AF344268

ENVIRONMENTAL BIOSENSORS

Edited by **Vernon Somerset**

INTECHOPEN.COM

Environmental Biosensors
Edited by Vernon Somerset

CBS, Edition 2016

Published by InTech
Janeza Trdine 9, 51000 Rijeka, Croatia

All chapters are distributed under the Creative Commons Non Commercial
Share Alike Attribution 3.0 license, which permits to copy, distribute, transmit,
and adapt the work in any medium, so long as the original work is properly cited.
After this work has been published by InTech, authors have the right to republish
it, in whole or part, in any publication of which they are the author, and to make
other personal use of the work. Any republication, referencing or personal use
of the work must explicitly identify the original source.

Statements and opinions expressed in the chapters are these of the individual contributors
and not necessarily those of the editors or publisher. No responsibility is accepted
for the accuracy of information contained in the published articles. The publisher
assumes no responsibility for any damage or injury to persons or property arising out
of the use of any materials, instructions, methods or ideas contained in the book.
Copyright © 2011 InTech

Publishing Process Manager Mirna Cvijic

Technical Editor Teodora Smiljanic

Cover Designer InTech Design Team

Image Copyright Rusty Dodson, 2010. Used under license from Shutterstock.com

Additional hard copies can be obtained from orders@intechopen.com

Environmental Biosensors,
Edited by Vernon Somerset
p. cm.
ISBN 978-953-307-486-3

Contents

Preface

Recent years have seen increasing interest in the application of simple, rapid, inexpensive and disposable electrochemical sensors for use in the fields of clinical, environmental or industrial analysis. Biosensors and chemical sensors represent analytical devices that utilise the sensitivity and selectivity of a biomaterial, chemical compound or a combination of both attached onto the surface of a physical transducer for sensing purposes. This book is an attempt to highlight current research issues in the field of biosensors on the topics of health, environment and biosecurity. It is divided into three sections with headings of current trends and developments; materials design and developments; and detection and monitoring. The book will provide valuable reference and learning material to other researchers and students in the field, with the references at the end of each chapter serving as guidance to further reading material. My sincere gratitude is expressed to the contributing authors for their hard work and dedication.

Vernon S. Somerset, Ph.D
Senior Researcher
Water Ecosystems and Human Health RG
Natural Resources and the Environment (NRE)
Cape Town,
South Africa

Part 1

Current Trends and Developments

Biosensors for Environmental Applications

Lívia Maria da Costa Silva,
Ariana Farias Melo and Andréa Medeiros Salgado
*Laboratory of Biological Sensors/EQ/UFRJ Biochemical Engineering Department,
Chemistry School, Technology Center, Federal University of Rio de Janeiro, Rio de Janeiro
Brazil*

1. Introduction

The increasing number of potentially harmful pollutants in the environment calls for fast
and cost-effective analytical techniques to be used in extensive monitoring programs.
Additionally, over the last few years, a growing number of initiatives and legislative actions
for environmental pollution control have been adopted in parallel with increasing scientific
and social concern in this area [1-4]. The requirements for application of most traditional
analytical methods to environmental pollutants analysis, often constitute an important
impediment for their application on a regular basis. The need for disposable systems or tools
for environmental applications, in particular for environmental monitoring, has encouraged
the development of new technologies and more suitable methodologies. In this context,
biosensors appear as a suitable alternative or as a complementary analytical tool. Biosensors
can be considered as a subgroup of chemical sensors in which a biological mechanism is
used for analyte detection [1,3-4].
A biosensor is defined by the International Union of Pure and Applied Chemistry (IUPAC)
as a self-contained integrated device that is capable of providing specific quantitative or
semi-quantitative analytical information using a biological recognition element (biochemical
receptor), which is retained in contact direct spatial with a transduction element [5].
Biosensors should be distinguished from bioassays where the transducer is not an integral
part of the analytical system [4-7]. Biosensing systems and methods are being developed as
suitable tools for different applications, including bioprocess control, food quality control,
agriculture, military and in particular, for medical applications.
Biosensors are usually classified according to the bioreceptor element involved in the
biological recognition process (e.g., enzymes, immunoaffinity recognition elements, whole-
cells of micro-organisms, plants or animals, or DNA fragments), or according to the
physicochemical transducer used (e.g., electrochemical, optical, piezoelectrical or thermal).
The main classes of bioreceptor elements that are applied in environmental analysis are
whole cells of microorganisms, enzymes, antibodies and DNA. Additionally, in the most of
the biosensors described in the literature for environmental applications electrochemical
transducers are used [5].
For environmental applications, the main advantages offered by biosensors over
conventional analytical techniques are the possibility of portability, miniaturization, work
on-site, and the ability to measure pollutants in complex matrices with minimal sample

preparation. Although many of the developed systems cannot compete yet with conventional analytical methods in terms of accuracy and reproducibility, they can be used by regulatory authorities and by industry to provide enough information for routine testing and screening of samples [1,3,8].

Biosensors can be used as environmental quality monitoring tools in the assessment of biological/ecological quality or for the chemical monitoring of both inorganic and organic priority pollutants. In this review article we provide an overview of biosensor systems for environmental applications, and in the following sections we describe the various biosensors that have been developed for environmental monitoring, considering the pollutants and analysis that are usually mentioned in the literature.

2. Heavy metals

Heavy metals are currently the cause of some of the most serious pollution problems. Even in small concentrations, they are a threat to the environment and human health because they are non-biodegradable. People are constantly been exposed to heavy metals in the environment [8-9]. The dangers associated with heavy metals are due to the ubiquitous presence of these elements in the biosphere, their bioavailability from both natural and anthropogenic sources, and their high toxicity. Thus, there are several cases described in the literature where exposure of populations to these pollutants has resulted in severe damage to their health, including a significant amount of deaths. The metal contaminants most commonly observed in the environment are: lead, chromium, zinc, mercury, cadmium and copper [10-11].

Conventional analytical techniques for heavy metals (such as cold vapour atomic absorption spectrometry, and inductively coupled plasma mass spectrometry) are precise, but suffer from the disadvantages of high cost, the need for trained personnel and the fact that they need to be performed in the laboratory [9]. For the reasons cited earlier, biosensors are being developed and utilized for monitoring heavy metal concentrations in environmental samples [9,12-14]. Furthermore, their biological basis makes them ideal for toxicological measurement of heavy metals, while conventional techniques can only measure concentrations [15].

Many of the bacterial biosensors developed for analysis of heavy metals in environmental samples, make use of specific genes responsible for bacterial resistance to these elements, such as biological receptors. Bacterial strains resistant to a number of metals such as zinc, copper, tin, silver, mercury and cobalt have been isolated as possible biological receptors [16-18]. The metal resistance of these genes is induced only when the element reaches the bacteria's cytoplasm. The specificity of this resistance mechanism contributes to the construction of cell biosensors for detection of metals from the fusion of these resistance genes with genes encoding bioluminescent proteins, for example, luciferin. In this case, the production of light, which can be measured by luminometers and photometers, indicates the presence of a heavy metal in the sample [9,17].

Enzymatic methods are also commonly used for metal ion determination, since these can be based on the use of a wide range of enzymes that are specifically inhibited by low concentrations of certain metal ions [19-20]. Domínguez-Renedo et al. [21] developed enzymatic amperometric biosensors for the measurement of $Hg+2$, based on the inhibitory action of this ion on urease activity. They used screen-printed carbon electrodes as support and screen-printed carbon electrodes modified with gold nanoparticles. The same enzyme was used by Kuswandi [22] in the development of a simple optical fibre biosensor for the determination of various heavy metal ions: Hg(II), Ag(I), Cu(II), Ni(II), Zn(II), Co(II) and Pb(II). Durrieu and Tran-Minh [23] developed an optical biosensor to detect lead and cadmium by inhibition of alkaline phosphatase present on the external membrane of Chlorella vulgaris microalgae, used as biological recognition element. Moreoever, a biosensor with microalgae Tetraselmis chui was developed for the voltammetric measurement of Cu+2 for Alpat et al. [24]. The developed algae-based biosensor was also successfully applied to the determination of copper (II) in real sample and the results were confirmed when compared to those obtained by atomic absorption spectrophotometric method. Table 1 lists some examples of biosensors developed for heavy metal determination.

Analyte	Recognition biocomponent	Transduction system	Matrix	Ref.
Zinc, copper, cadmium and nickel	*Pseudomonas fluorescens* 10586s pUCD607 with the *lux* insertion on a plasmid	Optical (luminometer)	Soil	[25]
Cadmium	DNA	Electrochemical	Standard solutions	[26]
Cadmiun	Phytochelatins	Optical (localized surface plasmon resonance)	Standard solutions	[27]
Mercury, cadmium and arsenic	Urease enzyme	Electrochemical	Standard solutions	[28]
Nickel ions	*Bacillus sphaericus* strain MTCC 5100	Electrochemical	Industrial effluents and foods	[29]
Zinc, copper, cadmium, nickel, lead, iron and aluminum	*Chlorella vulgaris* strain CCAP 211/12	Electrochemical	Urban waters	[30]
Cadmiun	*Escherichia coli* RBE23-17	Electrochemical	Wastewater	[31]
Zinc, cobalt and copper	*Pseudomonas sp.* B4251, *Bacillus cereus* B4368 and *E. coli* 1257	Electrochemical	Water	[32]
Mercury (II) and lead (II) ions	DNA	Optical	Water	[33]
Copper (I) and (II) ions	DsRed (red fluorescent) protein	Optical	Standard solutions	[34]
Cadmium, copper and lead	Sol-gel-immobilized-urease	Electrochemical	Synthetic effluents	[35]

Table 1. Examples of biosensors developed for heavy metal determination.

3. Biochemistry Oxygen Demand (BOD)

Biochemical oxygen demand (BOD or BOD$_5$) is a parameter widely used to indicate the amount of biodegradable organic material in water. Its determination is time consuming, and consequently it is not suitable for online process monitoring. Fast determination of BOD could be achieved with biosensor-based methods. Most BOD sensors rely on the measurement of the bacterial respiration rate in close proximity to a transducer, commonly of the Clark type (an amperometric sensor developed by Clark in 1956 for measuring dissolved oxygen) [2].

BOD biosensors are the most common commercial biosensors for environmental monitoring. The first commercial BOD sensor was produced by the Japanese company Nisshin Electric in 1983 and a number of other commercial BOD biosensors based on microbial cells are being marketed by Autoteam GmbH, Medingen GmbH and Dr. Lange GmbH in Germany; Kelma (Belgium); Bioscience, Inc. and US Filter (USA) [2,4,12].

Nakamura and Karube [36] developed a system for measuring BOD from cells of recombinant *Escherichia coli* with *Vibrio fisheri* genes lux AE. With this system the real time analysis of multiple samples was possible. These handy devices have been marketed primarily for food and pharmaceutical industries. Moreover, an optical biosensor for parallel multi-sample determination of biochemical oxygen demand in wastewater samples has been developed by Kwok *et al.* [37]. The biosensor monitors the dissolved oxygen concentration in artificial wastewater through an oxygen sensing film immobilized on the bottom of glass sample vials. Then, the microbial samples were immobilized on this film and the BOD value was determined from the rate of oxygen consumption by the microorganisms in the first 20 minutes.

4. Nitrogen compounds

Nitrites are widely used for food preservation and for fertilization of soils. However, continuous consumption of these ions can cause serious implications on human health, particularly because it can react irreversibly with hemoglobin [38]. The increasing levels of nitrate found in groundwater and surface water are of concern because they can harm the aquatic environment. In line with this, the regulations for treatment of urban wastewater in order to reduce pollution, including pollution by nitrates, from sewage treatment works of industrial and domestic, have been implemented [2].

Chen *et al.* [39] developed a biosensor for amperometric determination of nitrite using cytochrome c nitrite reductase (ccNiR) from *Desulfovibrio desulfuricans* immobilized and electrically connected on a glassy carbon electrode by entrapment into redox active [ZnCr-AQS] double layered hydroxide containing anthraquinone-2-sulfonate (AQS). The instrument showed a fast response to nitrite (5 seconds) with a linear range between concentrations of nitrite 0.015 and 2.35 µM and a detection limit of 4nM.

A highly sensitive, fast and stable conductimetric enzymatic biosensor for the determination of nitrate in waters was described in Wang *et al.* [40-41]. Conductimetric electrodes were modified by methyl viologen mediator mixed with nitrate reductase from *Aspergillus niger* by cross-linking with glutaraldehyde in the presence of bovine serum albumin and Nafion® cation-exchange polymer, allowing retention of viologen mediator. A linear calibration curve in the range of 0.02 and 0.25 mM with detection limits of 0.005 mM nitrate was obtained. When stored in pH 7.5 phosphate buffer, the sensors showed good stability over two weeks. Moreover, Khadro *et al.* [42], developed an enzymatic conductimetric biosensor for the determination of nitrate in water, validated and used for natural water samples. The instrument was based on a methyl viologen mediator mixed with nitrate reductase from *Aspergillus niger* and Nafion® cation-exchange polymer dissolved in a plasticized PVC membrane deposited on the sensitive surface of interdigitated electrodes. When stored in phosphate buffer pH 7.5 at 4°C, the sensor showed good stability over 2 months.

5. PCBs

Polychlorinated biphenyls (PCBs) are toxic organic compounds [43-44] that are ubiquitous environmental pollutants, even though their production was banned in several countries many years ago [44]. It is currently assumed that food is the major source of the PCB exposure since PCBs are highly lypophilic and accumulate in the food chain, so foods of animal origin are an important source of exposure. The level of PCBs in the environment depends on the matrix where it originated [43,45].

There are 209 polychlorinated biphenyl congeners that persist worldwide in the environment and food chain. These congeners are divided into three classes based upon the orientation of the chlorine moieties, *i.e.*, coplanar, mono-*ortho* coplanar, and non-coplanar [43]. Conventional techniques used for the analysis of PCBs are generally based on gas chromatography coupled with mass spectrometry (GC.MS) [43,45]. Alternative techniques based, for example, on immunoassays, are inexpensive and rapid screening tools for sample monitoring in laboratory and field analysis. Moreover, immunoassays are simple, sensitive, reliable, and relatively selective for PCBs testing. Among several immunoassay techniques, the enzyme-linked immunosorbent assay (ELISA) combined with colorimetric end-point detection are the most popular. Another interesting approach is the use of immunosensor technology [45]. Imunosensors are a class of biosensors that use as biological recognition elements, antibodies or antigens [45]. Pribyl *et al.* [44] developed a novel piezoelectric immunosensor for determination of PCB congeners in the range of concentrations usually found in real matrices (soil). The presented method allows one to carry out analysis of extracts directly without any additional purification steps. Moreover Gavlasova *et al.* [45] had a successful application of a sol-gel silica entrapment of viable *Pseudomonas sp.* P2 cells for constructing low-cost sensors for environmental monitoring using real soil.

6. Phenolic compounds

A considerable number of organic pollutants, which are found widely distributed in the environment, have phenolic structures. Phenols and their derivatives, are well known because of their high toxicity and are common compounds in industrial effluents, coming from the activities related to production of plastics, dyes, drugs, antioxidants, polymers, synthetic resins, pesticides, detergents, desinfectants, oil refinery and mainly pulp and paper [46].

Several substituted phenols, such as chloro- and nitrophenols, are highly toxic to humans and aquatic organisms [47-48]. These two groups of substituted phenols are the main degradation products of organophosphorus pesticides and chlorinated phenoxyacids. Even at small concentrations (< 1 ppm), phenolic compounds affect the taste and odor of drinking water and fish [47]. Many of these compounds have toxic effects in animals and plants, because they easily penetrate the skin and cell membranes, determining a wide range of genotoxicity, mutagenicity, hepatotoxic effects, and affect the rate of biocatalyzed reactions, and the processes of respiration and photosynthesis [2]. Thus, phenols and specially their chlorinated, nitro and alkyl derivatives have been defined as hazardous pollutants due to their high toxicity and persistence in the environment, and are found in the list of hazardous substances and priority pollutants of the EC (European Commission) and the U.S. Environmental Protection Agency (EPA).

The toxic pollutants in wastewater usually interact with DNA, leading to the damage to human health, but these very interactions between these toxic pollutants and DNA can be used in electrochemical DNA biosensors, generating a response signal, thus providing an effective approach for rapid screening of pollutants. Based on this principle, various electrochemical DNA sensors for environmental monitoring have been proposed. It has been demonstrated that DNA-based devices hold great promise for environmental screening of toxic aromatic compounds and for elucidating molecular interactions between intercalating pollutants and DNA. Using a disposable electrochemical DNA biosensor made by immobilizing double stranded DNA onto the surface of a disposable carbon screen-printed electrode, toxicants in water and wastewater samples have been successfully detected, which correlates well with the classic genotoxicity tests based on bioluminescent bacteria [49-53]. Parellada *et al.* [54] developed an amperometric biosensor with tyrosinase (a polyphenol oxidase with a relatively wide selectivity for phenolic compounds) immobilized in a hygrogel on a graphite electrode, which correlated satisfactorily with the official method for the determination of the phenol index in environmental samples. Chlorophenols have been also detected with a chemiluminescence fibre-optic biosensor adapted a flow injection analysis (FIA).

Phenolic compounds widely distributed in the environment as organic pollutants can be oxidized by conventional carbonaceous electrodes generally in high voltage (0.8 V *vs.* ECS). Under such over-voltage conditions, these compounds can dimerize and produce other electroactive species (radicals), resulting in higher than expected electrical current levels. In other cases, there may be adsorption or formation of polymeric products with consequent passivation of the electrode, leading to the observation of peaks with intensities very below those expected. In these cases the high applied potential may increase background current levels, and consequently the level of noise. Thus, by the use of electrodes modified with oxidase enzymes, coupled with the principle of biochemical oxidation followed by electrochemical reduction, one can undo or minimize these variables. Enzymes commonly used in the manufacture of biosensors are the laccase, tyrosinase and peroxidases [55-56].

Analyte	Recognition enzyme	Transduction system	Matrix	Ref.
Binary mixtures: phenol/clorophenol, catechol/phenol, cresol/clorocresol and phenol/cresol	Laccase and tyrosinase	Amperometric multicanal	wastewater	[37]
m-cresol or catechol	DNA	Amperometric	wastewater	[30]
Phenol	Mushroom tissue (tyrosinase)	Amperometric	wastewater	[38]
phenol, p-cresol, m-cresol and catechol	Polyphenol oxidase	Amperometric	wastewater	[39]

Table 2. Some of the most commonly used biosensors in phenolic compounds

Regarding the selectivity of biosensors based on peroxidase, greater sensitivity to the compounds 2-amino-4-chlorophenol and 4-chloro-3-methylphenol was observed among 20 compounds tested [57]. While the action of tyrosinase is confined largely to phenol and *ortho*-benzenediols the laccases are able to oxidize various phenolic substrates, including phenols and diphenols (*ortho*-, *meta*- and *para*-benzenediols), phenols, catecholamines, *etc.*, Liu *et al.* [58] developed a biosensor for phenol based on the immobilization of tyrosinase on the surface of modified magnetic $MgFe_2O_4$ nanoparticles. Table 2 presents some examples of biosensors used in the detection of phenolic compounds in wastewater matrices.

7. Endocrine disruptors and hormones

Endocrine disruptors, exogenous compounds that alter the endogenous hormone homeostasis, have been systematically discharged in the environment during the last years [62]. These contaminants have been related to the decrease of human sperm numbers and increased incidence of testicular, breast and thyroid cancers. These endocrine disruptors can act by the following mechanisms: a) inhibition of enzymes related to hormone synthesis; b) alteration of free concentration of hormones by interaction with plasmatic globulins; c) alteration in expression of hormone metabolism enzymes; d) interaction with hormone receptors, acting as agonists or antagonists; e) alteration of signal transduction resulting from hormone action. The importance of the identification of endocrine disruptors involves characterization of environmental contaminants and inquiry of new substances discharged in the environment.

Natural and synthetic hormone residues can be found in the environment as a result of human or animal excretion due to population growing and more intensive farming. Hormones such as estradiol, estrone and ethynylestradiol have been found in water at ng/L levels, but even at these low concentrations, some of them may have endocrine-disrupting activity in aquatic or even terrestrial organisms. Estrone, progesterone and testosterone, along with other organic pollutants, have been determined with a fully automated optical immunosensor in water samples, reaching limits of detection up to sub-ng/L [63].

An electrochemical biosensor for progesterone in cow's milk was developed and used in a competitive immunoassay by Xu *et al.* [64]. The sensor was fabricated by depositing anti-progesterone monoclonal antibody (mAb) onto screen-printed carbon electrodes (SPCEs) which were coated with rabbit anti-sheep IgG (rIgG). This sensor was operated following the steps of competitive binding between sample and conjugate (alkaline-phosphatase-labelled progesterone) for the immobilised mAb sites and measurements of an amperometric signal in the presence of *p*-nitrophenyl phosphate using either colorimetric assays or cyclic voltammetry [64].

8. Organophosphorus compounds (OP)

Organophosphorus (OP) compounds are a group of chemicals that are widely used as insecticides in modern agriculture for controlling a wide variety of insect pests, weeds, and disease-transmitting vectors [65].

8.1 Pesticides

A pesticide, as defined by the EPA, is any substance or mixture of substances intended for preventing, destroying, repelling, or lessening the damage of any pest, [65]. Of all the environmental pollutants, pesticides are the most abundant, present in water, atmosphere, soil, plants, and food [2].

Concerns about the toxicity, ubiquity and persistence of pesticides in the environment have led the European Community to set limits on the concentration of pesticides in different environmental waters. Directive 98/83/EC on the quality of water for human consumption has set a limit of 0.1 µg/L for individual pesticides and of 0.5 µg/L for total pesticides. Enzymatic sensors, based on the inhibition of a selected enzyme are the most extensively used biosensors for the determination of these compounds [66].

Parathion (*O,O*-diethyl-*O*-4-nitrophenyl thiophosphate), is a broad-spectrum OP pesticide having a wide range of applications against numerous insect species on several crops. Parathion is also used as a preharvest soil fumigant and foliage treatment for a wide variety of plants, both in the field and in the greenhouse. Parathion is highly toxic by all routes of exposure — ingestion, skin adsorption, and inhalation — all of which have resulted in human fatalities. Like all pesticides, parathion irreversibly inhibits AChE [67]. Table 3 presents a few examples of biosensors for determination of pesticides, including parathion.

8.2 Herbicides

For the detection of herbicides such as the phenylureas and triazines, which inhibit photosynthesis, biosensors have been designed with membrane receptors of thylakoid and chloroplasts, photosystems and reaction centers or complete cells such as unicellular alga and phenylureas and triazines, in which mainly amperometric and optical transductors have been employed [68]. Table 3 also presents some examples of biosensors used in the detection of herbicides.

Analyte	Type of interaction	Recognition biocatalyzer	Transduction system	Ref.
Pesticides				
Simazina	Biocatalytic	Peroxidase	Potentiometric	[47]
Isoproturon	Biocatalytic	Antibody encapsulate	Immunosensor immunoreaction	[48]
Parathion	Biocatalytic	Parathion hydrolase	Amperometric	[50]
Paraoxon	Biocatalytic	Alkaline phosphatase	Optical	[46,50]
Carbaril	Biocatalytic	Acetilcolinesterase	Amperometric	[51]
Herbicides				
2,4-Diclorofenoxiacetic	Immunoanalysis	Acetilcolinesterase	Amperometric	[52]
Diuron, Paraquat,	Biocatalytic	Cyanobacterial	Bioluminescence	[53]

Table 3. Biosensors used in the detection of pesticides and herbicides.

8.3 Dioxins

Dioxins are potentially toxic substances for humans with a major impact on the environment that can reach the food chain accidentally, as contaminating residues present in water and soil. Dioxins are organosoluble, toxic, teratogenic and cancinogenic. They are unintended by-products of many industrial processes where chlorine and chemicals derived from it are produced, used and disposed of. Industrial emissions of dioxin to the environment can be transported over long distances by airstreams and, less importantly, by rivers and sea currents. Consequently, dioxins are now widely present all over the globe. It is estimated that even if production completely stopped today, the current environmental levels would take years to diminish. This is because dioxins are persistent, taking years to centuries to deteriorate, and can be continuously recycled in the environment [69]. Biosensors for detection and monitoring of these pollutants in the area would be extremely useful.

9. Commercial activities

About 200 companies worldwide were working in the area of biosensors and bioelectronics at the turn of the century. Some of these companies are still involved in biosensor fabrication/marketing, whereas others just provide the pertinent materials and instruments for biosensor manufacture. Most of these companies are working on existing biosensor technologies and only a few of them are developing new technologies [76]. The application of new biodevices to real-world environmental samples is a must in the final steps of development. However, despite of the great number of newly developed biosensors, most literature references overlook the real-world step and only report applications of the biosensor in either distilled water or buffered solutions.

Most of the reviewed systems still have some way to go before application to real samples can be made, and the study of matrix effects, stability issues and careful comparison with established methods are crucial steps in this approach. Even though commercial returns from environmental biosensors are substantially less than from medical diagnostics, public concern and government funding have generated a major research effort aimed at the application of biosensors to the measurement of pollutants and other environmental hazards [77].

10. Other applications

Many biosensors reported for environmental applications show the potential to be developed for either single-sample formats for field screening applications or continous formats for field monitoring applications. A discussion concerning the cleanup of a Superfund site may provide examples of the scope and kinds of environmental screening and monitoring problems for which biosensors could provide unique solutions

Most of the work on metal-specific bacterial sensors has been done by using a liquid suspension of viable sensor bacteria. However, a more advanced approach is to use these bactéria in the biosensor system, e.g., by immobilising the cells onto optical fibres connected to a photo detection device. This type of fibre-optic biosensors have been previously constructed for Cu [76], genotoxicants, or general toxicity of industrial effluents [78]. These fibre-optic sensors can easily be brought to the field and used for on-line monitoring.

Ivask *et al* (2006) developed fibre-optic biosensors for the analysis of environmental samples, e.g., soils and sediments *in situ*. For that, the existing recombinant luminescent *Escherichia coli* MC1061 (pmerRluxCDABE) and MC1061 (parsluxCDABE) [79] responding specifically to Hg and As, respectively, were immobilised onto optical fibres in order to develop self contained biosensors. The system was optimised for the Hg biosensor and the derived protocol was used for analysing bioavailable Hg and As in natural soil or sediment suspensions. The biosensors consist of alginate-immobilised recombinant bacteria emitting light specifically in the presence of bioavailable Hg or As in a dosedependent manner. The biosensors alongside with the non-immobilised Hg and As sensor bacteria were successfully applied for the analysis of bioavailable fractions of Hg and As in soil and sediment samples from the Aznalcollar mining area (Spain).

Haron *et al* (2006) developed three layer waveguiding silicon dioxide (SiO2)/silicon nitride (Si3N4)/SiO2 structure on silicon substrate was proposed as an optically efficient biosensor for calibration of heavy metal ions in drinking water. Total attenuated reflection (ATR) in portable and miniaturized SiO2/Si3N4/SiO2 waveguides was successfully exploited in the present investigation for development of a highly sensitive biosensors and the detection limit as low as 1 ppb was achieved for Cd2+ and Pb2+ ions.Composite membranes containing both biologically active components, e.g. enzymes, and organic chromophores (indicators) were formed using the polyelectrolyte selfassembly deposition technique. The latter are sensitive to small changes in local pH caused by enzyme reactions, and thus provided a transuding function. The difference in inhibition of enzymes urease and acetylcholine esterase by the heavy metal ions established the possibility of designing a sensor array for discrimination between different types of water pollutants and the device is light, portable and robust

11. Conclusions

Most biosensor systems have been tested only on distilled water or buffered solutions, but more biosensors that can be applied to real samples have appeared in recent years. In this context, biosensors for potential environmental applications continue to show advances in areas such as genetic modification of enzymes and microorganisms, improvement of recognition element immobilization and sensor interfaces.

12. Acknowledgements

This work has been supported by CNPq (National Council for Scientific and Technological Development), CAPES (Coordination for the Improvement of Higher Level Personnel) and the FAPERJ (Foundation for Research of the State of Rio de Janeiro).

13. References

K.R. Rogers and C.L. Gerlach, *Environmental biosensors: A status report*, Environ. Sci. Technol. 30 (1996), pp. 486-491.

S. Rodriguez-Mozaz, M-P. Marco, M.J.L. Alda and D. Barceló, *Biosensors for environmental applications: future development trends*, Pure Appl. Chem. 76 (2004), pp. 723-752.

K.R. Rogers, *Recent advances in biosensor techniques for environmental monitoring*, Anal. Chim. Acta. 568 (2006), pp. 222-231.

S. Rodriguez-Mozaz, M-P. Marco, M.J.L. Alda and D. Barceló, *A global perspective: Biosensors for environmental monitoring*, Talanta. 65 (2005), pp. 291-297.

D.R. Thévenot, K. Toth, R.A. Durst and G.S. Wilson, *Electrochemical biosensors: Recommended definitions and classification*, Pure Appl. Chem. 71 (1999), pp. 2333-2348.

P. Leonard, S. Hearty, J. Brennan, L. Dunne, J. Quinn, T. Chakraborty, and R. O'Kennedy, *Advances in biosensors for the detection of pathogens in food and water*, Enzyme Microb. Technol. 32 (2003), pp. 3-13.

C. Ziegler and W. Gopel, *Biosensor development*, Curr. Opin. Chem. Boil. 2 (1998), pp. 585-591.

M. Sharpe, *It's a bug's life: Biosensors for environmental monitoring*, J. Environ. Monit. 5 (2003), pp. 109-113.

N. Verma and M. Singh, *Biosensors for heavy metals*, BioMetals. 18 (2005), pp. 121-129.

P.R.G. Barrocas, A.C.S. Vasconcellos, S.S. Duque, L.M.G. Santos, S.C. Jacob, A.L. Lauria-Filgueiras and J.C. Moreira, *Biossensores para o monitoramento da exposição a poluentes ambientais*, Cad. Saúde Colet. Rio de Janeiro. 16 (2008), pp. 677-700.

Brian, R. Babai, K. Levcov, J. Rishpon and E.Z. Ron, *Online and in situ monitoring of environmental pollutants: Electrochemical biosensing of cadmium*. Environ. Microbiol. 2 (2000), pp. 285-290.

S.F. D'Souza, *Microbial biosensors*, Biosens. Bioelectron. 16 (2001), pp. 337-353.

J. Davis, D.H. Vaughan and M.F. Cardosi, *Elements of biosensor construction*, Enzyme Microb. Technol. 17 (1995), pp. 1030-1035.

Mulchandani and A.S. Bassi, *Principles and applications of biosensors for bioprocess monitoring and control*. Crit. Rev. Biotechnol. 15 (1995), pp. 105-124.

K. Riedel, G. Kunze and A. Konig, *Microbial sensors on a respiratory basis for wastewater monitoring*, Adv. Biochem. Eng. Biotechnol. 75 (2002), pp. 81-118.

S. Magrisso, Y. Erel and S. Belkin, *Microbial reporters of metal bioavailability*, Microbial. Biotechnol. 1 (2008), pp. 20-330.

S. Ramanathan, M. Ensor and S. Daunert, *Bacterial biosensors for monitoring toxic metals*, Trends Biotechnol. 15 (1997), pp. 500-506.

I.V.N. Rathnayake, M. Megharaj, N. Bolan and R. Naidu, *Tolerance of heavy metals by gram positive soil bacteria*, World Acad. Sci. Eng. Technol. 53 (2009), pp. 1185-1189.

P.R.B.O. Marques and H. Yamanaka, *Biossensores baseados no processo de inibição enzimática*, Quim. Nova. 7 (2008), pp. 1791-1799.

Aminea, H. Mohammadia, I. Bourais and G. Palleschi, *Enzyme inhibition-based biosensors for food safety and environmental monitoring*, Biosens. Bioelectron. 21 (2006), pp. 1405-1423.

O. Domínguez-Renedo, M.A. Alonso-Lomillo, L. Ferreira-Gonçalves and M.J. Arcos-Martínez, *Development of urease bades amperometric biosensors for the inhibitive determination of Hg (II)*, Talanta 79 (2009), pp. 1306-1310.

B. Kuswandi, *Simple optical fibre biosensor based on immobilised enzyme for monitoring of trace heavy metal ions*, Anal. Bioanal. Chem. 376 (2003), pp. 1104-1110.

C. Durrieu and C. Tran-Minhw, *Optical algal biosensor using alkaline phosphatase for determination of heavy metals*, Environ. Res. Sect. B. 51 (2002), pp. 206-209.

S.K. Alpat, S. Alpar, B. Kutlu, O. Ozbayrak and H.B. Buyukisik, *Development of biosorption-based algal biosensor for Cu(II) using Tetraselmis chuii*, Sens. Actuat. B Chem. 128 (2007), pp. 273-278.

S.P. Mcgrath, B. Knight, K. Killham, S. Preston and G.I. Paton, *Assessment of the toxicity of metals in soils amended with sewage sludge using a chemical speciation technique and a lux-based biosensor*, Environ. Toxicol. Chem. 18 (1999), pp. 659-663.

E.L.S Wong, E. Chow and J.J. Gooding, *The electrochemical detection of cadmium using surface-immobilized DNA*, Electrochem. Commun. 9 (2007), pp. 845-849.

T-J. Lin and M-F Chung, *Detection of cadmium by a fiber-optic biosensor based on localized surface plasmon resonance*, Biosens. Bioelectron. 24 (2009), pp. 1213-1218.

P. Pal, D. Bhattacharyay, A. Mukhopadhyay and P. Sarkar, *The detection of mercury, cadium, and arsenic by the deactivation of urease on rhodinized carbon,*. Environ. Eng. Sci. 26 (2009), pp. 25-32.

N. Verma and M.A. Singh, *Bacillus sphaericus based biosensor for monitoring nickel ions in industrial effluents and foods*, J. Autom. Methods Manag. Chem. 2006 (2006), pp. 1-4.

D. Claude, G. Houssemeddine, B. Andriy and C. Jean-Marc, *Whole cell algal biosensors for urban waters monitoring*, Novatech 7(3) (2007), pp. 1507-1514.

Biran, R. Babai, K. Levcov, J. Rishpon, and E.Z. Ron, *Online and in situ monitoring of environmental pollutants: Electrochemical biosensing of cadmium*, Environ. Microbiol. 2 (2000), pp. 285-290.

T.G. Gruzina, A.M. Zadorozhnyaya, G.A. Gutnik, V.V. Vember, Z.R. Ulberg, N.I. Kanyuk and N.F. Starodub, *A bacterial multisensor for determination of the contents of heavy metals in water*, J. Water Chem. Technol. 29 (2007), pp. 50-53.

M.R. Knecht and M. Sethi, *Bio-inspired colorimetric detection of Hg^{2+} and Pb^{2+} heavy metal ions using Au nanoparticles*, Anal. Bioanal. Chem. 394 (2009), pp. 33-46.

J.P. Sumner, N.M. Westerberg, A.K. Stoddard, T.K. Hurst, M. Cramer, R.B. Thompson, C.A. Fierke and R. Kopelman, *DsRed as a highly sensitive, selective, and reversible fluorescence-based biosensor for both Cu^+ and Cu^{2+} ions*, Biosens. Bioelectron. 21 (2006), pp. 1302-1308.

R. Ilangovan, D. Daniel, A. Krastanov, C. Zachariah and R. Elizabeth, *Enzyme based biosensor for heavy metal ions determination*, Biotechnol. Biotechnol. Eq. 20 (2006), pp. 184-189.

H. Nakamura and I. Karube, Current *research activity in biosensors*, Anal. Bioanal. Chem. 377 (2003), pp. 446-468.

N-Y. Kwok, S. Dongb and W. Loa, *An optical biosensor for multi-sample determination of biochemical oxygen demand (BOD)*, Sens. Actuat. B Chem. 110 (2005), pp. 289-298.

M.J. Moorcroft, J. Davis and R.G. Compton, *Detection and determination of nitrate and nitrite: A review*, Talanta 54 (2001), pp. 785-803.

H. Chen, C. Mousty, S. Cosnier, C. Silveira, J.J.G. Moura, and M.G. Almeida, *Highly sensitive nitrite biosensor based on the electrical wiring of nitrite reductase by [ZnCr-AQS] LDH*. Electrochem. Commun. 9 (2007), pp. 2240-2245.

X. Wang, S.V. Dzyadevych, J.M. Chovelon, N. Jaffrezic-Renault, C. Ling and X. Siqing, *Development of conductometric nitrate biosensor based on Methyl viologen/Nafion® composite film*, Electrochem. Commun. 8 (2006), pp. 201-205.

X. Wang, S.V. Dzyadevych, J.M. Chovelon, N. Jaffrezic-Renault, C. Ling and X. Siqing, *Conductometric nitrate biosensor based on Methyl viologen/Nafion®/Nitrate reductase interdigitated electrodes*, Talanta 69 (2006), pp. 450-455.

B. Khadro, P. Namour, F. Bessueille, D. Leonard and N. Jaffrezic-Renault, *Enzymatic conductometric biosensor based on PVC membrane containing methyl viologen/nafion®/nitrate reductase for determination of nitrate in natural water samples*, Sens. Mater. 20 (2008), pp. 267-279.

S. Centi, B. Rozum, S. Laschi, I. Palchetti and M. Mascini, *Disposable electrochemical megnetic beads-based immunosensors*, Chem. Anal. 51 (2006), pp. 963-975.

J. Pribyl, M. Hepel and P. Skládal, *Piezoelectric immunosensors for polychlorinated biphenyls operating in aqueous and organic phases*, Sens. Actuat. B Chem. 113 (2006), pp. 900-910.

P. Gavlasova, G. Kuncova, L. Kochankova and M. Mackova, *Whole cell biosensor for polychlorinated biphenyl analysis based on optical detection*, Int. Biodeterior. Biodegrad. 62 (2008), pp. 304-312.

J.H.T. Luong, K.B. Male and J.D. Glennon, *Biosensor technology: technology push versus market pull*, Biotechnol. Adv. 26 (2008), pp. 492-500.

S. Rodriguez-Mozaz, M.J.L. Alda and D. Barceló, *Biosensors as useful tools for environmental analysis and monitoring*, Anal. Bioanal. Chem. 386 (2006), pp. 1025-1041.

L. Su-Hsia and J. Ruey-Shin, *Adsorption of phenol and its derivatives from water using synthetic resins and low-cost natural adsorbents: A review*, J. Environ. Manag. 90 (2009), pp. 1336-1349.

J.F. Stara, D. Mukerjee, R. McGaughy, P. Durkint and M.L. Dourson, *The current use of studies on promoters and cocarcinogens in quantitative risk assessment*, Environ. Health Perspect. 50 (1983), pp. 359-368.

A.P. Wezel and A. Opperhuizen, *Narcosis due to environmental pollutants in aquatic organisms: residue-based toxicity, mechanisms, and membrane burdens*, Crit. Rev. Toxicol. 25 (1995), pp. 255-279.

Y. Zheng, C. Yang, W. Pu and J. Zhang, *Carbon nanotube-based DNA biosensor for monitoring phenolic pollutants*, Microchim. Acta 166 (2009), pp. 21-26.

G. Bagni, D. Osella, E. Sturchio and M. Mascini, *Deoxyribonucleic acid (DNA) biosensors for environmental risk assessment and drug studies*, Anal. Chim. Acta 573-574 (2006), pp. 81-89.

J. Parellada, A. Narvaez, M.A. Lopez, E. Dominguez, J.J. Fernandez, V. Pavlov and I. Katakis, *Amperometric immunosensors and enzyme electrodes for environmental applications*, Anal. Chim. Acta 362 (1998), pp. 47-57.

A.J. Blasco, M.C. Rogerio, M.C. Gonzalez and A. Escarpa, *"Electrochemical Index" as a screening method to determine "total polyphenolics" in foods: A proposal*, Anal. Chim. Acta 539 (2005), pp. 237-244.

N. Duran, M.A. Rosa, A. D'annibale and L. Gianfreda, *Applications of laccases and tyrosinases (phenoloxidases) immobilized on different supports: a review*, Enzyme Microb. Technol. 31(7) (2002), pp. 907-931.

E.S. Gil, L. Muller, M.F. Santiago and A.T. Garcia, *Biosensor based on brut extract from Laccase (Pycnoporus Sanguineus) for environmental analysis of phenolic compounds*, Portugaliae Electrochim. Acta 27 (2009), pp. 215-225.

Z. Liu, Y. Liu, H. Yang, Y. Yang, G. Shen and G. Yu, *A phenol biosensor based on immobilizing tyrosinase to modified core-shell magnetic nanoparticles supported at carbon paste electrode*, Anal. Chim. Acta 533(1) (2005), pp. 3-9.

S.S. Rosatto, R.S. Freire, N. Duran and L.T. Kubota, *Biossensores amperométricos para determinação de compostos fenólicos em amostras de interesse ambiental*, Quim. Nova 24 (2001), pp. 77-86.

L.M.C. Silva, A.M. Salgado and M.A.Z. Coelho, *Agaricus bisporus as a source of tyrosinase for phenol detection for future biosensor development*, Environ. Technol. 31 (2010) pp. 611-616.

H. Xue and Z. Shen, *A highly stable biosensor for phenols prepared by immobilizing polyphenol oxidase into polyaniline-polyacrylonitrile composite matrix*, Talanta 57 (2002), pp. 289-295.

N.V. Moraes, M.D. Grando, D.A.R. Valério and D.P. Oliveira, *Exposição ambiental a desreguladores endócrinos: alterações na homeostase dos hormônios esteroidais e tireoideanos*, Braz. J. Toxicol. 21 (2008), pp. 1-8.

R.H. Waring and R.M. Harris, *Endocrine disrupters: A human risk?*, Mol. Cell. Endocrinol. 244 (2005), pp. 2-9.

Y.F. Xu, M. Velasco-Garcia and T.T. Mottram, *Quantitative analysis of the response of an electrochemical biosensor for progesterone in milk*, Biosens. Bioelectron. 20 (2005), pp. 2061-2070.

M. Badihi-Mossberg, V. Buchner and J. Rishpon, *Electrochemical biosensors for pollutants in the environment*, Electroanalysis 19 (2007) pp. 2015-2028.

I.E. Tothill, *Biosensors developments and potential applications in the agricultural diagnosis sector*, Comput. Electron. Agric. 30 (2001), pp. 205-218.

L.S. Cock, A.M.Z. Arenas and A.A. Aponte, *Use of enzymatic biosensors as quality indices: A synopsis of present and future trends in the food industry*, Chilean J. Agric. Res. 69 (2009), pp. 270-280.

P.D. Patel, *(Bio)sensors for measurement of analytes implicated in food safety: A review*, Trends Anal. Chem. 21 (2002), pp. 96-115.

M.N. Velasco-García and T. Mottram, *Biosensor technology addressing agricultural problems*, Biosyst. Eng. 84 (2003), pp. 1-12.

M.F. Yulaev, R.A. Sitdikov, N.M. Dmitrieva, E.V. Yazynina, A.V. Zherdev and B.B. Dzantiev, *Development of a potentiometric immunosensor for herbicide simazine and its application for food testing*, Sens. Actuat. 75(1-2) (2001), pp. 129-135.

P. Pulido-Tofiño, J.M. Barrero-Moreno, and M.C. Pérez-Conde, *Sol–gel glass doped with isoproturon antibody as selective support for the development of a flow-through fluoroimmunosensor*, Anal. Chim. Acta. 429(2) (2001), pp. 337-345.

L.D. Mello and L.T. Kubota, *Review of the use of biosensors as analytical tools in the food and drink industries*, Food Chem. 77 (2002), pp. 237-256.

M. Del Carlo, M. Mascini, A. Pepe, G. Diletti and D. Compagnone, *Screening of food samples for carbamate and organophosphate pesticides using an electrochemical bioassay*, Food Chem. 84(4) (2004), pp. 651-656.

E.P. Medyantseva, M.G. Vertlib, M.P. Kutyreva, E.I. Khaldeeva, G.K. Budnikov and S.A. Eremin, *The specific immunochemical detection of 2,4-dichlorophenoxyacetic acid and 2,4,5-trichlorophenoxyacetic acid pesticides by amperometric cholinesterase biosensors*, Anal. Chim. Acta 347 (1997), pp. 71-78.

C.Y. Shao, C.J. Howe, A.J.R. Porter, and L.A. Glover, *Novel cyanobacterial biosensor for detection of herbicides*, Appl. Environ. Microbiol. 68 (2002), pp. 5026-5033.

J.H.T. Luong, K.B. Male and J.D. Glennon, *Biosensor technology: Technology push versus market pull*, Biotechnol. Adv. 26 (2008), pp. 492-500.

S. Rodriguez-Mozaz, M.J.L. Alda and D. Barceló, *Biosensors as useful tools for environmental analysis and monitoring*, Anal. Bioanal. Chem. 386 (2006), pp. 1025-1041.

Corbisier, P., van der Lelie, D., Borremans, B., Provoost, A., de Lorenzo, V., Brown, N.L., Lloyd, J.R., Hobman, J.L., Cs¨oregi, E., Johansson, G., Mattiasson, B., 1999. Anal. Chim. Acta 387, 235–244.

Horsburgh, A.M., Mardlin, D.P., Turner, N.L., Henkler, R., Strachan, N., Glover, L.A., Paton, G.I., Killham, K., 2002. Biosens. Bioelectron. 17 (6/7), 495–501.

Hakkila, K., Green, T., Leskinen, P., Ivask, A., Marks, R., Virta, M., 2004. J. Appl. Toxicol. 24, 333–342

2

Advances in
Aptamer-based Biosensors for Food Safety

Maureen McKeague, Amanda Giamberardino and Maria C. DeRosa
Carleton University, Department of Chemistry Ottawa, Ontario,
Canada

1. Introduction

The presence of unsafe levels of chemical compounds, toxins, and pathogens in food constitutes a growing public health problem that necessitates new technology for the detection of these contaminants along the food continuum from production to consumption. While traditional techniques that are highly selective and sensitive exist, there is still a need for simpler, more rapid and cost-effective approaches to food safety testing. Within this context, the field of food safety biosensors has emerged. Biosensors consist of a specific molecular recognition probe targeting an analyte of interest and a means of converting that recognition event into a measurable signal. As molecular recognition is the foundation of biosensing, there has been increased focus on the development of new molecular recognition probes for food-safety related molecular targets. Antibodies have been the gold standard in molecular recognition for several decades and have been incorporated widely into biosensors and assays relating to food (Ricci et al., 2007). Despite their applicability to food monitoring, they are not without their disadvantages, which are primarily linked to the requirement that antibody generation is an *in vivo* process. For example, highly toxic substances are not conducive to antibody generation. Furthermore, the batch to batch reproducibility of antibody generation can be less than satisfactory. Compounding these limitations is the fact that antibodies have short shelf-lives and can be challenging to chemically modify for incorporation into a biosensor platform. Nevertheless, the affinity and specificity of antibodies for their molecular targets make them convenient receptors for biosensing strategies.

Many of the disadvantages described above could be avoided with a molecular recognition probe of synthetic origin that would still maintain the required specificity and affinity. Because of their *in vitro* selection and production, the relatively new technology of aptamers has emerged as a viable alternative for use in biosensor platforms (Mascini, 2009). This chapter will focus on the recent literature in aptamers for food safety related targets, as well as the biosensor platforms in which these probes have been incorporated.

2. Aptamers in biosensing

Aptamers are single-stranded oligonucleotides that fold into distinct three-dimensional conformations capable of binding strongly and selectively to a target molecule. As molecular recognition probes, aptamers have binding affinities and specificities that are comparable to, and in some cases even surpass those of monoclonal antibodies. For example, aptamers have been selected that have dissociation constants (Kd) in the nanomolar or picomolar range. Aptamers are discovered using an *in vitro* process called Systematic Evolution of Ligands by EXponential enrichment (SELEX), a procedure where target-binding oligonucleotides are selected from a random pool of sequences through iterative cycles of affinity separation and amplification (see Fig. 1) (Ellington & Szostak, 1990; Tuerk & Gold, 1990). The SELEX process begins with a large, random oligonucleotide library whose complexity and diversity can be tailored through its distribution and number of random nucleotide regions (Luo et al., 2010). These sequences are exposed to the molecule of interest and those with an affinity for the target are separated from non-binding sequences. The elution of the binding sequences from the target and the polymerase chain reaction (PCR) amplification of those binders yields an enriched pool for subsequent, more stringent, selection rounds. After several rounds, this pool is cloned, sequenced, and characterized to find aptamers with the desired properties.

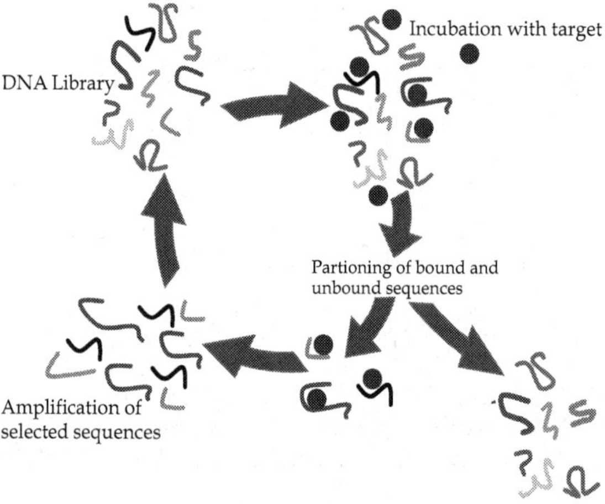

Fig. 1. Schematic overview of the SELEX procedure for the selection of aptamer sequences.

Targets for which aptamers can be developed are varied and range from small molecules (Huizenga & Szostak, 1995), to proteins and even whole cells. The *in vitro* nature of the selection process allows for the discovery of aptamers for even non-immunogenic or highly toxic substances. In addition to this advantage, aptamer technology offers several other benefits over antibodies. First, high-purity aptamers can be chemically synthesized at a low cost and can be easily modified with dyes, labels, and surface attachment groups without affecting their affinities. Second, aptamers are more chemically stable under most environmental conditions, have a longer shelf life, and can be reversibly denatured without loss of specificity. These properties make aptamers attractive in the development of low-cost, robust diagnostics and biosensors.

An examination of the development of aptamers and aptamer-based biosensors for food safety related targets can be found below. The chapter will be divided into two main parts: aptamers for small molecule food contaminants and aptamers for food safety-related pathogens.

Target Class	Target	DNA/RNA	K_d (nM)	Biosensor Platform	LOD	Ref
Antibiotics	Chloramphenicol	RNA	2100	-	-	(Burke et al., 1997)
Tetracyclines	Tetracycline	RNA	1000	-	-	(Berens et al., 2001)
		DNA	64	Electrochemical	10 nM	(Niazi, Lee Gu, 2008)
		DNA	-	Electrochemical	1 ng/mL	(Zhang et al., 2010)
	Oxytetracycline	DNA	10	-	-	(Niazi, Lee, Kim et al., 2008)
				Electrochemical	-	(Y. S. Kim et al., 2009)
				Colorimetric	-	(Y. S. Kim et al., 2010)
Aminoglycosides	Streptomycin	RNA	1000	-	-	(Wallace & Schroeder, 1998)
	Tobramycin	RNA	30-100	-	-	(Goertz, Colin Cox Ellington, 2004a)
		RNA	20000	-	-	(Morse, 2007)
		RNA	9	-	-	(Y. Wang & Rando, 1995)
		2'-OMe-RNA		Electrochemical (Impedence)	0.7 µM	(González-Fernández et al., 2011)
	Kanamycin	RNA	180	-	-	(Kwon et al., 2001)
		RNA	10-30	-	-	(Goertz, Colin Cox Ellington, 2004b)
	Neomycin	RNA	Low nM	-	-	(Goertz, Colin Cox Ellington, 2004b)
		RNA	1800	-	-	(Cowan et al., 2000)
				Electrochemical (Impedence)	"sub µM"	(de-los-Santos-Alvarez et al., 2007)
		2'-OMe-RNA		Surface Plasmon Resonance (SPR)	10 nM	(de-los-Santos-Álvarez et al., 2009)
Mycotoxins	Ochratoxin A	DNA	200	-		(Cruz-Aguado & Penner, 2008a)
				Fluorescence Polarization	5 nM	(Cruz-Aguado & Penner, 2008b)
				Electrochemical	30 pg/mL	(Kuang et al., 2010)
				Electrochemiluminescent	0.007 ng/mL	(Z. Wang et al., 2010)
				Colorimetric	20 nM	(Yang et al., 2010)

Target Class	Target	DNA/ RNA	K_d (nM)	Biosensor Platform	LOD	Ref
				Strip (Fluorescence)	1.9 ng/mL	(L. Wang et al., 2011a)
				Strip (Chromato-graphic/ Absorbance)	0.18 ng/mL	(L. Wang et al., 2011b)
				Electrochemical	0.07 ng/mL	(Bonel et al.,)
				Fluorescence (quenching)	22 nM	(Sheng et al.,)
	Fumonisin B1	DNA	100			(McKeague et al., 2010)
Food packaging	Bisphenol A	DNA	-	Electrochemical (Sandwich Carbon Nano-tube sensor)	10 fM	(Lee et al., 2011)
		DNA	-	-	-	(Okada et al., 2003)
Adulterants	Sulfor-hodamine	DNA	190	-	-	(Wilson & Szostak, 1998)
		RNA	70	-	-	(Holeman et al., 1998)
	Melamine	DNA		Resonance Scattering	10-15 ng/L	(Z. Jiang et al., 2011)
Pesticides *Herbicides*	Atrazine	DNA	2000	-	-	(Sinha et al., 2010)
Fungicides	Malachite Green	RNA	<1000	-	-	(Grate & Wilson, 1999)
			600	Fluorescence	-	(Stead et al., 2010)
Inorganic ions	Arsenic (3+, 5+)	DNA	5 (3+), 7 (5+)			(M. Kim et al., 2009)
	Mercury (2+)	DNA	-	Colorimetric	0.6 nM	(Li et al., 2009)
				Resonance Scattering	34 fg/mL	(Z. Jiang et al., 2010)
				Fluorescence	220 µM	(Xu et al.,2010)

Table 1. Table of aptamers and aptamer-based biosensor platforms for small-molecule targets.

2.1 Aptamers for small molecule food safety targets

Targets that can be classified under this category include high priorities for food analysis such as pesticides, toxins, veterinary drugs and other contaminants that may be present in a wide array of food products. Table 1 lists aptamers and aptamer-based biosensor platforms that have been developed for these targets. Section 2.1 will briefly highlight these systems and three promising biosensor platforms that target small molecules will be discussed in Section 2.2.

2.1.1 Antibiotics

In addition to their use in the treatment of bacterial diseases in humans and animals, antibiotics are important in animal husbandry because they significantly enhance growth when added to animal feed. However, the accumulation of antibiotics in food-producing animals is a potential cause of the increasing occurrence of antibiotic resistance. For this reason, European Union (EU) legislation has forbidden the practice of adding antibiotics to animal feed since 2006. As a result, fast, sensitive methodologies are being developed and used by food-safety control laboratories to ensure the control of antibiotic residues in live animals and animal products (Cháfer-Pericás et al., 2010).

While several antibiotic families are used in veterinary medicine and are tested in foods, only a handful of aptamers have been developed that recognize them. As seen in Table 1, RNA aptamers have been developed against chloramphenicol and several aminoglycosides such as streptomycin, kanamycin, tobramycin and neomycin. In particular, several aptamers have been developed for members of the tetracycline family. An RNA and DNA aptamer exist that recognize tetracycline as well as a DNA aptamer for oxytetracyline. Several of these have been integrated into electrochemical based testing systems. Overall, these antibiotic aptamers have a wide range of affinities for their targets, having dissociation constants from micromolar to low nanomolar. However, the detection systems that have been developed that use these aptamers all display detection limits in the nanomolar range.

2.1.2 Mycotoxins

Toxic fungal metabolites known as mycotoxins can contaminate a wide range of agricultural commodities and are high priority targets for the development of new molecular recognition probes and biosensors. It is estimated that at least 25% of the grain produced worldwide is contaminated with mycotoxins. Problematically, even small concentrations of mycotoxins can induce significant health problems including vomiting, kidney disease, liver disease, cancer and death (Cheli et al., 2008). The first mycotoxin aptamer was developed for Ochratoxin A (OTA). This toxin is produced by *Aspergillus ochraceus* and *Penicillium verrucosum* and is one of the most abundant food-contaminating mycotoxins in the world (De Girolamo et al., 2011). OTA is a nephrotoxic toxin, with strong carcinogenic effects on rodents, as well as documented teratogenic and immunotoxic effects in humans (Cruz-Aguado & Penner, 2008a). Since its development in 2008, this aptamer has been integrated into several biosensor detection systems including electrochemical, electrochemiluminescent, colorimetric and fluorescent platforms as well as chromatographic and fluorescent test strips. More recently, an aptamer for fumonisin B_1 (FB_1) has recently been developed. Fumonisin mycotoxins are produced by *Fusarium verticillioides* and *F. proliferatum* species, fungi that are ubiquitous in corn (maize). Insect damage and some other environmental conditions result in the accumulation of fumonisins in corn-based products worldwide. FB_1 is a nephrotoxin in all species tested, a carcinogen in rodents and a reproductive toxicant in rodents and likely in humans (Bolger et al., 2001). Both the FB_1 and the OTA DNA aptamers bind to their target with nanomolar dissociation constants.

2.1.3 Heavy metals

Inorganic metals contaminate foodstuffs including fish and fish products, meat and meat products, milk and milk products, eggs, fats and oils as well as animal feeds. Arsenic is a toxic carcinogen found in many parts of the world. It can exist in four oxidation states (-3, 0,+3, and+5); however, arsenate [As(V)] and arsenite [As(III)] are the most abundant.

Human exposure can occur through direct ingestion, such as drinking arsenic contaminated water, and through indirect ingestion such as consuming crops grown from arsenic-accumulated soils. Both acute and chronic health effects may result, the more serious effects include cancer, skin lesions, arsenicosis and cardiovascular diseases (M. Kim et al., 2009). Kim *et al.* developed a high affinity DNA aptamer for arsenic that can bind to arsenate [(As(V)] and arsenite [As(III)] with a dissociation constant of 5 and 7 nM respectively (M. Kim et al., 2009).

Mercury ion (Hg^{2+}) is highly toxic and a widespread contaminant. It is a potent neurotoxin that damages the central nervous and endocrine system. In addition, fish and shellfish concentrate mercury in their bodies, often in the form of methylmercury. The presence of mercury in fish can be a health issue, particularly for women who are or may become pregnant, nursing mothers, and young children (Xu et al., 2010). Routine detection of mercury is central for evaluating the safety of aquatically derived food supplies. As shown in Table 1, several label-free DNA aptamer based sensors have been developed for the direct detection of Hg^{2+}.

2.1.4 Food packaging

Contaminants migrating from food packaging have been detected in several food related matrices including foodstuffs, food stimulants, and food contact materials and articles. Bisphenol A (BPA), known to be an endrocrine disruptor since the 1930s, is used as a monomer compound in polycarbonate plastic products. Major concerns regarding the use of bisphenol A in consumer products were reported by the media in 2008. In 2010, a report from the United States Food and Drug Administration (FDA) raised further concerns regarding BPA exposure to fetuses, infants and young children. In September 2010, Canada became the first country to declare BPA as a toxic substance and to ban its use in baby bottles (US FDA, 2010). As seen in Table 1, several aptamers and aptamer platforms have been developed in response to these concerns. For example, Lee *et al.* developed an aptamer-sandwich based carbon nanotube sensor able to detect BPA at very low concentrations (Lee et al., 2011).

2.1.5 Adulterants

After the first report in 2003 indicating the illegal presence of the dye Sudan I in some foods in the European Union (EU), there have been many reports of the presence of illegal dyes in chili powder, curry powder, processed products containing chili or curry powder, sumac, curcuma and palm oil. These dyes are often both genotoxic and carcinogenic (EFSA, 2005). One such illegal dye is rhodamine B. While there is no known aptamer that recognizes rhodamine B, there is both a DNA and RNA aptamer that recognizes and binds to sulforhodamine B (Holeman et al., 1998; Wilson & Szostak, 1998).

Melamine is another chemical adulterant that is sometimes illegally added to food samples to increase the apparent protein content. Codex guidelines set the maximum amount of melamine allowed in powdered infant formula to 1 mg/kg and the amount allowed in other foods to 2.5 mg/kg. An aptamer-based resonance scattering assay has been developed for melamine detection. The oligo-T sequences with melamine binding affinity, however, were not determined through SELEX, but rather designed based on the idea that multiple hydrogen bonding interactions could support formation of an aptamer-melamine complex. While the design of this aptamer raises some concern about potential specificity problems, the authors confirmed that common metal ions, amino acids and proteins do not interfere with the assay (Z. Jiang et al., 2011).

2.1.6 Pesticides

Malachite Green (MG) has been used in aquaculture since 1936. It is a fungicide approved for use in aquarium fish in Canada; however, it is not approved for use in fish intended for human consumption. MG has been used for the treatment of external fungal and parasitic infections on fish eggs, fish and shellfish. It is also a very popular treatment against the common fresh water fish disease, ichthyophthirius (Stead et al., 2010). However, it has been determined that eating fish contaminated with malachite green poses a significant health risk. Scientific evidence indicates that a metabolite of malachite green (LMG), leucomalachite green, may be a genotoxic carcinogen that persists in fish tissues. A malachite green aptamer was developed in 1999 (Grate & Wilson, 1999). This aptamer has been recently used by Stead *et al.* as the recognition element in a fluorescence-based screening assay for MG detection in fish tissue (Stead et al., 2010). (See Section 2.2.1)

Atrazine is one of the most heavily used herbicides in Canada and the United States, as it is used as a pre- and post-emergence weed control agent for corn and rapeseed. Atrazine is a persistent environmental pollutant and widespread contamination of groundwater has been reported in the United States. It is associated with birth defects, menstrual problems and cancer when consumed by humans (Health Canada, 1993). Although it has been excluded from a re-registration process in the European Union, it is still one of the most widely used herbicides in the world. Sinha *et al.* used a combination of *in vitro* and *in vivo* selection to develop a synthetic riboswitch (an aptamer sequence coupled to an expression platform that can regulate a gene upon binding) to atrazine (Sinha et al., 2010).

2.2 Small molecule food safety targets case studies

As seen in Table 1, there are several biosensor platforms that are used for the detection of small molecular food contaminants. Here, we highlight a few recent sensors from the literature that have been used and tested directly in food samples.

2.2.1 Fluorescence-based detection of malachite green in fish tissue

Malachite green (MG) is a triphenylmethane dye used as a fungicide, antiparasitic and antiprotozoan agent in the treatment of farmed fish. Leucomalachite green (LMG) is the primary metabolite and is predominant to MG *in vivo* and highly persistent in edible fish tissue. Despite the potential health risk MG and LMG pose to humans, surveillance programs have identified their continuing incidence in farmed fish samples. For example, there was over 100 notifications under the EU Rapid Alert System for Food and Feed (RASFF) regarding the illegal use of MG between 2003 and 2007 (Stead et al., 2010). Current screening methods must be able to detect total MG and LMG concentrations at or below 2µg/kg. Unfortunately, typical detection methods for MG are time consuming and include high-performance liquid chromatography HPLC with visible detection or liquid chromatography–mass spectrometry LC-MS. Prior to the novel aptamer-based procedure, there has been limited progress in the development of rapid screening methods. This screening procedure, developed by Stead et al., is the first reported use of an RNA aptamer based detection system for the detection of chemical residues in food.

In 1999, Grate et al. developed a 38-mer RNA aptamer (see Fig. 2) that bound to MG with a dissociation constant (K_d) of less than 1µM (Grate & Wilson, 1999). The K_d value for the aptamer produced for use in this study was estimated to be approximately 600 nM, assuming a 1:1 binding model.

While MG has a low fluorescence quantum yield due to its facile vibrational de-excitation, a greater than 2000 times fluorescence enhancement was noted when bound by the aptamer.

Electrostatic forces and base stacking upon binding with the aptamer causes the aromatic rings of MG to adopt a more coplanar, rotationally stabilized structure, leading to this effect.

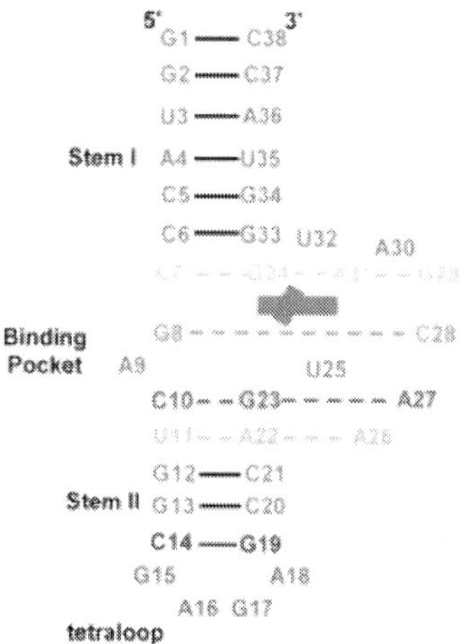

Fig. 2. Secondary stucture of the malachite green-RNA complex with the MG binding pocket indicated. Yellow indicates a base quadruple, red and green mark base triples, cyan the G8-C28 base pair and purple a GNRA tetraloop. The base pair adjacent to the tetraloop is shown in dark blue, the other stem base pairs are colored grey (Flinders et al., 2004). Reproduced with permission from Wiley-VCH.

The fluorescence signal associated with the RNA-MG complex was determined by excitation at a wavelength of 618nm and measuring emission at 643nm. The binding interaction produced a clear linear relationship with fluorescence intensity over the range of 0-40nM of MG. The pH operational range was determined to be between pH 5 and 9 for the RNA-MG complex, where the optimal fluorescent signals being observed between pH 6 and 7.5. A concentration of 5-10mM of Mg2+ and the presence of some monovalent cation (either Na+ or K+) was found to be required for the stability of the complex. Formation of this complex was time and temperature dependent. Maximum binding was achieved after 16 minutes at ambient temperature and remained stable for an hour. At temperatures outside this range, the formation of the ligand binding pocket was likely disrupted and signals were not fully recoverable. However, the complex at 0°C, once formed, was the most stable, surviving for about 4 weeks.

In order to perform the MG assay, fish tissues were extracted using acetonitrile. Since total LMG and MG is required for food safety detection, the ability of the RNA aptamer to detect LMG was investigated. However, addition of LMG to the RNA aptamer resulted in no change in fluorescence intensity. Due to this apparent lack of binding, an oxidation procedure was incorporated following extraction to ensure LMG was oxidized to the MG form. A final clean-up was then performed using solid-phase extraction (SPE) cartridges prior to analysis. Testing was performed on several spiked fish tissue samples including *Salmo trutta* and *Oncorhynchus mykiss* spp. and *Salmo salar*. Strong fluorescent signals were obtained in the positive samples compared to the blank controls. The threshold concentration of MG at which the signal was detectable was 5nmol/L (2ng/mL), a value sufficient for RASFF testing standards. It was estimated that a batch of 20 samples can be prepared and analyzed within 4 hours and it is possible to automate the SPE steps. Therefore, this novel RNA aptamer testing system has proven to be a successful rapid detection and semiquantitative method for direct MG detection in fish tissue.

2.2.2 Aptamer-based strip assay for toxin detection

Ochratoxins are dangerous by-products of several fungal species, mainly in the *Aspergillus* and *Penicillium* genera, which can contaminate foods and beverages including cereals, beans, nuts, spices, dried fruits, coffee, cocoa, beer, and wine. Ochratoxin A (OTA) is nephrotoxic and carcinogenic and poses a serious threat to the health of both humans and animals. The International Agency for Research on Cancer has classified OTA as a possible carcinogen to humans. Based on risk assessment performed by the Joint Food and Agriculture Organization/WHO Expert Committee on Food Additives (JECFA), over 50% of human exposure to OTA is a result of exposure from cereals and cereal products. Regulatory limits for OTA exist in several countries, and testing of products is carried out at central testing laboratories. For example, the European Commission established a maximum limit of 5 µg/kg for raw cereal grains and 3 µg/kg for all cereal-derived products (De Girolamo et al., 2011). In 2008, Cruz-Aguado et al. developed the first mycotoxin aptamer for OTA. The selected DNA aptamer bound to OTA with a nanomolar dissociation constant and displayed high selectivity (Cruz-Aguado & Penner, 2008a).

With this sequence, Wang et al. developed a chromatographic strip assay method for rapid OTA detection (L. Wang et al., 2011b). The aptamer-based strip assay was based on the competition for the aptamer modified with gold nanoparticles (GNPs) as the visual reporter between ochratoxin A and DNA probes. GNPs were prepared and thiolated aptamers (Aptamer: 5'-GAT CGG GTG TGG GTG GCG TAA AGG GAG CAT CGG ACA AAA AAA AAA AAA AAA AAA-SH-3') were self-assembled on the nanoparticle surface. DNA probes (Test line DNA probe 1: 5'-Biotin-CTA GCC CAC ACC CAC CGC ATT TCC CTC GTA GCC TGT-3' Control line DNA probe 2: 5'-Biotin-TTT TTT TTT TTT TTT TTT-3.) were conjugated to streptavidin using the 5' biotin. The strip was assembled as follows (see Fig. 3). The nitrocellulose membrane, glass-fiber membrane (conjugated pad), sample pad and absorbent pad were layered in sequence and pasted onto a plastic backing plate. The streptavidin–DNA probe 1 and streptavidin–DNA probe 2 conjugates were deposited onto the strip to form the test line and control line. Finally the GNP–aptamer probe was added to the glass fiber membrane.

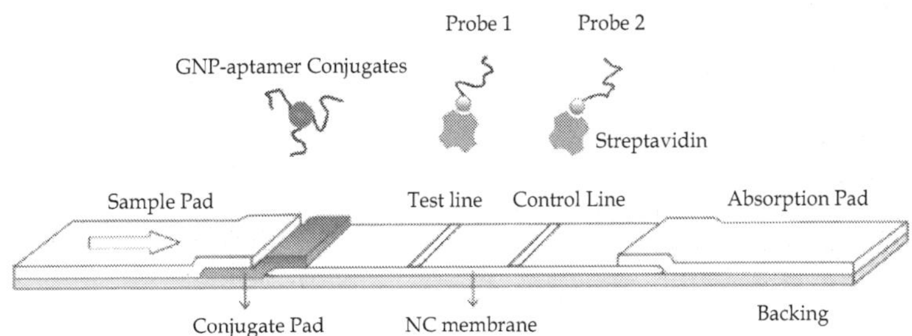

Fig. 3. Construction of the aptamer-based strip. Reproduced with permission from Elsevier.

This strip relied on the competitive reaction between the DNA probe 1 (test line) and the target OTA. In the presence of OTA, the aptamer-GNP could not hybridize to the DNA probe 1 in the test line, thus causing the red color intensity to become weaker. Regardless of the concentration of OTA, the aptamer–GNP probe could hybridize with DNA probe 2 in the control line to ensure the validity of the detection test (see Fig. 4).

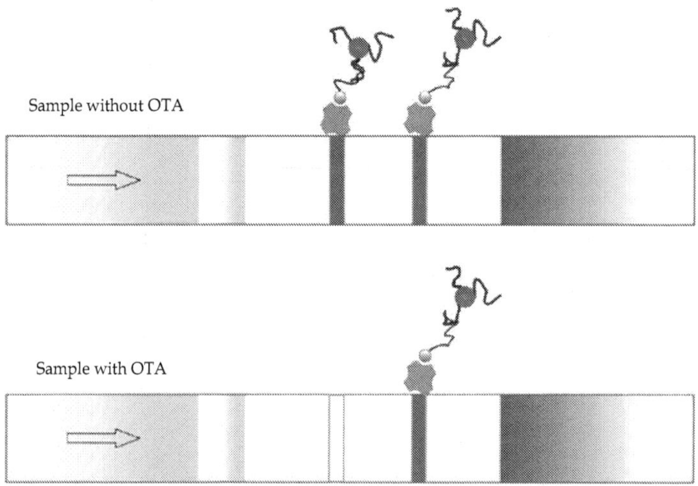

Fig. 4. Schematic of the detection principle of the strip. Reproduced with permission from Elsevier.

When using this strip for qualitative purposes, it was found to have a visual limit of detection (LOD) of 1ng/mL. However, it is possible to use this strip for semi-quantitative purposes. Using a scanning reader, a quantitative calibration curve was constructed. Based on this, the quantitative LOD was 0.18ng/mL which is better than the antibody-based strip method and comparable to the ELISA methods used for detection. Specificity of this strip assay was also tested using the mycotoxins deoxynivalenol, fumonisin B_1, zearalenone and microcystin-LR and was found to be specific for OTA. These fabricated strips were stable after 30 days. In addition, OTA analysis was performed on spiked wine samples. The recoveries were from 96% to 110% which met the detection requirements. Overall, this semi-quantitative assay proved to be rapid (less than 10 minutes), inexpensive and reliable in the detection of OTA.

2.2.3 Tetracyline determination in milk using an aptamer-modified electrode

Tetracycline is a naturally occurring, broad-spectrum polyketide antibiotic produced by species of *Streptomyces*. Tetracycline has been a popular and economically valuable drug for the last six decades as it can be easily isolated by fermentation. It is used against many bacterial infections due to its ability to inhibit protein synthesis. For these reasons, tetracyline has been widely used in human therapy, aquaculture and veterinary medicine. It is also extensively used as an animal growth promoter (Berens et al., 2001). As a result, tetracycline has been detected in food products, such as meat, milk, eggs and chicken.

Therefore, there is an increase in demand to discover novel methods for fast tetracycline detection that are applicable to food matrices. Zhang et al. developed one such method using a DNA aptamer modified glassy carbon electrode (Zhang et al., 2010). While there are several aptamers that have been developed for tetracycline (as seen in table 1), the authors performed a new selection to develop DNA aptamers for this assay. These modified glassy carbon electrodes were amino-functionalized using 3-aminopropyltriethoxysilane chemistry, allowing the aptamers to be tethered to the surface using 1-Ethyl-3-[3-dimethylaminopropyl]carbodiimide hydrochloride/ N-hydroxysulfosuccinimide (EDC/NHS) chemistry. The cyclic voltammetry (CV) measurements were performed using the aptamer-modified electrode with a standard three electrode system in the presence of $K_3Fe(CN)_6$ as a redox reporter. In the absence of tetracycline, a $Fe_3(CN)_6{}^{3-}/Fe_2(CN)_6{}^{4-}$ redox couple was measured due to its interaction with the electrode surface. In the presence of tetracyclines, the aptamer binds to the target, blocking the electrode surface and interrupting the redox system. This leads to a decrease in peak current. Peak current vs. concentration was measured and plotted as a standard curve for samples in phosphate buffered saline as well as directly in milk (see Fig. 5). A linear relationship between the tetracycline concentration and the current was found between 0.1–100ng/mL, and a LOD of 1 ng/mL was achieved. There was no significant response from penicillin, demonstrating the specificity of the system. Finally, six concentrations of tetracyclines in milk samples were analyzed using the electrode biosensor. The measured concentration vs. the actual concentration showed less than a 10% coefficient of variation, indicating the biosensor produces precise results. This biosensor proved to be a reliable, sensitive, inexpensive and specific detection system for tetracycline in milk samples. In addition, detection time was only 5 minutes; therefore, this system could prove useful in food safety testing.

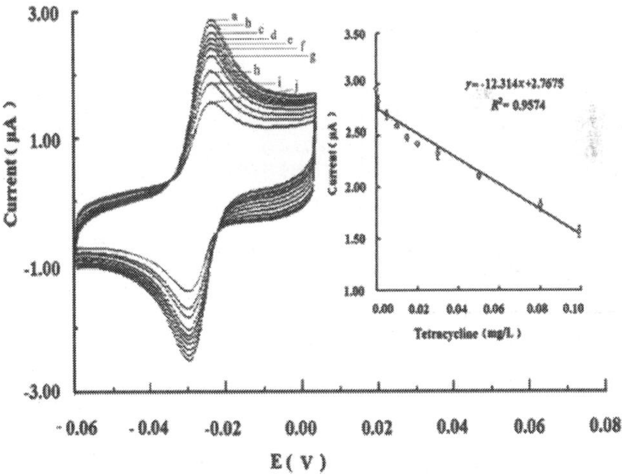

Fig. 5. Cyclic voltammograms of the tetracycline milk determination. The tetracycline concentrations are (a) 0mg/L (b) 0.001 mg/L (c) 0.005 mg/L (d) 0.01 mg/L (e) 0.015 mg/L (f) 0.02 mg/L (g) 0.03 mg/L (h) 0.05 mg/L (i) 0.08 mg/L (j) 0.1 mg/L. The inset is the plot of the tetracycline concentration versus the peak current at 0.02V. Reproduced with permission from RSC.

2.3 Aptamers for pathogenic food safety targets

Foodborne pathogens are implicated in millions of illnesses and thousands of deaths on a yearly basis. In addition to the clear health risk associated with contaminated foods, there is also the often devastating economic impact to the food producer that must be considered. As a result, rapid, accurate and sensitive assays and biosensors for bacteria, viruses, and other similar targets are of significant interest. Table 2 lists aptamers and aptamer-based biosensor platforms that have been developed for these targets.

Target Class	Target	DNA/RNA	K_d	Biosensor Platform	LOD	Ref
Bacteria	*Bacillus Thuingiensis* (spores)	DNA	-	Fluorescence (Quantum dots)	1000 cfu/mL	(Ikanovic et al., 2007) (Bruno & Kiel, 2002)
	Campylobacter Jejuni	DNA	-	Fluorescence (Magnetic Bead/Quantum dot sandwich)	10-250 cfu in food matrix; 2.5 cfu in buffer	(Bruno et al., 2009)
			-	Capillary Electrophoresis Immunoassay	6.4×10^6 cells/mL	(Stratis-Cullum et al., 2009)
	Escherichia Coli DH5α strain	RNA	-	Single-walled Carbon Nanotube (SWNT) Field Effect Transistor (FET)	-	(So et al., 2008)
	CECT 675 cells	RNA	-	Electrochemical (SWNT Potentiometry)	6 cfu/mL (milk) 26 cfu/mL (apple juice)	(Zelada-Guillen et al., 2010)
	Crook's strain (8739)	DNA	-	Fluorescence Resonance Energy Transfer (FRET) and Surface Plasmon Resonance (SPR)	30 *E. Coli* units/mL	(Bruno et al., 2010)
	ETEC K88 fimbriae protein	DNA	25-44 nM	Fluorescence Assay	-	(Li et. al., 2011)
	Francisella tularensis - japonica (bacterial antigen)	DNA	25 ng (K_a)	Aptamer-Linked Immobilized Sorbent Assay (ALISA)	1.7×10^3 bacteria/mL	(Vivekananda & Kiel, 2006)
	Listeria monocytogenes (internalin A protein)	DNA	84 nM	Fibre-optic sensor (Aptamer and Antibody-modified)	1000 cfu/mL	(Ohk et al., 2010)
				Magnetic beads	1 cfu/mL	(Yamamoto et al., 2010)

Target Class	Target	DNA/ RNA	K_d	Biosensor Platform	LOD	Ref
	Salmonella Typhimurium (outer membrane proteins)	DNA	-	Magnetic beads	<10 cfu/g (spike and recovery)	(Joshi et al., 2009)
	(Type IVB pilus protein)	RNA	9 nM	-	-	(Pan et al., 2005)
	Staphylococcus aureus	DNA	480 ± 210 nM	Confocal microscopy of infected samples	-	(Cao et al., 2009)
	Yersenia (Yop51)	RNA	18 nM, 28 nM	-	-	(Bell et al., 1998)
Viruses	Apple-Stem pitting virus	DNA	PSA-H protein: 8 nM; MT32 protein: 55 nM	Surface Plasmon Resonance	250 nM	(Lautner et al., 2010)
				Double Oligonucleotide Sandwich Enzyme-Linked Oligonucleotide Assay (DOS-ELONA)	100 ng/mL	(Balogh et al., 2010)
Other	Botulinum neurotoxin	DNA	3-51 nM	-		(Tok & Fischer, 2008)
	Egg-white lysozyme (allergen trigger)	DNA	3 nM	-	-	(Tran et al., 2010)
				Electrochemical (voltammetry)	0.5 µg/mL	(Cheng et al., 2007) (Kirby et al., 2004)
		RNA	-	Fluorescence (microarray)	70 fM	(Collett et al., 2005)

Table 2. Table of aptamers and aptamer-based biosensor platforms for pathogen and macromolecular targets.

Section 2.3 will describe these systems briefly, while three biosensor platforms targeting pathogens in food will be discussed in Section 2.4.

2.3.1 Bacteria

Foodborne illnesses and outbreaks are commonly caused by bacteria. Visible symptoms of infection typically do not appear until about 12-72 hours after food consumption. Thus, it is important to be able to detect these pathogens rapidly, at very low levels in complex food matrices, as early as possible in the food continuum to minimize the risk of illness and to avoid product recalls. In response to these needs, a number of aptamers and aptamer-based sensor platforms have been developed and demonstrate much potential in food safety applications.

Some strains of the gram-positive bacteria *B. Cereus* are responsible for foodborne illnesses causing severe nausea, vomiting and diarrhea. While the presence of large numbers of *B. cereus* (greater than 10^6 organisms/g) in a food can lead to illness, the US FDA recommends that *B. cereus* in infant formula does not exceed 100 cfu/g. Ikanovic et *al.* selected a DNA aptamer for *Bacillus thuingiensis* (BT) spores with eight rounds of selection. The aptamer was then applied to produce a novel, solution-based sensor by using quantum-dot reporter molecules for detection via fluoresence spectroscopy. BT spores are closely related to the food poisoning agent *B. cereus* so detection of BT demonstrates potential cross-over applications to its foodborne relative.

C. jejuni, a species of gram-negative bacteria commonly found in animal feces, is one of the most common causes of human gastroenteritis and has been linked with subsequent development of Guillain-Barré syndrome. A dose as low as 400-500 organisms can cause infection. DNA aptamers for *Campylobacter jejuni* have been incorporated into sensing platforms (Bruno et al., 2009; Stratis-Cullum et al., 2009). The former used a sandwich assay, comprised of aptamers conjugated to both magnetic beads and red quantum dots, in order to detect the bacterial protein in food matrices (see section 2.4). The latter implements a capillary electrophoresis immunoassay to characterize the bacterial cells.

Detection of *E. Coli* O157:H7 is of high priority in food surveillance as infection with this gram-negative bacteria can lead to hemorrhagic diarrhea, and occasionally kidney failure, especially in vulnerable populations such as infants and the elderly. Sensitive methods are particularly important here as the infectious dose can be as low as 10-100 organisms. Several DNA and RNA aptamer-based platforms have been developed for strains of food-borne *E.coli.*, which serve as models for O157:H7 detection. One example is an aptamer-modified single-walled carbon nanotube in both field-effect transistors (So et al., 2008) and label-free potentiometry measurements (see section 2.4) (Zelada-Guillen et al., 2010) to provide rapid, reusable biosensors. Additionally, a FRET-based assay using a DNA aptamer for *E. coli.* has been developed (Bruno et al., 2010).

Francisella tularensis–japonica is a bacterial pathogen that causes tularemia in humans and can be transmitted through contact with infected animals. With an estimated infectious dose of 1 bacillus, signs of the infection in the form of skin lesions occur within 3-5 days. Other signs include, fever, lethargy, anorexia and potentially death. A DNA aptamer cocktail for this bacteria was selected with 10 rounds of SELEX. Upon testing in a sandwich Aptamer-Linked Immobilized Sorbent Assay (ALISA) and dot blot analysis, the aptamer cocktail, which contained 25 unique sequences, exhibited good binding specificity in its ability to recognize only the bacteria in comparison to *Bartonella henselae*, pure chicken albumin or chicken lysozyme (Vivekananda & Kiel, 2006). Thus, it appears that this novel anti-tularemia aptamer cocktail may find application as a detection reagent for *F. tularensis* (Jones et al, 2005).

Listeria Monocytogenes is a virulent foodborne bacteria that causes hundreds of deaths in the US annually. The FDA, USDA and EU have all implemented a zero-tolerance rule for *L. monocytogenes* in ready-to-eat foods, leading to a substantial effort to develop very sensitive biosensors for the pathogen. Aptamer-based sensors have been developed to detect components of the *Listeria monocytogenes* bacteria. One platform uses magnetic beads conjugated to DNA aptamers for target detection and also investigated therapeutic applications for this platform. Another platform currently developed involves an antibody-aptamer functionalized fibre optic sensor that detects *L. monocytogenes* cells from contaminated ready-to-eat meat products such as sliced beef, chicken and turkey (Yamamoto et al., 2008, 2009, 2010).

Salmonella enterica is among the most commonly encountered bacteria and a prominent cause of foodborne illness around the world. Most frequently found in contaminated beef and poultry, the infectious dose for the enteric subspecies *typhimurium* is approximately 10^4 bacilli (Srinivasan et al, 2004). It is a notable target of interest as it has been demonstrated to have a multi-drug resistance. Joshi and co-workers (2009) have reported a DNA aptamer-based platform for capturing and detecting *S. typhimurium* outer membrane proteins at low levels using magnetic beads. Conversely, an RNA aptamer was selected for the IVB Pili protein of S. *typhimurium,* which is linked to the pathogenesis of the bacteria. The 70-nt sequence was determined to have low nanomolar binding constant a stem-loop secondary structure. Further studies found that the aptamer could also significantly inhibit the protein into human monocytic cells (THP-1) (Pan et al., 2005), demonstrating its potential as a therapeutic.

The U.S. FDA estimates an infectious dose of $10^5\text{-}10^6$ bacilli for *Staphylococcus aureus* (Schmid-Hempel & Frank, 2007). This bacteria is well known for causing a wide range of illnesses from minor wound infections to major diseases such as pneumonia. Staphylococcal food poisoning can cause symptoms such as nausea, vomiting, retching and abdominal cramping. Whole-bacteria subtractive SELEX was performed on *Staphylococcus aureus* to yield an aptamer cocktail of five sequences (Cao et al., 2009). As determined by competition experiments and flow cytometry, the sequences were effective in binding to various strains of the bacteria in a mixture.

Yersinia enterocolitica, a rare foodborne bacteria, has a relatively higher infectious dose than the above mentioned bacteria ($\sim10^9$ bacteria) but nevertheless produces similar clinical symptoms including diarrhea, mild fever and abdominal pain. It is often found in undercooked pork but also in unpasteurized milk and untreated water. One promising study comes from Bell *et al.* (1998) where two aptamers were selected for the PTPase enzyme, Yop51, which has been considered a virulent determinant for *Yersinia.* The sequences were characterized with nanomolar binding and found to effectively inhibit the enzyme, which have important implications in viral replication. It was found that the two sequences shared a 21-residue motif and bound specifically to Yop51 over a homologous PTPase called rat PTP1.

2.3.2 Viruses

Despite the existence of pathogenic foodborne viruses, few virus-targeting aptamers have been developed. Aptamers for food viruses such as Apple Stem-Pitting Virus (ASPV) have also demonstrated their significance in the food industry. Although these viruses may not necessarily be contracted by humans, they compromise the integrity and appearance of crops in agricultural development, which ultimately impacts their marketing value and can result in a loss of revenue. This aptasensor example (Lautner et al., 2010) and its applicability will be discussed in section 2.4.

2.3.3 Other macromolecular targets

There are other targets that do not fall under the classification of bacteria or viruses that are relevant for food safety detection. One example is the botulism neurotoxin. This protein is associated with *Clostridium botulinium,* a common contaminant can in canned foods that have been improperly prepared (e.g. heated before canning). The neurotoxin has a very lethal dose (~ng of bacteria) and is considered one of the most poisonous substances known. Tok & Fischer (2008) selected a DNA aptamer for two botulinum neurotoxin-related targets, BoNT-toxoid and BoNT Hc-peptide. Using single microbead SELEX, the study resulted in high-affinity sequences that competitively bound to the target over corresponding antibodies. Another noteworthy example that has been studied over the past few years is egg-white lysozyme, a protein considered to be an allergen-trigger in consumers with egg allergies. A few promising aptasensor platforms developed so far include label-free voltammetric assays (Cheng et al, 2007), Faradaic impendance spectroscopy (Rodriguez et al., 2005) and RNA microarray technology (Collett et al., 2005).

2.4 Pathogen case studies

As seen in Table 2, there are several biosensor platforms that may be applicable to real-world food testing. Here, we highlight a few recent aptamer-based sensors from the literature that have been used and tested directly in food samples.

2.4.1 Real-time potentiometric detection of bacteria in complex samples

One example of a feasible aptasensor for use in pathogen detection in food is the work presented by Zelada-Guillén and co-workers. They report the development of an electrochemical aptasensor using single-walled carbon nanotubes (SWNT) to selectively detect a non-pathogenic strain of *E. coli* cells to serve as a model for pathogenic strain O157:H7. Their motive for this aptasensor was to provide a simple, rapid and label-free means of detecting the target with comparably high sensitivity and selectivity.

The aptamer-based sensor was prepared by depositing SWNT on a polished glassy carbon surface. Amino-modifed aptamers for *E. coli* CECT 675 were tethered to the SWNT carboxyl groups via EDC/NHS chemistry. For *E. coli* CECT 675 testing in phosphate buffered saline (PBS), electromagnetic field (EMF) response was recorded to provide insight into the aptasensor's performance in terms of its stability and response. A rapid increase in EMF signal was observed from 4 cfu/mL to ~10^4 cfu/mL with 50% of the response achieved within seconds. The signal remained stable over almost 2 hours. Upon regeneration of the sensor, the EMF baseline remained the same and it was determined that regeneration could be done up to five times before the detection limit and sensitivity were compromised. The selectivity tests showed no cross-reactivity indicating that the aptamer not only had preference for its target but it could also discriminate the particular *E.coli* strain. In order to detect the bacteria in real beverage samples, the ionic strength of the sample needed to be controlled through a prefiltration step to prevent a false-positive response. The biosensor was tested on apple juice and skim milk samples that had been spiked with with *E. coli* CECT 675 and were compared to control samples of the liquid matrices (Fig. 6). Specificity was confirmed through controls of buffer, juice and milk contaminated with *E. coli* CECT 4558, *Lactobacillus casei* CECT 4180 and *Salmonella enterica* CECT 409. The aptasensor was exposed to increasing concentrations of bacteria. EMF measurements were made with a change in concentration every 10 seconds. Simultaneous measurements using the plate count method and the MacConkey agar test were conducted for experimental validation. The limit of detection for CECT 675 in milk was determined to be 6 cfu/mL and in apple juice was 26 cfu/mL. These detection limits are comparable with European regulated limits demonstrating the sensor's applicability in food testing. Incorporating O157 aptamers in a similar biosensor format therefore holds a great deal of potential.

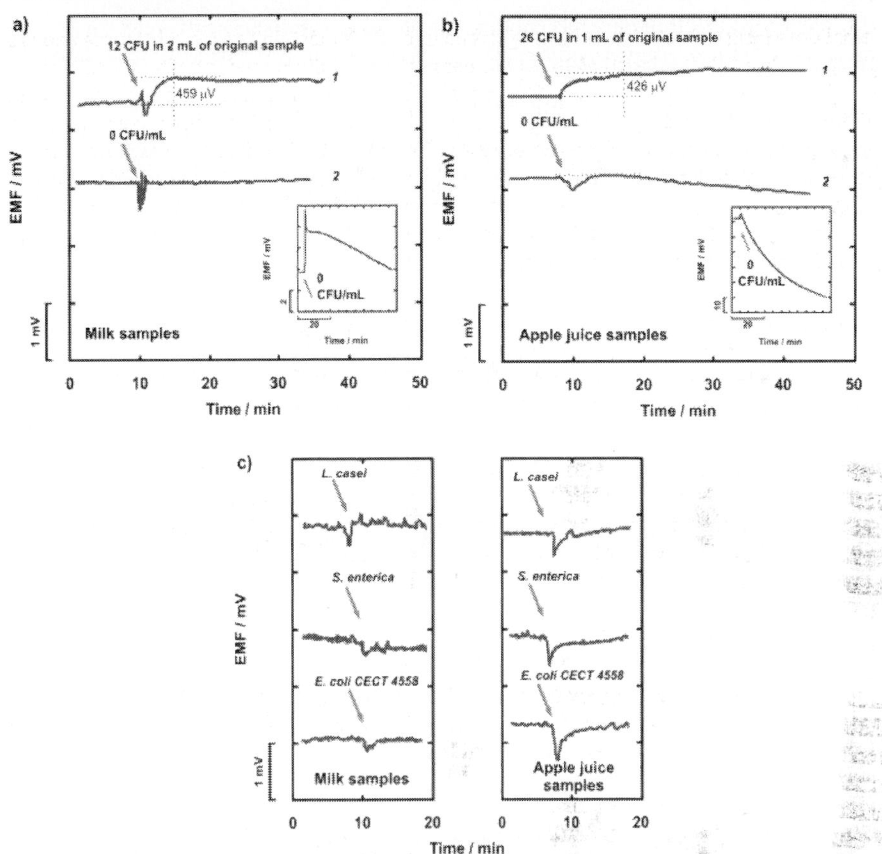

Fig. 6. Detection of microorganisms in liquid matrices via potentiometry. (a) Sample of skim milk containing 12 cfu of *E. coli* CECT 675 (1) compared to uninfected milk (2). (b) Sample of apple juice containing 26 CFU of *E. coli* CECT 675 (1) compared to uninfected juice. Insets in both graphs demonstrate the importance of pretreatment of these samples to reduce false positive results caused by other charged species in the samples. (c) Selectivity studies in milk (left) and apple juice (right) samples with 10^3 CFU/mL of various bacteria (*L. casei, S. enterica, E. coli* CECT 4558). Reproduced with permission from ACS.

2.4.2 Plastic-adherent DNA aptamer-magnetic bead and quantum dot sandwich assay for *Campylobacter* detection

Another example of a promising aptasensor for food safety is presented by Bruno *et al.* where they have developed a sandwich assay for *Campylobacter jejuni*. This concept of a an assay built from aptamer functionalized magnetic beads and quantum dots is based on their previous assay developed for BT spores (found in Table 2). The schematic of the assay is illustrated in Fig. 7. Two amino functionalized aptamers, designated C2 and C3 were prepared. C2 was conjugated to magnetic beads to behave as the capture probe, while C3 was conjugated to red CdSe/ZnS quantum dots to behave as the reporter probe. Equivalent amounts of the two aptamer particles were added to a 1cm polystyrene cuvette. Live and heat-killed *C. jejuni* bacteria were transferred to the cuvettes and with the facilitation of a magnet, the probes were led towards the inner face of the cuvettes allowing them to adhere to the surface. This eliminated the need for washing steps and allowed fluoresence to be measured in one plane with low background signal (Fig. 7). Moreover, the assay components could remain on the wall of the cuvette for up to a few weeks in sterile buffer at ambient temperature.

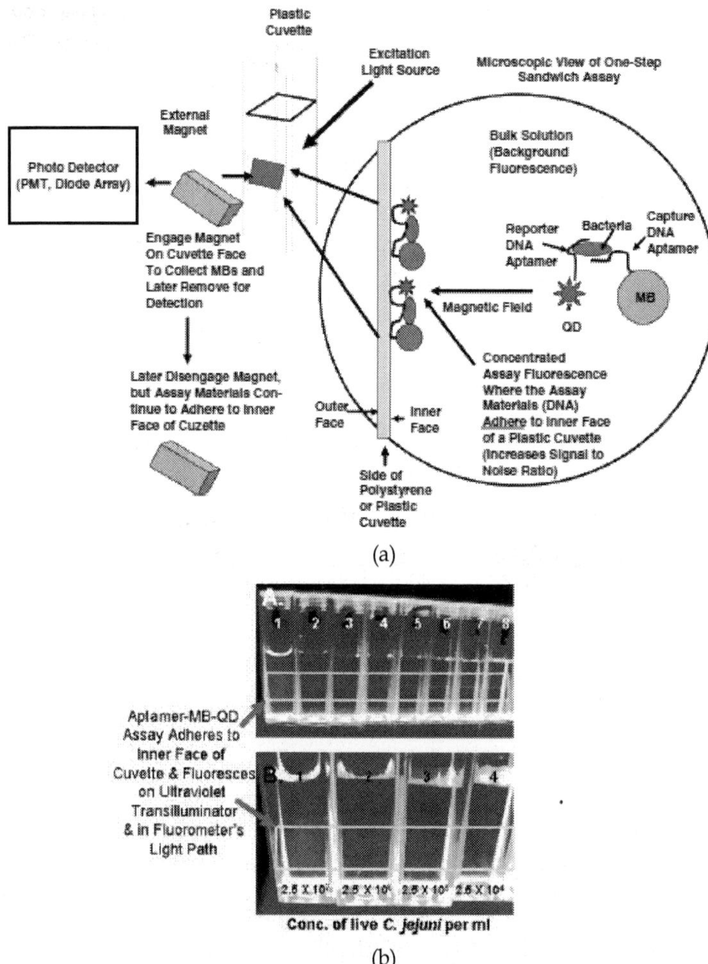

Fig. 7. (a) Schematic illustrating concept of sandwich assay for Campylobacter jejuni detection. (b) Assay of 10-fold serial dilutions of C.jejuni in binding buffer. Reproduced with permission from Springer.

Fluoresence emission from the quantum dot conjugated aptamer was measured and the detection limit in binding buffer was found to be 2.5 cfu/mL for both live and heat-killed *C. jejuni* bacteria (Fig. 8). Bacterial detection was also attempted in diluted food matrices such as 2% milk, chicken juice and ground beef wash. The detection limit for both live and heat-killed *C. jejuni* bacteria was determined to be in the range of 10-250 cfu/mL. In this case, a portable fluorometer could be effectively used as the fluorescence detector allowing this biosensor to serve as a practical field-based detection kit.

This aptasensor was also previously found to have very low cross-reactivity with other bacteria such as *E. coli* O157:H7, *L.monocytogenes* and *S.typhimurium*. In the present study, the aptasensor was further tested against different species of *Campylobacter* such as *C. coli* and *C. lari* in addition to other bacterial genera. The resulting cross-reactivity was observed to be very low for species outside the *Campylobacter* family but was fairly high for those within the genus, indicating potential use for detecting these other two pathogens.

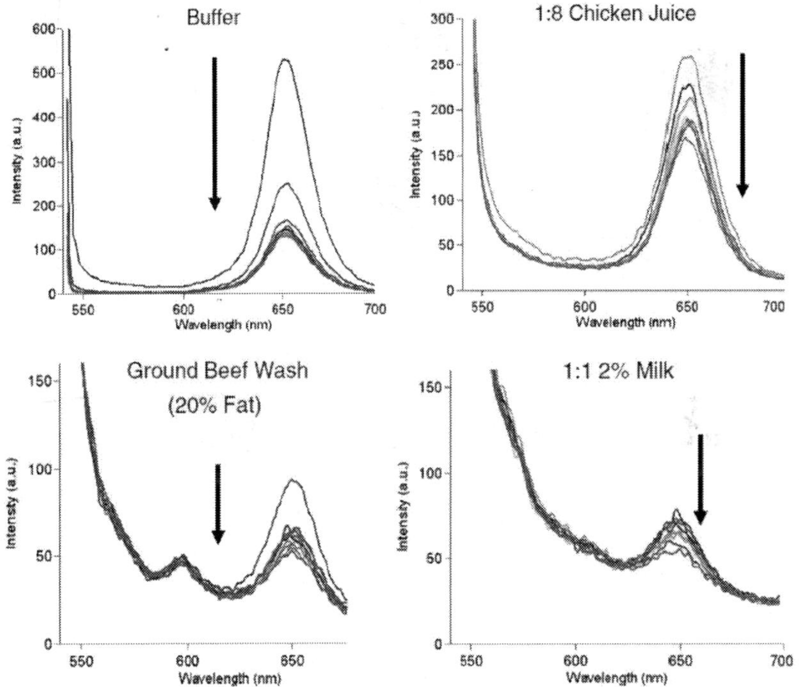

Fig. 8. Detection of C.jejuni bacteria in various food matrices using spectrofluorometry. Tenfold serial dilutions (using binding buffer as diluent) from 2.5 x 10^6 cfu/mL to 0 bacteria, indicated by the direction of the arrows, were measured. Reproduced with permission from Springer.

2.4.3 Aptamer-based biochips for label-free detection of plant virus coat proteins by SPR imaging

Apart from human health as a factor that is accounted for in food biosensors, food quality and integrity are also of importance. The appearance of food can influence the consumer's decision to buy based on their perception of what "looks safe". Apple-stem pitting virus (ASPV) is a widely spread filamentous virus associated with growth disorders of pome fruits (e.g. apples and pears). Characteristics of the virus include yellowed veins and darkened spotted areas that appear charred. Current methods in detecting the virus such as real-time PCR are effective yet involve pre-treatment of samples and may become less sensitive when evolved strains of the virus are present. Thus, a label-free aptamer-based sensor for ASPV detection was proposed by Lautner and coworkers. Their sensor is composed of thiolated aptamers for ASPV coat proteins, PSA-H and MT32, tethered to a gold sensor chip surface (Fig. 9). Similarly, other sensor chips were prepared with random sequences for comparison. The sensor was tested with different concentrations of PSA-H and MT32 in PBS using Surface Plasmon Resonance (SPR) which allowed the determination of (K_d) for each aptamer. These were 55nM for the MT32 aptamer, 8nM for the PSA-H aptamer.

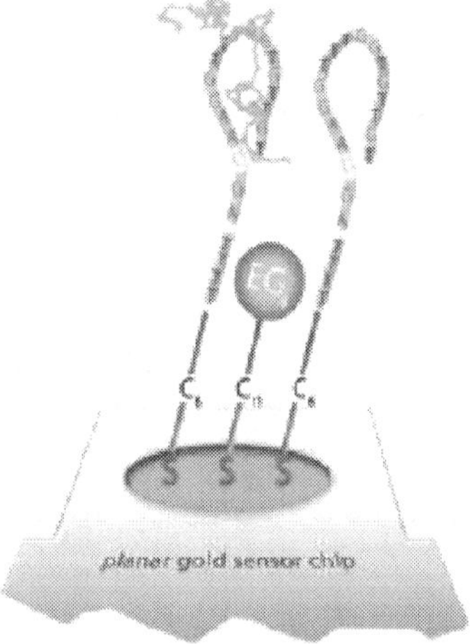

Fig. 9. Schematic of aptamer-based biochip for ASPV detection. Reproduced with permission from RSC.

Calibration curves were measured with the aptamer-functionalized surfaces and increasing concentrations of MT32 in apple leaf extract. The measurements on the graph were made relative to the surface prepared with a same-length random sequence. Specificity of the aptamer to its respective target was demonstrated and a nearly linear response was obtained. For detection of real ASPV-infected samples, ASPV-positive and ASPV-negative plant extracts were tested. Fig. 10 shows the response curves for the samples at different dilutions. The SPR signal predictably decreases upon a decreasing concentration of virus, indicating that this is a feasible method for virus detection in real samples.

3. Conclusion

Aptamers are proving to be effective molecular recognition probes for the analysis of high priority food contaminants. Aptamer-based biosensor platforms generally show sensitivities and specificities that could allow them to be competitive with existing detection methods.

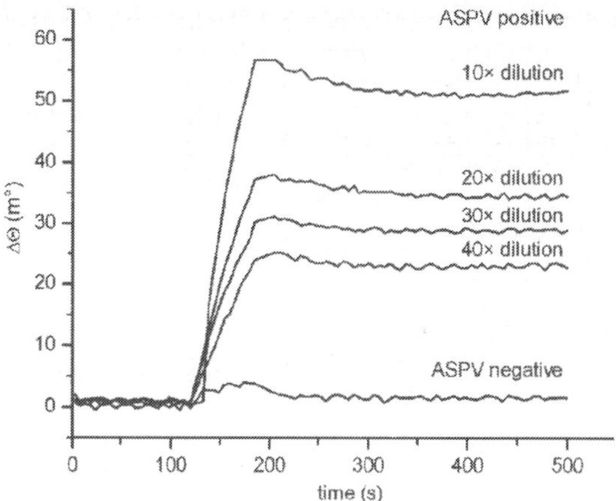

Fig. 10. SPR response curves of the MT32-based aptasensor detecting ASPV in apple leaf extracts at serial dilutions of total protein content. Reproduced with permission from RSC.

The *in vitro* nature of aptamer selection allows for a theoretically limitless variety of targets for which aptamers can be generated. This area of research is still in its infancy and the examples provided here represent only a small fraction of the potential aptamers which could be developed. In terms of small molecule targets, more focus should be paid to mycotoxins, given their prevalence and deleterious health effects. Similarly, aptamers for viral food pathogens are underrepresented, and should be investigated, particularly those related to gastroenteritis, such as rotaviruses and noroviruses. While several existing aptamer platforms have been tested in complex food matrices, it remains to be seen how many of these systems will perform under real-world conditions. Nevertheless, aptamer technology presents an opportunity for the development of robust, low cost, specific biosensors for food-safety applications.

4. Acknowledgment

The authors thank the Natural Sciences and Engineering Council of Canada (NSERC), the Canadian Foundation for Innovation (CFI), the Ontario Research Fund (ORF) and an Ontario Early Researcher Award for funding.

5. References

Balogh, Z.; Lautner, G.; Bardoczy, V.; Komorowska, B.; Gyurcsanyi, R. E. & Meszaros, T. (2010). Selection and versatile application of virus-specific aptamers. *The FASEB Journal : Official Publication of the Federation of American Societies for Experimental Biology,* 24, 11, (Nov), pp.4187-4195, 1530-6860; 0892-6638

Bell, S. D.; Denu, J. M.; Dixon, J. E. & Ellington, A. D. (1998). RNA molecules that bind to and inhibit the active site of a tyrosine phosphatase. *The Journal of Biological Chemistry,* 273, 23, (Jun 5), pp.14309-14314, 0021-9258; 0021-9258

Berens, C.; Thain, A. & Schroeder, R. (2001). A tetracycline-binding RNA aptamer. *Bioorganic & Medicinal Chemistry,* 9, 10, (Oct), pp.2549-2556, 0968-0896; 0968-0896

Bolger, M.; Coker, R. D.; DiNovi, M.; Gaylor, D.; Gelderblom, W.; Olsen, M., . . . Speijers, G. J. A. (2001). Fumonisins. WHO/IPCS safety evaluation of certain mycotoxins in food. *WHO Food Additives Series*, 47, pp.557-680.

Bonel, L.; Vidal, J. C.; Duato, P. & Castillo, J. R.An electrochemical competitive biosensor for ochratoxin A based on a DNA biotinylated aptamer. *Biosensors and Bioelectronics*, In Press, Corrected Proof, 0956-5663

Bruno, J. G.; Carrillo, M. P.; Phillips, T. & Andrews, C. J. (2010). A novel screening method for competitive FRET-aptamers applied to E. coli assay development. *Journal of Fluorescence*, 20, 6, (Nov), pp.1211-1223, 1573-4994; 1053-0509

Bruno, J. G., & Kiel, J. L. (2002). Use of magnetic beads in selection and detection of biotoxin aptamers by electrochemiluminescence and enzymatic methods. *BioTechniques*, 32, 1, (Jan), pp.178-80, 182-3, 0736-6205; 0736-6205

Bruno, J. G.; Phillips, T.; Carrillo, M. P. & Crowell, R. (2009). Plastic-adherent DNA aptamer-magnetic bead and quantum dot sandwich assay for campylobacter detection. *Journal of Fluorescence*, 19, 3, (May), pp.427-435, 1573-4994; 1053-0509

Burke, D. H.; Hoffman, D. C.; Brown, A.; Hansen, M.; Pardi, A. & Gold, L. (1997). RNA aptamers to the peptidyl transferase inhibitor chloramphenicol. *Chemistry & Biology*, 4, 11, (11), pp.833-843, 1074-5521

Cao, X.; Li, S.; Chen, L.; Ding, H.; Xu, H.; Huang, Y., . . . Shao, N. (2009). Combining use of a panel of ssDNA aptamers in the detection of staphylococcus aureus. *Nucleic Acids Research*, 37, 14, (Aug), pp.4621-4628, 1362-4962; 0305-1048

Cháfer-Pericás, C.; Maquieira, Á. & Puchades, R. (2010). Fast screening methods to detect antibiotic residues in food samples. *TrAC Trends in Analytical Chemistry*, 29, 9, (10), pp.1038-1049, 0165-9936

Cheli, F.; Pinotti, L.; Campagnoli, A.; Fusi, E.; Rebucci, R. & Baldi, A. (2008). Mycotoxin analysis, mycotoxin-producing fungi assays and mycotoxin toxicity bioassays in food mycotoxin monitoring and surveillance. *Italian Journal of Food Science*, 20, 4, pp.447-462, 1120-1770

Cheng, A. K.; Ge, B. & Yu, H. Z. (2007). Aptamer-based biosensors for label-free voltammetric detection of lysozyme. *Analytical Chemistry*, 79, 14, (Jul 15), pp.5158-5164, 0003-2700; 0003-2700

Collett, J. R.; Cho, E. J.; Lee, J. F.; Levy, M.; Hood, A. J.; Wan, C. & Ellington, A. D. (2005). Functional RNA microarrays for high-throughput screening of antiprotein aptamers. *Analytical Biochemistry*, 338, 1, (Mar 1), pp.113-123, 0003-2697; 0003-2697

Cowan, J. A.; Ohyama, T.; Wang, D. & Natarajan, K. (2000). Recognition of a cognate RNA aptamer by neomycin B: Quantitative evaluation of hydrogen bonding and electrostatic interactions. *Nucleic Acids Research*, 28, 15, (Aug 1), pp.2935-2942, 1362-4962; 0305-1048

Cruz-Aguado, J. A., & Penner, G. (2008a). Determination of ochratoxin a with a DNA aptamer. *Journal of Agricultural and Food Chemistry*, 56, 22, (Nov 26), pp.10456-10461, 1520-5118 (Electronic)

Cruz-Aguado, J. A., & Penner, G. (2008b). Fluorescence polarization based displacement assay for the determination of small molecules with aptamers. *Analytical Chemistry*, 80, 22, (Nov 15), pp.8853-8855, 1520-6882; 0003-2700

De Girolamo, A.; McKeague, M.; Miller, J. D.; DeRosa, M. C. & Visconti, A. (2011). Determination of ochratoxin A in wheat after clean-up through a DNA aptamer-based solid phase extraction column. *Food Chemistry*, In Press, Corrected Proof, 0308-8146

de-los-Santos-Alvarez, N.; Lobo-Castanon, M. J.; Miranda-Ordieres, A. J. & Tunon-Blanco, P. (2007). Modified-RNA aptamer-based sensor for competitive impedimetric assay of neomycin B. *Journal of the American Chemical Society*, 129, 13, (Apr 4), pp.3808-3809, 0002-7863; 0002-7863

de-los-Santos-Álvarez, N.; Lobo-Castañón, M. J.; Miranda-Ordieres, A. J. & Tuñón-Blanco, P. (2009). SPR sensing of small molecules with modified RNA aptamers: Detection of neomycin B. *Biosensors and Bioelectronics*, 24, 8, (4/15), pp.2547-2553, 0956-5663

Ellington, A. D., & Szostak, J. W. (1990). In vitro selection of RNA molecules that bind specific ligands. *Nature*, 346, 6287, (Aug 30), pp.818-822, 0028-0836; 0028-0836

European Food Safety Authority; (2005). Opinion of the Scientific Panel on Food Additives, Flavourings, Processing Aids and Materials in Contact with Food on a request from the Commission to Review the toxicology of a number of dyes illegally present in food in the EU. *The EFSA Journal.* 263, 1-71 (September 2005) available at http://www.efsa.europa.eu/en/efsajournal/pub/263.htm

Flinders, J.; DeFina, S.; Brackett, D.; Baugh, C.; Wilson, C.; Dieckmann, T. (2004). Recognition of planar and nonplanar ligands in the malachite green-RNA aptamer complex. *ChemBioChem*, 5, 1, (Jan 5), pp. 67-72.

Goertz, P. W.; Colin Cox, J. & Ellington, A. D. (2004a). Automated selection of aminoglycoside aptamers. *Journal of the Association for Laboratory Automation*, 9, 3, (6), pp.150-154, 1535-5535

Goertz, P. W.; Colin Cox, J. & Ellington, A. D. (2004b). Automated selection of aminoglycoside aptamers. *Journal of the Association for Laboratory Automation*, 9, 3, (6), pp.150-154, 1535-5535

González-Fernández, E.; de-los-Santos-Álvarez, N.; Lobo-Castañón, M. J.; Miranda-Ordieres, A. J. & Tuñón-Blanco, P. (2011). Impedimetric aptasensor for tobramycin detection in human serum. *Biosensors and Bioelectronics*, 26, 5, (1/15), pp.2354-2360, 0956-5663

Grate, D., & Wilson, C. (1999). Laser-mediated, site-specific inactivation of RNA transcripts. *Proceedings of the National Academy of Sciences of the United States of America*, 96, 11, (May 25), pp.6131-6136, 0027-8424; 0027-8424

Health Canada; (1993). Guidelines for Canadian Drinking Water Quality - Technical Document on Atrazine available at http://www.hc-sc.gc.ca/ewh-semt/pubs/water-eau/atrazine/index-eng.php

Holeman, L. A.; Robinson, S. L.; Szostak, J. W. & Wilson, C. (1998). Isolation and characterization of fluorophore-binding RNA aptamers. *Folding and Design*, 3, 6, (11), pp.423-431, 1359-0278

Huizenga, D. E., & Szostak, J. W. (1995). A dna aptamer that binds adenosine and atp. *Biochemistry*, 34, 2, (JAN 17), pp.656-665, 0006-2960

Ikanovic, M.; Rudzinski, W. E.; Bruno, J. G.; Allman, A.; Carrillo, M. P.; Dwarakanath, S., . . . Andrews, C. J. (2007). Fluorescence assay based on aptamer-quantum dot binding to bacillus thuringiensis spores. *Journal of Fluorescence*, 17, 2, (Mar), pp.193-199, 1053-0509; 1053-0509

Jiang, Z.; Zhou, L. & Liang, A. (2011). Resonance scattering detection of trace melamine using aptamer-modified nanosilver probe as catalyst without separation of its aggregations. *Chemical Communications (Cambridge, England)*, (Jan 28), 1364-548X; 1359-7345

Jiang, Z.; Wen, G.; Fan, Y.; Jiang, C.; Liu, Q.; Huang, Z. & Liang, A. (2010). A highly selective nanogold-aptamer catalytic resonance scattering spectral assay for trace Hg2+ using HAuCl4-ascorbic acid as indicator reaction. *Talanta*, 80, 3, (1), pp.1287-1291, 0039-9140

Joshi, R.; Janagama, H.; Dwivedi, H. P.; Senthil Kumar, T. M.; Jaykus, L. A.; Schefers, J. & Sreevatsan, S. (2009). Selection, characterization, and application of DNA aptamers for the capture and detection of salmonella enterica serovars. *Molecular and Cellular Probes*, 23, 1, (Feb), pp.20-28, 0890-8508; 0890-8508

Kim, M.; Um, H. J.; Bang, S.; Lee, S. H.; Oh, S. J.; Han, J. H., . . . Kim, Y. H. (2009). Arsenic removal from vietnamese groundwater using the arsenic-binding DNA aptamer. *Environmental Science & Technology*, 43, 24, (Dec 15), pp.9335-9340, 0013-936X; 0013-936X

Kim, Y. S.; Kim, J. H.; Kim, I. A.; Lee, S. J.; Jurng, J. & Gu, M. B. (2010). A novel colorimetric aptasensor using gold nanoparticle for a highly sensitive and specific detection of oxytetracycline. *Biosensors & Bioelectronics*, 26, 4, (Dec 15), pp.1644-1649, 1873-4235; 0956-5663

Kim, Y. S.; Niazi, J. H. & Gu, M. B. (2009). Specific detection of oxytetracycline using DNA aptamer-immobilized interdigitated array electrode chip. *Analytica Chimica Acta*, 634, 2, (2/23), pp.250-254, 0003-2670

Kirby, R.; Cho, E. J.; Gehrke, B.; Bayer, T.; Park, Y. S.; Neikirk, D. P., . . . Ellington, A. D. (2004). Aptamer-based sensor arrays for the detection and quantitation of proteins. *Analytical Chemistry*, 76, 14, (Jul 15), pp.4066-4075, 0003-2700; 0003-2700

Kuang, H.; Chen, W.; Xu, D.; Xu, L.; Zhu, Y.; Liu, L., . . . Zhu, S. (2010). Fabricated aptamer-based electrochemical "signal-off" sensor of ochratoxin A. *Biosensors and Bioelectronics*, 26, 2, (10/15), pp.710-716, 0956-5663

Kwon, M.; Chun, S. M.; Jeong, S. & Yu, J. (2001). In vitro selection of RNA against kanamycin B. *Molecules and Cells*, 11, 3, (Jun 30), pp.303-311, 1016-8478; 1016-8478

Lautner, G.; Balogh, Z.; Bardoczy, V.; Meszaros, T. & Gyurcsanyi, R. E. (2010). Aptamer-based biochips for label-free detection of plant virus coat proteins by SPR imaging. *The Analyst*, 135, 5, (May), pp.918-926, 1364-5528; 0003-2654

Lee, J.; Jo, M.; Kim, T. H.; Ahn, J. Y.; Lee, D. K.; Kim, S. & Hong, S. (2011). Aptamer sandwich-based carbon nanotube sensors for single-carbon-atomic-resolution detection of non-polar small molecular species. *Lab on a Chip*, 11, 1, (Jan 7), pp.52-56, 1473-0189; 1473-0189

Li, L.; Li, B.; Qi, Y. & Jin, Y. (2009). Label-free aptamer-based colorimetric detection of mercury ions in aqueous media using unmodified gold nanoparticles as colorimetric probe. *Analytical and Bioanalytical Chemistry*, 393, 8, (Apr), pp.2051-2057, 1618-2650

Luo, X.; McKeague, M.; Pitre, S.; Dumontier, M.; Green, J.; Golshani, A., . . . Dehne, F. (2010). Computational approaches toward the design of pools for the in vitro selection of complex aptamers. *RNA (New York, N.Y.)*, 16, 11, (Nov), pp.2252-2262, 1469-9001; 1355-8382

Mascini, M. (2009). *Aptamers in bioanalysis*. Hoboken, New Jersey, USA: John Wiley & Sons, Inc.

McKeague, M.; Bradley, C. R.; De Girolamo, A.; Visconti, A.; Miller, J. D. & DeRosa, M. C. (2010). Screening and initial binding assessment of fumonisin B-1 aptamers. *International Journal of Molecular Sciences*, 11, 12, (DEC), pp.4864-4881, 1422-0067

Morse, D. P. (2007). Direct selection of RNA beacon aptamers. *Biochemical and Biophysical Research Communications*, 359, 1, (7/20), pp.94-101, 0006-291X

Niazi, J. H.; Lee, S. J. & Gu, M. B. (2008). Single-stranded DNA aptamers specific for antibiotics tetracyclines. *Bioorganic & Medicinal Chemistry*, 16, 15, (8/1), pp.7245-7253, 0968-0896

Niazi, J. H.; Lee, S. J.; Kim, Y. S. & Gu, M. B. (2008). ssDNA aptamers that selectively bind oxytetracycline. *Bioorganic & Medicinal Chemistry*, 16, 3, (2/1), pp.1254-1261, 0968-0896

Ohk, S. H.; Koo, O. K.; Sen, T.; Yamamoto, C. M. & Bhunia, A. K. (2010). Antibody-aptamer functionalized fibre-optic biosensor for specific detection of listeria monocytogenes from food. *Journal of Applied Microbiology,* 109, 3, (Sep), pp.808-817, 1365-2672; 1364-5072

Okada, K.; Senda, S.; Kobayashi, A.; Fukusaki, E.; Yanagihara, I. & Nakanishi, T. (2003). Aptamer Capable of Specifically Adsorbing to Bisphenol A and Method for Obtaining the Aptamer. US Patent Application Publication , (April 2003) US2003/0064530A1.

Pan, Q.; Zhang, X. L.; Wu, H. Y.; He, P. W.; Wang, F.; Zhang, M. S., . . . Wu, J. (2005). Aptamers that preferentially bind type IVB pili and inhibit human monocytic-cell invasion by salmonella enterica serovar typhi. *Antimicrobial Agents and Chemotherapy,* 49, 10, (Oct), pp.4052-4060, 0066-4804; 0066-4804

Ricci, F.; Volpe, G.; Micheli, L. & Palleschi, G. (2007). A review on novel developments and applications of immunosensors in food analysis. *Analytica Chimica Acta,* 605, 2, (12/19), pp.111-129, 0003-2670

Rodriguez, M. C.; Kawde, A. N. & Wang, J. (2005). Aptamer biosensor for label-free impedance spectroscopy detection of proteins based on recognition-induced switching of the surface charge. *Chemical Communications (Cambridge, England),* (34), 34, (Sep 14), pp.4267-4269, 1359-7345; 1359-7345

Schmid-Hempel, P., & Frank, S. A. (2007). Pathogenesis, virulence, and infective dose. *PLoS Pathogens,* 3, 10, (Oct 26), pp.1372-1373, 1553-7374; 1553-7366

Sheng, L.; Ren, J.; Miao, Y.; Wang, J. & Wang, E.PVP-coated graphene oxide for selective determination of ochratoxin A via quenching fluorescence of free aptamer. *Biosensors and Bioelectronics,* In Press, Corrected Proof, 0956-5663

Sinha, J.; Reyes, S. J. & Gallivan, J. P. (2010). Reprogramming bacteria to seek and destroy an herbicide. *Nature Chemical Biology,* 6, 6, (Jun), pp.464-470, 1552-4469; 1552-4450

So, H. M.; Park, D. W.; Jeon, E. K.; Kim, Y. H.; Kim, B. S.; Lee, C. K., . . . Lee, J. O. (2008). Detection and titer estimation of escherichia coli using aptamer-functionalized single-walled carbon-nanotube field-effect transistors. *Small (Weinheim an Der Bergstrasse, Germany),* 4, 2, (Feb), pp.197-201, 1613-6829; 1613-6810

Srinivasan, A.; Foley, J.; Ravindran, R. & McSorley, S. J. (2004). Low-dose salmonella infection evades activation of flagellin-specific CD4 T cells. *Journal of Immunology (Baltimore, Md.: 1950),* 173, 6, (Sep 15), pp.4091-4099, 0022-1767; 0022-1767

Stead, S. L.; Ashwin, H.; Johnston, B.; Dallas, A.; Kazakov, S. A.; Tarbin, J. A., . . . Keely, B. J. (2010). An RNA-aptamer-based assay for the detection and analysis of malachite green and leucomalachite green residues in fish tissue. *Analytical Chemistry,* 82, 7, (Apr 1), pp.2652-2660, 1520-6882; 0003-2700

Stratis-Cullum, D. N.; McMasters, S. & Pellegrino, P. M. (2009). Evaluation of relative aptamer binding to campylobacter jejuni bacteria using affinity probe capillary electrophoresis. *Analytical Letters,* 42, 15, pp.2389-2402, 0003-2719

Tok, J. B., & Fischer, N. O. (2008). Single microbead SELEX for efficient ssDNA aptamer generation against botulinum neurotoxin. *Chemical Communications (Cambridge, England),* (16), 16, (Apr 28), pp.1883-1885, 1359-7345; 1359-7345

Tran, D. T.; Janssen, K. P.; Pollet, J.; Lammertyn, E.; Anne, J.; Van Schepdael, A. & Lammertyn, J. (2010). Selection and characterization of DNA aptamers for egg white lysozyme. *Molecules (Basel, Switzerland),* 15, 3, (Mar 2), pp.1127-1140, 1420-3049; 1420-3049

Tuerk, C., & Gold, L. (1990). Systematic evolution of ligands by exponential enrichment: RNA ligands to bacteriophage T4 DNA polymerase. *Science,* 249, 4968, (Aug 3), pp.505-510, 0036-8075 (Print)

US Food and Drug Administration; (2010) Update on Bisphenol A for Use in Food Contact Applications. (January, 2010) Available at http://www.fda.gov/newsevents/publichealthfocus/ucm197739.htm

Vivekananda, J., & Kiel, J. L. (2006). Anti-francisella tularensis DNA aptamers detect tularemia antigen from different subspecies by aptamer-linked immobilized sorbent assay. *Laboratory Investigation; a Journal of Technical Methods and Pathology,* 86, 6, (Jun), pp.610-618, 0023-6837; 0023-6837

Wallace, S. T., & Schroeder, R. (1998). In vitro selection and characterization of streptomycin-binding RNAs: Recognition discrimination between antibiotics. *RNA (New York, N.Y.),* 4, 1, (Jan), pp.112-123, 1355-8382; 1355-8382

Wang, L.; Chen, W.; Ma, W.; Liu, L.; Ma, W.; Zhao, Y., . . . Xu, C. (2011a). Fluorescent strip sensor for rapid determination of toxins. *Chemical Communications (Cambridge, England),* 47, 5, (Feb 7), pp.1574-1576, 1364-548X; 1359-7345

Wang, L.; Ma, W.; Chen, W.; Liu, L.; Ma, W.; Zhu, Y., . . . Xu, C. (2011b). An aptamer-based chromatographic strip assay for sensitive toxin semi-quantitative detection. *Biosensors and Bioelectronics,* 26, 6, (2/15), pp.3059-3062, 0956-5663

Wang, Y., & Rando, R. R. (1995). Specific binding of aminoglycoside antibiotics to RNA. *Chemistry & Biology,* 2, 5, (5), pp.281-290, 1074-5521

Wang, Z.; Duan, N.; Hun, X. & Wu, S. (2010). Electrochemiluminescent aptamer biosensor for the determination of ochratoxin A at a gold-nanoparticles-modified gold electrode using N-(aminobutyl)-N-ethylisoluminol as a luminescent label. *Analytical and Bioanalytical Chemistry,* 398, 5, (Nov), pp.2125-2132, 1618-2650

Wilson, C., & Szostak, J. W. (1998). Isolation of a fluorophore-specific DNA aptamer with weak redox activity. *Chemistry & Biology,* 5, 11, (11), pp.609-617, 1074-5521

Xu, J. P.; Song, Z. G.; Fang, Y.; Mei, J.; Jia, L.; Qin, A. J., . . . Tang, B. Z. (2010). Label-free fluorescence detection of mercury(II) and glutathione based on Hg2+-DNA complexes stimulating aggregation-induced emission of a tetraphenylethene derivative. *The Analyst,* 135, 11, (Nov), pp.3002-3007, 1364-5528; 0003-2654

Yang, C.; Wang, Y.; Marty, J. & Yang, X. (2010). Aptamer-based colorimetric biosensing of ochratoxin A using unmodified gold nanoparticles indicator. *Biosensors and Bioelectronics,* In Press, Accepted Manuscript, 0956-5663

Yamamoto, C. & Sen, T; (2008). Nucleic acids ligands capable of binding to internalin B or internalin A. Internation Application Published under the patent cooperation treaty, (January, 2008) WO2008/011608A2.

Yamamoto, C. & Sen, T; (2009). Nucleic acids ligands capable of binding to internalin B or internalin A. US Patent Application Publication , (October, 2009) US20090264512A1.

Yamamoto, C. & Sen, T; (2010) Aptamers that bind to *Listeria* surface proteins. US Patent Application Publication , (January, 2010) US7645582B2.

Zelada-Guillen, G. A.; Bhosale, S. V.; Riu, J. & Rius, F. X. (2010). Real-time potentiometric detection of bacteria in complex samples. *Analytical Chemistry,* 82, 22, (Nov 15), pp.9254-9260, 1520-6882; 0003-2700

Zhang, J.; Zhang, B.; Wu, Y.; Jia, S.; Fan, T.; Zhang, Z. & Zhang, C. (2010). Fast determination of the tetracyclines in milk samples by the aptamer biosensor. *The Analyst,* 135, 10, (Oct), pp.2706-2710, 1364-5528; 0003-2654

3

Biosensors Applications in Agri-food Industry

Liliana Serna-Cock and Jeyson G. Perenguez-Verdugo
Facultad de Ingeniería, Universidad Nacional de Colombia Sede Palmira
Colombia

1. Introduction

The dynamic worldwide agri-food market has produced increased biological, chemical and physical threats to food products, and a bigger consumer demand on process control, quality and safety of these products. As a result, farmers and processors are using technological tools to allow for the quick, effective and efficient determination of hazards inherent to safety and quality of products.

Biosensors are an important option in the agricultural and food sectors to control production processes and ensure food quality and safety by reliable, fast and cost effective procedures. Biosensors are promising alternatives to conventional analytical tools since they offer advantages in size, cost, specificity, rapid response, precision and sensitivity.

The agricultural and food industry integrates the production, processing and commercialization of agricultural materials for specific markets and consumer demands. Each step in the agricultural and food production chain is susceptible to several threats since agricultural materials and food products are transported to different parts of the world, favoring the loss of quality and the transmission of diseases.

The biological, chemical and/or physical threats may be the result of environmental contamination or failures during food handling, processing, packing and distribution. Detect, correct and prevent these failures are the basis for the development of quality management systems to ensure food safety.

According to the World Health Organization (WHO), each year millions of people around the world are affected by foodborne diseases. Biological risks produced by bacteria, viruses and biotoxins, chemical substances (*e.g.*, additives), food nutrients used to add product value, pesticides, veterinary drugs residues, and processing operations and technologies that may generate polycyclic aromatic hydrocarbons (PAHs) derived from proteins decomposition, or mutagenic agents (*e.g.*, tryptophan) in cooked foods, constitute risks to consumers health.

On the other hand, consumer demands have contributed to dramatic changes in the production and commercialization of foods, leading producers to search for innovative products and technologies. Furthermore, recent food crises have caused concerns regarding the safety of food products, resulting in increased sanitary standards. Today, food safety and food quality are key elements that influence consumers purchasing behavior. However, consumer perception varies with ethnicity, purchasing power, education level and awareness. Consequently, there are consumer groups that make purchasing decisions based on product price, regardless of food handling practices and potential public health risks;

while other consumer groups, aware of the importance of quality, safety and traceability, demand foods with high quality and safety standards, such as healthy, fresh, organic and additives-free products and are willing to pay the added value.

Food-related conflicts, higher demands for novel and safe foods, and better consumer eating habits, have imposed challenges on food producers to minimize potential health risks and build consumer-producer relationships that offer confidence and security to customers. This requires the development of effective traceability technologies throughout the food chain. Within this frame, biosensor technologies surge as effective and efficient alternatives to control production processes and ensure food quality and safety.

Although the initial development in biosensor technologies was done in the biomedical field, the ability to detect, analyze and quantify molecules of different biological origins and the diversity of measuring principles, resulted in the creation of personalized biosensor designs to cover a wide range of technical needs. The various operating principles have allowed the creation of biosensors to analyze external substances in food products such as pesticides, fertilizers and dioxins residues, water and soil residues (accidentally incorporated to the food chain), genetically modified organisms, pathogenic microorganisms and their toxins, and food components such as antinutrients, allergens, drugs, additives, and hydrocarbons.

Regarding food quality, biosensors have been useful for the evaluation of food composition, particularly in food products from plant and animal origin that are transformed during postharvest and processing. The sensory properties and chemical composition will depend on factors such as storage temperature conditions, exposure to sunlight, oxygen level and best handling practices. However, if these factors are not well managed, catalytic and enzymatic processes can lead to the generation of substances that cause undesirable changes in texture, color, taste, smell and aroma. Also, biosensors can be used in products enriched with components such as vitamins, minerals, and antioxidants. They quantify the different food components to evaluate rancidity, maturity, decline and shelf life and detect substances used as indicators of food freshness.

Another agro-industrial application for biosensor is in controlling the various steps involved in the farm to consumer chain, which is vital in the development of modern industrial and manufacturing processes. Biosensors allow the manipulation and real time monitoring of variables such as pH, temperature, pressure, oxygen, flow and volatile substances, affecting productivity, profitability and safety of food processing operations. Biosensors can also quantify compounds found in low or high concentrations (*e.g.*, sugars, amino acids, alcohols, and acids), as they may become the limiting factors in a production process.

The detection of genetically modified organisms (GMO) is another useful application for biosensors, since several countries have laws regulating the commercialization of GMO products and their derivatives. Biosensors can also be useful in the implementation of hazard analysis and critical control points (HACCP) plans by verifying process developments and correcting errors in due time.

Biosensor technologies and nanotechnologies are being used together in many food and agricultural applications such as the development of nano-scale tools for biosafety, nano-scale compounds for food packaging, and nano-sensors for pathogen detection in animals and plants, among others. These are helpful tools in the detection and control of potential food contaminants by the agricultural and food industries.

In the first section of this chapter, we describe current biosensor applications for food security in two main areas: detection of foodborne pathogens and detection of chemical contaminants. Pathogens detection is important in the prevention of microbiological hazards acquired during the production, processing, storage, transportation and distribution of food products. We will highlight the use of biosensors for the detection of microorganisms responsible of large economic losses in the food industry due to productivity loss, medical costs and food product recalls, including *Salmonella, Listeria monocytogenes, Campylobacter, and Escherichia coli*. We give a comprehensive view on the role of biosensors in pathogens detection and discuss constraints such as pre-treatment requirements and analysis time. The detection of chemical contaminants outlines biosensor applications in the determination of contaminant residues including pesticides, fertilizers, heavy metals, food additives and antibiotics. Antibiotics are of great concern for consumer health since they create resistance to microorganisms. The importance of the appropriate HACCP plan implementation is discussed together with its impact in the meat, milk and honey processing industries. Examples of enzyme biosensors are presented, especially those with greater sensitivity and faster response time.

The second section covers biosensors as low-cost, easy to use analytical tools for the fresh product industry, with emphasis on food quality, as defined by food nutritional value, safety, and acceptability. Food quality is evaluated through organoleptic parameters (freshness, appearance, taste and texture), where food composition allows the verification of the type of elements present in a given food. In this case, biosensors are used for the quantification of starch, glucose, lactose, lecithin, and ethanol as indicators of quality and consumer acceptability. We also describe the use of biosensors for the evaluation of freshness and shelf life. The detection of polyphenols in olive oils, short-chain fatty acids in milk and dairy foods, amines formation in fish; rancidity evaluation; quantification of glucose and fructose as indicators of fruit ripening and firmness, are among the various applications described in the second section.

The last section describes the use of biosensors in controlling processing operations, including HACCP plans. The importance in productivity, performance and efficiency in biotechnological processes is discussed and the online identification and quantification of various compounds (e.g., amino acids, sugars and alcohols) is highlighted. These determinations allow the monitoring of high or low concentrations of components that significantly affect the processing of foods. In the case of sugars, we will describe amperometric biosensors used to determine glucose using glucose oxidase in fruit juices, lactose using ß-galactosidase and glucose oxidase, and lactulose using ß-galactosidase and fructose dehydrogenase (an enzyme used as indicator of excessive thermal processing of milk). We will also describe the use of amperometric biosensors for the determination of alcohol using the enzyme dehydrogenase from *Gluconobacter oxydans*. Emphasis will be given to the monitoring of fermentation processes, including the description of interfering compounds (e.g., ascorbic acid and tyrosine), procedures to remove such compounds, and correction methods that reduce the duration of the analysis, including filtration, evaporation and acid or alkaline hydrolysis. Biosensors characteristic such as specificity, analysis time, automation, and sensitivity will be discussed. Finally, the challenges facing the food industry and the new trends in biosensors technology are presented.

2. Biosensors in food security

Food Security refers to the availability and continuous, timely and permanent access to foods that meet quality and safety standards by the entire world population (FAO, 1996).

Food agro-industrialization is a strategy that allows guaranteeing the continuous availability of foods. However, physical, biological, and chemical hazards from preharvest to storage and final product marketing, may affect food quality and safety.

Physical pollutants affect the quality of food components, mainly influencing product appearance and causing rejection by consumers. To prevent physical damage, it is necessary to maintain control over every stage of the process, and manage logistics, including the compliance with personnel safety standards and the assurance of good infrastructure conditions in the food industry.

Biological hazards are primarily caused by microorganisms and their toxins, which may or may not affect the organoleptic properties of foods but nevertheless affect consumer health. Food poisoning and infections have a strong economic and social impact, causing loss of productivity due to morbidity and mortality (Vásquez de Plata, 2003).

Microorganisms involved in biological hazards that require specific processing interventions include bacteria belonging to the genus *Salmonella*, *Shigella*, *Escherichia coli*, *Listeria monocytogenes* and *Clostridium botulinum*. Microbiological analyses during processing and on final products have been traditionally used to assure the control of these pathogens. These methods are based on the detection of potential pathogenic microorganisms through their isolation in differential and selective growth media, requiring long response times (ranging from 48 hours to 5 days), and also involving high identification costs (Meng & Doyle, 2002).

Another traditional method for the detection of pathogens in food and human and animal biological tissues is through enzyme-linked immunosorbent assays (ELISA) (Arbault et al., 2000; Nowak et al., 2007; Deng et al., 2008; Cabrera et al., 2009). The ELISA test is based on qualitative or quantitative color changes, using an enzyme as a reaction biomarker between an antigen and an antibody. If the substrate of the enzyme is cromogenic, a strong color change will indicate a greater concentration of the analyte (Sismani et al., 2008). However, the reuse of antibody receptors linked to enzymes increases the cost of detection, as well as the limited sensitivity of antibodies increases the possibility of unreliable results in a rapid test (Turner et al., 2009).

Biosensors offer rapid and effective detection options to control biological hazards. Biosensor technologies are advantageous due to the potential for miniaturized, rapid, specific, and sensitive detection of biological hazards (Pathirana et al., 2000; Ropkins & Beck, 2000).

2.1 Biosensors used for the detection of bacteria
2.1.1 *Salmonella*

Salmonella are Gram-negative bacteria naturally found in the gastrointestinal tract of warm-blooded animals and humans (Nowak et al., 2007; Lu et al., 2009). When out of their natural habitat, these bacteria are able to survive in water and food products. The consumption of contaminated food products causes diseases such as enteric fever and salmonellosis.

The majority of food processing plants are not equipped with water purification systems, and *Salmonella* becomes the principal biological contamination hazard (White et al., 2002).

For the rapid detection of *Salmonella*, piezoelectric antigen-antibody biosensors are used. These sensors may apply the Langmuir-Blodgett (LB) monolayer method, which is based on the immobilization of amphiphilic antibodies. The antibodies are captured and distributed uniformly in a liquid having affinity to the antibody. Antibodies can be monoclonal or polyclonal; however, the use of monoclonal antibodies is ideal because it offers greater specificity. As shown in Figure 1, when submerging a support probe with a polar and a non-polar surface, little interaction with the antibody occurs. However, binding between the antibody and the polar surface (antibody film) will occur when the support emerges from the liquid. The support will have transduction resonance properties, i.e., when affinity with antigen takes place, the resonance frequency will be affected due to increased mass in the crystal support (Guntupalli et al., 2007).

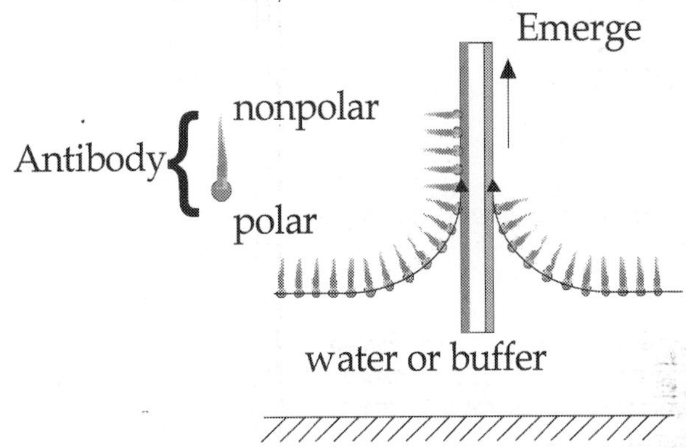

Fig. 1. Antibodies film formation in the support following the Langmuir-Blodgett technique when emerging from the solution.

The sensor platform used by Guntupalli et al. (2007) is made of magnetostrictive $Fe_{40}Ni_{38}Mo_4B_{18}$ and immobilized polyclonal rabbit antibodies using the LB method to detect *Salmonella typhimurium* by measuring changes in surface pressure by the Wilhelmy method. The system reaches equilibrium in the surface in order to reduce vibrations generated by external factors, in such a way that frequency and amplitude changes are only caused by mass variations in the crystal. This method was studied using different biosensor sizes and it was concluded that better changes in resonance frequency were achieved with small sensors. In these sensors, the resonance frequency decreases with an increase in the number of pathogen microorganisms bound to the antibody. As a consequence, to detect *Salmonella* it is convenient to use bigger biosensors since smaller ones would not allow enough contact area with the antigen, and low pathogen concentrations would not be detected. Figure 2 shows the common detection method for pathogens or bacterial cells by using a resonance magnetoelastic biosensensor.

Many *Salmonella* detection studies focus on this type of biosensors. The advantage of using these biosensors with antibodies as the recognition element is their high specificity, sensitivity, chemical stability and rapid response; however, as mentioned earlier, the low specificity achieved by polyclonal antibodies may be a drawback in some sensors. These biosensors are used in water sanitation control, which is required in all food transformation processes. They are also used in milk production and in the control of milk products, including milking equipment, pollution system, pumps, farms or livestock stables, processing industry and finished products.

2.1.2 *Listeria monocytogenes*

Listeria monocytogenes, a Gram positive, flagellate microaerophilic coccobacillus, can cause listeriosis associated to the consumption of fresh and processed foods such as meats,

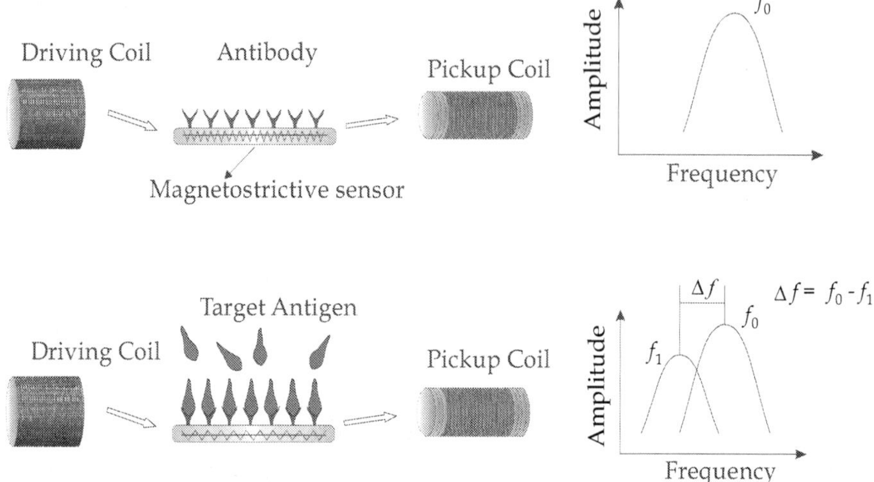

Δf is the shift in frequency due to antigen binding

Fig. 2. Detection of pathogens or bacterial cells using a magnetoelastic resonant biosensor by the change in amplitude of the resonant wave, due to frequency variation when the analyte binds to the receptor (Guntupalli et al., 2007).

shellfish, unpasteurized milk and vegetables (Sánchez et al., 2009). The disease produces little morbidity but important mortality in humans and animals, with symptoms such as muscle pain, nausea, diarrhea and chills. Complications are serious since it affects the central nervous system, causing seizures, headaches and even comatose state. Similarly to the bacteria genus *Salmonella*, *L. monocytogenes* has had negative repercussions at productive, economic and social levels (FAO/WHO, 2004).

Unlike *E. coli* and *Salmonella*, *Listeria monocytogenes* can grow at refrigeration temperatures, being only inactivated by high temperature and pasteurization treatments (FAO/WHO, 2004). Due to its ability to adapt to a wide range of pH and temperatures, it is associated with silage or poorly fermented foods. Organic acids, such as lactic acid produced during fermentation, inhibits the growth of *Listeria monocytogenes* at pH <5.2 being more effective than inorganic acids such as hydrochloric acid (HCl), where the bacteria can growth at pH 4.8 (Pellicer et al., 2009).

A rapid detection of this pathogen is essential. Biosensors offer advantages over traditional methods that need longer test time and specific growth media before the pathogen is detected (Geng et al., 2004). The use of optical biosensors allows the reaction between receptor and analyte, providing confidence, specificity and less time for detection. The excitation of atoms related to the receptor-analyte reaction, results in mass, energy, and physical changes that are captured by the biosensor. In this way, real time data is used to identify bacteria, viruses or toxins. Moreover, these biosensors allow the identification of pathogens at low concentrations (Geng et al., 2004).

A fiber-optic biosensor can be used for the detection of *Listeria monocytogenes* (Geng et al., 2004). It is based on the measurement of fluorescent light generated by a wave when there is a change in resonance. The biosensor uses monoclonal fluorescent antibodies that generate excitation of atoms and wave production when bound to the pathogen. The biosensor is made of a polystyrene support that immobilizes the antibodies. The antibodies are obtained by inducing the immune response of rabbits after inoculating with *Listeria monocytogenes* for three months (Leonard et al., 2004). The quick response from this biosensor and the specificity against *L. monocytogenes* were verified in the presence of bacterial pollutants common in foods such as *E. coli, Enterococcus faecalis,* and *S. enteric.* Its response in cold (4 °C) and salty conditions showed lower reaction rates due to "stress" of the monoclonal antibodies at low temperatures. The biosensor also showed a reduction in sensitivity at low *L. monocytogenes* concentrations when the medium contained other microbiota able to affect the antigen-antibody reaction (Geng et al., 2004). The fiber-optic biosensor detects concentrations from 10^7 to 10^8 CFU/ml (Geng et al., 2004). In situations where the product is suspected to contain *L. monocytogenes* counts below 10^7 CFU/ml, an enrichment step prior to the biosensor detection is required.

Although optical biosensors have been widely implemented for the detection of *L. monocytogenes,* biosensors using surface plasmon resonance (SPR) had shown better response. Although SPR sensors are more expensive than fiber optic sensors, they are the best alternative for a rapid - 5 to 15 minutes – detection (Van der Merwe, 2011). In addition, low concentrations of bacteria (10^5 *L. monocytogenes* CFU/ml) can be identified (Leonard et al., 2004), thus sample enrichment prior to analysis is not required.

SPR biosensors have disadvantages that are important to consider too, such as wear of the miniaturized and thin plate or support where the light wave contacts the sensor. The use of these biosensors for high-performance detection tests is not recommended. The efficiency, speed and sensitivity of the biosensor increases when the analytes are large molecules such as cells (Van der Merwe, 2011), since mass change on the surface is the basis for detection (Van der Merwe, 2011).

Fiber optic and SPR biosensors help preventing the spread of foodborne illnesses and ensure food safety in processed and minimally processed foods. In addition, their nano-scale dimensions offer advantages for handling, transporting, storing, processing, and distributing foods.

2.1.3 *Campylobacter*

The genus *Campylobacter,* are Gram negative bacteria, characterized for being microaerophiles (oxygen concentrations between 3-5%), comma-shaped or "S" flagellates, with motility (Ryan & Ray, 2010).

Infections caused by these bacteria are the main cause of gastroenteritis, being very frequent in children. Campylobacteriosis causes symptoms such as fever, abdominal pain, diarrhea, vomiting, and nausea. The disease has side effects including reactive arthritis and muscle pain due to alterations in the immune system (WHO, 2000). Campylobacter can be found in poultry and other birds, cattle, and even seafood (WHO, 2000). Poor cooking practices in processed meats and dairy products further favor the spread of this thermophilic bacterium. An efficient and opportune detection prevents the spread infections among animals and humans. Traditional microbiological methods for the identification of *Campylobacter* require 3 to 4 days (Wei et al., 2007). The difficult phenotypic characterization further complicates the proper identification of the bacteria (On, 2001). In this case, biosensors are also reliable for a rapid detection of the microorganism.

A SPR optical biosensor has great sensitivity for *Campylobacter,* being capable of rapidly detecting concentrations of *Listeria monocytogenes* having the specific antibody for

Campylobacter populations of at least 10^3 CFU/ml (Wei et al., 2007). To obtain these results, four antibodies from different manufacturers were used and reactivity, specificity and sensitivity tests were done, showing a maximum detection dilution of 1/200 for the chosen antibody. Reactivity tests were verified with an ELISA test, using inhibitors and denaturing agents to check whether the bacteria could be detected by antibodies even after inactivation. High temperature pretreatments did not result in changes in the detection capabilities of the biosensor. The specificity of the biosensor was also tested by the binding between the antibody and *Campylobacter* in the presence of other bacteria. Other bacteria exhibited temporary union with the antibody due to a weak association promoted by precipitation with glycine buffer. The sensitivity of the SPR biosensor was tested in pure bacterial populations ranging from 10^1 to 10^8 CFU/ml, with adequate detection starting at populations of 10^3 CFU/ml (Wei et al., 2007).

Campylobacteriosis illness is mainly spread by eating contaminated food, being a major problem in poultry production. Optical SPR biosensors are not the only detection methods, but other types of recognition elements such as DNA, RNA, enzymes, and tissues could be implemented; since interactions with other types of cells and/or macromolecules may probably occur.

2.1.4 *Escherichia coli*

Escherichia coli is a Gram negative bacilli normally inhabiting the intestine of humans and warm blooded animals (Darnton et al., 2007). An *E. coli* strain, known as 0157:H7, may cause inflammation in the small intestine, causing severe diarrhea (including bleeding) and kidney damage (Lin et al., 2010). These bacteria are also responsible for the hemolytic uremic syndrome (Waswa et al., 2007), which is an infectious and contagious disease characterized by kidney failure, microangiopathic hemolytic anemia, thrombocytopenia, coagulation defects and other neurologic disorders.

The infection caused by *E. coli* results from ingesting food, mainly fresh fruits and vegetables, and/or contaminated water. Infections may also be caused by eating poorly cooked animal foods or foods that were washed with contaminated water (WHO, 2005).

Animal derived foods such as raw milk can have a high *E. coli* count, either due to inadequate feeding of animals or diseases such as bovine mastitis, resulting from poor cleaning of processing equipment and/or inadequate sanitation practices.

E. coli grows at temperature ranging from 7 to 50 °C, and has the ability to grow in acid foods with pH down to 4.4 and with low water activity. Therefore, a minimum of 70 °C in the food is required to inactivate the bacteria (WHO, 2005).

When considering the effects that infections may produce in human and animals through food and water, traditional detection methods for *E. coli* that require 24 to 48 hours are considered time consuming and costly. Therefore, fast, simple, efficient and cost-effective methods are needed, and the development of biosensors may play a key role in detecting small *E. coli* concentrations (Gfeller et al., 2005).

There are different methods for a rapid detection using biosensors with detection limits down to 10^3 *E. coli* CFU/ml in less than 10 hours and confirmation using PCR (Tims & Lim, 2003). Other techniques use biosensors for detecting bacteria by flow cytometry followed by inmunomagnetic separation. This technique can detect down to 4 cells/g of *E. coli* in beef, in a 7 hours time frame (Seo et al., 1998).

E. coli detection in shorter time has been done using amperometric biosensors, through the detection of hydroxyl radicals produced by *E. coli* oxygen reduction during aerobic metabolism (Tang et al., 2006). This technique uses the covalent immobilization of enzymes lactase and peroxidase on indium and tin oxides, and the oxidation of salicylic acid to polyphenolic compounds by hydroxyl radicals. Enzymes can act on polyphenolic compounds, where the enzyme lactase in presence of oxygen produces quinones and peroxide residues that serve as electron donors and the enzyme peroxidase catalyzes the reaction on other polyphenolic compounds. The latter reaction also generates quinones, producing a reversible reaction measured by the amperometric biosensor. The amount of polyphenolic compounds generated through the microbial metabolic cycle depends on the concentration of *E. coli* cells.

2.2 Biosensors used for the detection of contaminant residues and pesticides

Pesticides (herbicides, fungicides, and insecticides) are used worldwide due to their wide range of activity. The presence of pesticide residues and metabolites in food, water and soil currently represents one of the major issues in environmental chemistry research (Mostafa, 2010). Due to their increasing use in agriculture, pesticides are among the most important environmental pollutants. However, the existing analytical methods for the determination of organophosphate pesticides and N-methyl carbamates are complex or not existent for some compounds. High Performance Liquid Chromatography (HPLC) is an appropriate technique for the determination of these compounds since it preserves pesticide stability. However, to set the adequate sensitivity for the method, several pretreatment steps are required, adding time and cost (Hiemstra & De Kok, 1994). Due to the restrictions in conventional methodologies, the development of biosensors for direct and indirect pesticide detection is of particular interest. Serna et al. (2009) discussed the use of most common enzymatic biosensors for the detection of pesticides, fertilizers, and heavy metals. Table 1 shows the biosensors that have been reported in the literature since 2004 for direct detection of pesticides.

Enzymes like cholinesterase (AChE, BChE), organophosphorus-hydrolase (OPH), and urease are used in the design of electrochemical biosensors for pesticides detection. Analytical devices, based on OPH and cholinesterase inhibition, have been widely used for the detection of carbamates (Zhang et al., 2005) and organophosphate compounds (OP) (Pavlov et al., 2005). OPH, an organophosphotriester hydrolyzing enzyme, has broad substrate specificity and is able to hydrolyze a number of OP pesticides such as paraoxon, parathion, coumaphos, diazinon, and dursban. Organophosphorus acid anhydrolase catalyzed hydrolysis of OP compounds generates two protons as a result of the cleavage of the P-O, P-F, substitution or P-CN bonds and an alcohol, which in many cases is chromophoric and/or Electroactive (Mostafa, 2010). The resulting reaction ions are detected through potentiometric biosensors. Another way to detect the action of the enzyme OPH is by monitoring, through amperometric biosensors, oxidation and reduction reactions occurred by the hydrolysis of the substrate.

OPH was used in a biosensor for the detection of paraoxon and parathion. The transducer structure of the sensor chip consists of a pH sensitive electrolyte-insulator- semiconductor (EIS) structure that reacts to pH changes caused by the OPH catalyzed hydrolysis of the organophosphate compounds (Schöning et al., 2003). Conductimetric AChE biosensors have been used to evaluate the toxicity of methyl parathion and its photodegradation products in water (Dzyadevych et al., 2002). Similarly, biosensors using immobilized enzymes for the detection of OP parathion hydrolase immobilized in a coal electrode have been developed. The organophosphorus parathion hydrolase hydrolyzes parathion on p-nitrophenol and it is detected by its anodic oxidation (Schöning et al., 2003). Immobilized cells of *Flavobacterium sp* have been used for the detection of methyl parathion. Whole cells of *Flavobacterium sp.* were immobilized by trapping in glass fiber filter and were used as biocomponents along with optic fiber system. *Flavobacterium* sp. has the enzyme organophosphorus hydrolase, which hydrolyzes the methyl parathion into detectable product *p*-nitrophenol (Kumar et al., 2006). Biosensors based on BChE are very sensitive in the detection of OP pesticide mixtures such as clorfenvinfos and diazinon. Immunochemical techniques, including the piezoelectric immunosensors, are gaining acceptance as alternative or complementary methods for the analysis of pesticides. Immunosensors have great potential for monitoring herbicides in drinking water (Yokoyama et al., 1995; Székács et al., 2003), dioxins concentration in real environmental samples (Park et al., 2006), and the detection of polychlorinated terphenyls (biphenyls) (Pribyl et al., 2006) and atrazine (Pribyl et al., 2003) among others.

Piezoelectric immunosensors "can be considered as alternatives comparable to other well-established immunochemical methods such as ELISA. In contrast to the ELISA techniques that require labeling of reactants and about two hours per assay, this technique can be automated, does not use markers and analysis results are available in a few minutes" (Ocampo et al., 2007).

An inmunosensor is a biosensor using antigen-antibody reactions as the basis for detection. Usually, the analyte contains the antigen or hapten and the antibody is attached to the receptor. It is important to note that pesticides are generally small molecules that do not produce immune response, and then it is necessary to unite these compounds to inmunogenic molecules like proteins (haptens), keeping intact the chemical composition of the compound to be analyzed. The design of the specific hapten is decisive in the development of immunoassays, since it is responsible for determining the recognition characteristics of the antibody (Haasnoot et al., 2000).

The piezoelectric inmunosensor uses a crystal quartz micro scale, in which the antigen, a conjugate or an antibody, is immobilized to the glass surface. The antigen-antibody reaction is detected by the oscillating piezoelectric crystal quartz that resonates at a fundamental frequency and has a linear behavior with respect to its surface mass density. The operation of piezoelectric resonators in liquid phases includes changes in the mass induced frequency and effects induced by the change in viscosity or density of the surrounding liquid layers (Kim et al., 2007; Fohlerová et al., 2007).

There are also other potentially toxic substances for humans, such as contaminated residues present in water and soil, with a major impact on the environment accidentally reaching the food chain (Serna et al., 2009). Some of these residues are by-products from diverse industrial processes (dioxins) used as dielectric or hydraulic fluid agents (polychlorinated biphenyls or PCBs) or generated from fossil fuels or wood burning (polycyclic aromatic hydrocarbons or PAHs), benzene, toluene and xylene (named BETX) and derived phenolics. Immuno-, enzymatic, and bio- sensors with complete cells are used for the detection of these organic compounds (Patel, 2002).

2.3 Biosensors used for the detection of heavy metals

Heavy metals are toxic substances that accumulate in the organism and cause metabolic alterations since there is no way of metabolizing or excreting them. We may accumulate heavy metals when eating foods of animal origin, since animals have greater contact with poorly treated water, can graze close to industries, and even eat foods treated with water contaminated with heavy metals.

Diseases caused by the ingestion if heavy metals include cardiovascular and respiratory problems, infertility, irritations, inhibition of some hormonal activities, malfunction of the principal organs, and death.

Devices have been designed to determine the concentration of heavy metals such as arsenic, cadmium, mercury, and lead, in water and soil samples. These devices incorporate genetically modified microorganisms and enzymes such as urease, cholinesterase, glucose oxidase, alkaline phosphatase, ascorbate oxidase and peroxidase (Tsai et al., 2003), incorporated to electrochemical and optical transduction systems.

Bi-enzymatic biosensors are used to detect enzymes inhibition in water. Enzymes like alkaline phosphatase and acetyl cholinesterase are inhibited by the presence of heavy metals, carbamates and organophosphates. Conductometric biosensors use immobilized *Chlorella vulgaris* microalgae as bioreceptors. Chouteau et al. (2005) immobilized algae inside bovine serum albumin membranes reticulated with glutaraldehyde vapors deposited on interdigitated conductometric electrodes. Local conductivity variations caused by algae alkaline phosphatase and acetylcholinesterase activities could be detected.

In addition, it is possible to detect the presence of cadmium through transduction systems based on fiber optics. The biosensor detects the inhibition of the enzyme urease, can sense down to 0.1 g/l of cadmium in milk. The device can be constructed from whole cells of *Bacillus badius* with phenol red as an indicator co-immobilized onto circular plastic discs with sol-gel approach and fiber optic transducer system. Urea is added to the plastic disc to detect the inhibition of enzymatic reactions (Verma et al., 2010).

Other methodologies used in the detection of heavy metals are shown in Table 1.

Analyte	Type of interaction	Recognition biocatalyzer	Transduction system	References
Pesticides				
Methyl Parathion	Biocatalytic	*Sphingomonas*sp. *Flavobacterium*sp Methyl parathion hydrolase. Acetyl Cholinesterase	Fiber optic Amperometric. (square wave voltammetric). Electrochemical	Kumar & D'Souza, 2010 Kumar et al., 2006 Chen et al., 2010 Gong et al., 2009
Organophosphorus	Biocatalytic	Organophosphorus hydrolase	Amperometric	Deo et al., 2005
Triazophos	Biocatalytic	Acetyl Cholinesterase	Amperometric	Du et al., 2007; Du et al, 2010
Monocrothopos, Malathion, Metasystox and Lannate	Biocatalytic	Acetyl Cholinesterase	Electrochemical	Dutta et al, 2008
Chlorpyrifos–Oxon, Chlorfenvinphos, Pirimiphos-methyl, Malathion,Carbofuran, Methomyl and Carbendazim	Biocatalytic	Acethyl Cholinesterase	Amperometric	Hildebrandt et al., 2008

Analyte	Type of interaction	Recognition biocatalyzer	Transduction system	References
Acetylcholine	Biocatalytic	Choline oxidase and Acethyl Cholinesterase	Amperometric	Shimomura et al., 2009
Fertilizers				
Nitrate	Biocatalytic	Nitrate reductase	Potentiometric	Sohail & Adeloju, 2008
Nitrite	Biocatalytic	Nitrite reductase	Voltammetric Amperometric	Almeida et al., 2007 Silveira et al., 2010
Phosphate	Biocatalytic	pyruvate oxidase	Amperometric	Rahman et al, 2006
		Maltose phosphorylase.	Conductometric	Zhang et al., 2008
		pyruvate oxidase	Voltammetric and Amperometric	Akar et al., 2010
Urea	Biocatalytic	Urease	Amperometric Potentiometric	Kuralay et al., 2006 Trivedi et al., 2009
Heavy metals				
Hg, Ag, Pb and Cd	Biocatalytic	Invertase and glucose oxidase	ultramicroelectrode	Bagal-Kestwal et al., 2008
Cadmium, copper, chrome, nickel, zinc	Biocatalytic	Ureasa	Optical	Verma et al., 2010
Lead		bovine serum albumin	piezoelectric quartz crystal impedance	Yin et al., 2007
Lead	Direct structure-competitive detection mode	Monoclonal antibody	Gold nanoparticle-modified optical fiber	Lin & Chung, 2008
Mercury		Envanescent wave DNA-based biosensor	Fiber optic sensor	Long et al., 2011
Copper and mercury	Biocatalytic	Glucose oxidase	Amperometric	Ghica & Brett, 2008; Jian-Xiao et al., 2009

Table 1. Most important biosensors used in the detection of pesticides, fertilizers and other pollutants.

2.4 Biosensors as indicators of product acceptability

Food quality involves nutritional and organoleptic characteristics important for consumers such as freshness, appearance, taste and texture. The food sensory basis is essential for the industry (Perez et al., 2007; Vadivambal & Jayas, 2007).

A method to determine food freshness is by assessing food composition of products such as meats, fish, fruits and vegetables. During storage, compounds that provide aroma and abnormal flavors or may be harmful to consumer may be synthesized, indicating in most cases microbial growth and insufficient food safety (Serna et al., 2009).

Food freshness is negatively affected by storage time, incorrect packaging design, inadequate temperature and oxygen management during the handling of fruits and vegetables in modified atmosphere storage conditions.

Ethanol and methanol have been used as indicators of food freshness and quality of alcoholic beverages. Their determination is done by colorimetric, refractometric, chromatographic and spectrophotometric methods. Some of these techniques require expensive equipments and all require considerable time. These drawbacks can be overcome with the use of biosensors.

Biosensors that use whole cells or enzymes have been used for the detection of alcohol (Valach at al., 2009). Alcohol enzymatic biosensors described in the literature are mainly based on alcohol dehydrogenase and alcohol oxidase, and less commonly on catalase. Ethanol biosensors using alcohol oxidase as biorecognition element are the most abundant. Alcohol oxidase is an oligomeric enzyme responsible for the oxidation of low molecular weight primary alcohols, using molecular oxygen as the electron acceptor and producing acetaldehyde and hydrogen peroxide. Due to the strong oxidizing character of oxygen, the oxidation of alcohols by alcohol oxidase is irreversible. The reaction may be followed by measuring either O_2 decline or H_2O_2 increase using optical or electrochemical detections.

Smyth et al. (1999) used a biosensor with immobilized enzymes alcohol oxidase and alcohol peroxidase and a chromogen, to detect injuries caused by low O_2 in lettuce, cauliflower, broccoli and cabbage lightly processed and packed in a modified atmosphere. By measuring the ethanol accumulation in the free space, results obtained with this biosensor are similar to those obtained using gas chromatography. This biosensor could also be used to monitor ethanol during the storage of apples in a controlled atmosphere, the decay in potato tubers (Castillo et al., 2003), or in any other application where ethanol accumulation can be associated with quality loss. Similar research has been done to detect organic acids and sugars as indicators of fruit and vegetables maturity (Cañas & Macias, 2004).

Hnaien et al. (2010) pioneered the construction of a conductimetric biosensor for the detection of ethanol in foods. The biosensor uses the enzymes ethanol oxidase and catalase co-immobilized on the surface of interdigitated thin-film electrodes. The transduction mode has the following advantages: (a) thin-film electrodes are suitable for miniaturization and large scale production using inexpensive technology, (b) it does not require any reference electrode and the differential mode measurements allow cancellation of many interferences, (c) it is not sensitive to light, (d) the driving voltage can be sufficiently low to significantly reduce power consumption, and (e) large spectrum of compounds of different nature can be determined on the basis of various reactions and mechanisms (Jaffrezic & Dzyadevych, 2008).

Similarly, Asav & Akyilmaz (2010) developed an amperometric biosensor for the determination of glucose and ethanol. The biosensor is based on the co-immobilization of alcohol oxidase and glucose oxidase on the same electrode by formation of self-assembled monolayers on a gold disc electrode. Measurements are based on monitoring decrease in current on reduction potential of tetrathiafulvalene (at 0.1V vs. Ag/AgCl) by using a cyclic voltammetry method. Cyclic voltammograms were taken at a potential range between −0.1 and 0.4V vs. Ag/AgCl and correlations between decreases in biosensor responses and glucose oxidase or alcohol oxidase activity were monitored.

Multiple compounds giving unpleasant flavors and aromas in foods can be potentially detected by biosensors. This is the case of the 2,4,6-trichloroanisole, a compound (Varelas et al., 2011) related to corks in wine bottles causing considerable losses to the wine industry. Varelas et al. (2011) developed a rapid novel biosensor system based on a bioelectric recognition assay. The sensor measured the electric response of cultured membrane-engineered fibroblast cells suspended in an alginate gel matrix due to the change of their membrane potential in the presence of the analyte.

Fish freshness has been identified by the quantification of total percent inosine and hypoxanthine generated during fish postmortem changes (Hamada-Sato et al, 2005). Other process indicator compounds can be detected through biosensors, such as lactulose, a disaccharide formed during the thermal processing of milk that allows differentiating between UHT vs. sterilized in container milks.

The amount of starch, glucose, lactose, lecithin, and ethanol present in food, are indicators of quality and consumer acceptability. These compounds also serve as indicators of processing steps completion in manufactured food products. Biosensors technology for substance detection significantly reduces analysis time, and improves specificity, reliability and test sensitivity. These properties allow for real time decision making during food processing. A listing of biosensors used to evaluate food composition is presented by Serna et al. (2009).

Product quality standards are also determined by sensory analysis. During sensory evaluation, a set of techniques to measure properties such appearance, smell, taste and texture when chewing or eating food are used to determine human sensations. However, standardization and accreditation methods for sensory quality analysis are still needed for the certification of food products.

Using expert panels, Pérez et al. (2007) proposed a set of accredited methods for the sensory evaluation of the "Idiazabal" cheese that could be generalized to any type of food and drink as a reference for sensory accreditation. The use of biosensors instead of expert panels in sensory analysis seems to be a good alternative. Electronic and bioelectronic tongues, which could be known as taste sensors, are the advanced and emerging analytical technologies simulating the taste detection modality of the human tongue by means of electrochemical sensors or biosensor arrays.

Pioggia et al. (2007) characterized five compounds with different chemical characteristics and determined taste perceptions, using an electrical impedance biosensor. Five sensors of three different types based on carbon nanotubes or carbon black dispersed in polymeric matrices and doped polythiophenes composed the array. Fifty different solutions eliciting the five basic tastes (sodium chloride, citric acid, glucose, glutamic acid and sodium dehydrocholate for salty, sour, sweet, umami and bitter, respectively) at 10 concentration levels comprising the human perceptive range were analyzed. The impedentiometric composite sensor array was shown to be sensitive, selective and stable for using it in an electronic tongue.

Taste substances are getting through the biological membrane of gustatory cells in the taste buds of tongue. Taste is perceived when the information on taste substances is transduced into an electrical signal, which is transmitted along the nerve fiber to the brain. Electronic and probably bioelectronic tongues are known as two promising tools for the taste assessment of the foodstuffs. In a review article, Ghasemi-Varnamkhasti et al. (2011) concluded that bioelectronic tongues would be a useful tool for process control and although, application of electronic tongue has been reported for alcoholic beers, up to now, no research has been published on the flavor evaluation of non alcoholic beer by means of bioelectronics tongue.

3. Potential use of biosensors in hazard analysis and critical control points (HACCP)

A quality management system should include quality control, quality assurance and a continuous improvement system. In quality management of agri-food processes, hazard analysis and critical control points (HACCP) is considered as the most effective system to ensure food safety. HACCP can also be used as a food quality control system (Ropkins & Beck, 2000).

Well implemented HACCP systems can be used to improve processing efficiency and the quality of food products in order to meet the client needs, and even as defense arguments in litigations. Although HACCP systems are great tools, microbiological contamination of products from biological origin will continue to be the greatest threat in food production. As mentioned before, bacteria counts can be detected by microbial cultures, which are very effective, but require expensive and time consuming techniques to yield reliable results. Many other methods have been proposed for the rapid detection, isolation, identification and enumeration of bacteria, including among others, impedimetric detection, automated cells counting, and immunoassays.

Biosensors have been adapted to detect or quantify analytes in systems in-line (Rasooly, 2001). The HACCP system can be used to verify that a given process is under control, since high biosensor sensitivity allows the detection of pathogenic microorganisms, pesticides, herbicides and other contaminants in hours or minutes (Luo et al., 2009; Mostafa, 2010).

In a HACCP system setup, the ATP bioluminescence method is very suitable for monitoring bacteria in-line, since it does not need prior bacteria growth. The ATP bioluminescence assay is based on the detection of ATP present in living organisms through the luciferin-luciferase enzymatic system.

Luo et al. (2009) developed a low-cost biosensor to detect bacterial counts, with easy operation and rapid response, based on adenosine 50-triphosphate (ATP). The biosensor is composed of a key sensitive element and a photomultiplier tube as a detector element. The disposable sensitive element consists of a sampler, a cartridge where intracellular ATP is chemically extracted from bacteria, and a microtube where the extracted ATP reacts with the luciferin–luciferase reagent to produce bioluminescence. The bioluminescence signal is transformed into relevant electrical signal by the detector and further measured with a homemade luminometer.

In HACCP systems implemented for fisheries, histamine can be used as a quality indicator for the evaluation of fresh fish. Hamada-Sato et al. (2005) considered that raw fish freshness could be estimated by the "K" freshness value. The K value is the total percent of inosine and hypoxanthine to that of ATP-related compounds, and it has shown a good relationship with fish muscle changes. Postmortem adenine phosphate nucleotides (ATP, ADP, and AMP) contents decline rapidly and inosine monophosphate (IMP) contents increase sharply at 0°C within about 24 h after death. Furthermore, inosine and hypoxanthine contents increase when IMP contents decline. Hamada-Sato et al. (2005) simplified the K value to KI since it represents "the ratio (%) of total amount of inosine and hypoxanthine to that of IMP, inosine and hypoxantine".

A biosensor to determine the KI value has already been developed and it is commercially available (Tokaseiki Ltd.). The biosensor consists of two enzyme reactors, two oxygen electrodes, a peristaltic pump, an A/D converter, a microcomputer and a data logger. Inosine in fish muscle is converted into hypoxanthine by nucleoside phosphorylase in enzyme reactor A, followed by the oxidation of hypoxanthine by xanthine oxidase (XOD) in the same reactor. The total amount of inosine and hypoxanthine was determined from the output of the electrode corresponding to the amount of oxygen consumed by hypoxanthine oxidation. IMP is converted into inosine by nucleotidase (NT) in enzyme reactor B, followed by the same reaction in enzyme reactor A.

The total amount of IMP, inosine and hypoxanthine was determined from the output of the electrode corresponding to the amount of oxygen consumed by hypoxanthine oxidation. Fifty microliters of the sample solution were injected through the injection port, and after about 3 min, the obtained results were displayed on the screen (Okuma et al., 1992).

Similarly, the concentration of amino acids (e.g., lysine) obtained by fermentation and used as supplements in animal feed have been controlled with the enzyme lysine oxidase. Amperometric biosensors have been used to measure lactic acid used to control the acidity and crust formation in cheese, assessed using the enzyme lactate oxidase. These biosensors can be integrated into a HACCP system to monitor process acceptability.

4. Conclusions

The use of biosensors technology for food safety will facilitate complying with international quality and safety standards, allowing for the efficient, safe and reliable detection and quantification of pathogenic microorganisms involved in food borne illnesses and inorganic contaminants that threaten consumer health. However, the detection of small concentrations of chemical and biological polluting substances in products for human and animal consumption is still needed.

The selectivity, specificity, and rapid response depend on the biosensor's reception and transduction systems, since they are based on recording reactions that generate physical, chemical and/or immunological changes. This form of reading that allows for a rapid response is ideal in the agri-food process control.

Although biosensors technology is approximately 50 years old, there are several fields that are still under study. A continue search for new recognition elements which comply with minimum wear characteristics and are absent of inhibitory substances that block analyte detection must be done. In the same way, receptors with greater stability for the analyte are also needed.

The detection of pathogenic microorganisms with the use of biosensors shows advantages when compared to traditional methods. In addition to easy handling, biosensors provide timely detection and on line process control. The complex metabolism of pathogenic bacteria and the defense reactions produced by the guest organisms are also used to build biosensors. Furthermore, antigen-antibody interactions are efficient and selective, and the detection speed will depend on the sensitivity of the transduction system and on the method used for signal amplification.

Pesticides, fertilizers, and heavy metals can be quickly detected in small quantities with biosensors, facilitating *in situ* implementation in pre- and post-harvest processes. Transduction, electrochemical and optical methods provide better detection sensitivity in these cases.

In the agri-food industry, biosensors have been useful to assess the freshness of raw materials such as meat, fish, fruits and vegetables. In these cases, biosensors detect compounds that provide abnormal flavors and aromas, indicating microbial growth and food safety problems. However, although biosensors developed in recent years are efficient and some of them successfully used in the industry, their use is limited by the need for receiver renovation and calibration and the adequate control of system variables.

5. References

Akar, S., Tek, V., Bange, A., Lagos, L., Gill, P., Munroe, N. & Thundat, T. (2010). Development of a biosensor for detection of phosphate species in uranium contaminated ground water and wastewater sediments. *WM2010 Conference*, Phoenix, AZ, USA, March 7-11, 2010

Almeida, M., Silveira, C. & Moura, J. (2007). Biosensing nitrite using the system nitrite redutase/Nafion/methyl viologen—A voltammetric study. *Biosensors and Bioelectronics*, Vol. 22, No. 11, pp. 2485-2492

Arbault, P., Buecher, V., Poumerol, S. & Sorin, M. (2000). Study of an ELISA method for the detection of *E. coli* O157 in food. *Progress in Biotechnology*, Vol. 17, pp. 359-368.

Asav, E. & Akyilmaz, E. (2010). Preparation and optimization of a bienzymic biosensor based on self-assembled monolayer modified gold electrode for alcohol and glucose detection. *Biosensors and Bioelectronics*, Vol. 25, No. 5, pp. 1014–1018

Bagal-Kestwal, D., Karve, M., Kakadeb, B. & Pillai, V. (2008). Invertase inhibition based electrochemical sensor for the detection of heavy metal ions in aqueous system: Application of ultra-microelectrode to enhance sucrose biosensor's sensitivity. *Biosensors and Bioelectronics*, Vol. 24, No. 4, pp. 657–664.

Cabrera, L., Witte, J., Victor. B., Vermeiren, L., Zimic, M., Brandt, J. & Geysen, D. (2009). Specific detection and identification of African trypanosomes in bovine peripheral blood by means of a PCR-ELISA assay. *Veterinary Parasitology*, Vol. 164, No. 2-4, pp. 111–117.

Cañas, A. & Macias, M. (2004). Desarrollo de un sistema sensor para la cuantificación de glucosa en jugos de frutas. *Revista de la Sociedad Química de Mexico*, Mexico, Vol. 48, No. 8, pp. 106-110

Castillo, J., Gáspár, S., Sakharov, I. & Csöregi, E. (2003). Bienzyme biosensors for glucose, ethanol and putrescine built on oxidase and sweet potato peroxidase. *Biosensors and Bioelectronics*, Vol. 18, No. 5-6, pp. 705-714.

Chen, S., Huang, J., Du, D., Li, J., Tu, H., Liu, D. & Zhang, A. (2010). Methyl parathion hydrolase based nanocomposite biosensors for highly sensitive and selective determination of methyl parathion. *Biosensors and Bioelectronics*, Article in Press, doi:10.1016/j.bios.2011.04.025.

Chouteau, C., Dzyadevych, S., Durrieu, C. & Chovelon, J. (2005). A bi-enzymatic whole cell conductometric biosensor for heavy metal ions and pesticides detection in water samples. *Biosensors and Bioelectronics*, Vol. 21, No. 2, pp. 273–281

Darnton, N., Turner, L., Rojevsky, S. & Berg, H. (2007). On Torque and Tumbling in Swimming *Escherichia coli*. *Journal of Bacteriology*, Vol. 189, No. 5, pp. 1756–1764. ISBN 0021-9193

Deng, L., Xu, Y. & Huang, J. (2008). Developing a double-antigen sandwich ELISA for effective detection of human hepatitis B core antibody. *Comparative Immunology, Microbiology & Infectious Diseases*, Vol. 31, No. 6, pp. 515-526

Deo, R., Wang, J., Block, I., Mulchandani, A., Joshi, K., Trojanowicz, M., Scholz, F., Chen, W. & Lin, Y. (2005). Determination of organophosphate pesticides at a carbon nanotube/organophosphorus hydrolase electrochemical biosensor. *Analytica. Chimica Acta*, Vol. 530, No. 2, pp. 185–189.

Du, D., Huang, X., Cai, J. & Zhang, A. (2007). Amperometric detection of triazophos pesticide using acetylcholinesterase biosensor based on multiwall carbon nanotube–chitosan matrix. *Sensors and Actuators B: Chemical*, Vol. 127, No.2, pp. 531–535

Du, D., Wang, M., Cai, J. & Zhang, A. (2010). Sensitive acetylcholinesterase biosensor based on assembly of β-cyclodextrins onto multiwall carbon nanotubes for detection of organophosphates pesticide. *Sensors and Actuators B: Chemical*, Vol. 146, No. 1, pp. 337–341

Dutta, K., Bhattacharyay, D., Mukherjee, A., Setford, S., Turner, A. & Sarkar, P. (2008). Detection of pesticide by polymeric enzyme electrodes. *Ecotoxicology and Environmental Safety*, Vol. 69, No. 3, pp. 556–561

Dzyadevych, S., Soldatkin, A., Arkhypova, V., El'skaya, A. & Chovelon, J. (2005). Early-warning electrochemical biosensor system for environmental monitoring based on enzyme inhibition. *Sensors and Actuators B: Chemical*, Vol. 105, No. 1, pp. 81–87

FAO (1996). Plan de Acción de la Cumbre Mundial sobre la Alimentación, párrafo 1. In: *Declaración de Roma sobre la Seguridad Alimentaria Mundial y Plan de Acción de la Cumbre Mundial sobre la Alimentación*. Cumbre Mundial sobre la Alimentación, 13-17 de noviembre de 1996, FAO, 43 p., Roma, Italia,

FAO/WHO (2004). Risk assessment of Listeria monocytogenes in ready to eat foods. *Microbiological Risk Assessment Series, No. 4, Interpretive Summary*. ISBN 92 4 156261 7 (WHO), Available from
http://www.who.int/foodsafety/publications/micro/mra_listeria/en/index.html

Fohlerová, Z., Skládal, P. & Turánek, J. (2007). Adhesion of eukaryotic cell lines on the gold surface modified with extracellular matrix proteins monitored by the piezoelectric sensor. *Biosensors and Bioelectronics*, Vol. 22, No. 9-10, pp. 1896-1901

Geng, T., Morgan, M. & Bhunia A. (2004). Detection of Low Levels of *Listeria monocytogenes* Cells by Using a Fiber-Optic Immunosensor. *Applied and Environmental Microbiology*, Vol. 70, No. 10, pp. 6138-6146, ISBN 0099-2240

Gfeller, K., Nugaeva, N. & Hegner, M. (2005). Micromechanical oscillators as rapid biosensor for the detection of active growth of *Escherichia coli*. *Biosensors and Bioelectronics*, Vol. 21, No. 3, pp. 528-533

Ghasemi-Varnamkhasti, M., Mohtasebi, S., Rodrıguez-Mendez, M., Siadat, M., Ahmadi, H. & Razavi, S. (2011). Electronic and bioelectronic tongues, two promising analytical tools for the quality evaluation of non alcoholic beer. *Trends in Food Science & Technology*, Vol. 22, No. 5, pp. 245-248

Ghica, M. & Brett, C. (2008). Glucose oxidase inhibition in poly(neutral red) mediated enzyme biosensors for heavy metal determination. *Microchimica Acta*, Vol. 163, No. 3-4, 185-193.

Gong, J., Wang, L. & Zhang, L. (2009). Electrochemical biosensing of methyl parathion pesticide based on acetylcholinesterase immobilized onto Au–polypyrrole interlaced network-like nanocomposite. *Biosensors and Bioelectronics*, Vol. 24, No. 7, pp. 2285-2288

Guntupalli, R., Hu, J., Lakshmanan, R., Huang, T., Barbaree, J. & Chin, B. (2007). A magnetoelastic resonance biosensor immobilized with polyclonal antibody for the detection of *Salmonella typhimurium*. *Biosensors and Bioelectronics*, Vol. 22, No. 7, pp. 1474-1479

Haasnoot, W., Du Pre, J., Cazemier, G., Kemmers-Voncken, A., Verheijen, R. & Jansen, B. (2000). Monoclonal antibodies against a sulfathiazole derivative for the immunochemical detection of sulphonamides. *Food and Agricultural Immunology*, Vol. 12, No. 2, pp. 127-138

Hamada-Sato, N., Usui, K., Kobayashi, T., Imada, C. & Watanabe, E. (2005). Quality assurance of raw fish based on HACCP concept. *Food Control*, Vol. 16, No. 4, pp. 301–307

Hiemstra, M. & De Kok, A. (1994). Determination of N-methylcarbamate pesticides in environmental water samples using automated on-line trace enrichment with exchangeable cartridges and high-performance liquid chromatography. *Journal of Chromatography*, Vol. 667, No., pp. 155-166

Hildebrandt, A., Bragos, R., Lacorte, S. & Marty, J. (2008). Performance of a portable biosensor for the analysis of organophosphorus and carbamate insecticides in water and food. *Sensors and Actuators B: Chemical*, Vol. 133, No. 1, pp. 195–201

Hnaien, M., Lagarde, F. & Jaffrezic-Renault, N. (2010). A rapid and sensitive alcohol oxidase/catalase conductometric biosensor for alcohol determination. *Talanta*, Vol. 81, No. 1-2, 222–227

Jaffrezic-Renault, N. & Dzyadevych, S. (2008). Conductometric Microbiosensors for Environmental Monitoring. *Sensors*, Vol. 8, No. 4, pp. 2569-2588

Jian-xiao, L., Xiang-min, X., Lin, T. & Guang-ming, Z. (2009). Determination of trace mercury in compost extract by inhibithion based glucose oxidase biosensor. *Transactions of Nonferrous Metals Society of China*.Vol. 19, No. 1, pp.235-240

Kim, N., Park, I. & Kim, D. (2007). High-sensitivity detection for model organophosphorus and carbamatepesticide with quartz crystal microbalance-precipitation sensor. *Biosensors and Bioelectronics*, Vol. 22, No. 8, pp. 1593-1599

Kumar, J., Kumar, S. & D'Souza, J. (2006). Optical microbial biosensor for detection of methyl parathion pesticide using *Flavobacterium* sp. whole cells adsorbed on glass fiber filters as disposable biocomponent. *Biosensors and Bioelectronics*, Vol. 21, No. 11, pp. 2100–2105

Kumar, J. & D'Souza, S. (2010). An optical microbial biosensor for detection of methyl parathion using *Sphingomonas* sp. immobilized on microplate as a reusable biocomponent. *Biosensors and Bioelectronics*, Vol. 26, No. 4, pp. 1292–1296

Kuralay, F., Ozyoruk, H. & Yıldız, A. (2006). Amperometric enzyme electrode for urea determination usingimmobilized urease in poly(vinylferrocenium) film. *Sensors and Actuators B: Chemical*, Vol. 114, No. 1, pp. 500–506

Leonard, P., Hearty, S., Quinn, J. & O'Kennedy, R. (2004). A generic approach for the detection of whole *Listeria monocytogenes* cells in contaminated samples using surface plasmon resonance. *Biosensors and Bioelectronics*, Vol. 19, No. 10, 1331–1335

Lin, H., Lu, Q., Ge, S., Cai, Q. & Grimes, C. (2010). Detection of pathogen *Escherichia coli* O157:H7 with a wireless magnetoelastic-sensing device amplified by using chitosan-modified magnetic Fe_3O_4 nanoparticles. *Sensors and Actuators B: Chemical*, Vol 147, No. 1, pp. 343–349

Lin, T. & Chung, M. (2008). Using monoclonal antibody to determine lead ions with a localized surface plasmon resonance fiber-optic biosensor. *Sensors*, Vol. 8, No. 1, pp. 582-593.

Long, F., Gao, C., Shi, H., He, M., Zhu, A., Klibanov, A. & Gu, A. (2011). Reusable evanescent wave DNA biosensor for rapid, highly sensitive, and selective detection of mercury ions, *Biosensors and Bioelectronics*, Article in Press, doi: 10.1016/j.bios.2011.03.022

Lu, Y., Yang, W., Shi, L., Li, L., Alam, M., Guo, S. & Miyoshi, S. (2009). Specific Detection of Viable *Salmonella* Cells by an Ethidium Monoazide-Loop Mediated Isothermal

Amplification (EMA-LAMP) Method. *Journal of Health Science*, Vol. 55, No. 5, pp. 820–824

Luo, J., Liu, X., Tian, Q., Yue, W., Zeng, J., Chen, G. & Cai, X. (2009). Disposable bioluminescence-based biosensor for detection of bacterial count in food. *Analytical Biochemistry*, Vol. 394, No. 1, pp. 1–6

Meng, J. & Doyle, M. (2002). Introduction. Microbiological food safety. *Microbes and Infection*, Vol. 4, No. 4, pp. 395–397

Mostafa, G. A. (2010). Electrochemical Biosensors for the detection of pesticides. The Open Electrochemistry Journal. Vol. 2, pp. 22-42.

Nowak B., Müffling T., Chaunchom S. & Hartung, J. (2007). *Salmonella* contamination in pigs at slaughter and on the farm: A field studyusing an antibody ELISA test and a PCR technique. *International Journal of Food Microbiology*, Vol. 115, No. 3, pp. 259–267

Ocampo A., March, C. & Montoya, A. (2007). Inmunosensores piezoeléctricos: revisión general y su aplicación en el análisis de pesticidas. *Revista EIA*, Escuela de Ingeniería de Antioquia, Medellín, Colombia, No. 7, pp. 97-110, ISSN 1794-1237

Okuma, H., Takahashi, H., Yazawa, S., Sekimukai, S. & Watanabe, E., 1992. Development of a system with double enzyme reactors for the determination of fish freshness. *Analytica Chimica Acta*, Vol. 260, No. 1, pp. 93–98

On, S. (2001). Taxonomy of *Campylobacter*, *Arcobacter*, *Helicobacter* and related bacteria: current status, future prospects and immediate concerns. *Journal of Applied Microbiology*, Vol. 90, No. S6, pp. 1S-15S.

Park, J., Kurosawa, S., Aizawa, H., Hamano, H., Harada, Y., Asano, S., Mizushima, Y. & Higaki, M. (2006). Dioxin immunosensor using anti-2,3,7,8-TCDD antibody which was produced with mono 6-(2,3,6,7-tetrachloroxanthene-9-ylidene) hexyl succinate as a hapten. *Biosensors and Bioelectronics*, Vol. 22, No. 3, pp. 409-414

Patel, P. (2002). (Bio)sensors for measurement of analytes implicated in food safety: a review. *Trends in Analytical Chemistry*, Vol. 21, No. 2, pp. 96-115

Pathirana, S., Barbaree, J., Chin, B., Hartell, M., Neely, W. & Vodyanoy, V. (2000). Rapid and sensitive biosensor for *Salmonella*. *Biosensors and Bioelectronics*, Vol. 15, No. 3-4, pp. 135–141

Pavlov, V., Xiao, Y. & Willner, I. (2005). Inhibition of the acetylcholine esterase-stimulated growth of au nanoparticles: nanotechnology-based sensing of nerve gases. *Nano Letters*. Vol. 5, No.4, pp. 649–653

Pellicer, K., Hoyo, G., Brocardo, M., Aliverti, V., Aliverti, F. & Copes, J. (2009). Effect of Chlorhidric and Lactic Acid on the Development of Thirty Strains of Listeria spp. from Foods Stuffs. *Revista de la Facultad de Ciencias Veterinarias*, Universidad Central de Venezuela, Vol. 50, No. 1, pp. 19-22, ISBN 0048-7724

Pérez Elortondo, F., Ojeda, M., Albisu, M., Salmerón, J., Etayo, I. & Molina, M. (2007). Food quality certification: An approach for the development of accredited sensory evaluation methods. *Food Quality and Preference*, Vol. 18, No. 2, pp. 425–439

Pioggia, G., Di Francesco, F., Marchetti, A., Ferro, M. & Ahluwalia, A. (2007). A composite sensor array impedentiometric electronic tongue Part I. Characterization. *Biosensors and Bioelectronics*, Vol. 22, No. 11, pp. 2618-2623

Pribyl, J., Hepel, M. & Skládal, P. (2006). Piezoelectric immunosensors for polychlorinated biphenyls operatingin aqueous and organic phases. *Sensors and Actuators B: Chemical*, Vol. 113, No. 2, pp. 900-910.

Pribyl, J., Hepel, M., Halámek, J. & Skládal, P. (2003). Development of piezoelectric immunosensors for competitive and direct determination of atrazine. *Sensors and Actuators B: Chemical*, Vol. 91, No. 1-3, pp. 333-341

Rahman, M., Park, D., Chang, S., McNeil, C. & Shim, Y. (2006). The biosensor based on the pyruvate oxidase modified conducting polymer for phosphate ions determinations. *Biosensors and Bioelectronics*, Vol. 21, No. 7, pp. 1116–1124

Rasooly, A. (2001). Surface plasmon resonance analysis of Staphylococcal enterotoxin B in food. *Journal of Food Protection*, Vol. 64, No. 1, pp. 37–43

Ropkins, K. & Beck, A. (2000). Evaluation of worldwide approaches to the use of HACCP to control food safety. *Trends in Food Science and Technology*, Vol. 11, No. 1, pp. 10–21

Ryan, K & Ray, C. (Eds). (2010) *Sherris medical microbiology,* Fifth edition. McGraw Hill Companies, Inc. 1026 p. ISBN 978-0-07-160402-4, USA

Sánchez, J., Jiménez, S., Navarro, R. & Villarejo, M. (2009). Patógenos Emergentes en la Línea de Sacrificio de Porcino. Fundamentos de Seguridad Alimentaria. Ediciones Díaz de Santos, 209 p. ISBN 978 84 7978 922 0, España

Schöning, M., Arzdorf, M., Mulchandani, P., Chen, W. & Mulchandani, A. (2003). Towards a capacitive enzyme sensor for direct determination of organophosphorus pesticides: fundamental studies and aspects of development. *Sensors,* Vol. 3, No. 6, 119-127

Serna, L., Zetty A. & Ayala, A. (2009). Use of enzymatic biosensors as quality indices: a synopsis of present and future trends in the food industry. *Chilean Journal of Agricultural Research*, Vol. 69, No. 2, pp. 270-280

Seo, K., Brackett, R. & Frank, J. (1998). Rapid detection of *Escherichia coli* O157:H7 using immunomagnetic flow cytometry in ground beef, apple juice, and milk. *International Journal of Food Microbiology*, Vol. 44, No. 1-2, pp. 115–123.

Shimomura, T., Itoh, T., Sumiya, T., Mizukami, F. & Ono, M. (2009). Amperometric biosensor based on enzymes immobilized in hybrid mesoporous membranes for the determination of acetylcholine. *Enzyme and Microbial Technology*, Vol. 45, No. 6-7, 443-448

Silveira, C., Gomes, S., Araújo, A., Montenegro, B., Todorovic, S., Viana, A., Silva, R., Moura, J. & Almeida, M. (2010). An efficient non-mediated amperometric biosensor for nitrite determination. *Biosensors and Bioelectronics*, Vol. 25, No. 9, pp. 2026–2032

Sismani, C., Kousoulidou, L. & Patsalis, P. (2008). *Molecular biomethods handbook.* 2nd Edition.Humana Prees, 1124 p., ISBN 1603273700, Totowa, NJ

Smyth, A., Talasila, P. & Cameron, A. (1999). An ethanol biosensor can detect low-oxygen injury in modified atmosphere packages of fresh-cut produce. *Postharvest Biology and Technology*, Vol. 5, No. 2, pp. 127-134

Sohail, M. & Adeloju, S. (2008). Electroimmobilization of nitrate reductase and nicotinamide adenine dinucleotide into polypyrrole films for potentiometric detection of nitrate. *Sensors and Actuators B: Chemical*, Vol. 133, No. 1, pp. 333–339

Székács, A., Trummer, N., Adányi, N., Váradi, M. & Szendrö, I. (2003). Development of a non-labeled immunosensor for the herbicide trifluralin via OWLS detection. *Analytica Chimica Acta*, Vol. 487, No. 1, pp. 31–42

Tang, H., Zhang, W., Geng, P., Wang, Q., Jin, L., Wu, Z. & Lou, M. (2006). A new amperometric method for rapid detection of *Escherichia coli* density using a self-assembled monolayer-based bienzyme biosensor. *Analytica Chimica Acta*, Vol. 562, No. 2, pp. 190–196

Tims, T. & Lim, D. (2003). Confirmation of viable *E. coli* O157:H7 by enrichment and PCR after rapid biosensor detection. *Journal of Microbiological Methods*, Vol. 55, No. 1, pp. 141- 147

Trivedi, U., Lakshminarayana, D., Kothari, I., Patel, N., Kapse, H., Makhija, K., Patel, P. & Panchal, C. (2009). Potentiometric biosensor for urea determination in milk. *Sensors and Actuators B: Chemical*, Vol. 140, No. 1, pp. 260–266

Tsai, H., Doong, R., Chiang, H. & Chen, K. (2003). Sol-gel derived urease- based optical biosensor for the rapid determination of heavy metals. *Analytica Chimica Acta*, Vol.481, No. 1, pp. 75-84.

Turner, N., Subrahmanyam, S. & Piletsky, S. (2009). Analytical methods for determination of mycotoxins: A review. *Analytica Chimica Acta*, Vol. 632, No. 2, pp. 168–180

Vadivambal, R. & Jayas, D. (2007). Changes in quality of microwave-treated agricultural products: a review. *Biosystems Engineering*, Vol. 98, No. 1, pp. 1-16

Valach, M., Katrlik, J., Sturdik, E. & Gemeiner, P. (2009). Ethanol *Gluconobacter* biosensor designed for flow injection analysis: Application in ethanol fermentation off-line monitoring. *Sensors and Actuators B: Chemical*, Vol. 138, No. 2, pp. 581–586

Van der Merwe, P. (May 2011) Surface Plasmon Resonance. Available from http://users.path.ox.ac.uk/~vdmerwe/Internal/spr.PDF.

Varelas, V., Sanvicens, N., Pilar-Marco, M. & Kintzios, S. (2011). Development of a cellular biosensor for the detection of 2,4,6-trichloroanisole (TCA). *Talanta*, Vol. 84, No. 3, pp. 936–940

Vasquez de Plata, G. (2003). La Contaminación de los Alimentos, un Problema por Resolver. *Salud UIS*, Vol. 35, pp. 48-57

Verma, N., Kumar, S. & Kaur, H. (2010). Fiber Optic Biosensor for the Detection of Cd in Milk. *Journal of Biosensors and Bioelectronics*, 1:102. doi:10.4172/2155-6210.1000102.

Waswa, J., Irudayaraj, J. & DebRoy, C. (2007). Direct detection of *E. Coli* O157:H7 in selected food systems by a surface plasmon resonance biosensor. *Food Science and Technology*, Vol. 40, No. 2, pp. 187–192

Wei, D., Oyarzabal, O., Huang, T., Balasubramanian, S., Sista, S. & Simonian, A. (2007). Development of a surface plasmon resonance biosensor for the identification of *Campylobacter jejuni*. *Journal of Microbiological Methods*, Vol. 69, No. 1, pp. 78–85

White, D., Zhao, S., Simjee, S., Wagner, D. & McDermott, P. (2002). Antimicrobial resistance of foodborne pathogens. *Microbes and Infection*, Vol. 4, No. 4, pp. 405–412

WHO (2000). Factsheet No 255: *Campylobacter*. World Health Organization, Available from http://www.who.int/mediacentre/factsheets/fs255/en

WHO (2005). Factsheet No 125: Enterohaemorrahic *Escherichia coli* (EHEC). World Health Organization, Available from
http://www.who.int/mediacentre/factsheets/fs125/en/

Yin, J., Wei, W., Liu, X., Kong, B., Wu, L. & Gong, S. (2007). Immobilization of bovine serum albumin as a sensitive biosensorfor the detection of trace lead ion in solution by piezoelectric quartz crystal impedance. *Analytical Biochemistry*, Vol. 360, No. 1, pp. 99–104

Yokoyama, K., Ikebukuro, H., Tamiya, E., Karube, I., Ichiki, N. & Arikawa, Y. (1995). Highly sensitive quartz crystal immunosensors for multisample detection of herbicides. *Analytica Chimica Acta*, Vol. 304, No. 2, pp. 139-145

Zhang, Y., Muench, S., Schulze, H., Perz, R., Yang, B., Schmid, R. & Bachmann, T. (2005). Disposable biosensor test for organophosphate and carbamate insecticides in milk. *Journal of Agricultural and Food Chemistry*, Vol. 53, No. 13, pp. 5110–5115

Zhang, Z., Jaffrezic-Renault, N., Bessueille, F., Leonard, D., Xia, S., Wang, X., Chen, L. & Zhao, J. (2008). Development of a conductometric phosphate biosensor based on tri-layer maltose phosphorylase composite films. *Analytica Chimica Acta*, Vol. 615, No. 1, pp. 73–79

4

Modified Cholinesterase Technology in the Construction of Biosensors for Organophosphorus Nerve Agents and Pesticides Detection

Monika Hoskovcová and Zbyněk Kobliha
University of Defence, NBC Defence Institute
Czech Republic

1. Introduction

Organophosphorus compounds are a wide-spread group of agents which can be used among others as pesticides, especially insecticides, or chemical warfare agents. These extremely toxic compounds irreversibly inhibit the enzymes of hydrolases class, which have the catalytic capability of hydrolyzing their specific neurotransmitters in synaptic clefts of the nervous system. The organophosphorus pesticides and carbamates are agents used in agriculture very often, because of their relatively low stability. Many of them, however, have very high acute toxicity for warm-blooded animals.

The group of organophosphorus compounds also includes nerve agents with their dominant position among chemical warfare agents. The discovery of their effects was made more or less accidentally in the 30-ies of the last century during the research of fluoroorganic compounds in IG Farben, a German company, by Dr. Gerhard Schrader. The first synthesized agent was ethyl-(dimethylamido)phosphorocyanidate, designated as trilon 83 or tabun. However, it was not the first known agent with a cholinergic effect. Back in 1854 a French chemist, Phillip de Clermont, synthesized the first organophosphate – tetraethyl pyrophosphate (de Clermont, 1855, as cited in Holmstedt, 2000). Gradually more agents with N-P, P-CN or C-F bonds were synthesized in order to produce insecticides and later also nerve agents (Holmstedt, 2000). After verifying the effects of tabun for warm-blooded animals, a synthesis of other, even more toxic, agents followed (Pitschmann, 1999). In 1939 isopropyl-methylphosphonofluoridate was discovered, also called trilon 46 or sarin, and in 1944 (3,3-dimethylbutane-2-yl)-methylphosphonofluoridate, the so-called soman, followed. These agents belong to the so-called G-series of nerve agents. The origin of a new V-series dates back to the end of 50-ies. This group of agents has increased toxicity when penetrating the skin. The principal representative is *S*-[2-(diisopropylamino)ethyl]-*O*-ethyl-methylphosphonothioate, also known as VX, and *S*-[2-(diethylamino)ethyl]-*O*-isobutyl-methylphosphonothioate with the code designation R-33. In the 70-ies up to 90-ies another group of nerve agents was discovered in the former USSR within the so-called Foliant program. These are compounds based on phosphorylated and phosphonylated oximes and amidases. According to unauthorized sources they reach at least the toxicity of VX, some are even 5 to 8 times more toxic. In a case like this there could be a problem with their detection.

The limit of detection of the most of current technical means is, with the exception of means based on biochemical cholinesterase reaction, unsatisfactory.

The inhibition of enzyme efficacy happens even at very low concentrations of nerve agents or pesticides. Even trace amounts of these agents can represent a considerable hazard to health for living organisms. Detection through the biochemical method is characterized by high sensitivity and enables setting of about $1.10^{-5} - 1.10^{-7}$ mg.ml^{-1} in the solution or 1 l of air depending on the type of inhibitor. This sensitivity is up to five times higher compared to methods based on chemical or physical principle (Halámek et al., 2009). For this reason a number of biosensors, which provide a suitable alternative to classical analytical methods, has been developed based on the biochemical reaction. It is an analytical device which consists of a biologically active material such as an enzyme, an antibody or a binding protein and a converter which converts the biochemical reaction to an output signal which can be quantified. These converters can be based on amperometric, potentiometric, chemiluminescent, piezoelectric or optical principles and on a semiconductor or an ion-selective effect of transistor techniques. Their development has been important especially in the last two decades. They meet the demands for a quick and cheap analysis with high sensitivity (Kenar, 2010; Rekha et al., 2000).

This chapter deals with a general overview and characteristics of methods for detection and identification of nerve agents and pesticides based on the cholinesterase reaction and its modifications. It shows some practical applications of biosensors and indicates some new trends in the field leading to increasing the selectivity for identification of individual nerve agents.

2. Cholinesterases

The hydrolases, which are used for the construction of biosensors, are basically two, acetylcholinesterase (AChE, EC 3.1.1.7), also called a genuine or specific cholinesterase, and butyrylcholinesterase (BuChE, EC 3.1.1.8), the so-called false or also nonspecific cholinesterase. The task of ACHE is the hydrolysis of a specific neurotransmitter acetylcholine (ACh) in a neurosynaptic cleft and thus the termination of irritating cholinergic receptors or a neuromuscular junction. The physiological importance of BuCHE in a body is not, compared to AChE, so essential. Its physiological function has not been exactly identified yet (Masson et al., 2009). However, it has been found out that it plays some role in metabolism of local anaesthetics, novocaine and cocaine (Carmona et al., 2000). Its specific substrate is butyrylcholine (BuCh). Besides natural substrates also a synthetic substrate, e.g. 4-aminophenyl acetate, can be used for both enzymes. Both enzymes basically have a similar three-dimensional structure but there is a difference in their substrate specificity and sensitivity to inhibitors (Hanin & Dudas, 2001). Hydrolytic reactions catalyzed by cholinesterases proceed in a narrow space of the active center where the so-called esteratic and anionic site can be found. The esteratic site is made up of the so-called catalytic triad a part of which is the serine residuum which reacts with the substrate but also with organophosphates. The presence of substrate in the active center of AChE stabilizes the configuration of the catalytic triad. The size of the substrate molecule and its positive charge at the nitrogen atom influence the size of the bond to anionic site. It has been experimentally found out that this bond is reduced by relocating the positive charge (Kua et al., 2002). The decomposition of a neurotransmitter results in ACh bond to the active site of AChE enzyme which is placed in the hole of the active center, about 20 Å deep (Sussman et al., 1991). The tertiary nitrogen in ACh molecule is bonded to the anionic site represented by a glutamate ion. The oxygen from the acyl part of the substrate molecule is bonded in the oxy anion hole. Thereby the ACh molecule is anchored in the active center. The hydroxyl serine group, which binds the acetyl part of ACh, participates in splitting of ester. Then the substrate is split up into acetyl and choline. The effect of hydrolysis by means of a water molecule, present in the cleft of the active center, causes spontaneous reconstruction of enzyme activity. An acylating group, which can be active within another choline acetylation, splits off from the enzyme molecule (Somani et al., 1992). The mechanism of acetylcholine hydrolysis catalyzed by the acetylcholinesterase enzyme is shown in the reaction scheme 1.

$$CH_3C(O)OCH_2CH_2 \overset{+}{N}(CH_3)_3 + H_2O \xrightarrow{\text{AChE}} CH_3COOH + HOCH_2CH_2 \overset{+}{N}(CH_3)_3$$

Reaction scheme 1. Mechanism of acetylcholine hydrolysis catalyzed by the acetylcholinesterase enzyme.

Nerve agents, even at very low concentrations, inhibit their hydrolytic capacity through a bond in this active center. The inhibition takes place in several steps, as shown in the reaction scheme 2. In the first one the inhibitor is bonded to a hydroxyl serine group in the esteratic center of AChE. Thus the reversible enzyme-inhibitor complex arises. Further serin is covalently phosphorylated. An electronegative leaving group is released from the molecule of organophosphorus toxic agent. This bond is still reversible. A competing process is the ageing of AChE, the phosphorylated AChE switches to its dealkylated form. It is a process where the alcohol is split off by means of water and the alkoxyl group bonded to phosphor is replaced by the hydroxyl group. This state is irreversible. The result of enzyme inactivation is accumulation of acetylcholine in the nerve ending and development of toxic exposures demonstrated in the form of nicotine, muscarinic and central effects (Watson et al., 2009).

Reaction scheme 2. Mechanism of inhibition of the acetylcholinesterase enzyme.

The enzyme inhibition will prove itself in reducing its activity depending on the degree of inactivation. This reduction can be measured objectively and it is a significant factor for the construction of biosensors. The basis for measuring the enzyme activity is the reaction with a suitable substrate. AChE catalyzes the hydrolysis of acetylcholine substrate, or its sulfur isoester – acetylthiocholine; BuChE hydrolyzes then the catalysis of butyrylcholine or butyrylthiocholine. Subsequently you can observe the decrease in substrate, the increase in products or an unreacted substrate. Quite usual is the observation of concentration of hydrolysis products, i.e. an appropriate acid or choline. We shall not forget pH-metric methods based on the measurement of concentration of an arising acid in combination with acid-base indicators and buffers or finished with alkalimetric titration or also with determination of CO_2 (Tomeček & Matoušek, 1961). The methods of manometric, fluorometric, colorimetric and spectrophotometric evaluation or electrochemical methods such as potentiometry, amperometry, coulometry or voltammetric indication are also worth mentioning. Some benefits can be seen in using the so-called chromogenous substrates which brought modification and expansion of possibilities of the cholinesterase method. Those are used for example with photometric methods; we observe changes in coloring caused by the reaction of hydrolysis products with chromogenous reagents.

The advantage of enzymatic methods for identification of organophosphorus compounds is their high sensitivity and specificity. The demands on making an analysis are, compared to common analytical methods, lower as well as their cost. For these reasons they are applied more and more often, e.g. when checking the environment contamination.

The detection of nerve agents and pesticides based on organophosphates can be done not only on the basis of choline esterases but there is also a new trend, a non-inhibitory method based on organophosphate hydrolase (OPH, EC 3.1.8.1) or organophosphorus acid anhydrolase (OPAA, EC 3.1.8.2). OPH hydrolyses compounds with P-O, P-F, P-S and P-CN bonds. There were tests of some pesticides, such as paraoxon and parathion, and nerve agents, e.g. sarin or VX (Donarski et al., 1989; as cited in Wang et al., 1999; Joshi et al., 2006). So the organophosphorus compounds act for OPH as a substrate. These types of biosensors offer, same as the cholinesterase ones, using optical, fluorescent, potentiometric or amperometric converters (Karnati et al., 2007; Mulchandani, et al. 1999a, 1999b; Rogers et al., 1999; Wang et al., 2003). Also a field version for in-situ monitoring has been developed (Wang et al., 1999; Mulchandani et al., 2001). The OPAA enzyme selectively hydrolyses organophosphorus compounds containing the P-F bond; for compounds with other common bonds occurring in nerve agents and pesticides the substrate activity is low. A biosensor with this enzyme was successfully tested for example for diisopropyl-fluorophosphate. Besides measuring of pH changes using a glass electrode, pH-sensitive field effect transistor (pH-FET), also an ion-selective electrode for fluoride ions can be used (Simonian et al., 2001).

3. Cholinesterase biosensors

The inhibitory reactions can proceed in the solution; a disadvantage, however, is the limited use of enzyme and thus also the increase of cost or enzyme immobilization onto a suitable material and connection with an appropriate converter. The first biosensor based on the cholinesterase inhibition through organophosphorus compounds was constructed by G. Guibault back in 1962 (Guibault et al., 1962; as cited in Arduini et al., 2010). Since that time many biosensors have been developed for identification of nerve agents and pesticides.

Cholinesterase biosensors may be classified according to a converter they use. Most often electrochemical converters are used, further piezoelectric and optical converters and surface plasmon resonance (SPR) (Arduini et al., 2010; Lin et al., 2006). This chapter deals particularly with electrochemical and optical methods.

A common disadvantage of cholinesterase biosensors is the irreversible character of inhibition in the course of analysis. In many cases it is impossible to recover a biosensor like that and to use it once again. This limits the life of biosensor to 10 – 15 identifications. It is necessary to recover the enzyme continuously, which can cause considerable difficulties, especially in field conditions. Some procedures have been developed, for example using nucleophilic reagents, the so-called cholinesterases reactivators (Dăneţ et al., 2003; Gulla et al., 2002; Gyurcsányi, et al., 1999; Tušarová et al., 1999) or surface regeneration using self-assemble layers with a piezoelectric biosensor (Makower et al., 2003).

3.1 Electrochemical biosensors

Electrochemical methods are based on a potentiometric, amperometric or conductometric method of measurement. Most often amperometric and potentiometric devices are used.

Electrochemical biosensors make use of mono- and multienzymatic systems. The monoenzymatic systems use the above-mentioned AChE or BuChE hydrolases, the bienzymatic ones use then the cholinesterase-choline oxidase (ChE-ChO) system (Rekha et al., 2000), the trienzymatic ones use cholinesterase, cholinoxidase and peroxidase (POD) (Ghindilis et al., 1996).

3.1.1 Amperometric biosensors

Amperometric biosensors are the most widely used type of electrochemical sensors and a good substitution for the potentiometric ones. They provide a linear output signal depending on the concentration of analyte. There are several ways how to evaluate the enzyme activity by means of amperometric methods. The first one includes the bienzymatic cholinesterase-cholinoxidase (ChE-ChO) system with an oxygen or peroxide amperometric converter (Campanella, 2007). The choline from the hydrolytic enzyme catalysis is not electrochemically active; that is why cholinoxidase is used by means of which the choline oxidizes to betaine while developing the hydrogen peroxide, as shown in the reaction scheme 3. This results in detection of the occurrence of hydrogen peroxide or oxygen consumption.

$$(CH_3)_3 \overset{+}{N} CH_2CH_2OH + O_2 + H_2O \xrightarrow{\text{ChO}} (CH_3)_3 \overset{+}{N} CH_2CH_2COOH + 2H_2O_2$$

Reaction scheme 3. Reaction mechanism of the bienzymatic amperometric ChE-ChO system.

The second way includes the modification using ACh sulfur isoester, acetylthiocholine, which is electrochemically active (Pohanka, 2009). With applied voltage in the reaction mixture the resulting thiocholine undergoes an oxidation-reduction reaction according to the reaction mechanism in scheme 4. These systems were tested not only for pesticides and nerve agents but also for toxins with a cholinergic effect (Kandimalla & Ju, 2006; Pohanka et. al., 2008). Also a bienzymatic amperometric biosensor based on BuChE and sulfhydryl oxidase (SOX) has been developed even if it has not been tested for organophosphorus esterases inhibitors so far. Butyrylthiocholine catalytically hydrolyzes while developing the thiocholine which acts as a SOX substrate in the subsequent oxidation (see the reaction scheme 5). The measurement was made through the subsequent detection of oxygen consumption which is proportional to BuTCh concentration. SOX catalyzes the formation of disulfide bridges between sulfhydryl groups (Teksoy, 2007).

$$CH_3C(O)SCH_2CH_2 \overset{+}{N}(CH_3)_3 + H_2O \xrightarrow{\text{AChE}} CH_3COOH + HSCH_2CH_2 \overset{+}{N} H(CH_3)_3$$

$$2HSCH_2CH_2\overset{+}{N}(CH_3)_3 \xrightarrow{\text{+410mV}} (CH_3)_3\overset{+}{N}CH_2CH_3-S-S-CH_2CH_2\overset{+}{N}(CH_3)_3 + 2H^+ + 2e^-$$

Reaction scheme 4. Reaction mechanism of the enzyme acetylthiocholine hydrolysis and subsequent oxidation through applied voltage.

$$CH_3(CH_2)_2C(O)SCH_2CH_2\overset{+}{N}(CH_3)_3 + H_2O \xrightarrow[\text{SOX}]{\text{BuChE}} CH_3(CH_2)_2COOH + HSCH_2CH_2\overset{+}{N}(CH_3)_3$$

$$2(CH_3)_3\overset{+}{N}CH_2CH_2SH + O_2 \longrightarrow (CH_3)_3\overset{+}{N}CH_2CH_3-S-S-CH_2CH_2\overset{+}{N}(CH_3)_3 + H_2O_2$$

Reaction scheme 5. Reaction mechanism of the bienzymatic ChE-SOX system.

Besides natural substrates it is possible to use also the synthetic 4-aminophenyl acetate. The enzyme-catalyzed hydrolysis gives rise to an electroactive 4-aminophenol which is oxidized on the electrode surface (see the reaction scheme 6) (La Rosa et al., 1994). The advantage lies in simplicity of using the system with this substrate and a low applied potential.

$$H_2N-\!\!\bigcirc\!\!-O-\underset{\underset{O}{\|}}{C}-CH_3 \;+\; H_2O \xrightarrow{\text{AChE}} H_2N-\!\!\bigcirc\!\!-OH \;+\; CH_3COOH$$

Reaction scheme 6. Reaction mechanism of the enzyme 4-aminophenyl acetate hydrolysis.

As for cholinesterases in amperometric measurements the above-mentioned AChE or BuChE are used; they hydrolyze the substrate to an appropriate choline (acetyl or butyrylcholine), in case of 4-aminophenyl acetate it is 4-aminophenol and an acid (acetic or butyric). These reactions, however, are complicated by occurrence of byproducts deposited on the electrode or by a spontaneous substrate hydrolysis itself. Limiting is also the necessity of using high working voltage which limits the use of biosensors for samples with electroactive molecules. In case of cholinesterase biosensors it is possible to lower the voltage and interferences of electroactive sample components by means of redox transmitters such as ferophtalocyanine, cobalt phtalocyanine, Prussian blue (ferrihexacyanofferate), 7,7,8,8-tetracyanoquinodimethane (TCNQ) in combination with carbon electrodes, screen electrodes or using the carbon nanotube (CNT) (Arduini et al., 2007; Ciucu et. al., 2003; Hart et al., 1997; Hernandez et al., 2000; Lin et al., 2004; Ricci et al., 2003; Ivanov et al., 2003a; Nunes et al., 2004; Shulga & Kirchhoff, 2007; Skládal, 1992; Skládal & Mascini, 1992; Sun & Wang, 2010; Suprun et al., 2005; Wring et al., 1989) or electrochemical decomposition of the arising peroxide using a gold or platinum amperometric electrode (Yao, 1983).

CNTs are commonly used for electrochemical biosensors. They occur in two forms – single-walled (SWCNT) and multi-walled (MWCNT). They are characterized by excellent chemical stability, good structural and mechanical properties and electrical conductivity. And these are exactly the qualities which make them an ideal means for construction of biosensors. They provide an improved electrochemical detection of enzymatic-developed thiocholine not only thanks to the reduction of over-voltage but also thanks to higher sensitivity and stability (Liu & Lin, 2006; Zhang & Gorski, 2005; as cited in Du et al., 2007). However, they are insoluble in most of the organic solvents and water solutions. Nevertheless, CNT biosensors for detection of pesticides with an immobilized enzyme have been developed, e.g. MWCNT with chitosan matrix (Du et al, 2007) or a glass carbon electrode (GCE) modified by MWCNT with chitosan matrix with immobilized AChE (Sun et al., 2010) or a glass electrode modified by CNT for detection of pesticides and nerve agents (Liu et al., 2005).

The principle of thiocholine substrate splitting is used for example by a biosensor with platinum electrode with immobilized AChE and Ag/AgCl reference electrode in the midst of which a potential of +410 mV is applied (Martorell et al., 1994), at Pt anode with a potential of +410 mV vs. Ag/AgCl (Marty et al., 1993) or Ti-Au-Pt electrode with +700 mV versus a saturated calomel electrode, etc. Thiocholine is oxidized at the anode. The platinum electrode is not suitable for detection of sulfhydryl compounds because of the necessity of using too high over-voltage. For this reason the aforementioned carbon electrodes are used, chemically modified by cobalt phtalocyanine or other modifiers, which leads to reduction of the applied potential and lowering of interferences caused by electrochemical impurities (Halbert & Baldwin, 1985; Ciucu & Baldwin, 1992; as all cited in Ciucu et al., 2003; Skládal, 1992). An ideal tool for monoenzymatic systems based on the acetylthiocholine hydrolysis is also the detection using screen electrodes, for example in connection with cobalt phtalocyanine which transmits the oxidation of thiocholine (Bucur et al., 2005).

Another application makes use of a working electrode with graphite compound and Ag/AgCl reference electrode and applied voltage of +700 mV (Turdean, 2002). The amperometric biosensor for detection of methyl paraoxon, carbofuran and phoxim has been developed based on the modified platinum electrode with immobilized AChE at gold nanoparticles and a silk fibroin.

The hydrogen peroxide can be detected via amperometry at +650 mV versus the reference Ag/AgCl electrode (Andreescu & Marty, 2006). An approach of a ceramic microelectrode with cholinoxidase for measurement of changes to the extracellular choline with a detection limit of 300 nM and an ability to measure sudden ACh changes in brain and its hydrolysis to choline has been developed (Burmeister et al., 2003; Parikh et al., 2004; Parik et al., 2006; as all cited in Philips, 2005). For detection of organophosphorus pesticides a biosensor has been constructed, containing BuChE and CHO layered onto a nylon membrane of a hydrogen peroxide electrode (Campanella et al., 1992). Another system is based on detection of oxygen consumption in this reaction using Clark's electrode (Mizutani & Tsuda, 1982). After enzyme inhibition there is reduction in the production of hydrolytic products and thus also cutting of oxygen consumption which is proportional to the enzyme activity. The disadvantage of this system is the required incubation period and enzyme consumption for every other identification.

An amperometric biosensor, which uses the AChE enzyme and 4-aminophenyl acetate as a substrate, has been described. AChE is immobilized onto a glass carbon electrode (La Rosa et al., 1994; Pariente et al., 1993) or a screen electrode (Andreescu et al., 2002a). The enzymatic-hydrolyzed 4-aminophenol can be detected also via voltametry. The disadvantage of screen electrodes is their low stability in organic solvents because of partial solubility of the printed layer.

In the paper (Mitchell, 2004) a multienzyme biosensor for amperometric detection of ACh and choline (Ch) in vivo is described. This biosensor makes use of AChE, ChO and ascorbic acid oxidase (AAO) immobilized onto a platinum-iridium wire with an electropolymer layer Poly(*m*-(1,3)-phenylenediamine).

For detection of organophosphorus pesticides and carbamates in water and food a portable amperometric biosensor has been assembled, consisting of a screen electrode with 10 mU of immobilized AChE connected with potentiostat and a portable computer. It enables detection even in a pepper extract (Hildebrandt et al., 2008). Pesticides present in water samples were successfully detected for example by ChE-ChO biosensor through a peroxide electrode (Bernabei et al., 1993) or in real samples in some organic solvents which can influence the enzyme activity (Palchetti et al., 1997). The sensor for detection of organophosphorus pesticides in river, sea and waste water, based on immobilized ChO and esterase in the solution, uses the Pt anode vs. Ag/AgCl cathode with +650 mV (Palleschi et al., 1992).

The development of a new technology for immobilization based on a silicate sol-gel matrix enabled creation of very stable AChE biosensors with nanoparticles. For example an AChE amperometric biosensor with gold nanoparticles (AuNP) in sol-gel matrix with a three-electrode detection system has been developed, including the Pt wire as an auxiliary electrode, a saturated calomel reference electrode and a modified glass carbon working electrode with AChE, an ATCh substrate. The flow of arising products after the catalytic hydrolysis and substrate oxidation is detected (Du et al., 2008). Further use of the sol-gel technology based on an alternative non-silica material is offered by the AChE biosensor with ZnO matrix and SPE electrodes (Sinha et al., 2010).

Most of the amperometric biosensors are tested for detection of pesticides, less of them for nerve agents. For detection of nerve agents for example a portable biosensor, which is a part of soldier's individual equipment, has been developed (Arduini et al., 2007).

Generally it can be said that the improvement of amperometric biosensors is based on two approaches, i.e. enhancement of the catalytic capacity and transfer of electrons through various modifications of working electrodes and at a lower potential. There is also a description of multisensors for current detection and identification of mixtures of cholinesterase insecticides based on various options of one enzyme (Bachmann & Schmidt, 1999; Bachmann et al., 2000), various enzymes (Danzer & Schwedt, 1996; Istamboulie et al., 2009; Kok & Hasirci, 2004) using Artificial Neural Networks (ANN).

3.1.2 Potentiometric biosensors

As it has been already mentioned, an appropriate organic acid arises in the course of hydrolysis, which is the basis for potentiometric measurements. The simplest one is based on detecting the acid increment by means of a pH-metric device with continuous recording. This configuration, however, is not that suitable for the use in field conditions (Miao et al., 2010).

Potentiometric detectors were also constructed for both hydrolases substrates, Ach and BuCh. For this reason they require different substrate sensors – acetylcholine or butyrylcholine selective electrodes which are based on measurement of the redox potential, either with transmitters or without them. Most often, however, we use pH sensors based on detection of pH changes, caused by system acidification after hydrolysis (Zhang et al., 2009). Glass pH electrodes are the basis for potentiometric measurements but they are expensive, fragile and unfit for miniaturization and require conditiation before using. A substitution may be electrodes based on metal, metal/metal oxide or metal oxide – Au, Pd/PdO, Ir/IrO$_2$, RuO$_2$ (Gyurcsányi, et al., 1999; Koncki & Mascini, 1997; Reybier et al., 2002; Tran-Minh et al., 1990). Furthermore, we use pH ISFET electrodes (Ion-sensitive field effect transistors) which represent a sensitive, cheap, available and simple monitoring device suitable for miniaturization (Dzyadevych et al., 2004; Dzyadevych et al., 2006), a light addressable potentiometric sensor (LAPS) and electrodes modified by polymers (Snejdarkova et al. 2004; Ivanov et al., 2003b).

Most biosensors based on the pH detection are handicapped thanks to an extended signal response and lower substrate sensitivity. The response extension is given by the necessity of perfusing the reaction layer or by membrane strength and permeability. The basic limit of sensitivity of potentiometric electrodes is given by sensitivity of the pH converter. This jump in potential versus pH does not mostly exceed 59 mV per pH unit. The ISFET sensitivity is 44-54 mV/pH (Cara et al., 1985; as cited in Karyakin et al., 1996). The disadvantages can be partially eliminated by various modifications of electrodes. For this purpose for example a potentiometric acetylcholinesterase biosensor with an antimony pH electrode has been assembled, having – contrary to the glass ones – an easily recoverable surface for immobilization, low electrical stability and variability of shapes and sizes. For the electrode an immobilization technique of covalent cross-linking with glutaraldehyde vapors in the vacuum was used (Gyurcsányi, et al., 1999).

Potentiometric measurements make use of the above-mentioned ion-selective electrodes (ISE) which serve as a polymer matrix for transmission of reagent substances but also as a converter for this type of electrochemical biosensors. In the paper (Ding & Qin, 2009) there is a presentation of a detection system where the butyrylcholine substrate is released through ISE membrane and thus the activity of BuChE enzyme or its application for detection of the presence of organophosphorus pesticides in the sample is detected. Constantly releasing the substrate under the zero-current conditions from an external solution to a sample solution provides a measurable signal in situ. Detecting the enzyme activity is then a consequence of disturbing the ion flux induced by membrane potential after the enzyme catalysis.

As far as potentiometric substrate sensors for detection of pesticides are concerned, for example a butyrylcholine sensor, placed on a plasticized polyvinyl chloride membrane by a cationic ion exchanger tetrakis (3,5-bis[2-methoxy-hexafluoro-methyl]phenyl) borate, has been described. The advantage is high hydrophobicity of the ion exchanger used. Compared to other biosensors there has been an improvement of potential stability, life cycle and detection limit (Imato & Ishibashi, 1995). A biosensor of flow injection type for identification of pesticides is described in the paper (Lee et al., 2002). It consists of AChE immobilized onto controlled porous glass and of a detector with tubular H^+ of a selective membrane electrode. The organophosphate is oxidized, which leads to increasing the sensitivity of cholinesterase biosensors. Oxidized forms show a higher inhibitory activity.

The potentiometric system without transmitters uses the peroxidase enzyme. The peroxidase (POD) catalyzes the reaction of the arising peroxide through electro-reduction by a mechanism of direct electron transfer from an electrode to a substrate molecule through the active enzyme site (Ghindilis et al., 1996; Ghindilis et al., 1997). This principle was used to test the screen enzyme electrode for detection of organophosphorus pesticides. ChE along with CHO and POD are co-immobilized on the electrode surface. The enzyme activity is specified based on individual enzymatic reactions, i.e. butyrylcholine hydrolysis to choline, choline oxidation and peroxide formation, and peroxide electro-reduction by means of POD according to the reaction scheme 7 (Espinosa et al., 1999).

Even the potentiometric biosensors can be used to monitor waste substances in the living environment, for example in waste waters (Espinosa et al., 1999; Evtugyn et al., 1997).

$$\text{choline} + O_2 \xrightarrow{\text{ChO}} \text{betaine} + H_2O_2$$

$$H_2O_2 + 2e^- + 2H^+ \xrightarrow{\text{POD}} + 2H_2O$$

Reaction scheme 7. Reaction mechanism of the POD detector.

Over fifty years there has been a research conducted in the field of reactivation of inhibited cholinesterases in vitro and in vivo. For this purpose the so-called reactivators are practically used – strong nucleophilic reagents which recover the function of an inhibited enzyme. The compounds based on mono- and bispyridinium aldoximes or ketoximes proved to be efficient. Based on this also the potentiometric sensor was built up. It works on a principle of the reaction of ketoxime and organophosphorus chemical warfare agent under the development of phosphorylated ketoxime and separation of the leaving group (here CN⁻). The cyanide ion is detected by a silver electrode (Moll et al., 1976; as cited in Oh & Masel, 2007). Later on this system was reevaluated and optimized not only for organophosphates containing a cyanide group in their structure but also for their simulators using the cyanide ion-selective electrode (Oh & Masel, 2007).

When comparing amperometric and potentiometric sensors, the amperometric ones are rated as quicker and more sensitive. The potentiometric ones, however, are simpler and more suitable for a field analysis.

3.1.3 Conductometric biosensors

The conductometric measurement is based on measuring the conductivity changes which are directly proportional to the occurrence of ions in the measured solution (Dzyadevych et al., 2005). The conductometric biosensors have a big advantage because they do not require the reference electrode and the converter provides the possibility of miniaturization (Miao et al., 2010).

It is impossible to analyze various mixtures of analytes in real samples using just one biosensor type. This problem led to construction of the so-called multibiosensor. For this purpose two converters were linked – the potentiometric one with a pH-sensitive transistor and the conductometric one with thin-films interdigitated electrodes, and three enzymes – urease, AChE and BuChE. This group enables simultaneous identification of some heavy metals and pesticides. Working parameters and experimental conditions are similar as in the case of individual sensors. The result of hydrolytic reactions is the urease consumption or proton production (with AChE and BuChE), which causes the increase in pH and changes to conductivity at membranes with a subsequent possibility of potentiometric or conductometric way of detection. This system represents a simple way of measurement and it is applicable for the control of drinking water (Arkhypova et al., 2001).

3.2 Optical biosensors
3.2.1 Colorimetric and spectrophotometric biosensors

Probably the first colorimetric method is Hestrin's quantitative method for identification of an unreacted ACh and esters similar to it. The identification is based on reaction of ACh with hydroxylamine in the alkaline environment in the presence of ferric chloride when the hydroxylamine reacts with ACh while producing the acetylhydroxamic acid and developing the brown-purple complex of ACh-acetylhydroxamic product which can be identified via spectrophotometry at 540 mµ (Hestrin, 1949).

One of the oldest procedures for detection of the presence of nerve agents is measuring the degree of inhibition of cholinesterases using a chromogenous substrate via the so-called Ellman's method designed for identification of thiols (Ellman, 1959; Ellman et al., 1961). It is the most often used method in clinical biochemistry to include its employment in field individual equipment kits (Wilson, et al., 1997; Capacio et al., 2008). Its principle is splitting of hydrolysis-thiocholine (acetyl- or butyrylthiocholine) while releasing the appropriate acid and thiocholine according to the reaction scheme 8. The thiocholine molecule contains the -SH group which can be detected by means of the so-called Ellman's reagent, 5,5´-dithiobis(2-nitrobenzoic) acid (DTNB). Splitting up the substrate into thiocholine is indicated by yellow coloring which is caused by development of a reduced form of the Ellman's reagent. This enables spectrophotometric detection at 412 nm.

Various modifications of Ellman's reaction are used in a number of detection means such as detection tubes or a detection biosensor, Detehit (Tušarová & Halámek, 2001). This means is utilized by the Armed Forces of the Czech Republic not only as an individual equipment item but also as a means of chemical survey and monitoring. It is a strip consisting of a zone with a cotton cloth with immobilized and stabilized acetylcholinesterase and a detection paper with acetylthiocholine iodide and 5,5´-dithiobis(2-nitrobenzoic) acid. The acetylcholinesterase is immobilized in the form of a stable enzyme chimera with polysaccharide cellulose. The enzyme remains the solid phase and its use is polyvalent. An advantage is the possibility of removing the surplus of organophosphate and reactivator by simply flushing the cloth with water.

This simple colorimetric biosensor can be used, after exposition through nucleophilic reagents, for selective inhibitor identification (Halámek et al., 2009; Hoskovcová et al., 2009). The nucleophilic reagents restore the function of the enzyme by unbinding the inhibitor from its active center under the formation of the so-called phosphorylated oxime. This exposition, however, must happen even before the development of the so-called non-reactivable form of the phosphorylated enzyme. In clinical practice compounds based on pyridinium aldoximes are used (Milatović & Jokanović, 2009). Their efficacy varies and depends on many factors such as structure, number and placement of functional groups and cationic points, number of pyridinium nuclei etc., but also the type of the inhibitor. Thanks to those influences there is no broad-spectrum reactivator which would be able to react effectively with an enzyme inhibited by any nerve agent. Exactly the different efficacy of these oxime compounds compared to characteristic enzyme-inhibitor complexes results in the already mentioned change to the intensity of color of the biosensor. From the original white back to yellow owing to arising products of the substrate hydrolysis. The measurement of coloring intensity is done by detecting the reflectance of the color surface of an impregnated cloth (Tušarová et al., 1999; Hoskovcová et al., 2009).

For observing the activity of enzymes also other chromogens have been studied, e.g. indolephenylacetate, 2,6-dichlorindolephenyl acetate or β-naphthyl acetate, which are characteristic through blue coloring (Kramer & Gamson, 1958; No et al., 2007), also 2-azobenzene-1-naphthylacetate with developing a red product (Epstein et al., 1957) or reaction with bromothymol blue which, due to pH changes in hydrolysis, switches from the originally blue-green color to the yellow one (Limperos & Ranta, 1953). There has been also a description of substitution of the Ellman's reagent by a more stable compound 5-(2-aminoethyl)dithio-2-nitrobenzoate (Zhu et al., 2004).

Using the redox indicator of Guinea green B enables the indication of presence of the esterases inhibitor in automatic detectors. The basic principle is the ability of nerve agents to slow down or stop the biochemical reaction (see the reaction scheme 9). This reaction applies BuTCh and Guinea green B whereas the normal course of the enzyme-catalyzed hydrolysis will result in discoloring of indicator. In the presence of nerve agents there is no change to coloring (Český obranný standard, 2007).

$$CH_3C(O)SCH_2CH_2\overset{+}{N}(CH_3)_3 + H_2O \xrightarrow{\text{AChE}} CH_3COOH + (CH_3)_3\overset{+}{N}CH_2CH_2SH$$

$$2\,(CH_2)_3\overset{+}{N}CH_2CH_2S^- + O_2N\!-\!\!\bigcirc\!\!-\!S\!-\!S\!-\!\!\bigcirc\!\!-\!NO_2 \quad (HOOC \ldots COOH)$$

$$(CH_3)_3\overset{+}{N}CH_2CH_2\text{-}S\text{-}S\text{-}CH_2CH_2\overset{+}{N}(CH_3)_3 + 2\ O_2N\!-\!\!\bigcirc\!\!-\!SH \quad (HOOC)$$

$$H^+ \updownarrow OH^-$$

Reaction scheme 8. Ellman's method for identification of acetylthiocholine hydrolysis products.

For colorimetric evaluation a pH indicator strip has been developed, working in the range of pH 6.0 up to 7.7. The acetic acid originated from hydrolysis will be marked dark red in the detector. The method is comparable to the Ellman's method but in contrast to it there is an advantage of characteristic coloring which is visible even under the artificial light (Pohanka, et al. 2010a). Same as the previous means it can function without any other instrumentation and it offers a simple and rapid semiquantitative evaluation of contamination by cholinesterases inhibitors.

Another modification of the Ellman's method is a spectrophotometric FIA system developed for detection of pesticides in water. Here the acetylcholinesterase is immobilized on amidated glass pearls. The advantage of the system is the reactivation of an inhibited enzyme through 2-PAM reactivator (Dănet, 2003). There has been also a procedure described where DTNB is substituted by a chemiluminescent substance – 1,2-dioxethane.

The test is based on a chemically induced electron luminescent exchange (Sabelle et al., 2002).

The simple spectrophotometric identification has been described using a different chromogenous substrate, indophenyl acetate, which is hydrolyzed to a color product by enzymatic catalysis at pH 8. The production rate of this product depends on the enzyme activity (Kramer & Gamson, 1958). There has been also a description of the method of spectrophotometric identification of pesticides in the analyte through measuring of color products of α-naphthol acetate. The substrate is catalytically hydrolyzed by AChE to α-naphthol which reacts with *p*-nitrobenzendiazonium fluoroborate while developing color products (Leon-Gonzales & Townshend, 1990).

green

$$2 \ (CH_3)_3NCH_2CH_2SH$$

colourless

$$(CH_3)_3NCH_2CH_2-S-S-CH_2CH_2N(CH_3)_3$$

Reaction scheme 9. Reaction mechanism of biochemical reaction modified by Guinea green B

Without the presence of chromogen the biosensors work based on non-linear optics. Conformation changes of the applied enzyme are measured, for example at adsorption of the produced thiocholine (Lin et al. 2006).

Also the system of spectrophotometric identification of cholinesterase inhibitors by means of an integrated acetyl-butyrylcholinesterase surface has been described. The interaction of these esterases with a reversible inhibitor, tetraphenyl porphyrin, leads to production of characteristic maxima at 446 and 421 nm in the absorption spectrum. The exposition through competitive inhibitors to a porphyrin-enzyme complex leads to reduction in values of absorbance intensity for both characteristic porphyrin-enzyme peaks, which is, however, not the case of noncompetitive inhibitors. The reduction of absorbance is caused by dissociation of porphyrin from active enzyme sites (Brandy et al., 2003).

It seems to be interesting to use the so-called chromo-fluorogenous reagents with gold nanoparticles. This nanotechnology was successfully tested for example for detection of paraoxon. The system consists of AChE, HAuCl$_4$, gold nanoparticles and acetylthiocholine iodide. The plasmon absorbance band of nanoparticles has higher intensity (blue) after increasing the concentration of aceylthiocholine. The gold nanoparticles enlarge with increasing the substrate concentration, which can be caused by increased production of thiocholine that acts as a reduction substance for AuCl$_4$, while producing the metal gold which is deposited onto gold nanoparticles. These gold nanoparticles increase their size proportionally to the concentration of thiocholine whereby changes to plasmon absorbance are made. The gold nanoparticles are deposited onto a glass surface. In reaction of enzyme with substrate the blue coloring will be shown (λ = 570 nm), after adding an inhibitor the plasmon band will be reduced and the glass plate will turn pink (Pavlov et al., 2005).

3.2.2 Fluorescent biosensors

This type of optical biosensors has been described especially in recent years. It is based on the principle of detecting fluorescent changes caused by the shift of pH in the course of substrate hydrolysis (Díaz & Peinado, 1997).

Various papers describe a greater number of fluorescent substances which are suitable for detection of nerve agents as well as organophosphorus pesticides and their simulators (Dale & Rebek, 2006; Bencic-Nagale et al., 2006) but many of them do not correspond to modification of the cholinesterase method; that is why they will not be included in this chapter. Probably the first fiber optic cholinesterase biosensor with a fluorescent substance was constructed with immobilized AChE with fluorescein isothiocyanate (FITC) on quartz fibers (FITC-AChE) (Rogers et al., 1991). In the course of ACh hydrolysis, depending on the pH change, a fluorescent signal under the influence of FITC-AChE is produced in the system. It is present in the evanescent zone of the fiber surface and is quenched by produced protons, which enables detecting the enzyme activity. Totally different is the fiber optic biosensor based on covalently bonded AChE on isothiocyanate glass mixed with thymol blue (immobilized on aminopropylic glass) and subsequently measuring the reflectance of color changes of a pH-sensitive indicator in the course of enzyme-catalyzed hydrolysis at 600 nm (Andres & Narayanaswamy, 1997).

Fiber-optic biosensors provide a lot of benefits such as the possibility of miniaturization, the possibility of remote sensing, in situ monitoring without the necessity of direct electric supply, and except for that they minimize undesirable interactions between the fluorescent detector and the sample.

The sol-gel technology meant a significant advancement for the construction of optical biosensors. Very often it is used just for the optical fluorescent biosensors. These biosensors contain a pH-sensitive fluorescent indicator encapsulated along with the enzyme in the sol-gel network. Biorecognition elements are protected by polymeric material which enables the substrate to diffuse easily to the enzyme and to be bonded subsequently. For detection of organophosphorus pesticides we have used for example the ChE immobilized in the sol-gel silicone matrix along with indoxyl-acetate substrate which is not a fluorescent substance but its hydrolysis gives rise to a highly fluorescent indoxyl. In the presence of the AChE inhibitor this fluorescence is reduced (Díaz & Peinado, 1997). In the paper (Tsai & Doong, 2000) altogether nine fluorescent substances were used for optimization of the fiber-optic biosensor on the sol-gel principle for detection of organophosphorus pesticides whereas the best results were achieved with FITC-dextran.

The Ellman's reagent was substituted by fluorescent substances such as fluorescein-5-maleimide or methylcoumarin maleimide analoga (Parvari et al., as cited in Capacio et al.,

2008). Same as DTNB they react with thiocholine substrates. Coumarinylphenylmaleimide (CPM) forms with thiocholine a CPM-thiocholine product which is continually monitored by laser-induced fluorescence. After inhibitor exposition the fluorescent signal is reduced due to decreased concentration of thiocholine originated from the hydrolytic reaction catalyzed by AChE, and it results in negative peak depending on the type of inhibition (Hadd et al., 1999).

Another fluorescent biosensor for detection of pesticides applies AChE and a pH-sensitive fluorescent indicator·– pyranine, immobilized onto nanoliposoms. Reducing the fluorescent signal of the pH indicator is proportional to the concentration of pesticides tested. This biosensor was tested for detection of the total toxicity in samples of drinking water (Vamvakaki & Chaniotakis, 2007; as cited in Miao et al., 2010).

Using the fluorescent substances for detection of cholinesterases inhibitors has found its way also to analytical kits. Such an analysis is quick and reliable and it enables the acquisition of results even in the field conditions (Technical Bulletin 296, 1996; Product information K 015-F1, 2004; Product information A 12217, 2009).

3.3 Piezoelectric biosensors

Besides the already mentioned methods we also use biosensors with piezoelectric converters for detection of organophosphates and carbamates. This type of converters can be used for biosensors with AChE immobilized onto the crystal face which serves as a frequency control element for the connection of oscillator. This device works based on the principle of changes to resonance frequency of the crystal. The frequency rises as a result of changes to oscillation, thickness or elasticity of the film applied on the crystal (O´Sullivan & Guilbault, 1999), i.e. in case of piezoelectric biosensors the change of resonance frequency depends directly on the mass of molecules bonded to the sensitive surface (Skládal, 1995; Grate et al., 1993; as all cited in Skládal & Macholán, 1997).

In the paper (Abad et al., 1998) 3-indolylacetate, which is transformed to indigo as a result of enzymatic reactions, was used as a substrate. This insoluble pigment is entrapped on the surface of the crystal, which results in frequency changes. The rate and extent of enzymatic reaction are proportional to those frequency changes related to thickness of the QCM (Quartz crystal microbalance) surface, caused by the product of enzymatic reaction.

Another piezoelectric biosensor works, instead of detecting the enzyme activity – i.e. a substrate transformation, based on the principle of measuring the AChE bond to a reversible inhibitor, benzoylecgonine-1,8-dioxaoctane, which is immobilized onto the layer of 11-mercaptomonoundecanoic acid on the gold surface of the sensor. The AChE bond to the inhibitor is detected by ·a mass-sensitive QCM detector. In case of presence of the organophosphate in a sample the enzyme bond to an immobilized reversible inhibitor is reduced. The reduction in mass changes is proportional to the concentration of a free inhibitor in a sample. This biosensor can be practically applied for control and quick analysis of the presence of organophosphates in water streams (Halámek et al., 2005).

4. Enzyme immobilization

The most important step in preparation of a biosensor is immobilization and stabilization of enzyme on the working surface. It can be an electrode or just a cloth or paper (Tušarová & Halámek, 2001). The enzyme immobilization determines the possibilities and construction of a biosensor. The selection of an appropriate way of immobilization is important to keep the catalytic activity of the enzyme and the mechanism for signal transmission. This process is given by interactions between the enzyme and the surface of the given material which serves as a carrier. The principal requirement is to keep the given system sufficiently stable and efficient and to preserve its sensitivity and selectivity. An unsubstitutable role is played also by the cost of the whole system.

Carrying matrices, which are used for enzyme immobilization, can be included in groups of natural polymers such as polysaccharides or proteins, synthetic polymers such as polystyrene, polyacrylates, methacrylates etc. or inorganic carriers such as minerals, active carbon, fiber glasses or porous metal oxides (Pohanka et al., 2010 b; Doretti et al., 1998).

The basic methods for enzyme immobilization include physical adsorption, covalent bond, self-assembled monolayer (SAM), physical entrapment and affinity precipitation (Andreescu & Marty, 2006). The AChE immobilized for purposes of detection of organophosphorus substances, however, should not be entrapped by high affinity.

The physical adsorption of an enzyme onto the surface is supported by van der Waal's forces, thus the bond is very weak. This way of immobilization is the simplest one within the whole range of techniques used. It is characterized by disadvantages such as a weak operational stability and limited storing capacities.

The most often used way of immobilization is the covalent bond. This method commonly uses the modifications of converters with a bifunctional cross-linker such as glutaraldehyde, carbodiimide/succinimide or aminopropyltriethoxysilanes, albumin, chitosan etc., which are scaled on the supporting side activated by amino, carboxyl or hydroxyl groups and on the other side with biomolecules. This technique means increasing the stability of enzyme but the disadvantage is a great amount of bioreagent and lower reproducibility (Andreescu & Marty, 2006; Gyurcsányi et al., 1999; Kandimalla & Ju, 2006; Li et al., 1999; Nunes et al., 2004). The most often applied immobilization technique is the cross-linking by means of glutaraldehyde. The cross-linking of enzyme with glutaraldehyde on an electropolymeric polyethyleneimine film on the electrode surface was used for example for potentiometric measurement in the paper (Reybier et al., 2002). Polyethylenimine as a carrier for ChE was used also for a screen electrode in the paper (Montesinos et al., 2001), chitosan with MWNT (Kandimalla & Ju, 2006) or with a glassy carbon electrode (Sun & Wang, 2010); the bienzymatic AChE-ChO system is immobilized through glutaraldehyde to a glass electrode modified by Au-Pt nanoparticles (Upadhyay et al., 2009) etc.

The self-assembled monolayer exploits the placement of a thin layer directly onto the converter. The layer is made up of strong adsorption of alkenesilanes or disulfides, sulfides and thiols onto the metallic surface of the converter. Very often gold surfaces are used (Somerset et al., 2009). However, also other electrode materials are exploited, such as carbon nanotubes (Liu & Lin, 2006). The stability of a layer is ensured through formation of long chains of *n*-alkylthiols or silanes. This immobilization technique enables the orientation and space control of the enzyme and degradation of diffusion barriers. A disadvantage is the repeated layering of biomolecules (Andreescu & Marty, 2006, Arduini et al., 2010).

The mechanical entrapment is possible through a photopolymeric monomer, sol-gel matrix or nanoparticles and magnetic microparticles (Pohanka et al., 2009). The enzyme, transmitters and possible additives can be entrapped in one layer (Andreescu & Marty, 2006). The entrapment in photopolymeric matrix includes mixing the enzyme with monomer with subsequent polymerization under the neon lamp (Andreescu et al., 2002b, c). The photopolymeric matrix is a material suitable for the use with various enzymes and for various types of electrodes. The sol-gel procedure enables the entrapment of organic molecules in an inorganic material. Compared to other immobilization techniques it enables the anchorage of a great amount of enzyme, thermal and chemical stability, the possibility for selection of size and shape of pores for penetration of a substrate or inhibitors and its preparation is simple. The so-called microencapsulation in pores of the sol-gel matrix is not dependant on protein properties, significantly reduces neither the activity nor the affinity of biomolecules, because the enzyme is not covalently bonded to the matrix. Very often the sol-gel silicate matrices are used for construction of fluorescent fiber-optic biosensors (Díaz & Peinado, 1997) or as a carrier for biosensors based on reactions of nanoparticles (Du et al., 2008). The sol-gel technology was described even based on non-silicate matrices, e.g. with aluminium oxide, titanium, vanadium peroxide, zirconium dioxide or zinc oxide (Daigle & Leech, 1997; Elessi et al., 1997; Li et al., 1998; Meulenkamp, 1998; as all cited in Sinha et al., 2010; Sinha et al., 2010).

Affinity tags represent a modern trend in enzyme immobilizations. This technique is focused on oriented and site-specific immobilization of enzymes. One of the trends is the formation of a bioaffinity bond between supporting and functional groups of proteins. This can be achieved through affinity interactions of functional groups of the activated electrode surface (lectins, (strept)avidin, sugars and metal chelates) and affinity tags of ChE. ChEs contain a limited number of those groups which are placed far enough from active sites and they are suitable for formation of an affinity bond (Andeescu & Marty, 2006). There was a description of an immobilization method based on an affinity bond through metal chelate (nitriloacetic acid in the complex with nickel ions) and a hexa-histidine bond with genetically modified AChE (Andreescu et al., 2001, Andreescu et al., 2002c). Another way of affinity interactions includes the bond between avidin and biotin which was used for enzyme multilayer membrane sets on the surface of a quartz plate and electrode. The enzyme is immobilized in turns onto individual layers. This set can be used for both mono- and bienzymatic biosensor (Chen et al., 1998). The streptavidin-biotin was used for AChE immobilization onto a carbon nanotube (CNT) (Gao et al., 2009). As for lectins, the use of concanavalin A is described. In the paper (Ivanov et al., 2010) there was a test of a biosensor with position-specified immobilized AChE through albumin, glutaraldehyde and lectin concanavalin A on a hybrid polymer membrane in combination with MWCN; in the paper (Bucur et al., 2004) an affinity bond using concanavalin A and a screen electrode is applied.

The biosensor selectivity and sensitivity depend on the conditions of immobilization. Very often we discuss the rule saying "a lower enzyme activity or a thinner enzymatic membrane means higher sensitivity or a lower detection limit of the inhibitor identified". According to the study (Ivanov et al., 2000) a thinner film with quicker reaction will be produced thanks to direct enzyme immobilization onto the electrode surface; however, this does not mean increasing the sensitivity of biosensor. It is caused by relative saturation of the enzyme layer with a substrate or an inhibitor and it depends on diffusion of reactants to this membrane. Buffering properties of the membrane material may contribute to relative accumulation of products of the catalyzed hydrolysis. This compensates for the lower enzyme activity and causes increasing the response of biosensor. It is important that the accumulation intensifies the changes to enzyme activity thanks to the contact of enzyme with inhibitor. The above-said results in reduction of a detection limit and a relative sensitivity reduction compared to biosensors without diffusion limits or mass transfers to the membrane surface. The influence of diffusion factors can be affected also by pre-concentrating of the pesticide at a membrane. Then the identification of inhibitor depends on its hydrophobicity (Ivanov et al., 2000).

Membranes placed onto the enzymatic layer reduce the relative saturation of this layer with substrate and thus they increase the linear extent of the identified substrate concentration. When applying the enzyme inhibitors, they do not provide any bigger advantage because the maximum inhibitor sensitivity requires substrate saturation of the membrane. One of the possibilities for membrane protection is the nafion layer placed between the working surface of the sensor and the enzymatic layer. This configuration means reducing the response of biosensor on one hand but on the other hand it provides higher stability of the carbon layer and greater stability of the thin layer of immobilized ChE and it does not reduce the biosensor sensitivity (Gogol et al., 2000).

The metal nanoparticles can be of great importance for increasing the stability and speed of the response of biosensors. The nanoparticles act as transmitters of an electron transfer from an enzyme molecule to an electrode. Very often gold nanoparticles are used because of their biocompatibility with enzymes (Marinov et al., 2010; Pavlov et al., 2005; Yin et al., 2009). They roughen the surface and subsequently increase interactions on the electrode (Shulga & Kirchoff, 2007). Also other nanoparticles were used, Au-CdS, placed on a gold electrode with AChE (Zayats et al., 2003; as cited in Shulga & Kirchoff, 2007). The use of nanoparticles also includes carbon, zircon and CdS particles (Ion et al., 2010; Liu et al., 2008).

5. Conclusion

Many substances with an anti-cholinergic effect can be, besides extensive industrial applications, used in the military or misused for terrorist purposes as it is shown by events from the recent past when in 1988 the nerve agents were used by Iraq against the Kurdish minority or in 1995 during the terrorist attack in Tokyo subway, conducted by the Aum Shinrikyo sect. Every year millions of tons of those substances/agents are produced for the needs of agriculture, medicine, industry and many other branches. They are quite toxic and for living organisms they mean a considerable risk even at very low concentrations. For these reasons the importance of early detection and analysis is growing. Many analytical methods (GC, MS, GC-MS, HPLC etc.) require a long preparation of samples such as homogenization, extraction, purification or derivatization. Some of these procedures may lead to sample devaluation. An alternative for those lengthy and laborious procedures is offered by biosensors. They meet the requirements for a rapid detection of nerve agents or organophosphorus pesticides for a quick adoption of protective measures to conceal, evacuate the inhabitants and conduct decontamination with checking its efficiency. The above requirements also include an easy handling and low cost. Their other indisputable benefits include a low sample volume, a partial protection of the immobilized enzyme against undesirable physical phenomena, a possibility for monitoring of kinetic and dissociation constants and the type of inhibition. Newly the biosensors are used for evaluation of new AChE reactivators (Pohanka et al., 2007).

6. References

Abad, J.M.; Pariente, F.; Hernández, L.; Abruña, H.D. & Lorenzo, E. (1998). Determination of organophosphorus and carbamate pesticides using a piezoelectric biosensor. *Analytical Chemistry*, Vol. 70, No. 14, pp. 2848-2855. ISSN 0003-2700

Andreescu, S.; Magearu, V.; Lougarre, A.; Fournier, D. & Marty, J.-L. (2001). Immobilization of enzyme on screen-printed sensors via an histidine tail. Application to the detection of pesticides using modified choinesterase. *Analytical Letters,* Vol. 34, No. 4, pp. 529-540, ISSN 0003-2719

Andreescu, S.; Noguer, T.; Magearu, V. & Marty, J.-L. (2002a). Screen-printed electrode based on AChE for the detection of pesticides in presence of organic solvents. *Talanta,* Vol. 57, No. 1, pp. 169-176, ISSN 0039-9140

Andreescu, S.; Avramescu, A.; Bala, C.; Magearu, V. & Marty, J.-L. (2002b). Detection of organophosphorus insecticides with immobilized acetylcholinesterase - Comparative study of two enzyme sensors. *Analytical and Bioanalytical Chemistry,* Vol. 374, No. 1, pp. 39-45, ISSN 1618-2642

Andreescu, S.; Barthelmebs, L. & Marty, J.-L. (2002c). Immobilization of acetylcholinesterase on screen-printed electrodes: Comparative study between three immobilization methods and applications to the detection of organophosphorus insecticides. *Analytica Chimica Acta,* Vol. 464, No. 2, pp. 171-180, ISSN 0003-2670

Andreescu, S. & Marty, J.-L. (2006). Twenty years research in cholinesterase biosensors: From basic research to practical applications. Review. *Biomolecular Engineering,* Vol. 23, pp. 1-15, ISSN 1389-0344

Andres, R.T. & Narayanaswamy, R. (1997). Fibre-optic biosensor based on covalently immobilized acetylcholiesterase and tymol blue. *Talanta,* Vol. 44, No. 8, pp. 1335-1352, ISSN 0039-9140

Arduini, F.; Amine, A.; Moscone, D. Ricci, F. & Palleschi, G. (2007). Fast, sensitive and cost-effective detection of nerve agents in the gas phase using a portable instrument and an electrochemical biosensor, *Analytical and Bioanalytical Chemistry,* Vol. 388, No. 5-6, pp. 1049-1057, ISSN 1618-2642

Arduini, F.; Amine, A.; Moscone, D. & Palleschi, G. (2010). Biosensors based on cholinesterase inhibition for insecticides nerve agents and aflatoxin B_1 detection (review). *Microchimica Acta,* Vol. 170, No. 3-4, pp. 193-214, ISSN 0026-3672

Arkhypova, V.N.; Dzyadevych, S.V.; Soldatkin, A.P.; El´skaya, A.V.; Jaffrezic-Renault, N.; Jaffrezic, H. & Martelet, C. (2001). Multibiosensor based on enzyme inhibition analysis for determination of different toxic substances. *Talanta,* Vol. 55, No. 5, pp. 919-927, ISSN 0039-9140

Bachmann, T.T. & Schmidt, R.D. (1999). A disposable multielectrode biosensor for rapid simultaneous detection of the insecticides paraoxon and carbofuran at high resolution. *Analytical Chimica Acta,* Vol. 401, No. 1-2, pp. 95-103, ISSN 0003-2670

Bachmann, T.T.; Leca, B.; Vilatte, F.; Marty, J.-L.; Fournier, D. & Schmid, R.D. (2000). Improved multianalyte detection of organophosphates and carbamates with disposable multielectrode biosensors using recombinant mutants of Drosophila acetylcholinesterase and artificial neural networks. *Biosensors and Bioelectronics,* Vol. 15, No. 3-4, pp. 193-201, ISSN 0956-5663

Bencic-Nagale, S.; Sternfeld, T. & Walt, D.R. (2006). Microbead chemical switches: An approach to deection of reactive organohosphate chemical warfare agent vapors. *Journal of American Chemical Society,* Vol. 128, No. 15, pp. 5041-5048, ISSN 0002-7863

Bernabei, M.; Chiavarini, S.; Cremisini, C. & Palleschi, G. (1993). Anticholinesterase activity measurement by a choline biosensor: Application in water analysis. *Biosensors and Bioelectronics,* Vol. 8, No. 5, pp. 265-271, ISSN 0956-5663

Brandy, J.; White, B.J.; Legako, J.A. & Harmon, H.J. (2003). Spectrophotometric detection of cholinesterase inhibitors with an integrated acetyl-/butyrylcholinesterase surface. *Sensors and Actuators, B: Chemical*, Vol. 89, No. 1-2, pp. 107-111, ISSN 0925-4005

Bucur, B.; Danet, A.F. & Marty, J.-L. (2004). Versatile method of cholinesterase immobilisation via affinity bonds using Concanavalin A applied to the construction of a screen-printed biosensor. *Biosensors and Bioelectronics*, Vol. 20, No. 2, pp.217-225, ISSN 0956-5663

Bucur, B.; Dondoi, M.; Danet, A. & Marty J.L. (2005). Insecticide identification using a flow injection analysis systém with biosensors based on variol cholinesterases. *Analytica Chimica Acta*, Vol. 539, pp. 195-201, ISSN 0003-2670

Campanella, L.; Cocco, R., Sammartino, M.P. & Tomassetti, M. (1992). A new enzyme inhibition sensor for organophosphorus pesticides analysis. *The Science of the Total Environment*, Vol. 123-124, pp. 1-16, ISSN 0048-9697

Campanella, L.; Lelo, D.; Martini, E. & Tomassetti, M. (2007). Organophosphorus and carbamate pesticide analysis using an inhibition tyrosinase organic phase enzyme sensor; comparison by butyrylcholinesterase + choline oxidase opee and application to natural waters. *Analytical Chimica Acta*, Vol. 587. No. 1., pp. 22-32, ISSN 0003-2670

Capacio, B.R.; Smith, J.R.; Gordon, R.K.; Haigh, J.R.; Barr, J.R. & Lukey, B.J. (2008). Clinical detection of exposure to chemical warfare agents, In: *Chemical Warfare Agents: Chemistry, Pharmacology, Toxicology and Therapeutics*, Romano, J.A., Lukey, B.J. & Salem, H., pp. 501-548, Taylor & Francis Group, ISBN-13: 978-1-4200-4661-8

Carmona, G.N.; Jufer, R.A.; Goldberg, S.R.; Gorelick, D.A.; Greig, N.H.; Yu, Q.S.; Cone, E.J. & Schindler, C.W. (2000). Butyrylcholinesterase accelerates cocaine metabolism: in vitro and in vivo effects in nonhuman primates and humans. *Drug Metabolism and Disposition*, Vol. 28, No. 3, pp. 367-371, ISSN 0090-9556

Ciucu, A.A.; Negulescu, C. & Baldwin R.P. (2003). Detection of pesticides using an amperometric biosensor based on ferophthalocyanine chemically modified carbon paste electrode and immobilized bienzymatic system. *Biosensor and Bioelectronics*, Vol. 18, pp. 303-310, ISSN 0956-5663

Český obranný standard. (2007). Základní metody vhodné pro detekci OL a PTL, In: *Automatické signalizátory bojových otravných látek a průmyslových škodlivin*, pp. 16-23. Úřad pro obrannou standardizaci, katalogizaci a státní ověřování jakosti, Praha. Czech Republic

Dale, T.J. & Rebek, J. jr. (2006). Fluorescent sensor for organophosphorus nerve agent mimics. *Journal of American Chemical Society*, Vol. 128, No. 14, pp. 4500-4501, ISSN 0002-7863

Danzer, T. & Schwedt, G. (1996) Development of biosensor system and evaluation of inhibition studies. *Biosensors and Bioelectronics*, Vol. 11, No. 8, pp. 18-19, ISSN 0956-5663

Dăneţ, A.F.; Bucur, B.; Cheregi, M.-C.; Badea, M. & Şerban, S. (2003). Spectrophotometric determination of organophosphoric insecticides in a FIA system based on AChE inhibition. *Analytical Letters*, Vol. 36, No. 1, pp. 59-73, ISSN 0003-2719

Ding, J. & Qin, W. (2009). Polymeric membrane ion-selective electrode for butyrylcholinesterase based on controlled release of substrate. *Electroanalysis*, Vol. 21, No. 17-18, pp. 2030-2035, ISSN 1040-030003-271997

Díaz, N.A.; Peinado, R.M.C. (1997). Sol-gel cholinesterase biosensor for organophosphorus pesticide fluorimetric analysis. *Sensors and Actuators, B: Chemical*, Vol. 38-39, No. 1-3, pp. 426-431, ISSN 0925-4005

Doretti, L.; Gattolin, P.; Burla, A.; Ferrara, D.; Lora, S. & Palma, G. (1998). Covalently immobilized choline oxidase and cholinesterases on a methacrylate copolymer for disposable membrane biosensors. *Applied Biochemistry and Biotechnology - Part A Enzyme Engineering and Biotechnology*, Vol. 74, No. 1, pp. 1-12, ISSN 0273-2289

Du, D.; Huang, X.; Cai, J. & Zhang. A. (2007). Comparison of pesticide sensitivity by electrochemical test based on acetylcholinesterase biosensor. *Biosensors and Bioelectronics*, Vol. 23, No. 2, pp. 285-289, ISSN 0956-5663

Du, D.; Chen, S.; Cai, J. & Zhang, A. (2008). Electrochemical pesticide sensitivity test using acetylcholinesterase biosensor based on colloidal gold nanoparticle modified sol-gel interface. *Talanta*, Vol. 74, No. 4, pp. 766-772, ISSN 00399140

Dzyadevych, S.V.; Arkhypova, V.N.; Martelet, C.; Jaffrezic-Renault, N.; Chovelon, J.-M.; El'skaya, A.V. & Soldatkin, A.P. (2004). Potentiometric biosensors based on ISFETs and immobilized cholinesterases. *Electroanalysis*, Vol. 16, No. 22, pp. 1873-1882, ISSN 1040-0397

Dzyadevych, S.V.; Arkhypova, V.N.; Martelet, C.; Jaffrezic-Renault, N.; Chovelon, J.-M.; El'skaya, A.V. & Soldatkin, A.P. (2006). Potentiometric biosensors based on ISFETs and immobilised cholinesterases. *International Journal of Applied Electromagnetics and Mechanics*, Vol. 23, No. 3-4, pp. 229-244, ISSN 1383-5416

Dzyadevych, S.V.; Shulga, A.A.; Soldatkin, A.P.; Hendi, A.M.N.; Jaffrezic-Renault, N. & Martelet, C. (2005). Conductometric biosensors based on cholinesterases for sensitive detection of pesticides. *Electroanalysis*, Vol. 6, pp. 752-758, ISSN 1040-0397

Ellman, G.L. (1959). Tissue sulfhydryl groups. *Archives of Biochemistry and Biophysics*, Vol. 82, pp. 70-77

Ellman, G.L.; Courtney, K.D.; Andres, V. jr. & Featherstone, R.M. (1961). A new and rapid colorimetric determination of acetylcholiesterase activity. *Biochemical Pharmacology*, Vol. 7, pp. 88-95, ISSN 0006-2952

Epstein, J.; Demek, M. & Wolff, V.C. (1957). Enzymatic method for estimation of sarin in dilute aqueous solution. *Analytical Chemistry*, Vol. 29, No. 7, pp. 1050-1053, ISSN 0003-2700

Espinosa, M.; Atanasov, P. & Wilkins, E. (1999). Development of a disposable organophosphate biosensor. *Electroanalysis*, Vol. 11, No. 14, pp. 1055-1062, ISSN 1040-0397

Evtugyn, G.A.; Rizaeva, E.P.; Stoikova, E.E.; Latipova, V.Z. & Budnikov, H.C. (1997). The application of cholinesterase potentiometric biosensor for preliminary screening of the toxicity of waste waters *Electroanalysis*, Vol. 9, No. 14, pp. 1124–1128, ISSN 1040-0397

Chen, Q.; Kobayashi, Y.; Takeshita, H.; Hoshi, T. & Anzai, J.-I. (1998). Avidin-biotin system-based enzyme multilayer membranes for biosensor applications: optimization of loading of choline esterase and choline oxidase in the bienzyme membrane for acetylcholine biosensors. *Electroanalysis*, Vol. 10, No. 2, pp. 94-97, ISSN 1040-0397

Ghindilis, A.L.; Morzunova, T.G.; Barmin, A.V. & Kurochkin, I.N. (1996). Potentiometric biosensor for cholinesterase inhibitor analysis based on mediatorless bioelectrocatalysis. *Biosensors and Biocatalysis,* Vol. 11, No. 9, pp. 873-880, ISSN 0956-5663

Ghindilis, A.L.; Atanasov, P. & Wilkins, E. (1997). Enzyme-catalyzed direct electron transfer: Fundamentals and analytical applications. *Electroanalysis,* Vol. 9, No. 9, pp. 661–674, ISSN 1040-0397

Gao, Y.; Kyratzis, I.; Taylor, R.; Huynh, C. & Hickey, M. (2009). Immobilization of acetylcholinesterase onto carbon nanotubes utilizing streptavidin-biotin interaction for the construction of amperometric biosensors for pesticides. *Analytical Letters,* Vol. 42, No. 16, pp. 2711-2727, ISSN 0003-2719

Gogol, E.V.; Evtugyn, G.A.; Marty, J.-L.; Budnikov, H.C. & Winter, V.G. (2000). Amperometric biosensors based on nafion coated screen-printed electrodes for the determination of cholinesterase inhibitors. *Talanta,* Vol, 53, No. 2, pp. 379-389, ISSN 0039-9140

Gulla, K.C.; Gouda, M.D.; Thakur, M.S. & Karanth, N.G. (2002). Reactivation of immobilized acetyl cholinesterase in an amperometric biosensor for organophosphorus pesticide. *Biochimica et Biophysica Acta,* Vol. 1597, No. 1, pp. 133-139, ISSN 01674838

Gyurcsányi, R.E.; Vágföldi, Z.; Tóth, K. & Nagy, G. (1999). Fast response potentiometric acetylcholine biosensor. *Electroanalysis,* Vol. 11, No. 10-11, pp. 712–718, ISSN 1040-0397

Hadd A.G.; Jacobson, S.C. & Ramsey, M. (1999). Microfluidic assay of acetylcholinesterase inhibitors. *Analytical Chemistry,* Vol. 71, No. 22, pp. 5206-5212, ISSN 0003-2700

Halámek, J.; Přibyl, J.; Makower, P.; Skládal, P. & Scheller, F.W. (2005). Sensitive detection of organophosphorus in river water by means of piezoelectric biosensor. *Analytical and Bioanalytical Chemistry,* Vol. 382, No. 8, pp. 1904-1911, ISSN 1618-2650

Halámek, E.; Kobliha, Z. & Pitschmann, V. (2009). Organophosphorus chemical warfare agents, In: *Analysis of chemical warfare agents,* E. Halámek, Z. Kobliha & V. Pitschmann, pp. 63-69, University of Defence, Institute of NBC Defence, ISBN 978-80-7231-658-8, Brno, Czech Republic

Hanin, I. & Dudas, B. (2000). Measurement of cholinesterase activity, In: *Cholinesterases and Cholinesterase Inhibitors,* Giacobini, E., pp. 139-144, Martin Dunitz, ISBN 1-85317-910-8, U.K.

Hart, A.L.; Collier, W.A. & Janssen D. (1997). The response of green-printed enzyme electrodes containing cholinesterases to organo-phosphates in solution and from commercial formulations. *Biosensors and Bioelectronics,* Vol. 12, No. 7, pp.645-654, ISSN 0956-5663

Hernandez, S.; Palchetti, I. & Mascini, M. (2000). Determination of anticholinesterase activity for pesticide monitoring using a thiocholine sensor. *International Journal of Envirnmental Analytical Chemistry,* Vol. 58, No. 3-4, pp. 263-278, ISSN 0306-7319

Hestrin, S. (1949). The reaction of acetylcholine and other carboxylic acid derivatives with hydroxylamine, and its analytical application. *Journal of Biological Chemistry,* Vol. 180, No. 1, pp. 249-261, ISSN 0021-9258

Hildebrandt, A.; Bragós, R.; Lacorte, S. & Marty, J.L. (2008). Performance of a portable biosensor for the analysis of organophosphorus and carbamate insecticides in water and food. *Sensors and Actuators, B: Chemical,* Vol. 133, No. 1, pp. 195-201, ISSN 0925-4005

Holmstedt, B. (2000). Cholinesterase inhibitors: an introduction, In: *Cholinesterases and Cholinesterase Inhibitors*, Giacobini, E., pp. 1-8, Martin Dunitz, ISBN 1-85317-910-8, U.K.

Hoskovcová, M.; Kasalová, I.; Halámek, E. & Kobliha, Z. (2009). Suggestion for selective differentiating of nerve agents G and V type with utilization of modified Ellman´s method. *Enviromental Chemistry Letters*, Vol. 7, No. 3, pp. 277-281, ISSN 1610-3653

Imato, T. & Ishibashi, N. (1995). Potentiometric butyrylcholine sensor for organophosphate pesticides. *Biosensors and Bioelectronics*, Vol. 10, No. 5, pp. 435-441, ISSN 0956-5663

Ion, A.C.; Ion, I.; Culetu, A.; Gherase, D.; Moldovan, C.A.; Iosub, R. & Dinescu, A. (2010). Acetylcholinesterase voltammetric biosensors based on carbon nanostructure-chitosan composite material for organophosphate pesticides. *Materials Science and Engineering C*, Vol. 30, No. 6, pp. 817-821, ISSN 0928-4931

Istamboulie, G., Cortina-Puig, M., Marty, J.-L., Noguer, T. (2009). The use of Artificial Neural Networks for the selective detection of two organophosphate insecticides: Chlorpyrifos and chlorfenvinfos. *Talanta*, Vol. 79, No. 2, pp. 507-511, ISSN 0039-9140

Ivanov, A.; Evtugyn, G.; Gyurcsányi, R.E.; Tóth, K. & Budnikov, H.C. (2000). Comparative investigation of electrochemical cholinesterase biosensors for pesticide determination. *Analytica Chimica Acta*, Vol. 404, No. 1, pp. 55-65, ISSN 0003-2670

Ivanov, A.; Evtugyn, G.; Budnikov, H.; Ricci, F.; Moscone, D. & Palleschi, G. (2003a). Cholinesterase sensors based on screen-printed electrodes for detection of organophosphorus and carbamic pesticides. *Analytical and Bioanalytical Chemistry*, Vol. 377, No. 4, pp. 624-631, ISSN 1618-2642

Ivanov, A.N.; Evtugyn, G.A.; Lukachova, L.V.; Karyakina, E.E.; Budnkov H.C.; Kiseleva, S.G.; Orlov A.V.; Karpacheva, G.P. & Karyakin A.A. (2003b). New polyaniline based potemtiometric biosenzor for pesticides detection. *IEEE Sensors Journals - Analycal Chemistry Department*, Vol. 3, No. 3, pp. 333-340, ISSN 1530-437X

Ivanov, Y.; Marinov, I.; Gabrovska, K.; Dimcheva, N. & Godjevargova, T. (2010). Amperometric biosensor based on a site-specific immobilization of acetylcholinesterase via affinity bonds on a nanostructured polymer membrane with integrated multiwall carbon nanotubes. *Journal of Molecular Catalysis B: Enzymatic*, Vol. 63, No. 3-4, pp. 141-148, ISSN 1381-1177

Joshi, K.A.; Prouza, M.; Kum, M.; Wang, J.; Tang, J.; Haddon, R.; Chen, W. & Mulchandani, A. (2006). V-type nerve agent detection using a carbon nanotube-based amperometric enzyme electrode. *Analytical Chemistry*, Vol. 78, No. 1, pp. 331-336, ISSN 0003-2700

Kandimalla, V.B. & Ju, H. (2006). Binding of acetylcholinesterase to multiwall carbon nanotube-cross-linked chitosan composite for flow-injection amperometric detection of an organophosphorus insecticide. *Chemistry – A European Journal*, Vol. 12, pp. 1074-1080, ISSN 1521-3765

Karnati, C.; Du, H.; Ji, H.F.; Xu, X.; Lvov, Y.; Mulchandani, A.; Mulchandani P. & Chen W. (2007). Organophosphorus hydrolase multilayer modified microcantilevers for organophosphorus detection. *Biosensors and Bioelectronics*, Vol. 22, pp. 2636-2642, ISSN 0956-5663

Karyakin, A.A.; Bobrova, O.A.; Lukachova, L.V. & Karyakina, E.E. (1996). Potentiometric biosensors based on polyaniline semiconductor films. *Sensors and Actuators, B: Chemical*, Vol. 33, No. 1-3, pp. 34-38, ISSN 0925-4005

Kenar, L. (2010). The use of biosensors for the detection of chemical and Biological Weapons. *Turkish Journal of Biochemistry*, Vol. 35, No. 1, pp. 72-74. ISSN 0250-4685

Kok, F.N. & Hasirci, V. (2004). Determination of binary pesticide mixtures by an acetylcholinesterase- choline oxidase biosensor. *Biosensors and Bioelectronics*, Vol. 19, No. 7, pp. 661-665, ISSN 0956-5663

Koncki, R. & Mascini, M. (1997). Screen-printed ruthenium dioxide electrodes for pH measuremnts. *Analytica Chimica Acta*, Vol. 351, No. 1-3, pp. 143-149, ISSN 0003-2670

Kramer, D.N. & Gamson R.M. (1958). Colorimetric determination of acetylcholinesterase activity. *Analytical Chemistry*, Vol. 30, No. 2, pp.251-254, ISSN 0003-2700

Kua, J.; Zhang, Y. & McCammon, A. (2002). Studying enzyme binding specifity in acetylcholinesterase using a combined molecular dynamics and multiple docking approach. *Journal of American Chemical Society*, Vol. 124, No. 28, pp. 8260-8267, ISSN 0002-7863

La Rosa, C.; Pariente, F.; Hernández, L. & Lorenzo, E. (1994). Determination of organophosphorus and carbamic pesticides with an acetylcholinesterase amperometric biosensor using 4-aminophenyl acetate as substrate. *Analytica Chimica Acta*, Vol. 295, No. 3, pp. 273-282, ISSN 0003-2670

Lee, H.-S.; Kim, A.Y.; Cho, A.Y. & Lee, T.Y. (2002). Oxidation of organophosphorus pesticides for the sensitive detection by a cholinesterase-based biosensor. *Chemosphere*, Vol. 46, No. 4, pp. 571-576, ISSN 0045-6535

Leon-Gonzales, M.E. & Townshend, A. (1990). Flow-injection determination of paraoxon by inhibition of immobilized acetylcholinesterase. *Analytica Chimica Acta*, Vol. 236, pp. 267-272, ISSN 0003-2670

Li, Y.-G.; Zhou, Y.-X.; Feng, J.-L.; Jiang, Z.-H. & Ma, L.-R. (1999). Immobilization of enzyme on screen-printed electrode by exposure to glutaraldehyde vapour for the construction of amperometric acetylcholinesterase electrodes. *Analytica Chimica Acta*, Vol. 382, No. 3, pp. 277-282, ISSN 003-2670

Limperos, G. & Ranta, K.E. (1953). A rapid screening test for the determination of the approximate cholinesterase activity of human blood. *Science*, Vol. 117, No 3043, pp. 453-455, ISSN 0036-8075

Lin, T.J.; Huang, K.T. & Liu, C.Y. (2006). Determination of organophosphorus pesticides by a novel biosensor based on localized surface plasmon resonance. *Biosensors and Bioelectronics*, Vol. 22, No. 4, pp. 513-518, ISSN 0956-5663

Lin, Y., Lu, F., Wang, J. (2004). Disposable carbon nanotube modified screen-printed biosensor for amperometric detection of organophosphorus pesticides and nerve agents. *Electroanalysis*, Vol. 16, No. 1-2, pp. 145-149, ISSN 1040-0397

Liu, G.; Richers, S.L.; Mellen, M.C. & Lin, Y. (2005). Sensitive electrochemical detection of enzymatically generated thiocholine at carbon nanotube modified glasssy carbon electrode. *Electrochemistry Communications*, Vol. 7, No. 11, pp. 1163-1169, ISSN 1388-2481

Liu, G. & Lin Y. (2006). Biosensor based on self-assembling acetylcholinesterase on carbon nanotubes for flow injection/amperometric detection of organophosphate pesticides and nerve agents. *Analytical Chemistry*, Vol. 78, No. 3, pp. 835-843, ISSN 0003-2700

Liu, G.; Wang, J.; Barry, R.; Petersen, C.; Timchalk, C.; Gassman, P.L. & Lin, Y. (2008). Nanoparticle-based electrochemical immunosensor for the detection of phosphorylated acetylcholinesterase: An exposure biomarker of organophosphate pesticides and nerve agents. *Chemistry - A European Journal*, Vol. 14, No. 32, pp. 9951-9959, ISSN 1521-3765

Makower, A.; Halámek, J.; Skládal, P.; Kernchen, F. & Scheller, F.W. (2003). New principle of direct real-time monitoring of the interaction of cholinesterase and its inhibitors by piezoelectric biosensor. *Biosensors and Bioelectronics*, Vol. 18, No. 11, pp. 1329-1337, ISSN 0956-5663

Marinov, I.; Ivanov, Y.; Gabrovska, K. & Godjevargova, T. (2010). Amperometric acetylthiocholine sensor based on acetylcholinesterase immobilized on nanostructured polymer membrane containing gold nanoparticles. *Journal of Molecular Catalysis B: Enzymatic*, Vol. 62, No. 1, pp. 66-74, ISSN 1381-1177

Martorell, D.; Céspedes, F.; Martínez-Fàbregas, E. & Alegret, S. (1994). Amperometric determination of pesticides using a biosenzor based on a polishable graphite-epoxy biocomposite. *Analytica Chimica Acta*, Vol. 290, No. 3, pp. 343-348, ISSN 0003-2670

Marty J.-L.; Mionetto, N.; Noguer, T.; Ortega F. & Roux C. (1993). Enzyme sensors for the detection of pesticides. *Biosensors and Bioelectronics*, Vol. 8, No. 6, pp. 273-280, ISSN 0956-5663

Masson, P.; Carletti, E. & Nachon, F. (2009). Structure, activities and biomedical applications of human butyrylcholinesterase. *Protein and Peptide Letters*, Vol. 16. No. 10, pp. 1215-1224, ISSN 0929-8665

Miao, Y.; He, N. & Zhu, J.-J. (2010). History and new developments of assays for cholinesterase activity and inhibition. *Chemical Reviews*, Vol. 110, No. 9, pp. 5216-5234, ISSN 0009-2665

Milatović, D. & Jokanović, M. (2009). Pyridinium oximes as cholinesterase reactivators in the treatment of OP poisoning, In: *Hadbook of Toxicology of Chemical Warfare Agents*, Gupta, R.C., pp. 985-996, Elsevier, ISBN 978-0-12-374484-5, U.K.

Mitchell, K.M. (2004). Acetylcholine and choline amperometric enzyme sensors characterized in vitro and in vivo. *Analytical Chemistry*, Vol. 76, No. 4, pp. 1098-1106, ISSN 0003-2700

Mizutani, F. & Tsuda, K. (1982). Amperometric determination of cholinesterase with use of an immobilized enzyme electrode. *Analytica Chimica Acta*, Vol. 139, No. 1, pp. 356-362, ISSN 0003-2670

Montesinos, T.; Peréz-Munguia, S.; Valdez, F. & Marty, J.-L. (2001). Disposable cholinesterase biosensor for the detection of pesticides in water-miscible organic solvents. *Analytica Chimica Acta*, Vol. 431, No. 2, pp. 231-237, ISSN 0003-2670

Mulchandani, A.; Mulchandani P. & Chen, W. (1999a). Amperometric thick-film strip electrodes for monitoring organophosphate nerve agents based on immobilized organophosphorus hydrolase. *Analytical Chemistry*, Vol. 71, No. 11, pp. 2246-2249, ISSN 0003-2700

Mulchandani, A.; Mulchandani, P.; Kaneva, I. & Chen, W. (1999b). Biosensor for direct determination of organophosphate nerve agents. 1. Potentiometric enzyme electrode. *Biosensors and Bioelectronics*, Vol. 14, No. 1, pp. 77-85, ISSN 0956-5663

Mulchandani, A.; Chen, W.; Mulchandani, P.; Wang, J. & Rogers, K.R. (2001). Biosensors for direct determination of organophosphate pesticides. Review. *Biosensor and Bioelectronics*, Vol. 16, No. 4-5, pp. 225-230, ISSN 0956-5663

No, H.-Y; Kim, Y.A.; Lee, Y.T. & Lee, H.-S. (2007). Cholinesterase-based dipstick assai for the detection of organophosphate and carbamate pesticides. *Analytica Chimica Acta*, Vol. 594, No. 1., pp. 37-43, ISSN 0003-2670

Nunes, G.S.; Jeanty, G. & Marty J.-L. (2004). Enzyme immobilization procedures on screen-printed electrodes used for the detection of anticholinesterase pesticides. Comparative study. *Analytica Chimica Acta*, Vol. 523, No. 1, pp. 107-115, ISSN 0003-2670

Oh, I. & Masel, R.I. (2007). Electrochemical organophosphate sensor based on oxime chemismy. *Electrochemical and Solid-State Letters*, Vol. 10, No. 2, pp. J19-J22, ISBN 1099-0062

O'Sullivan, C.K. & Guilbault, G.G. (1999). Commercial quartz crystal microbalances – theory and applications. *Biosensors and Bioelectronics*, Vol. 14, No. 8-9, pp. 663-670, ISSN 0956-5663

Palchetti, I.; Cagnini, A.; Del Carlo, M.; Coppi, C.; Mascini, M. & Turner, A.P.F. (1997). Determination of anticholinesterase pesticides in real samples using a disposable biosensor. *Analytica Chimica Acta*, Vol. 337, No. 3, pp. 315-321, ISBN 0003-2670

Palleschi, G.; Bernabei, M.; Cremisini, C. & Mascini, M. (1992). Determination of organophosphorus insecticides with a choline electrochemical biosensor. *Sensors and Actuators: B. Chemical*, Vol. 7, No. 1-3, pp. 513-517, ISSN 0925-4005

Pariente, F.; Hernández L. & Lorenzo, E. (1993). 4-aminophenyl acetate as a substrate for amperometric esterase sensors. *Analytica Chimica Acta*, Vol. 273, No. 1-2, pp. 399-407, ISSN 0003-2670

Pavlov, V.; Xiao, Y. & Willner, I. (2005). Inhibition of the acetylcholine esterase-stimulated growth of Au nanoparticles: Nanotechnology-based sensing of nerve gases. *Nano Letters*, Vol. 5, No. 4, pp. 649-653, ISSN 1530-6984

Philips, J.W. (2005). Acetylcholine repase fro central nervous system: A 50-year retrospective. *Critical Review in Neurobiology*, Vol. 17. No. 3-4, pp. 161-217, ISSN 0892-0915

Pitschmann, V. (1999). Mezi světovými válkami. In: *Historie chemické války*. ed. V. Pitschman, pp. 41-45, Military System Line, s.r.o. ISBN 80-902669-0-8, Praha, Czech Republic

Pohanka, M.; Jun, D. & Kuca, K. (2007). Amperometric biosensor for evaluation of competitive cholinesterase inhibition by the reactivator HI-6. *Analytical Letters*, Vol. 40, No. 12 , pp. 2351-2359, ISSN 0003-2719

Pohanka, M.; Kuča, K. & Jun, D. (2008) Aflatoxin assay using an amperometric sensor strip and acetylcholinesterase as recognition element. *Sensor Letters*, Vol. 6, No. 3, pp. 450-453, ISSN 1546-198X

Pohanka, M. (2009). Cholinesterase based amperometric biosensors for assay of anticholinergic compounds. *Interdisciplinary Toxicology*, Vol. 2, No. 2, pp. 52-54, ISSN 1337-6853

Pohanka, M.; Musílek, K. & Kuča, K. (2009). Progress of biosensors based on cholinesterase inhibition. *Current Medicinal Chemistry*, Vol. 16, No. 14. Pp. 1790-1798, ISSN 0929-8673

Pohanka, M.; Karasová-Žďárová, J.; Kuča, K.; Pikula, J.; Holas, O.; Korábečný, J. & Cabal, J. (2010a). Colorimetric dipstick for assay of organophosphate pesticides and nerve agents represented by paraoxon, sarin and VX, *Talanta*, Vol. 81, No. 1-2, pp. 621-624, ISSN 0039-9140

Pohanka, M.; Vlček, V.; Žďárová-Karasová, J.; Kuča, K. & Cabal, J. (2010b). Colorimetric detectors based on acetylcholinesterase and its construction. *Vojenské Zdravotnické listy*, Vol. 79, No. 1, pp. 9-14, ISSN 0372-7025

Product information A 12217 (2004). Amplex® Red Acetylcholine/Acetylcholinesterase Assay Kit. Molecular Probes, Inc., USA,
http://probes.invitrogen.com/media/pis/mp12217.pdf

Product information K015-F1 (2009). DetectX® Acetylcholinesterase Fluorescent Activity Kit. ArborAssay, USA,
http://www.arborassays.com/products/inserts/K015-F1.pdf

Rekha, K.; Thakur M.S. & Karanth N.G. (2000). Biosensors for the detection of organophosphorus pesticides. *Critical Review in Biotechnology*, Vol. 23, No. 3, pp. 213-235, ISSN 0738-8551

Reybier, K.; Zairi, S.; Jaffrezic-Renault, N. & Fahys, B. (2002). The use of polyethyleneimine for fabrication of potentiometric cholinesterase biosensors. *Talanta*, Vol. 56, No. 6, pp. 1015-1020, ISSN 0039-9140

Ricci, F.; Amine, A.; Palleschi, G. & Moscone, D. (2003). Prussian blue based green-printed biosensors with improvedcharacteristics of long-term lefetime and pH stability. *Biosensors and Bioelectronics*, Vol. 18, No. 2-3, pp. 165-174, ISSN 0956-5663

Rogers, K.R.; Cao, C.J.; Valdes, J.J.; Eldefrawi, A.T. & Eldefrawi, M.E. (1991). Acetylcholinesterase fiber-optic biosensor for detection of anticholinesterases. *Fundamental and Applied Toxicology*, Vol. 16, No. 4, pp. 810-820, ISSN 0272-0590

Rogers, K.R.; Wang, Y.; Mulchandani, A.; Mulchandani, P. & Chen, W. (1999). Organophosphorus hydrolase-based assay for organophosphate pesticides. *Biotechnology Progress*, Vol. 15, No. 3, pp. 517-521, ISSN 8756-7938

Sabelle, S.; Renard, P.Y.; Pecorella, K.; De Suzzoni-dézard S.; Créminon, C.; Grassi, J. & Mioskowski, C. (2002). *Journal ofiAmerican Chemical Society*, Vol. 124, No. 17, pp. 4874-4880, ISSN 0002-7863

Shulga, O. & Kirchhoff, J.R. (2007). An acetylcholinesterase enzyme electrode stabilized by an electrodeposited gold nanoparticle layer. *Electrochemistry Communications*, Vol. 9, No. 5, pp. 935-940, ISSN 1388-2481

Simonian, A.L.; Grimsley, J.K.; Flounders, A.W.; Schoeniger, J.S.; Cheng, T.-C.; DeFrank, J.J. & Wild, J.R. (2001). Enzyme-based biosensor for the direct detection of fluorine-containing organophosphates. *Analytica Chimica Acta*, Vol. 442, No. 1, pp. 15-23, ISSN 0003-2670

Sinha, R., Ganesana, M., Andreescu, S., Stanciu, L. (2010). AChE biosensor based on zinc oxide sol-gel for the detection of pesticides. *Analytica Chimica Acta*, Vol. 661, No. 2, pp. 195-199, ISSN 0003-2670

Skládal, P.; (1992). Detection of organophosphate and carbamate pesticides using disposable biosensors based on chemically modified electrodes and immobilized cholinesterase. *Analytica Chimica Acta*, Vol. 269, pp. 281-287, ISSN 0003-2670

Skládal, P. & Mascini, M. (1992). Sensitive detection of pesticides using amperometric sensors based on cobalt phtalocyanine-modified composite electrodes and immobilized cholinesterases. *Biosensors and Bioelectronics*, Vol. 7, No. 5, pp. 335-343, ISSN 0956-5663

Skládal, P. & Macholán, L. (1997). Biosensory – současný stav a perspektivy. *Chemické Listy*, Vol. 91, pp. 105-113, ISSN 0009-2770

Snejdarkova, M.; Svobodova, L.; Evtugyn, G.; Budnikov, H.; Karyakin, A.; Nikolesis, D.P. & Hianik, T. (2004). Acetylcholinesterase sensors based on gold electrodes modified with dendrimer and polyaniline. A comparative research. *Analytica Chimica Acta*, Vol. 514, No. 1, pp. 79-88, ISSN 003-2670

Somani S.M.; Solana, R.P. & Dube, S.N. (1992). Toxicodinamics of nerve agents, In: *Chemical Warfare Agents*, Somani S.M., pp. 67-123, Academic Press, Inc., ISBN 0-12-654620-7, USA

Somerset, S.; Baker, P. & Iwuoha, E.I. (2009). Mercaptobenzothiazole on-gold organic phase biosensors systems: 1. Enhanced organophosphate pesticide determination. *Journal of Environmental Science and Health Part B*, Vol. 44, No. 2, pp. 164-178, ISSN 0360-1234

Sun, X. & Wang, X. (2010). Acetylcholinesterase biosensor based on prussian blue-modified electrode for detecting organophosphorous pesticides. *Biosensors and Bioelectronics*, Vol. 25, No. 12, pp. 2611-2614, ISSN 0956-5663

Sun, X., Wang, X., Zhao, W. (2010). Multiwall carbon nanotube-based acetylcholinesterase biosensor for detecting organophosphorous pesticides, *Sensor Letters*, Vol. 8, No. 2, pp. 247-252, ISSN 1546-198X

Suprun, E.; Evtugyn, G.; Budnikov, H.; Ricci, F.; Moscone, D. & Palleschi, G. (2005). Acetylcholinesterase sensor based on screen-printed carbon electrode modified with prussian blue. *Analytical and Bioanalytical Chemistry*, Vol. 383, No. 4, pp. 597-604, ISSN 1618-2642

Sussman, J.L.; Harel, M.; Frolow, F.; Oefner, C.; Goldman, D.; Toker, L. & Silman, I. (1991). Atomic structure of acetylcholinesterase from Torpedo californica: A prototypic acetylcholine binding protein. *Science*, Vol. 253, No. 5022, pp. 872-879, ISSN 0036-8075

Technical Bulletin 296. (1996). Verification of nerve agent exposure-monitoring blood cholinesterase activity with the tTest-Mate ™ OP Kit, In: *Assay techniques for detection of exposure to sutur mustard, cholinesterase inhibitors, sarin, soman, GF, and cyanide.* pp. 3-1 – 3-3, Headquarters Department of the Army, Washington, DC, USA http://armypubs.army.mil/med/dr_pubs/dr_a/pdf/tbmed296.pdf

Teksoy, S.; Odaci, D. & Timur, S. (2007). A new bienzymatic biosensor based on butyrylcholine esterase-sulfhydryl oxidase enzymes. *Analytical Letters*, Vol. 40, No. 15, pp. 2840-2850, ISSN 0003-2719

Tomeček, I. & Matoušek, J. (1961). Organické fosforové otravné látky, In: *Analýza bojových otravných látek*, Tomeček, I. & Matoušek, J., pp. 103-135, SPN Praha, Czech Republic

Tran-Minh, C.; Pandey, P.C. & Kumaran, S. (1990). Studies on acetylcholine sensor and its analytical application based on the inhibition of cholinesterase. *Biosensors and Bioelectronics*, Vol. 5, No. 6, pp. 461-471, ISSN0956-5663

Tsai, H.-C. & Doong, R.-A. (2000). Optimization of sol-gel based fibre-optic cholinesterase biosensor for the determination of organophosphorus pesticides. *Water Science and Technology*, Vol. 42, No. 7-8, pp. 283-290, ISSN 0273-1223

Turdean, G.L.; Popescu, I.C.; Oniciu, L. & Thevenot, D.R. (2002). Sensitive detection of organophosphorus pesticides using a needle type amperometric acetylcholinesterase based bioelectrode. Thiocholine electrochemistry and immobilized enzyme inhibition. *Journal of Enzyme Inhibition and Medicinal Chemistry*, Vol. 17, No. 2, pp. 107-115, ISSN 1475-6366

Tušarová, I.; Halámek, E. & Kobliha, Z. (1999). Study on reactivation of enzyme-inhibitor complexes by oximes using acetylcholine esterase inhibited by organophosphate chemical warfare agents. *Enzyme and Microbial Technology*, Vol. 25, No. 3-5, pp. 400-403, ISSN 0141-0229

Tušarová, I. & Halámek, E. (2001). Biosenzor pro detekci a rozlišení inhibitorů cholinesteráz, způsob přípravy zóny biosenzoru s imobilizovanou cholinesterázou, způsob detekce inhibitorů cholinesteráz a způsob rozlišení inhibitorů cholinesteráz. *Patent of the Czech Republic*, ČR 288 576. 22.5.2001

Upadhyay, S.; Rao, G.R.; Sharma, M.K.; Bhattacharya, B.K.; Rao, V.K. & Vijayaraghavan, R. (2009). Immobilization of acetylcholineesterase-choline oxidase on a gold-platinum bimetallic nanoparticles modified glassy carbon electrode for the sensitive detection of organophosphate pesticides, carbamates and nerve agents. *Biosensors and Bioelectronics*, Vol. 25, No. 4, pp. 832-838, ISSN 0956-5663

Wang, J.; Chen, L.; Mulchandani, A.; Mulchandani, P. & Chen, W. (1999). Remote biosensor for in-situ monitoring of organophosphate nerve agents. *Electroanalysis*, Vol. 11, No. 12, pp. 866–869, ISSN 1040-0397

Wang, J.; Krause, R.; Block, K.; Musameh, M.; Mulchandani, A. & Schöning, M.J. (2003). Flow injection amperometric detection of OP nerve agents based on an organophosphorus-hydrolase biosensor detector. *Biosensors and Bioelectronics*, Vol. 18, No. 2-3, pp. 255-260, ISSN 0956-5663

Watson, A.; Opresko, D.; Young, R.; Hauschild; V.; King, J. & Bakshi, K. (2009). Organophosphorus Nerve agents, In: *Hadbook of Toxicology of Chemical Warfare Agents*, Gupta, R.C., pp. 43-67, Elsevier, ISBN 978-0-12-374484-5, U.K.

Wilson, B.W.; Sanborn, J.R.; O'Malley, M.A.; Henderson, J.D. & Billitti, J.R. (1997). Monitoring the pesticide-exposed worker. *Occupational Medicine*, Vol. 12, No. 2, pp. 347-363, ISSN 0885-114X

Wring, S.A.; Hart, J.P. & Birch, B.J. (1989). Development of an improved carbon electrode chemically modified with cobalt phtalocyanine as a reusable sensor for glutathione. *Analyst*, Vol. 116, pp. 1563-1570, ISSN 0003-2654

Yao, T. (1983) Flow injection analysis for cholinesterase in blood serum by use of a choline-sensitive electrode as an amperometric detector. *Analytica Chimica Acta*, Vol. 153, No. 1, pp. 169-174, ISSN 0003-2670

Yin, H.; Ai, S.; Xu, J.; Shi, W. & Zhu, L. (2009). Amperometric biosensor based on immobilized acetylcholinesterase on gold nanoparticles and silk fibroin modified platinum electrode for detection of methyl paraoxon, carbofuran and phoxim. *Journal of Electroanalytical Chemistry*, Vol. 637, No. 1-2, pp. 21-27, ISSN 0022-0728

Zhang, J.; Luo, A; Liu, P.; Wei, S.; Wang, G. & Wei, S. (2009). Detection of organophosphorus pesticides using potentiometric enzymatic membrane biosensor based on methylcellulose immobilization. *Analytical Sciences*, Vol. 25, No. 4, pp. 511-515, ISSN 0910-6340

Zhu, J.; Dhimitruka, I. & Pei, D. (2004). 5-(2-amino)dithio-2-nitrobenzoate as a more base-stable alternative to Ellman´s reagent. *Organic Letters*, Vol. 6, No. 21, pp. 3809-3812, ISSN 1523-7060

5

Biosensors Applications on Assessment of Reactive Oxygen Species and Antioxidants

Simona Carmen Litescu[1], Sandra A. V. Eremia[1],
Mirela Diaconu[1,2], Andreia Tache[1,2] and Gabriel-Lucian Radu[1,2]
[1]*Centre of Bioanalysis, National Institute for Biological Science,*
[2]*Politehnica University of Bucharest, Faculty of Applied Chemistry and Materials Science*
Romania

1. Introduction

The importance of the subject dealing with oxidative stress and antioxidants protection against oxidative stress was increasing in the last three decades, a proof of this assertion being the huge number of publications appeared since 2000 (meaning 150582 publications) devoted, for example, to antioxidant.

Addressing such a topic like Reactive Oxygen Species (ROS) and/or antioxidants (Aox) assessment it has to be clearly described the mean of both these terms and, concomitantly, the close inter-dependence between their actions and, respectively, effects. To support the assertion we would like to mention the fact that two of the four recommended steps in protocol for antioxidant assessment (Becker et al., 2004) are that regarding the evaluation of compound activity as inhibitor of lipid peroxidation in biological model systems and the study of the compound efficiency against relevant oxidative markers, these providing the evidence of strict correlation between ROS toxicity and antioxidant efficacy.

As consequence, in this point it could be found the main reason in the attempt to develop flexible (versatile) bio-analytical tools applicable both in ROS toxicity assessment and antioxidant analysis, such are the sensors/biosensors.

Reactive oxygen species (ROS) toxicity assessment was a subject of highly interest in all types of publications about the oxidative stress because in the last decades was proven an increasing occurrence of pathologies associated to ROS presence (Dalle-Donne et al. 2006, Butterfield et al. 2001). The ROS "attack" arises on specific receptors from a cellular component which is the oxidizable substrate, producing as result an oxidized molecular product. Generally, this is the key-event in several diseases evolution like Alzheimer's (Jaeger et al, 2008) cardiovascular diseases (Knopp et al. 2008) and others age-related diseases (Wang et al., 2008).

In order to status a common basis of the used terms the beginning of this chapter is devoted to a short description of each of them, with examples, further followed by the biosensors development and application on assessment of ROS, respectively antioxidants.

Up to date all reported biosensors employed various approaches, from direct analysis of compounds with characteristics antioxidants, to measuring the antioxidant enzymes activity and detection of free radicals. Most reported biosensors use immobilized enzymes in combination with electrochemical transducers, mainly amperometric devices (Mello & Kubota, 2007).

We are presenting in this chapter the use of biosensor in such determinations underlining with several experimental critical points that have to be tackled when a biosensor is developed and are implemented in ROS and antioxidants assessments. All examples that we are giving in this chapter are performed using some case study, significant for each category, namely for ROS, respectively antioxidants. It has to be highlighted from the very beginning that all examples are fully applicable only *in vitro* and useful as screening tools for providing information eventually helpful *in vivo* , but the *in vivo* determinations request other deep investigations strictly corroborated to metabolic pathways and pathological substantiation.

In the above mentioned context the versatility of sensors and biosensors application in both the analysis of antioxidants and evaluation of reactive oxygen species is defined by their extended use as analytical tool either in quantification of phytochemical compounds acting as antioxidants (especially phenolics and phenolic derivatives) and their known applications in the evaluation of antioxidant properties with respect to relevant oxidative markers. Moreover, several studies reported the electrochemical sensors application when it has to be evaluated the inhibition/ending of lipid oxidation.

2. Basics on reactive oxygen species and antioxidants

Reactive oxygen species (ROS) describe very reactive molecules containing oxygen, their high reactivity being given by the presence of unpaired electrons in the valence shell. As underlined by Halliwell (Halliwell, 2007 Halliwell Barry, Gutteridge J.M.C, Free Radicals in Biology and Medicine, Fourth edition, Oxford University Press, 2007), ROS is a general term that includes both oxygen radicals and several non-radical derivatives of oxygen, being overall acceptable that all oxygen radicals are ROS but not all ROS are oxygen radicals. A brief review of the most important ROS, mainly those with proved *in vivo* action, is given below, either radicals or non-radicals ROS.

The first produced free radical in the aerobic organisms is the superoxide radical $O_2^{\bullet-}$, a very reactive radical that afterwards generates hydrogen peroxide and can lead to lipid peroxidation, DNA and RNA damage, etc.

Another radical ROS is hydroperoxyl, $HO_2^{\bullet}$ which generally exists in traces in equilibrium with $O_2^{\bullet-}$ at physiological pH, its reactivity being mainly related to its higher capability to membrane cross than superoxide anion radical.

Hydroxyl radical OH• is the most reactive among ROS and can be formed by interaction of superoxide anion O_2•- and hydrogen peroxide H_2O_2 with cellular compounds through reactions like Fenton or Haber–Weiss.

Peroxyl radicals $RO_2^{\bullet}$ and alkoxyl radicals, $RO^{\bullet}$ are oxygen centered radicals formed by various routes, like reaction of carbon centered radicals with oxygen, or by decomposition of organic peroxides, being extremely important in lipid peroxidation reactions, especially as reaction product.

Hydrogen peroxide is a non-radical ROS, widespread *in vivo*, generated as product in various enzymatic reactions (that involving xanthine oxidase, superoxid dismutase, D-aminoacid oxidases etc.), mitochondria being one of the main sources for hydrogen peroxide on cellular level.

Singlet oxygen, existing normally in two states, is produced via photosensitization reactions or by decomposing peroxyl radicals, both reactions being very important *in vivo*.

Anyway, usually ROS are produced in metabolic reactions all the time, their level being maintained at certain level limits by the same metabolic reaction and by the action of so-called antioxidants defence. When an imbalance between ROS and antioxidants occurs, it results in the generation of oxidative stress, defined by Sies as: a disturbance in the pro-oxidant – antioxidant balance in favour of the pro-oxidants, leading to potential damages. Oxidative stress plays an important role in the pathogenesis of many diseases such as atherosclerosis, diabetes, hypertension, cancer and in the ageing process.

Halliwell gave the general accepted definition of the antioxidant as any substance that, when present in low concentrations compared to those of an oxidizable substrate, significantly delays or prevents oxidation of that substance (Halliwell 1990).

The antioxidants classifications support various points of view, depending on the antioxidant source, antioxidant action (mechanism) during radical chain reactions, or depending on antioxidant mechanisms as hydrogen or electron transfer reactions. As sources antioxidants can be divided into endogenous (internally synthesized, enzymatic ones -superoxide dismutase, catalase, glutathione peroxidase- or non-enzymatic - uric acid, bilirubine, albumine, glutathione etc.) and exogenous (diet-derived, like polyphenols, anthocyans, vitamin E, carotens, ascorbic acid etc.), while depending on their mechanism of action they can be divided into chain breaking antioxidants and preventive antioxidants (Somogyi, et al. 2007). Obviously, a classification criterion does not exclude another. Therefore, it has to be stressed that preventive antioxidants include enzymes such as superoxide dismutase that scavenge, for example, superoxide radical, blocking the initiation of chain reactions, while chain breaking antioxidants destroy free radicals after that they were formed, thus inhibiting the propagation of chain reactions, such as tocopherol (vitamin E) action against peroxyl radicals, during lipo-peroxidation.

The assessment of free radicals (FR) toxicity became very important due to the increasing degree and sources of pollution as long as the increasing occurrence of pathologies associated to the presence of free radicals in living organisms (Dalle-Donne et al. 2006, Butterfield et al. 2001).

3. How to design electrochemical sensors for ROS assessement. Case study HO•, superoxide and lipoperoxides radicals' assessment

When developing a protocol to assess free radicals toxicity it has to be taken into account that the FR "attack" takes place against a substrate producing an oxidized molecular product, this being generally the key-event in several diseases evolution. Usually, the amount of oxidized product is proportional to FR concentration and strictly related to the intensity of the damaging effect.

All existing methods able to evaluate effects of the ROS against cellular membrane components bear the drawback of the ROS short life-time, making difficult any attempt of direct assay; therefore it is usually preferred to monitor the molecular product of the oxidative stress reaction (Dalle-Donne et al. 2006, Butterfield et al. 2001). This monitoring approach is possible due to the fact that the degree of substrate oxidative damaging correlates strictly to the ROS concentration.

Generally highly sensitive procedures employed in such type of measurements are based on hyphenated techniques like high-performance chromatography (HPLC) with mass spectrometry (MS) detection, either coupled mass spectrometry detection (HPLC-MS/MS), or with mass – spectrometry detection with resolved time of flight signal (MS-ToF); the most complex such technique is that based on Matrix Assisted Lased Desorption-Ionization-Time of Flight detection (MALDI ToF) (Dalle-Donne et al. 2006). Another measuring procedure reported for the molecular product determination of the oxidative reaction is based on enzyme immunoassay (EIA) detection as reported, for example, by Kohno (Kohno et al., 2000) which realized a sandwich EIA suitable for the measurement of human oxidized LDL (Ox-LDL) in blood, using mouse monoclonal antibody specific for oxidized phosphatidylcholine as the capture antibody, and a horseradish peroxidase (HRP)-labeled goat anti-human apolipoprotein-B (Apo-B) IgG for detection. The use of such kind of detection devices is costly in order to achieve reasonable analytical information.

Therefore, the logical outcome is that, by using an electrochemical device as screening tool, it is possible to appear an alternative to the mentioned expensive analysis for determination of certain free radicals toxicity.

This approach is more interesting when is undoubtedly necessary to obtain biological significant information about free radicals toxicity or about antioxidant efficacy, especially considering the Halliwell criteria about how to assess an antioxidant (Halliwell, 2006).

Direct determination of ROS extent in biological systems requires highly sensitive methods, at least on nanomolar level since, for example, superoxide anion physiological level is about 10^{-10} molL^{-1} and it can be performed by different methods as fluorescence assays (Benov et al., 1998), electron spin resonance (ESR) (Roubaud et al., 1997) or chemiluminescence (Yao et al., 2004). In the same time, the exact amount of reactive oxygen from complex samples using various types of sensors and biosensors was reported, starting with fluorimetric sensors (Pastor et al., 2004) and ending with electrochemical ones, either amperometric – with mediated or direct electron transfer (Tian et al., 2002, 2005; Ohsaka et al., 1995, 2001, 2002; Campanella et al., 2000; Dharmapandian et al, 2010) or voltammetric (Fan et al., 2004; Cortina-Puig et al., 2009). From all these analytical tools the highly sensitive ones should be mentioned, able to perform the determination of nitric oxide with a detection limit of 20 pmolL^{-1} (Fan et al., 2004) while the fluorimetric biosensor for superoxide anion exhibited a quantification limit of 20 nmolL^{-1}(Pastor et al., 2004). There are data reporting the simultaneous detection of reactive species of oxygen and nitrogen in macrophage cells using multi-step amperometric method (Amatore et al., 2008, 2010).

In this work we are presenting new electrochemical devices -electrochemical sensors- as alternative for ROS toxicity screening analysis, based on the use of bio-mimetic systems as oxidizable substrates.

It has to be emphasized once again that the biomimetic systems were developed for applications addressing information on ROS toxicity for ROS concentration levels with physiological signification, but without a sensitivity competing with *in vivo* measurements, since hyphenated techniques are used for this purpose.

The biomimetic systems use two substrates highly susceptible to lipoperoxidation, namely human low-density lipoprotein (LDL) and phosphatidylcholine (PC).

3.1 Why lipoprotein and phospholipids as biomimetic systems?

Lipoproteins in general and especially low-density lipoproteins (LDL), are the main target of free radicals "attack" on cellular level because lipoperoxidation is considered to be responsible of main damage of both proteins and lipids from cellular membranes (Halliwell & Gutteridge, 2007). LDL are considered to have an important role in biological process that initiates and accelerates the development of cardiovascular diseases (Knopp et al. 2008) and influence Alzheimer's (Jaeger & Pietrzik 2008) and others diseases (Irshad, 2004; Staples et al. 2008; Parhami, 2003).

This is the main point from which we started the development of a new bio-mimetic model, thought to be able to help in understanding the structural and conformational modifications occurring on LDL subsequent to oxidative damage induction without using *in vivo* assessments. Other data supporting the use of LDL as bio-mimetic model are those related to the evidence that a key event in development of the various pathologies is the LDL oxidation step. The LDL oxidative modification has a unique pathway with respect to other lipoproteins (Parthasarathy et al., 2008), LDL being more sensitive to oxidation (Lam et al., 2004). LDL oxidation is dependent on the antioxidants within the cellular medium (Fierth et al., 2008).

In the same time, another peroxidation process, membrane phospholipid peroxidation, was incriminated in relation to oxidative damage occurring during pathological changes such as immuno-functional modulations, atherogeneses, and aging (Nagakawa et al., 1996) Phospholipid hydro-peroxide-the main product of membrane phospholipid peroxidation, may accumulate when the oxidative damage takes place in cellular membrane. In the same time, it was found that considerably elevated levels of phospholipid hydroperoxides occurred in blood cells of patients with Alzheimer's disease as compared to healthy volunteers. As a result, it seems very important to find out the level of ROS inducing such oxidative damage, and consequently, which are the effective antioxidants that can inhibit the formation of phospholipid hydroperoxides to prevent the disease. All data supporting phopholipid implication in key-events of important degenerative diseases made us to attempt to built up a second bio-mimetic system, based on another substrate highly susceptible to lipoperoxidation, namely phosphatidylcholine (PC).

Phosphatidylcholine, despite of a simple structure, is following in main steps the same oxidation pathway as LDL, generating lipoperoxides in presence of peroxyl radicals (see even figure 1).

3.2 Development of biomimetic systems to be used in sensors/biosensors construction

When dealing with development of biomimetic systems used in construction of sensors for ROS assessment, we decided to exemplify by the mean of the easiest methods available for oxidative substrates immobilization.

Therefore, **when oxidative substrates containing –SH groups** are used (as example LDL or DNA) two simple immobilization methods are available:

a. Direct deposition on solid support (Au sheet) by solution casting from oxidative substrate suspensions using suspensions containing a known mass of lipoprotein. Our protocol, already reported (Litescu et al., 2002) supposes the use of 60 ppm LDL suspension in KCl, 12 hours immobilization by solution casting on solid support, then further washed, dried and stored under vacuum, at 4°C.

b. Co-immobilization of substrate from suspensions on gold nanoparticles (25 mg mL^{-1}) and further attachment on conductive solid support. According to our experience, for LDL the optimal procedure consists in using of 1000 µg mL^{-1} LDL in 0.1 mol L^{-1} KCl allowed to immobilize 12 hours on AuNP then washed, dried under vacuum and stored under vacuum at 4°C.

When phospholipids are the oxidation substrate (example phosphatidylcholine, PC) the immobilization procedure consisted in suspending an exact amount of phospholipid in a solvent containing the best supporting electrolyte for electrochemical sensors, followed by chemisorption in controlled conditions on conductive support: inert atmosphere, optimum deposition time (able to ensure the appropriate thickness for the substrate layer and sensor operational stability), appropriate temperature. For example, our protocol used to build up a PC-based biomimetic system consists in the following: suspending of 72 mg of PC in 5 mL KCl 0.1 mol L^{-1}, suspending 24 mg of magnetic nanobeads (Fe_3O_4) in the previous solution, vortexed for 24 hours. After that PC-modified nanobeads ($PC-Fe_3O_4$) were separated, washed repeatedly, dried at 60°C for 30 minutes and stored at 4°C. The voltammetry experiments were performed applying a magnetic field in order to focus the same population of $PC-Fe_3O_4$ on the surface of working electrode.

It should be emphasized that the **main critical point when the biomimetic systems are used in the ROS toxicity evaluation** (and, consequently in assessment of the structural damaging induced by lipoperoxidation) consists in compulsory preservation of main functional groups availability toward ROS "attack". As consequence, checking the proper immobilization by surface analysis techniques (like Fourier Transformed Infrared Spectroscopy, XPS etc.) is essential at least in the first steps of elaboration of the protocols for sensors construction.

In the same time, it has to be mentioned a *sine qua non* condition that is mandatory when electrochemical sensors for ROS assessments are developed: it is necessary that ROS attack generate a significant change in electroactive properties of the used substrate. This means that the structural changes of the substrate have either gave birth to electroactive oxidation products or modify in a measurable way the substrate genuine electroactivity. Of course the amount of oxidized product (the amount of modified charge) is proportional to FR concentration and strictly related to the intensity of the damaging effect, raising an electrochemically measurable signal (amperometric, conductometric or voltametric).

3.2 Substrates oxidation

There are three simple possibilities to induce a fast and controlled lipo-peroxidation: heating, making the substrate (lipoprotein, phospholipids) to react with peroxyl radicals produced by azo-initiators or generating HO· radicals, or using the classical Fenton reaction.

In our protocol , thermally generated peroxyl radicals (ROO·) by the decomposition of AAPH reacted with the substrates (LDL, respectively PC) as shown in figure 1 (where LH is the unoxidized lipoprotein, while PC is the unoxidized phosphatidylcholine); subsequently are generated the lipoperoxides (LOO·, respectively PCOO·) electrochemically active (see also figure 2). The lipoperoxides reduction electrode process is further monitored electrochemically because the LDL, respectively PC, structural modifications induced by oxidation gave rise to a measurable signal.

In our experiments lipoperoxidation of the two substrates (LDL and PC) was initiated using peroxyl radicals obtained at a controlled rate by a known procedure: an aqueous solution of free radical azo-initiator, 2,2′-azobis (2methylpropionamidine) dihydrochloride, AAPH (83,8 mg to 10 mL of KCl 0.1 mol L^{-1}) was left at 37°C for 10 minutes, these conditions inducing the generation of ROS according to reactions given in figure below. It was demonstrated (Nikki, 1990) that aqueous solution of AAPH, 10 mmol L^{-1}, constantly generates, at 37°C, 1.36×10^{-6} mol L^{-1} sec^{-1} of free radical.

The generated ROS, ROO·, is further capable to induce lipids peroxidation.

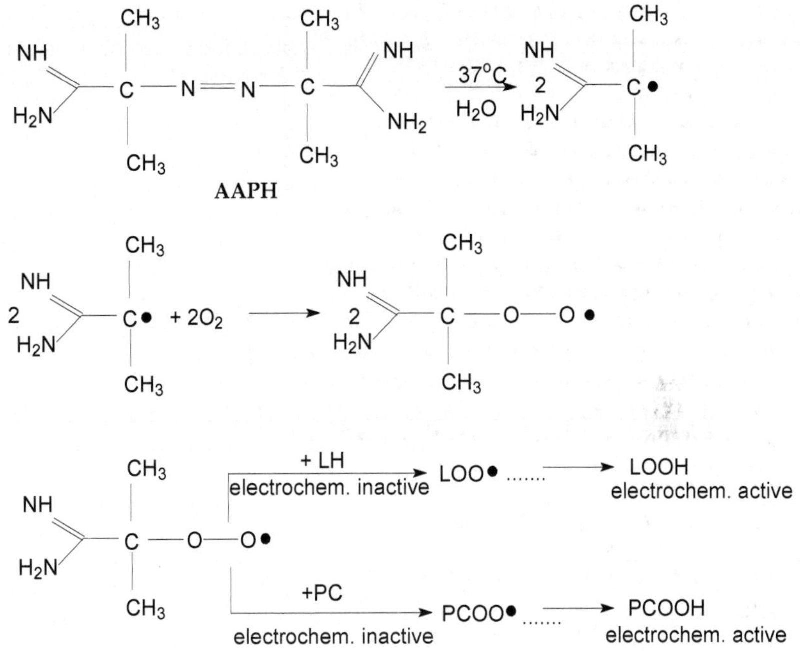

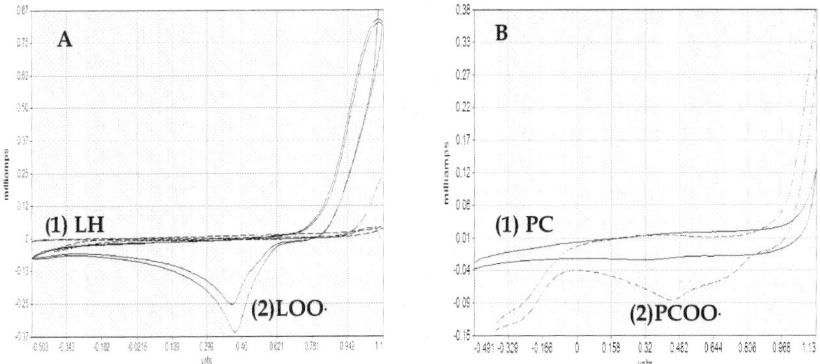

Fig. 1. Pathway of peroxidation induced by thermal decomposition of azo-initiator AAPH.

As noticed from figure 2, both phosphatidylcholine and LDL exhibit a similar behavior, proving no electroactive characteristics in solution or deposed, while an anodic reduction peak around + 0.395 ± 0.020 V raises in the presence of AAPH generated peroxidation.

Fig. 2. Cyclic voltammograms of LDL (A) and PC (B) un-oxidized (1) oxidized (2); $v = 100mVs^{-1}$ (WE=Au) KCl 0.1 molL^{-1}.

3.3 Applications of sensors based on biomimetic systems
3.3.1 Application of sensors based on biomimetic systems to ROS assessment

The variation of the current intensity of the lipoperoxides peak with the ROS concentration was proved to be linear for both substrates on a concentration range of physiological significance 10^{-7} - 1.6×10^{-6} molL^{-1}; the equation of the linear domain was $I(\mu A) = 160{,}14 \times C(\mu\,molL^{-1}) + 49.72$ ($R^2 = 0{,}9958$) for LDL and, respectively, $I(\mu A) = 211.15 \times C$ ($\mu\,molL^{-1}$) + 85.38 ($R^2 = 0{,}9636$) in the case of PC use, the obtained detection limit for both models calculated as $3 \times S/N$ being 4×10^{-7} mol L^{-1}.

In order to accomplish the goal of providing physiological significant data, the electrochemical device response has to be sensitive both to ROS concentration and to oxidizable substrate concentration. As consequence, chronoamperometric determinations using reduction potential of lipoperoxide were performed for various LDL concentrations and the same AAPH (in fact ROO·) concentration.

It was noticed that the lipoperoxides formation increased with the increasing of the substrate concentration in the range 200 to 500 µg mL^{-1}, concentrations higher than 500 µg mL^{-1} causing the electrode passivation (see table 1).

Substrate concentration (µg mL^{-1})	I peak ± SD nA, (E=0.395V±0.020)	
	WE= LDL/Au	WE = PC/Au
200	108 (± 26)	324 (± 32)
420	250 (± 18)	708 (± 29)
1000	302 (± 12)	600 (± 44)

Table 1. Dependence of LOO· signal on oxidizable substrate concentration (oxidation initiated by AAPH 10×10^{-3} molL^{-1}; results are the mean of 5 measurements)

These responses proved that the designed system is equally sensitive to substrate amount.
The obtained electrochemical data demonstrating the suitability of the built model in the study of lipoprotein ROS oxidation and assessment of the degree of ROS damage against lipoproteins biomimetic system were confirmed by FTIR and MALDI-ToF analysis. Subsequent to AAPH attack FTIR spectra performed on the LDL-Au surface confirmed the lipoperoxides formation: band corresponding to ester groups from lipid residues at 1740 cm^{-1} changed, and new HO absorption bands at 3600-3700 cm^{-1} and 917 cm^{-1} appear, these proving the lipoperoxides formation (FTIR data not shown) and a structural modification of the LDL molecule with respect to amide II absorption band that is modified. The MALDI-ToF analysis was also performed in order to obtain another confirmation of the modifications observed in the LDL structure, as result of oxidation by ROS attack. As could be observed from figure 3, the significant mass region for LDL is ranging between 1 and 10000 Da, because the signals in the mass region 10000 – 25000 were very weak and wide, while from 25000 to 80000 no signal was observed. The characteristic mass fragments for LDL itself are 1509 Da, 2569 Da, 4394 Da, 6420 Da, 6637 Da, 7647.5 Da and the 9432 Da. When the oxidation using AAPH initiator was performed it was noticed a 60% decrease in the intensity of the molecular fragments from 4394 Da, and for 1509 Da, while signals from 7647.5 Da and 9432 Da not only decreased significantly, but even are significantly shifted toward 7600 Da and, respectively 9492 Da, corresponding to mass modification multiples of peroxyl mass (Tache et al., in press).

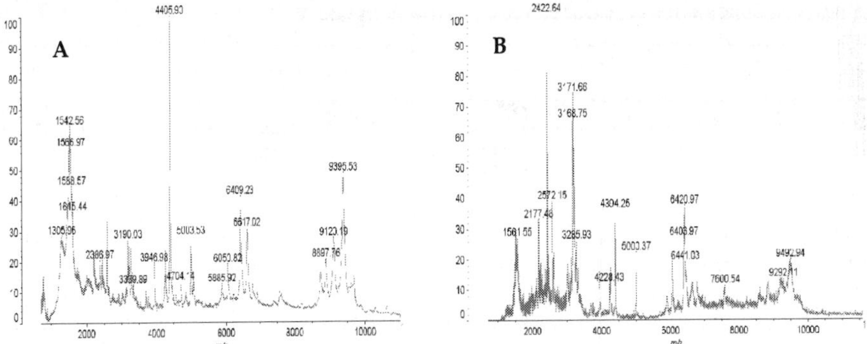

Fig. 3. MALDI-ToF spectra of un-oxidized LDL (A) and oxidized LDL (B), oxidation initiated by AAPH mmolL⁻¹

3.3.2 Application of sensors based on biomimetic systems to antioxidant efficacy assessments

As mentioned earlier a bio-analytical tool devoted to assessment of ROS has to be applicable both in ROS toxicity assessment and antioxidant analysis. As consequence, further was studied the potential application of realized sensors to assess the efficacy of few antioxidants against lipoperoxidation.

Taking into account the fact that an antioxidant is specific for a certain type of free radical that induces the oxidative stress (or, eventually to a narrow class of radicals) several antioxidants were tested against lipoperoxidation of LDL, using the LDL/Au sensor. Determinations were performed using an antioxidant concentration of 10^{-6} mol L⁻¹, the AAPH concentration being 10×10^{-3} mol L⁻¹ (respectively ROO· concentration 1.63×10^{-3} mol L⁻¹), and the modified electrode being incubated in the antioxidant solution for 20 minutes prior oxidation. The efficacy of antioxidants preservation was monitored in time, and the relative percent of lipoperoxide formation was calculated according to the formula (1), where %LOO is the percent of formed lipoperoxides, $i_{FR}^{LOO\bullet}$ is the current intensity of the peak corresponding to lipoperoxides formation after ROS attack, and $i_{ROS+Aox}^{LOO\bullet}$ is the current intensity of the same peak, when both ROS and antioxidant are in the measuring system (Litescu et al., 2002).

$$\%LOO\bullet = 100 - \left(\frac{i_{ROS}^{LOO\bullet}}{i_{ROS+Aox}^{LOO\bullet}} \right) \tag{1}$$

Essential oils from *Salvia* species (*Salvia*-EO) and astaxanthine (from *Haematoccocus pluvialis*) were used as lipo-soluble antioxidants and their efficacy against lipoperoxides formation was compared with that of two recognized lipophilic antioxidants, coenzyme Q10 (CoQ10) and vitamin E. An efficacy index against lipoperoxidation was established: astaxanthine > CoQ10 ≅ *Salvia*-EO > vitamin E.

Hydro-alcoholic extracts of the same *Salvia* species (Salvia extract) were used as antioxidants, and their efficacy was compared with that of known polyphenolic antioxidants: rosmarinic acid, caffeic acid and gallic acid, an index of efficacy being drawn: Caffeic acid > Rosmarinic acid > Salvia extract > Gallic acid.

It has to be stressed that when the substrate deposition is performed by solution casting, in optimal conditions, from suspensions containing a known mass of lipoprotein but without a controlled reproducibility of the deposition because of the unknown reproducibility of the deposition process, consequently affecting the data reproducibility.

Considering all obtained results it was proven that these new types of electrochemical sensors are applicable both to ROS assessment of and to assessment of antioxidant efficacy against lipoperoxidation. In the same time, the antioxidant efficacy depends strictly even on the nature of the antioxidant, the amount of antioxidant and on eventually occurring synergetic or antagonic effects exerted by a mixture of antioxidants. This is the reason of presenting further the development of biosensors devoted to a certain antioxidant class identification and quantification.

4. How to design biosensors for antioxidants quantitative determination. A case study- polyphenols antioxidants analysis

As mentioned in the introductory part, the majority of data published on antioxidant determination is based on biosensors that use immobilized enzymes (Mello & Kubota, 2007). In an overall acceptation of terms, biosensors are a sub-group of chemical sensors which could be defined as self-contained devices able to supply specific information. The provided analytical information is either quantitative or semi quantitative and is based on the use of a biological recognition element, which is in direct and spatial contact with a transduction element.

Several critical points have to be tackled when biosensors have to be developed:

a. the choice of the biological material and the choice of the transducer depend on the sample properties and on the type of physical magnitude to be measured.

b. the type and the nature of the bio-recognition component determine the degree of selectivity or specificity of the biosensor while the transducer correlates with biosensor sensitivity.

A general overview on designing of such type of biosensors involves following steps:

Quantification and identification of the compound of interest; this step concerns providing the appropriate bio-recognition element, able to supply accurate, sensitive, selective and reproducible information concerning samples composition. Appropriate bio-recognition elements are redox enzymes, the main advantage of using redox enzymes in amperometric biosensor construction being the value of the potential applied to monitor reduction or oxidation of the species at the electrode surface. This value generally occurs in -0.2V-0V range and allows reaching a minimum of possible electrochemical interferences (Mello et al., 2003)

Transduction, that supposes to transform the signal provided by the bio-recognition element into a measurable one, that could be current intensity (or charge), specific fluorescence and/or maximum absorbency.

Sensor performances assessments, that involves to provide the associate values of several performance parameters. This means to set up parameters like: the domain of applicability, selectivity/specificity, the linearity range of the sensor response, detection limit (LoD) and determination/quantitation limit (LoQ), accuracy of the determination, reproducibility, life time, operational stability, storage stability, and validation of the biosensors sensors response.

Validation of the built up biosensors, that means to validate the biosensors response with respect to "classic" methods of phytochemicals assessment.

4.1 Quantification and identification step; biological recognition element immobilization

This step is dealing with appropriate choice of bio-recognition element.

Polyphenols are one of the most important classes of antioxidants, naturally antioxidants, commonly occurring in fruits, vegetables and medicinal plants, and have been found to have a protective role against many chronic human diseases associated with oxidative stress. Polyphenols are divided in three large groups: phenolic acids (hydroxybenzoic acids and hydroxycinnamic acids), flavonoids (anthocyanidins, flavonols, flavononas, flavonas, isoflavonas and chalconas) and tannins (hydrolyzable tannins and condensed tannins) (Escarpa & Gonzales, 2001). Due to the importance of this class of compounds, many analytical strategies to evaluate the total phenolic content from plant extract and to establish phenolic profile have been reported, biosensors based on enzymes being proposed as an alternative device for total phenolic content (TPC) assessment. It has to be stressed that, in many cases, it is more important to measure the total content of polyphenols compounds than to determine each of them individually. The term 'total phenol' refers to all the phenols that are responsible for the total antioxidant capacity of a specific sample.

Biosensors based on various polyphenol-oxidases immobilisation are versatile devices in TPC assessment, due to class specificity. The main reaction consists in the oxidation of the substrate (phenols; poly-phenols) in the presence of enzymes molecules (phenol oxidases) the oxidation product – a quinone- being later reduced. While the tyrosinase biosensors are restricted to the monitoring of phenolic compounds with at least one free otho-position those based on laccase are applicable to a wider group of polyphenols, including ortho and para substitution or conjugated phenols with other functional groups.

The immobilisation of biological recognition element is considered as one of **the critical steps** that dictate the effectiveness of the enzymatic biosensor, due to the fact that biosensors performances -in terms of quantification and identification- are ensured by the preservation of the specific structure of the bio-component. For redox enzymes it is important that subsequent immobilization the active site remains available and, more, that immobilization did not affect the electron transfer from the enzyme active site and the electrode surface; this electron transfer could be affected by the insulating effect exerted from the protein structure. Data on laccase immobilization on different solid supports were reported. It has to be underlined that the immobilization matrix has to be, if possible, chemically inert with respect to biological element, stabile and with suitable conductive properties (there are cases when the immobilization matrix is, in fact, a conductive polymer).

Following, we exemplify two procedures of laccase immobilization on the surface of screen-printed working electrodes that lead to successful construction of versatile biosensors for polyphenols analysis. The first procedure is based on laccase embedding in an electrochemically generated chitosan matrix, the second one being based on Laccase entrapment in a Nafion stabilizing membrane:

Laccase immobilisation on solid modified supports via chitosan matrix. It was envisaged the immobilisation of *Laccase* on a stabilizing matrix deposed on the surface of a solid support;

the chosen solid support was gold. Two types of embedding matrix were used: one consisting only in chitosan (Chi), the other one involving the deposition of a composite matrix of Chi and multi-wall carbon-nanotubes (MWCNT). Chi and MWCNT-Chi films were electrodeposited on gold electrode surface using a -1.5 V controlled potential, deposition time was 5 minutes. Optimum conditions for electrodeposition of MWCNT –Chi film on a gold electrode were established taking into consideration the value of the layer capacity. Laccase immobilization was carried out by entrapment into the Chi-MWCNT nanocomposite film from multi-wall carbon nanotubes (MWCNT)-chitosan (Chi) solution containing 25U/mL enzyme during electrodeposition process (Diaconu et al., 2010).

Laccase immobilisation on screen-printed electrodes via Nafion membrane. Biosensors development was performed on the base of DROPSENS screen-printed electrochemical cell. The screen printed (either gold AuSPE or carbon CSPE) working electrodes of a three electrodes were modified by drop-casting from stock solution of 3 to 5µL of Lacccase solutions of exactly known activity, allowed to quickly dry, than followed by the immobilisation in Nafion membrane. It were used stock solutions of different Laccase activities, in order to obtain different set of biosensors that has different specific activities on the electrode surface, the units deposed ranging between 0.1U/electrode to 1 U/electrode. The Nafion membrane was obtained from aqueous/alcoholic solutions, 0.1%, 0.3 % or 0.5% perfluorinated Nafion. The cells were stored between measurements at 4°C (Litescu et al, 2010).

In order to check the efficiency of immobilisation the apparent Michaelis-Menten constant has to be determined and compared with the value of Michaelis-Menten constant for free enzyme; if the magnitude order is the same for both constants, then the immobilized enzyme preserve the affinity toward substrate exhibited by the free enzyme. The corresponding data related to performed immobilizations are presented further, in the section devoted to **biosensor performances assessment.**

4.2 Biosensors performances assessment

The assessment of biosensors performances is related to several specific characteristics, like: the domain of applicability, selectivity, the linearity range of the sensor response, detection limit (LoD) and the quantitation limit (LoQ), accuracy of the determination, reproducibility, life time, operational stability, storage stability (Thevenot et al., 2001).

When amperometric biosensors are developed the device performances evaluation is performed in optimal measuring conditions for the monitored enzymatic reaction. This means that first it is accomplished the electrochemical characterization of the product of the enzymatic reaction (for reaction in solution) in order to ascribe the corresponding oxidation and reduction potential peak values. It should be mentioned that the potential value differs according to the used conductive material. This electrochemical characterization is achieved by cyclic voltammetry (CV) experiments and it is important to be performed even for substrate, because, by this way, it is possible to solve a critical point by ascertain, from the very beginning, if substrate plays as electrochemical interferent in determination.

A suggestive example, related to laccase biosensor construction, in given in figure 4, where are presented the cyclic voltammograms of caffeic acid (a polyphenolic substrate for laccase) and that of the product of the enzymatic reaction.

After CV experiments it could be concluded which are the corresponding oxidation and reduction potentials. In the case of the considered example (figure 4) caffeic acid has an oxidation peak potential around +200 mV and a reduction one around 0 mV, while the quinone resulted from laccase catalyzed reaction has an oxidation peak potential around +63 mV and the reduction one is around -180 mV. The amount of enzymatic reaction product is direct proportional with the amount of substrate, in our case caffeic acid. As evident from figure one at about -200 mV the whole amount of quinone produced in enzymatic reaction is reduced. At this potential value, if the catalyst, namely the laccasse enzyme, is not in sufficient amount, correctly is not sufficiently active to ensure the complete substrate oxidation, then the substrate, caffeic acid, is able to electrochemically interfere in chronoamperometric determination of polyphenols. As consequence, the used enzyme amount, better activity is another critical point in biosensors development. By performing the same tests for several possible substrates it was established the optimal working potential, and further the performances characteristics for both constructive variants presented above. The investigated substrates were the polyphenolic antioxidants used to evaluate the LDL based sensors applicability to antioxidant efficacy assessment: caffeic acid, rosmarinic acid, gallic acid and, in addition chlorogenic acid.

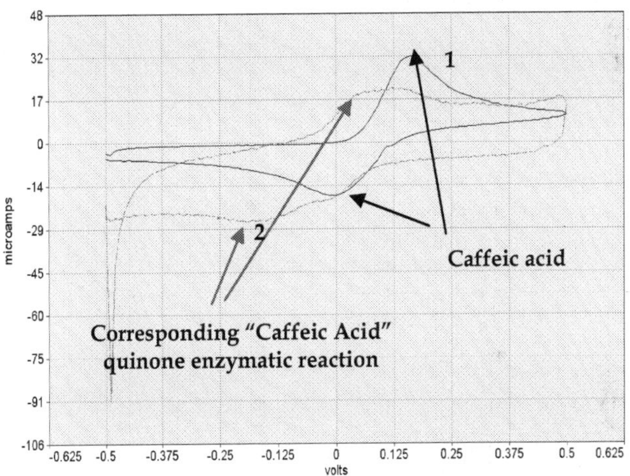

Fig. 4. Overlaid cyclic voltammograms of caffeic acid (1), respectively of caffeic acid in the presence of 1U Laccase (2); experiments performed on Au bare electrode

Obtain performance characteristics are given in table 2 and 3.

Substrate	Sensitivity nA/μmolL^{-1}	Linear range molL^{-1}	LoD molL^{-1}	K_M^{app} molL^{-1}
Caffeic acid	1446	10^{-6} –1.5×10^{-5}	7.7×10^{-7}	2.52×10^{-5}
Chlorogenic acid	1725	2×10^{-6} –1×10^{-5}	1×10^{-6}	2.41×10^{-5}
Gallic acid	72.3	2×10^{-6} –1×10^{-5}	1.5×10^{-6}	1.2×10^{-5}
Rosmarinic acid	788.6	10^{-6} –1.5×10^{-5}	4×10^{-7}	1.1×10^{-5}

Table 2. Performance assessment of Laccase-CHIT-MWCNT biosensor, buffer citrate-acetate pH=4.50, applied potential –0.200 V vs Ag/AgCl

Substrate	Sensitivity nA/μmolL^{-1}	Linear range molL^{-1}	LoD molL^{-1}	K_M^{app} molL^{-1}
Caffeic acid	245.3	3×10^{-6} –1.5×10^{-5}	2.5×10^{-6}	6.6×10^{-6}
Chlorogenic acid	255.0	2×10^{-6} –7×10^{-6}	2.8×10^{-6}	4.3×10^{-6}
Gallic acid	72.3	3×10^{-6} –1.5×10^{-5}	1.55×10^{-6}	4.12×10^{-5}
Rosmarinic acid	173.6	3×10^{-6} –1.5×10^{-5}	2.4×10^{-6}	4.3×10^{-6}

Table 3. Performance characteristics of laccase-Nafion based biosensor 1U/Au-SPE, pHʹ4.50, applied potential _0.200V vs. Ag=AgCl

The apparent Michaelis constant was determined using Lineweaver-Burk method, the obtained values proving that both devices could be applied to determination of the polyphenolic secondary metabolites (as shown in tables 2 and 3).

After these assays, the next to be established is biosensors stability. The operational stability of a biosensor response may vary depending on the sensor geometry, method of preparation, the used receptor and transducer. For operational stability determination, it has to be used a suitable analyte concentration (within dynamic range), the same type of biosensor contact (continuous or sequential) with the analyte solution, the same measuring parameters as temperature, pH, buffer composition, presence of organic solvents, and sample matrix composition. Even if some biosensors have been reported usable in laboratory conditions for more than one year, their practical lifetime is either unknown or limited to days or weeks when they are incorporated into industrial processes.

For storage stability assessment, significant parameters are the state of storage (dry or wet), the atmosphere composition, pH value, buffer composition, temperature and presence of additives. Whilst it is easy to determine the laboratory worktable stability of biosensors -either during storage and operational in the presence of analyte- the procedures for assessing their behavior during several days when biosensors were introduced in industrial reactors is very complex and difficult to handle. In both cases –lab or industrial set-ups-, it should be specified if lifetime is a storage (shelf) or operational (use) lifetime, the storage, respectively working conditions and specific substrate(s) concentration(s), as compared to the apparent Michaelis-Menten constant.

In general, after a raw assessment of the performances characteristics for biosensors, as in the above described protocols, several steps of biosensor optimization have to be performed, in terms of:

1. the amount (the activity) of the immobilized bio-recognition element
2. biosensor operational and storage stability

These two approaches drive out, in fact, due to necessity of reaching a compromise between: the amount of the immobilized units, the associated noise (because it is known that usually, with protein amount increasing the noise increases too, due to partially insulation owed to the protein itself) and the envisaged stability of the built up biosensors, both storage stability and operational one (because, in certain limits, the increase of the active units of enzyme on the working electrode generates a better stability).

1. re-evaluation of interferences, chemical and electrochemical interferences.
2. working electrode material, due to two main issues: the goal of diminishing of noise and interferences, and last but not least the costs of the built up biosensors

Continuing with the examples of laccase immobilization, different amounts of laccase were deposed on the working electrode surface and immobilized using Nafion membrane, the best responses, no matter the working electrode material, are obtained for an immobilized Laccase activity of 300 mU (Litescu et al., 2010).

In a simplistic way, it could be said that the biosensors operational stability is evaluated by repeated measurements performed using solutions of known concentration (within dynamic response range) of a certain standard substrate, while storage stability is evaluated by repeated measurements in time, at very well-established moments, in the same measuring conditions as optimum defined, using solutions of known concentration of a certain standard substrate.

The storage lifetime is defined as time necessary to decrease the biosensor sensitivity by a factor of 10 % or 50 %. The operational stability could be assessed as reproducibility or accuracy, and expressed as standard deviation of the measured signal for a certain number of determinations.

Laccase-Nafion biosensor operational and storage stability were checked in the following working conditions: MCIlvaine buffer pH = 4.50, 0.3 U Lacc/electrode (+0.1% Nafion), rosmarinic acid concentration level 5 μmolL^{-1}. The operational stability is fair up to 10 measurements, being very good for the first 7 measurements, with a RSD up to 4.00%, after that, for the next 3 determinations the stability decreased.

Operational stability of the Lacc/Chi-CNT electrode was checked for a rosmarinic acid concentration level of 5 μmolL^{-1}, in MCIlvaine buffer, pH = 4.50. Ten consecutive determinations gave a mean current of 890 nA with a relative standard deviation of 5.62 %. After 15 measurements a 10% decrease of the registered current was observed.

Another parameter to be defined is the **biosensor response selectivity**, two methods being generally used to determine biosensor selectivity. One method suppose the drawing of calibration curves for each of the possible interfering substances from sample matrix using the same measuring conditions as for the analyte determination and after that comparing the slopes of the curves with the slope of the analyte calibration curve, the selectivity being related to the ratio between slopes. Yet, for polyphenols analysis, considering the high-significance of the concentration range for phytochemicals, another method is the most important and has to be applied. This method consists in adding of the interfering substance at the expected concentration in the measuring cell which already contains a usual concentration of the analyte, the selectivity being expressed as the percentage of variation of the biosensor response.

The reliability of biosensors for given samples depends both on their selectivity and their reproducibility and it has to be determined under real operating conditions. This means that in the presence of possible interfering substance, the biosensor response should be directly related to the analyte concentration and should not vary with fluctuations of interfering substances concentrations.

For the same example of Laccase-Nafion biosensor, since the applied potential, -0.030V allows the avoidance of electrochemical interferences (is commonly accepted that the window of potential free of electrochemical interferences is ranging between -100 mV and +100 mV), the main problem that rest to be solved is that of chemical interferences. Supposing that the biosensor has to be applied for determination of polyphenols from *in vitro* cultivated plants, the main occurring interferences are those from growth media. Using the second method of selectivity determination and taking into account that an inhibition of 9-10 % of the enzyme activity itself was noticed when the Laccase activity was checked spectrophotometrically (against ABTS as unspecific substrate), it was concluded that no interference is taking place despite of the lowering of biosensors response with 10%.

4.3 Validation of the built up biosensors

Generally, when a new analytical method has to be validated, several principles should be satisfied and characteristic values of a number of parameters should be provided; we are mentioning as most important the following: method suitability to the analysis purpose, method specificity, precision and accuracy, repeatability, linear domain of response, limit of detection, limit of quantification, method traceability etc. Taking into account the peculiarity of applying analytical methods based on biosensor measurements, when discussing of validation for biosensor analysis the validation parameters addresses more the biosensor response than the overall method.

Using the same example of the built up laccase biosensors, the validation of the biosensors for phytochemical antioxidants analysis was supported by comparing the biosensor response with the high performance liquid chromatography-diode array-mass spectrometry (HPLC-DAD-MS) response obtained for the same samples, using the same measuring procedures. An in-house obtained example of feasibility of polyphenol-oxidases based biosensor application in polyphenols analysis is given in the table below and refers to two types of *Salvia* callus, *Salvia Maxima* and *Salvia verde*. The results are expressed as total rosmarinic acid equivalent, the main issue in validation step being precisely this one, namely to clearly define which is the main component of the sample in order to report data as main component equivalent.

Method used	Limit of Detection (mol L⁻¹)	Amount of polyphenolic content, RAEC (µg/ g fresh material)	
		S maxima	*S verde*
HPLC-DAD-MS	3.36×10^{-7}	103 µg/g	174 µg/g
Laccase – Nafion Biosensor	4.2×10^{-7}	97.8 µg/g	162.2 µg/g

Table 4. Determination of polyphenolic secondary metabolites as „total rosmarinic acid equivalent" in two callus samples

When the biosensors responses, expressed again in equivalent of rosmarinic acid, were compared with the response obtained by HPLC-DAD--MS for real samples of extracts of *Salvia officinalis* and *Mentha Piperita* good results were obtained, the biosensor response is about 94-95% from HPLC response (Litescu, 2010; Diaconu, 2011).

5. Conclusions

Application of sensors using lipoproteins in ROS determination is a feasible approach for lipoperoxides and peroxyl radicals, ensuring a fair measure sensitivity and specificity, but being strongly affected by the matrix complexity. In the same time such sensors are useful bio-analytical tools in *in vitro* assessing of the antioxidants efficacy against lipoperoxidation.

Based on reported data, comparing the results accuracy between biosensors based determinations and LC–DAD–MS determinations it could be concluded that the versatility of biosensors application in determination of phytochemical antioxidants content was proven by numerous publications, the critical point in ensuring a reliable result being the choice of the most suitable biological recognition element and of a transduction mode able to support the necessary measure sensitivity. Moreover, if biosensors designed for superoxide anion radical or hydrogen peroxide determinations are introduced in measuring solutions containing plant extracts, the antioxidants effects of phytochemicals could be assessed by the mean of radical scavenging monitoring.

6. Acknowledgments

The work was partially supported by the projects: FP7-SENSBIOSYN-232522/2009, BIOMONOXSIS 62053/2008 andUEFISCDI-DEI 241/2007.

7. References

Amatore C., Arbault S., Koh A.C.W.(2010). Simultaneous Detection of Reactive Oxygen and Nitrogen Species Released by a Single Macrophage by Triple Potential-Step Chronoamperometry. *Analitycal. Chemistry*,Vol 82, No 4, pp 1411-1419, ISSN 0003-2700.

Amatore C., Arbault S., Bouton C., Drapier J.-C., Ghandour H., Koh A.C.W. (2008) Real-Time Amperometric Analysis of Reactive Oxygen and Nitrogen Species Released by Single Immunostimulated Macrophages. *Chem BioChem*, Vol 9, No.9, pp 1472-1480, ISSN 1439-7633.

Benov L., Sztejnberg L., Fridovich I. (1998). Critical evaluation of the use of hydroethidine as a measure of superoxide anion radical. *Free Radical Biology & Medicine*, Vol. 25, pp 826-831, ISSN 0891-5849

Becker E. M., Nissen L.R., Skibsted L. H. (2004). Antioxidant evaluation protocols: Food quality or health effects, *European Food Research and Technology*, Vol. 219, pp. 561–S571, ISSN 1438-2377.

Butterfield D.A., Drake J., Pocernich C., Castegna A. (2001). Evidence of oxidative damage in Alzheimer's disease brain: central role for amyloid b- peptide. *Trends in Molecular Medicine*, Vol. 7, No 12, pp 548 - 554, ISSN: 1471-4914.

Campanella L., Persi L., Tomassetti M. (2000). A new tool for superoxide and nitric oxide radicals determination using suitable enzymatic sensors. *Sensors and Actuators B*, Vol. 68, No 1-3, pp 351, ISSN 0925-4005.

Cortina-Puig M., Muñoz-Berbel X., Rouillon R., Calas-Blanchard C., Marty J.-L. (2009). Development of a cytochrome *c*-based screen-printed biosensor for the determination of the antioxidant capacity of orange juices. *Bioelectrochemistry*, Vol 76, No 1-2, 76, ISSN:1567-5394.

Dalle-Donne I., Rossi R., Colombo R., Giustarini D., Milzani A. (2006). Biomarkers of Oxidative Damage in Human Disease. *Clinical Chemistry*, Vol. 52, No. 4, pp 601-623, ISSN 1874-2416.

Dharmapandian P., Rajesh S., Rajasingh S., Rajendran A., Karunakaran C. (2010). Electrochemical cysteine biosensor based on the selective oxidase–peroxidase activities of copper, zinc superoxide dismutase. *Sensors and actuators B*, Vol.148, No.1, pp 17-22, ISSN 0925-4005.

Diaconu M., Litescu S.C., Radu G.L.(2010) Laccase-MWCNT-chitosan biosensor-A new tool for total polyphenolic content evaluation from in vitro cultivated plants. *Sensors and Actuators B*, Vol. 145, No.2, pp 800-806, ISSN 0925-4005.

Diaconu M., Litescu S.C., Radu G.L. (2011).Bienzymatic sensor based on the use of redox enzymes and chitosan–MWCNTnanocomposite. Evaluation of total phenolic content in plant extracts.*Microchimica Acta*, Vol. 172, No 1-2, pp 177-184, ISSN 0026-3672

Escarpa A., Gonzalez M.C. (2001). An Overview of Analytical Chemistry of Phenolic Compounds in Foods. *Critical Reviews in Analytical Chemistry*,Vol. 31, No. 2, pp. 57–139, ISSN: 1040-8347

Fan C., Liu X., Pang J., Li G., Scheer H. (2004). Highly sensitive voltammetric biosensor for nitric oxide based on its high affinity with hemoglobin. *Analytica Chimica Acta*, Vol 523, No. 2,pp 225, ISSN 0003-2670

Halliwell B. (1990). How to characterize a biological antioxidant. *Free Radical Research Communications*, Vol. 9, No1, pp1-32, ISSN 8755-0199.

Halliwell B. (2006). Reactive Species and Antioxidants. Redox Biology Is a Fundamental Theme of Aerobic Life *Plant Physiology*, Vol. 141, pp. 312-322, ISSN 0032-0889

Hallliwell B, Gutteridge J.M.C. (2007). *Free Radicals in Biology and Medicine*, IVth edition, Oxford University Press, ISBN 978-0-19-856869-8, Oxford, Great Britain.

Irshad M. (2004). Serum lipoprotein (a) levels in liver disease caused by hepatitis. *Indian Journal of Medical Research*, Vol 120, pp 542-545, ISSN 0971-5916.

Jaeger S., Pietrzik C.U. (2008). Functional role of lipoprotein receptors in Alzheimer's disease. *Current Alzheimer Research*, Vol. 5, No 1, pp 15 -25, ISSN 1567-2050.

Knopp R.H., Paramsothy P., Atkinson B., Dowdy A. (2008). Comprehensive Lipid Management Versus Aggressive Low-Density Lipoprotein Lowering to Reduce Cardiovascular Risk. *American Journal of Cardiology*, Vol. 101, No 8 , pp S48- S57, ISSN: 0002-9149.

Kohno H., Sueshige N., Oguri K., Izumidate H., Masunari T., Kawamura M., Itabe

H.,Takano T., Hasegawa A., Nagai R. (2000). Simple and practical sandwich-type enzyme immunoassay for human oxidatively modified low density lipoprotein using antioxidized phosphatidylcholine monoclonal antibody and antihuman apolipoprotein-B antibody.*Clinical Biochemistry*, Vol. 33, No.4, pp 243-253, ISSN: 0009-9120.

Litescu S.C., Cioffi N., Sabbatini L, Radu G.L. (2002). Study of Phenol-Like Compounds Antioxidative Behavior on Low-Density Lipoprotein Gold Modified Electrode. *Electroanalysis*, Vol. 14, No.12, pp: 858-865, ISSN1040-0397.

Litescu S.C., Eremia SAV, Bertolli A., Pistelli L., Radu G.L. (2010). Laccase –Nafion based biosensor for the determination of polyphenolic secondary metabolites. *Analytical Letters*, Vol. 43, No. 7, pp 1089-1099, ISSN 0003-2719

Mello L.D., Kubota L.T, (2007). Biosensors as a tool for the antioxidant status evaluation, *Talanta*, Vol. 72, pp. 335–348, ISSN 0039-9140

Mello L. D., Sotomayor M.D. P. T., Kubota L. T. (2003). HRP-based amperometric biosensor for the polyphenols determination in vegetables extract. *Sensors and Actuators B*, Vol. 96, pp 636–645, ISSN 0925-4005

Nakagawa K, Fujimoto K., Miyazawa T, (1996) b-Carotene as a high-potency antioxidant to prevent the formation of phospholipid hydroperoxides in red blood cells of mice, *Biochimica and Biophysica Acta*, Vol.1299, No 1, pp 110-116, ISSN 0005-2760

Ohsaka T., Shintani Y., Matsumoto F., Okajima T., Tokuda K. (1995). Mediated electron transfer of polyethylene oxide-modified superoxide dismutase by methyl viologen *Bioelectrochemistry and. Bioenergetics.*, Vol 37, pp 73, ISSN 0302-4598.

Ohsaka T., Tian Y., Mao L., Okajima T. (2001). Superoxide Ion Sensor. *Analytical Sciences.*, Vol 17, pp. 379-381, ISSN: 0910-6340.

Ohsaka T., Tian Y., Shioda M., Kasahara S., Okajima T. (2002). A superoxide dismutase-modified electrode that detects superoxide ion. *Chemical Communications*, pp 990 - 991, ISSN 1359-7345.

Pastor I., Salinas-Castillo A., Esquembre R., Mallavia R., Mateo C R. (2010). Multienzymatic system immobilization in sol-gel slides: fluorescent superoxide biosensors development, *Biosensors & Bioelectronics* Vol. 25, No. 6, pp1526-1529, ISSN 0956-5663.

Parhami F. (2003). Possible role of oxidized lipids in osteoporosis: could hyperlipidemia be a risk factor?. *Prostaglandins, Leukotrienes and Essential Fatty Acids*, Vol 68, No 6, pp:373-378, ISSN 0952-3278.

Roubaud V., Sankarapandi S., Kuppusamy P., Tordo P., Zweier J.L.(1997). Quantitative Measurement of Superoxide Generation Using the Spin Trap 5-S(Diethoxyphosphoryl)-5-methyl-1-pyrroline-N-oxide *Analytical Biochemistry*, Vol. 247, No. 2, pp 404-411, ISSN 0003-2697

Somogyi A., Rosta K, Pusztai P., Tulassay Z., Nagy G, (2007) Antioxidant measurements, *Physiological Measurement*, Vol. 28, pp: R41–R55,ISSN: 0967-3334

Staples J., Taylor P., Magil A., Frohlich J., Johnston S., Koschinsky M. M., Chan-Yan C., Levin A.(2008). Progressive kidney disease in three sisters with elevated lipoprotein(a). *Nephrology Dialysis Transplantation*, Vol. 23, No.5, pp: 1756-1759, ISSN 0931-0509

Tache A, Cotrone S., Litescu S.C., Cioffi N, Torsi L., Sabbatini L, Radu G.L. (2011). Spectrochemical characterization of thin layers of lipoprotein self-assembled films on solid supports under oxidation process, Analytical Letters, in press, ISSN: 0003-2719

Thevenot D.R., Toth K, Durst R.A., Wilson G.S. (2001). Electrochemical biosensors: recommended definitions and classification. *Biosensors & Bioelectronics*, Vol 16, pp121-131, 0956-5663.

Tian Y., Mao L., Okajima T., Ohsaka T. (2005). A carbon fiber microelectrode-based third-generation biosensor for superoxide anion. *Biosensors & Biolelectronics*, Vol.21, No 4, pp: 557-564, ISSN 0956-5663.

Tian Y., Mao L., Okajima T., Ohsaka T.(2002). Superoxide Dismutase-Based Third-Generation Biosensor for Superoxide Anion. *Analytical Chemistry*, Vol 74, No 10, pp 2428-2434, ISSN 0003-2700.

Yao D., Vlessidis A.G., Gou Y., Zhou X., Zhou Y., Evmiridis N.P. (2004). Chemiluminescence detection of superoxide anion release and superoxide dismutase activity: modulation effect of *Pulsatilla chinensis. Analytical and Bioanalytical Chemistry*, Vol. 379, No. 1, pp 171-177, ISSN 1618-2642

Wang J., Hu B., Kong L., Cai H., Zhang C. (2008). Native, oxidized lipoprotein(a) and
    lipoprotein(a) immune complex in patients with active and inactive rheumatoid
    arthritis: Plasma concentrations and relationship to inflammation. *Clinica Chimica
    Acta*, Vol 390, No.1-2, pp 67-71, ISSN 0009-8981.

6

New Trends in
Biosensors for Water Monitoring

Florence Lagarde and Nicole Jaffrezic-Renault
University of Lyon
France

1. Introduction

Intensive industrialisation and farming associated to domestic uses of a growing number of chemicals have led to the release of many toxic compounds in the environment, causing an important pollution of aquatic ecosystems. In Europe, the Water Framework Directive WFD 2000/60/EC lays down the monitoring of a large number of substances, the so-called "priority substances", with the objective of restoring a good chemical and ecological status of all water bodies by 2015 (Allan et al., 2006). To implement effective monitoring and treatment programs, complementary analytical methods are required:
- low cost and high throughput screening methods for semi-quantitative determination of families of compounds and/or prediction of their harmful biological effects (overall toxicity, genotoxicity, estrogenicity),
- conventional methods based on chromatographic separation techniques (LC/MS, LC/MS/MS, GC/MS or ICP-MS), which are more time consuming, costful and require trained operators. These methods do not provide informations on water toxicity but allow the rescan of positive samples for more accurate analytes identification (Rodriguez-Mozaz et al., 2007).

Biological techniques, such as bioassays and biosensors, constitute the first category of methods. Many works in the past have been focused on the development of bioassays and have led to the commercialization of bacterial bioassays and immunoassays (Allan et al., 2006; Farre et al., 2005). In recent years, biosensors have received particular attention owing to their high sensitivity, low cost and possible easy adaptation for on-line measurements (Barcelo & Hansen, 2009).

A biosensor is an electronic device used to transform a biological interaction into an electrical signal. This device is based on the direct spatial coupling of the immobilised biologically active element, the so-called "bioreceptor", with a transducer that acts as detector and electronic amplifier. Different types of bioreceptors (enzymes, receptors, antibodies, DNA or microorganisms) combined with electrochemical, optical or mechanical transduction have been used for the elaboration of biosensors in view of water monitoring applications (Badihi-Mossberg et al., 2007; Rogers, 2006). To answer to the ever increasing requirements of water monitoring legislation, not only in terms of amount and reliability of informations provided, but also in terms of rapidity of response, selectivity, sensitivity and cost, tremendous efforts have been devoted in the last few years to improve the different elements contributing to the overall response of the biosensors, i.e. bioreceptor, transducer and cell/transducer interface. All these aspects will be addressed in the present chapter. New advances recorded in the field during the last five years will be more particularly emphasized.

2. Transduction modes

2.1 Electrochemical transduction

Electrochemical sensors are classified according to their transduction mode, which may be potentiometric, amperometric, conductimetric or impedimetric. In a general way, electrochemical transducers measure the electron transfers occuring between electroactive species (molecules or ions) present in a solution and an electrode, in well defined analytical conditions (Grieshaber et al., 2008). Over the past 10 years, electrochemical transduction technology has evolved significantly. Novel electrode materials such as boron doped diamond (BDD) have emerged as possible alternative materials to conventional gold, platinum or carbon (Luong et al., 2009). Owing to the recent advances in microfabrication techniques, it is also now possible to prepare microelectrodes of various sizes and geometries as well as to construct parallel arrays of microsensors on a same chip (Wei el al., 2009). Such systems are powerful tools able to answer to most of the environmental monitoring requirements such as rapidity of response, sensitivity, and parallel analysis of a large number of parameters and/or samples. Moreover, the small size is useful for the design of portable biosensors intended for on-field applications.

2.1.1 Potentiometric transduction

The two classical types of potentiometric transducers are ion selective electrodes (ISEs) and semiconductor-based field-effect devices (FEDs). The inherent miniaturization of ISEs and FEDs and their compatibility with advanced microfabrication technology make them very attractive for the integration into sensing arrays and microfluidic platforms and thus, the creation of miniaturized analytical systems suitable for environmental monitoring (Bakker et al., 2008; Bratov et al., 2010).

ISEs involve ion exchange equilibria at the interface between the solution and a membrane made of an ionic conducting material (inorganic solid electrolyte or organic liquid membrane). The nature of the membrane depends on exchanged ions, special glasses being typically used for H^+, ionic solid for halide ions, polymers including specific ionophores for other ions. Practically, potentiometric biosensors measure the difference of potential Ep between the selective electrode on which the bioreceptor is immobilised and a reference electrode when no significant current flows between them. Ep can be expressed by Nernst equation:

$$Ep = E^0 + RT/nF \ln a_{A^{n+}} \tag{1}$$

where E^0 is the selective electrode constant, $a_{A^{n+}}$ is the activity of A^{n+} ion

Significant efforts have been made during the past decade to improve the robustness of conventional ISEs, widen the range of ions detected and miniaturize the electrodes (Bakker et al, 2008; Tymecki et al., 2006).

FEDs belong to the second class of potentiometric transducers and include ion-sensitive field-effect transistors (ISFETs) and light-addressable potentiometric sensors (LAPSs). At present, only ISFET sensors measuring H^+ ions are commercially available. By deposition of enzymes or bacteria, it is possible to monitor enzymatic and metabolic reactions generating

H^+. LAPS devices are also extensively used to monitor cellular acidification in response to pollutants. Several recent reviews document the main features of these devices and their application to biosensing (C.-S. Lee et al., 2009; Schoening & Poghossian, 2006; Poghossian et al., 2009).

2.1.2 Conductometric transduction

Conductometric biosensors rely on the direct measurement of conductance variations in electrolytic media containing mobile electric charges. For that, an alternating voltage is applied between the working electrode, on which the bioreceptor is immobilised, and a reference electrode. The frequency value is chosen in order to minimize polarization effects. The conductance can be expressed by the following equation:

$$G = \gamma \frac{S}{l} \tag{2}$$

where γ (S. cm-1) is the specific conductance or conductivity, characteristics of the medium; S (cm2) is the working electrode surface; l (cm) is the distance between the electrodes.
Recent advances in the field have led to the production of miniaturized interdigitated electrodes that have been used to the elaboration of enzyme-based and cell-based biosensors for water monitoring (Hnaien et al., 2011; Jaffrezic-Renault & Dzyadevych, 2008). Enzymatic reactions between the pollutant and the bioreceptor induce a local change of conductivity due to the production of charged species.

2.1.3 Impedimetric transduction

Impedimetric transduction measures charge transfer processes occurring at electrode/electrolyte interfaces. Practically, measurement is performed using three electrodes, a working electrode modified by the bioreceptor, a reference electrode and an auxiliary electrode. A small amplitude sinusoidal voltage is imposed between reference and working electrodes and the resulting current generated between working and counter electrodes is measured. The applied voltage over measured current intensity ratio defines the impedance of the electrochemical system. Impedimetric data can be modelled by an equivalent electrical circuit from which electrical parameters that define charge transfer processes can be deduced (Katz & Willner, 2003). Impedimetric transduction is particularly well-suited to investigate reactions based on molecular affinity such as antigen-antibody or receptor-target interactions. Cell adhesion to the electrode surface is also expected to increase impedance value due to the insulative properties of the cell membrane. In the presence of cytotoxicants, morphological changes or functional alterations, and even death of the cells are also observed, inducing impedance variations. Therefore, these properties have been extensively exploited for water pollutants biosensing and toxicity assessment using electrodes modified with antibodies, receptors or cells.

2.1.4 Optical transduction

Optical transducers measure the effect of biological entities used as bioreceptors on light absorption, fluorescence, luminescence, refractive index or other optical parameters. In general, two different protocols can be implemented in optical biosensing. The first one requires a prelimary functionalisation of the bioreceptor or the target analyte with an optically active tag (labeling). Although this process produces highly sensitive biosensors, it is time-consuming and may interfere with the function of a biomolecule. In contrast, in the second protocole, target molecules are not labeled or altered, and are detected in their natural forms. This type of detection is relatively easy and cheap to perform (Fan et al., 2008). The most recent innovations in optical transduction applied to environmental biosensing are related to the development of new solid-state devices, microarrays and microfluidic systems for continuouous monitoring (Ligler et al., 2009)

2.1.4.1 Optical fibre sensors

An optical fibre is a waveguide that classically consists of a silica core (optical index: n_1) surrounded with a cladding of index n_2, slightly lower than n_1. The fiber is placed in a medium of index n_0. The light-guiding conditions are defined by:

$$n_0^2 \sin^2\theta_0 = n_1^2 - n_2^2 \qquad (3)$$

where θ_0 represents the numerical aperture of the fibre or the limit injection angle of the incident beam.

Fibre optical biosensors are all of extrinsic type. In some of them, called punctual biosensors, the physical or chemical effect is measured at the tip of the fibre on which the bioreceptor is deposited. The biosensor operates in reflection mode. In the so-called continuous biosensors, measurements are performed on a well defined length of the fibre. Cladding is removed in this zone and the bioreceptor layer is directly deposited on the core. This system can operate in reflection or in transmission modes. Bioreceptors are generally immobilised by adsorption or covalent attachment to a membrane, or recovered by a semi-permeable membrane. The most conventional fibre biosensors are based on absorption, fluorescence or luminescence detection (Goure & Blum, 2009). Others exploit the physical properties of the evanescent wave corresponding to the light power lost at the core-cladding interface. These biosensors, for which a stripping of the fibre is required, are more fragile than the massive optical systems based on the same principles and described in the following sections.

2.1.4.2 Mach-Zender interferometers

This type of sensor relies on the perturbation of the light propagating in one arm of an optical waveguide. In a typical Mach-Zehnder interferometer configuration, the light guide is divided into two branches via a Y-junction. A branch, functionalised with the biosensing element, is used as the sensitive arm, while the other is the reference branch (Fig. 1). The two branches recombine at the output, resulting in interference, and a photodetector measures the intensity. A change in the refractive index at the surface of the functionalised arm results in an optical phase change and a subsequent variation in the light intensity measured at the photodetector. This latter is proportional to cos ($\Delta n_2 k_o L$), where Δn_2 represents the refractive index change, k_o the amplitude of the wave vector and L the length of the sensitive region. These structures are made in glass or silicon and may be easily integrated into lab on chip laboratories (Sepulveda et al., 2006).

2.1.4.3 Surface plasmon resonance (SPR) sensors

These sensors are based on the physical principle of surface plasmon resonance (Hoa et al., 2007). The bioreceptor is deposited on a metal surface covering a glass support attached to the base of a prism (Kretschmann configuration). Interaction between the target and biorecognition molecules can be investigated in real time, with high precision and sensitivity, without specific labelling, through the measurement of the variations of

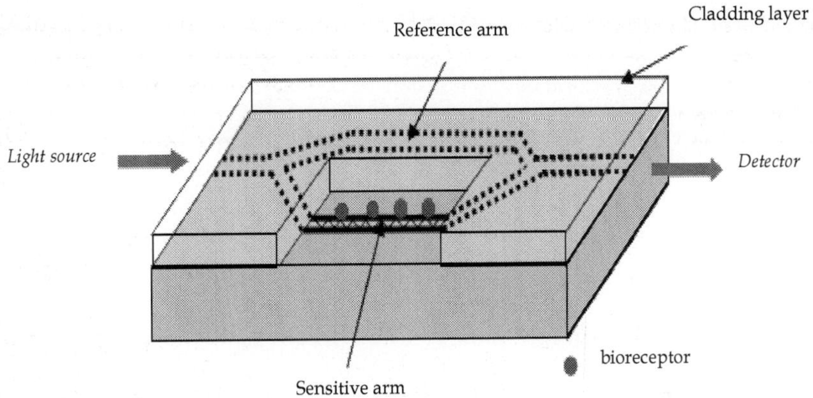

Fig. 1. Design of a Mach-Zehnder interferometer

refractive index near the interface. These sensors have been extensively used for the study of affinity interactions (eg antigen-antibody). SPRi systems allowing real-time and simultaeous imaging of several spots functionalised with different affinity systems are currently in full expansion (Scarano et al., 2010).

2.1.4.4 Optical waveguide light mode spectroscopy (OWLS)

This is a new detection technique based on evanescent field for *in* situ and label free investigation of surface processes at molecular level. It is based on accurate measurement of the resonance angle of a linearly polarized laser light, diffracted by a grating and coupled in a thin layer of the waveguide. Resonance coupling occurs at a specific angle characteristic of the refractive index of the medium covering the waveguide surface. The light is guided by total internal reflection on the edges where the detection is performed via photodiodes. By varying the light incidence angle, a spectrum is obtained, which allows the calculation of effective indices for both the electric and magnetic fields (Luppa et al., 2001).

2.1.4.5 Total internal reflection fluorescence (TIRF)

This technique has been used with planar waveguides and optical fibres as optical transducers in many biosensors. The light propagates along the waveguide, generating an evanescent wave on the surface of the optically denser part of the waveguide (quartz) as well as in the adjacent less dense medium (aqueous medium). The evanescent wave amplitude decreases exponentially with distance in the lower refractive index medium. The fluorescence of a fluorophore excited by the evanescent field can then be detected. Only fluorophores bound to the surface are excited. Real time kinetics of interaction of bioanalytes with molecules immobilised on the surface of the waveguide can be measured using TIRF. This is a rapid, nondestructive and sensitive technique used for the development of automated detection systems for environment monitoring (Tschemelak et al., 2005).

2.1.5 Mechanical transduction

Various mechanical methods have been used as detection in biosensors. These transducers have become increasingly popular over the years.

2.1.5.1 Transducers based on piezoelectric effect

A quartz crystal, to which a sinusoidal electric field is imposed, undergoes mechanical deformation due to the electrical potential appearing at its surface (piezoelectric effect). The crystal oscillates at its resonance frequency that depends on its structure (orientation, thickness…). Any change in mass (Δm) occuring at the crystal surface causes a proportional decrease in its resonance frequency (ΔF). This linear relationship is expressed quantitatively by the Sauerbrey equation:

$$\Delta F = -\frac{2F_0^2}{\sqrt{\mu_Q \cdot \rho_Q}} \cdot \frac{\Delta m}{A} \tag{4}$$

where Fo is fundamental frequency; A the geometric surface; μ_Q the shearing mode; ρ_Q the density of piezoelectric crystal

The Sauerbrey equation applies only for thin and rigid layers, excluding viscoelastic films, e.g. polymer or polyelectrolyte films. The most common transducer based on piezoelectric effect is the quartz crystal microbalance (QCM).

2.1.5.2 BioMEMS

BioMEMS (Bio-Micro-Electro-Mechanical-Systems) are mechanical systems of nanometer size that allow the translation of biomolecular interactions into mechanical data originating from the deflection of a cantilever. A biomembrane, attached to a silicon platform, constitutes the sensitive part of the device. This membrane, functionalised with specific molecules of interest (probes), vibrates in its fundamental mode by means of a piezoelectric patch. When the biomembrane is in contact with an aqueous solution containing species to be detected (target), the molecules are captured by the probe and membrane mass increases. By detecting the vibrations of this biomembrane, it is possible to measure the resonance frequency variations and thus estimate the quantity of biomolecules present in the solution. BioMEMS offer many advantages including rapidity, high sensitivity, low signal-to-noise ratio, ability for real time monitoring and possible integration of a large number of sensors on a small area (Hassen & Thundat, 2005).

2.1.6 Bioreceptor immobilisation

Immobilisation of the active biosensing element (enzymes, antibodies, cells ...) onto the transducer surface is a key point in the development of a biosensor. Apart from preserving the functionality of the biomaterial, the immobilisation method must ensure the accessibility of the cells towards target analytes as well as a close proximity between the bioreceptor and the transducer. The selection of an appropriate immobilisation method depends on the nature of the biological element and of the transducer, the physico-chemical properties of the analyte and the operating conditions of the biosensor. Several methods have been proposed in the literature, including chemical and physical methods (Fig. 2) (D'Souza, 2001). Physical methods include adsorption, retention into a membrane or entrapment within a polymeric network. Adsorption is based on the establishment of low energy interactions between the functional groups of the bioreceptor and of the substrate surface. This type of immobilisation offers the advantage of preserving bioreceptor properties but results in the formation of weak bondings that favours its desorption. To avoid leakage processes, biological elements can be covered by a thin polymer membrane that allows diffusion of the target molecule or entraped in a chemical or biological polymeric matrix. Sol-gel silica or

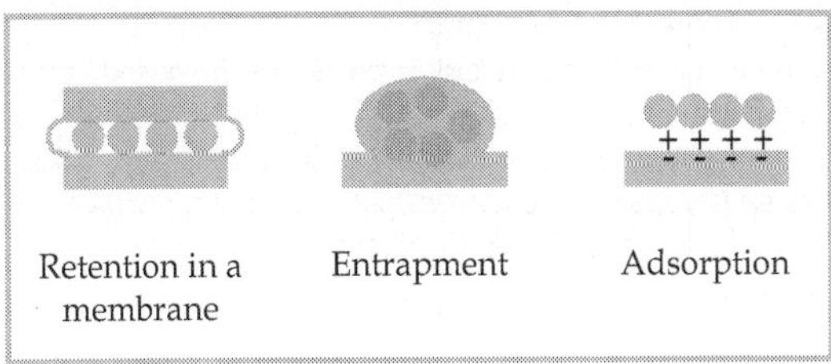

Physical methods

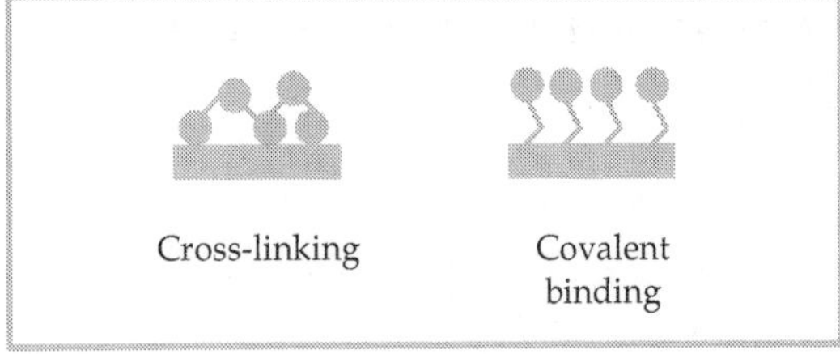

Chemical methods

Fig. 2. The different methods for bioreceptor immobilisation

hydrogels are typically used for that purpose. These polymers can efficiently protect the bioreceptor from external aggressions but may form a diffusion barrier that restricts the accessibility to the substrate and/or decrease light and electronic transfers to the transducer. The swelling properties of hydrogels may also limit their practical application in some cases. In chemical methods, biosensing elements may be attached directly to the transducer through covalent bindings or to an inert and biocompatible matrix through cross-linking using a bi-functional reagent. Proteinic supports, e.g. bovine serum albumin or gelatine are typically used to constitute the network with glutaraldehyde (GA) as cross-linking agent. This second technique is primarily used to attach enzymes or antibodies to the transducer, less often to immobilize microorganisms. Indeed, cross-linking involves the formation of covalent bindings between the functional groups located on the outer membrane of cells and GA. This mode of immobilisation is consequently not suited when cell viability is absolutely required or enzymes involved in the detection are expressed at the cell surface. Finally, covalent grafting is based on the reaction between functional groups of the biological element and previously activated groups of the transducer. Functional groups of the bioreceptor are typically ε-amines of lysine, carboxyl groups of aspartate or glutamate, sulfhydryl groups of cysteine and hydroxyphenolic groups of tyrosine, which belong to the side chains of aminoacids in proteins (enzymes, antibodies or external cellular proteins). To ensure the formation of covalent bondings with the transducer, this latter has to be functionalised first. Metal surfaces such as gold and silver can be functionalised with amine, hydroxyl or carboxyl groups through reaction with aminoalkanethiols, hydroxyalcanethiols and carboxyalcanethiols, respectively. Oxide surfaces are functionalised with organosilanes. More recently, metal electrodes on which films of functionalised conducting polymer (polypyrrole, polythiophene, polyaniline) are deposited electrochemically, have been used to immobilise active biomolecules through covalent bonding formation (Teles & Fouseca,

2008). In recent years, particular attention has been also paid to the use of nanomaterials, typically gold nanoparticles, magnetic beads, carbon nanotubes (CNTs) or quantum dots (QDs) for the elaboration of biosensors (Xu et al., 2006; X. Zhang et al., 2009). Their particular chemical and physical properties make them very attractive to improve bioreceptor stability as well as biosensor sensitivity.

2.2 Application to the determination of specific (groups of) pollutants
2.2.1 Enzyme-based biosensors

A large number of enzymes has been used in the construction of water pollution biosensors. They may be classified into different families corresponding to the type of reaction they catalyze, typically oxidation, reduction and hydrolysis. The enzyme is immobilised on a transducer that detects the consumption of a co-factor, e.g. oxygen in the case of oxidase enzymes, or the appearance of a product following enzymatic reaction. Hydrolase enzymes are generally associated with optical fibres, potentiometric or conductometric transducers to detect local changes in pH or in conductivity. For their part, reductase or oxidase enzymes are generally immobilised on an amperometric or conductometric transducer to record electronic transfers towards the electrode. These electronic transfers may be promoted by the use of redox mediators that allow the application of lower potentials and limit interferences from other electroactive species.

Tyrosinase is a copper monooxygenase that catalyzes the hydroxylation of monophenols and the oxidation of diphenols into reactive quinones. This reaction has been extensively exploited for the determination of phenolic compounds using tyrosinase-based amperometric and optical biosensors. In electrochemical systems, substances produced by the reduction of quinones at the electrode can be detected at a low potential value, in the absence of mediator. Various electrode materials, such as gold (S. Wang et al; 2010), platinum (Yildiz et al., 2007), carbon paste (De Albuquerque et Ferreira, 2007; Mita et al., 2007), glassy carbon (Carralero et al., 2006; J. Chen & Jin, 2010; Hervas Perez et al., 2006; Kochana et al., 2008; Kong et al., 2009; Y.-J. Lee et al., 2007; S. Wang et al., 2008; L. Wang et al., 2010) or BDD (Notsu et al., 2002; Zhao et al., 2009, Zhou & Zhi, 2006) have been used for that purpose. Very recently, Yuan et al. (Yuan et al., 2011) developed an amperometric biosensor using a carbon fiber paper (CFP) electrode. This biosensor exhibited short response times (10-20s) and very high sensitivities to phenolic compounds such as catechol, phenol, bisphenolA and 3-aminophenol, corresponding detection limits being 2, 5, 5 and 12 nM, respectively. A seen in Table 1, these values are much better than other figures reported in the literature and are 4 to 10 times lower than the values obtained in the same experimental conditions, by the same authors, using a commercial screen-printed carbon electrode (SCPE). In this work, tyrosinase was immobilised in photoreticulated polyvinylalcohool (PVA-SbQ) matrix. Many other modes of immobilisation have been proposed to stabilize tyrosinase on the transducer including, among others, entrapment into titania sol-gel (Kochana et al, 2008), polyacrylamide microgel (Hervas Perez et al., 2006), Fe_3O_4 or multi-walled carbon nanotubes (MWNT)-chitosane composites (Kong et al., 2009; Wang et al., 2008), MWNT-epoxy resin (Perez-Lopez et al., 2007), physical adsorption on ZnO nanorods (Zhao et al., 2009), or covalent binding (Zhou et al., 2006; Wang et al., 2010). Optical tyrosinase-based biosensors have been also reported (Abdullah et al., 2006; Jang et al., 2010). Table 1 presents some recent biosensors based on tyrosinase enzyme, with the type of transducer and immobilisation used, as well as the detection limits obtained for typical phenolic contaminants.

Pollutant	Enzyme	Transduction	Detection limit (μM)	Reference
Phenolic compounds				
17β-estradiol	tyrosinase	**Amperometry**		
		BDD electrode	10	Notsu et al., 2002
Bisphenol A	tyrosinase	**Amperometry**		
		BDD electrode	1	Notsu et al.
		CP electrode/SWCNT	0.02	Mita et al.
		Carbon fiber paper/PVA	0.005	Yuan et al.
2,4-dichloro-phenol	tyrosinase	Amperometry		
		GC electrode/MWNT+chitosane	0.06	Kong et al., 2011
Catechol	tyrosinase	**Amperometry**		
		GC electrode/MWNT/TiO$_2$/Nafion	0.087	Y.-J. Lee et al., 2007
		MWNT/epoxy resin	10	Perez-Lopez et al., 2007
		GC electrode/MWNT/TiO$_2$/Nafion	0.09	Kochana et al., 2008
		GC electrode/Fe$_3$O$_4$/chitosane	0.025	Sang et al., 2008
		GC/Teflon/Au nanoparticles	0.003	Carralero et al., 2006
		GC/polyacrylamide microgel	0.3	Hervas Perez et al., 2006
		Pt electrode/EDP	0.01	Yildiz et al., 2007
		GC electrode/ TiO$_2$	0.09	Kochana et al., 2008
		CP electrode/ CoPc	0.25	De Albuquerque et al., 2007
		GrC-acethylcellulose/ CoPc	0.45	De Albuquerque et al., 2007
		Carbon fiber paper/PVA	0.002	Yuan et al., 2011
		SPCE/PVA	20	Yuan et al., 2011
		Optical		
		Glass/SiO$_2$/Nafion	2.1	Abdullah et al., 2006
Phenol	tyrosinase	**Amperometry**		
		GC electrode/MWNT/TiO$_2$/Nafion	0.095	Y.-J. Lee et al., 2007
		Functionalised Au electrode	0.1	L. Wang et al., 2010
		GC electrode/palygorskite	0.05	J. Chen & Jin, 2010
		Functionnalised BDD electrode	0.2	Zhou et al., 2006
		GC/polyacrylamide microgel	1.4	Hervas Perez et al., 2006
		Pt/EDP	0.1	Yildiz et al., 2007
		GC electrode/ TiO$_2$	0.13	Kochana et al., 2008
		BDD electrode/ZnO nanorods	0.5	Zhao et al., 2009
		Carbon fiber paper/PVA	0.005	Yuan et al., 2011
		SPCE/PVA	20	Yuan et al., 2011
		Optical		
		Glass microarray/PEG-DA/QD	1	Jang et al., 2010
		Glass/SiO$_2$/Nafion	1.9	Abdullah et al., 2006
4-chlorophenol	tyrosinase	**Amperometric**		
		GC electrode/MWNT/TiO$_2$/Nafion	0.11	Y.-J. Lee et al., 2007
		GCelectrode+Teflon+Au nanoparticles	0.02	Carralero et al., 2006
		Functionnalized BDD electrode	0.1	Zhou et al., 2006
		GC/polyacrylamide microgel	0.03	Hervas Perez et al., 2006
		GC electrode/ TiO$_2$	0.17	Kochana et al., 2008
		BDD electrode/ZnO nanorods	0.4	J. Zhao et al., 2009

Pollutant	Enzyme	Transduction	Detection limit (μM)	Reference
Trophic pollutants				
Nitrates	Nitrate reductase	**Amperometry**		
		Pt electrode/Ppy	10	Sohail et al., 2009
		Conductometry		
		Au electrode /GA /Nafion	5	Xuejiang et al., 2006
Nitrites	Nitrite reductase	**Amperometry**		
		GC electrode/[ZrCr-AQS]-LDH/GA	0.004	H. Chen et al., 2007
		GC electrode/modified MV	0.06	Quan et al., 2010
		Conductometric		
		Au electrode /Nafion	0.05	Z. Zhang et al., 2009
Phosphates	Maltose phosphorylase	**Conductometric**		
		Au electrode /GA	1	Z. Zhang et al., 2008
	Pyruvate oxidase	**Amperometry**		
		Pt electrode/Nafion/PCS hydrogel	3.6	Gilbert et al., 2010
		SPC electrode/acetate cellulose	<300	Khadro et al., 2009
Organic Matter (proteic fraction)	Protéinase K + pronase	Conductometric	0.583 μg/L for TOC	
Organophosphorous pesticides				
Dichlorvos	tyrosinase (inhibition)	**Amperometric**		
		GC electrode/NQ/Nafion	0.17	
		GC electrode/NQ/o-PPD	0.19	Vidal et al., 2008
		GC electrode/Nafion	0.07	
methyl parathion	OPH	**Amperometric**		
		GC electrode/ MWCNT/Au/QD	0.004	Du et al., 2010
		GC electrode/ MWCNT	0.8	Deo et al., 2005
demeton-S	OPH	**Amperometric**		
		CSP electrode/MWCNT	1	Joshi et al., 2006
paraoxon	OPH	**Amperometric**		
		GC electrode/ MWCNT	0.15	Deo et al., 2005
		GC electrode/mesoporous C/C black	0.12	J.H. Lee et al., 2010
		Piezzoelectric		
		Microcantilever/LbL	0.1	Karnati et al., 2007
		SPR		
		Au/SiO$_2$	20	Luckarift et al., 2007
		OWLS		
		Glass/TiO$_2$ array	2.5	Ramanathan et al., 2007
		PMMA/sol gel	0.004	Zourob et al., 2007

AQS: anthraquinone sulfonate, BDD: boron doped diamond, CoPc: cobalt phtalocyanine, CP: carbon paste, CSP: carbon screen-printed, EDP: electrodeposition polymer, GA: glutaraldehyde, GC: glassy carbon, GrC: graphite carbon, Ppy: polypyrolle, LbL: layer by layer, LDH: Layered double hydroxide , MV: methylviologen, MWCNT: multi-walled carbon nanotubes,NQ: naphthoquinone, PCS: poly(carbamoyl) sulfonate , PVA: polyvinylalcohol, QD: quantum dot, SPC: screen-printed carbon, SWCNT: single-walled carbon nanotubes.

Table 1. Examples of enzymatic biosensors for the detection of chimical pollutants.

Another enzyme, organophosphate hydrolase (OPH), has been also commonly used for the development of electrochemical, optical and mechanical biosensors for organophosphorous pesticides detection, while nitrate reductase, nitrite reductase, maltose phosphorylase, pyruvate oxidase have been employed for the determination of trophic pollutants such as nitrates, nitrites or phosphates (Table 1).

2.2.2 Immunosensors

Immunosensors are based on highly selective antibody (Ab) - antigen (Ag) reactions. The immobilized sensing element can be either an Ab or an Ag which can be chemically modified (hapten). In the first case, analyte binding is measured directly. In the second case, the method is based on the competition between immobilized Ag, the analyte (Ag) and a fixed amount of Ab. All types of immunosensors can either be run as nonlabeled or labeled immunosensors. Label free immunosensors rely on the direct detection of antigen-antibody complex formation by measuring variations in electrical properties using electrochemical impedance spectroscopy (EIS), or changes in optical properties using SPR. The second type of immunosensors use signal-generating labels which allow more sensitive and versatile detection modes. Peroxidase, glucose oxidase, alkaline phosphatase, catalase enzymes and electroactive compounds such as ferrocene are the most common labels used for electrochemical detection, while fluorescent labels (rhodamine, fluorescein, Cy5, etc...) are employed for optical detection. Some recent examples are presented in Table 2. Over the two past decades, a large number of immunosensors targeting individual pollutants or groups of pollutants and based on these different configurations have been reported. Recent developments have been focused on label free immunosensors using EIS and SPR detection (Mitchell, 2010; Prodromidis, 2010) as well as on improvements in antibody design (Conry et al., 2009).

2.2.3 Cell-based biosensors

Many works have been focused on the development of cell-based biosensors. Bacteria, algae and yeasts are the main organisms used for that purpose. Various types of strains have been exploited, from commercial and well-characterized cells harbouring a broad range of substrates to genetically-engineered organisms specially constructed to detect specific molecules or groups of molecules, passing through environmental cells isolated from polluted sites offering higher robustness and more specific enzymatic properties. Cell membranes can be permeabilized in order to improve accessibility to internal enzymes. Compared to their individual components (enzymes, antibodies or DNA), cells are easier to produce in large quantities and are more tolerant to pH, ionic strength and temperature variations. Owing to the large number of enzymes and cofactors that the cells contain, a large variety of biosensors has been proposed for the detection of specific (groups of) analytes or for aquatic toxicity assessment, this latter application being addressed in section 2.3. Several reviews have been published on the topics (Lei et al., 2006), some of them being more specifically dedicated to yeast-based sensors (Baronian et al., 2004), genetically-modified bacteria sensors (Daunert et al., 2000; Girotti et al., 2008; Hansen & Sorensen, 2001; Van der Meer & Belkin, 2010; Woutersen et al., 2010), or electrochemical cell biosensors (Lagarde & Jaffrezic-Renault, 2011).

2.2.3.1 Electrochemical biosensors

Amperometry is the most common electrochemical transduction mode used in whole cell biosensors. It allows detecting oxygen consumption or production during respiration/

photosynthesis processes, consumption or production of specific compounds in course of analyte metabolisation, or induction of a specific enzyme activity by genetically modified microorganisms (Lagarde & Jaffrezic-Renault, 2011). Different microbial strains exhibiting a wide range of substrates have been used for the determination of biological demand of oxygen (BOD), an index of the amount of degradable organic compounds present in the sample (Nakamura, 2010; Ponomareva et al., 2011). Oxygen consumption during biological respiration is generally detected by means of conventional Clark type electrodes, but miniaturized systems based on small-size carbon screen-printed electrodes (SPEs) have been also proposed in recent years. In the same way, biosensors able to detect surfactants, phenolic derivatives, alcohols or organophosphorous pesticides have been constructed by immobilizing bacteria degrading specifically these groups of pollutants on classical oxygen,

Pollutant	Detection mode	Transduction	Detection limit (ng L^{-1})	Reference
Pesticides				
Isoproturon	Competitive/ Cy5.5 fluorescence labelling	TIRF	20	Tschmelak et al.,2005
Chlorpyrifos	Competitive/no labelling	SPR	55	Mauriz et al., 2006
DDT and derivated products	Competitive/no labelling	SPR	15	Mauriz et al., 2007
Atrazine	Direct	EIS	10	Hleli et al., 2006
	Competitive/ Cy5.5 fluorescence labelling	TIRF	10	Tschmelak et al. 2005
	Competitive	SPR	500	Farre et al., 2007
Picloram	Competitive HRP labelling	Amperometric	< 1	L. Chen et al., 2010
EDCs				
Testosterone	Competitive/labelling	TIRF	1.7	Tschmelak et al. 2006
	Competitive/no labelling	Amperometric	170	Eguilaz et al., 2010
Estradiol	Competitive	SPR	300	Ou et al., 2009
Bisphenol A	Direct	EIS	400	Rahman et al., 2007
	Competitive/labelling	TIRF	8	Marchesini et al., 2005: Tschmelak et al., 2005
Nonylphenols	Direct/HRP labelling	Amperometric	10000	Evtugyn et al., 2006
2,4-dichloro-phenoxyacetic acid	Competitive/Cy5.5 fluorescence labelling	TIRF	90	Long et al., 2008
	Competitive/no labelling	SPR	100	Kim et al., 2007
	Competitive (signal amplification)	SPR	8	Kim et al., 2007
Toxins				
Microcystin-LR	Competitive/Cy5.5 fluorescence labelling	TIRF	30	Long et al., 2008

Table 2. Examples of immunologic biosensors for the detection of chemical pollutants. EDC: endocrine disrupting compound

graphite carbon or carbon paste electrodes, more recently on SPEs (Lagarde & Jaffrezic-Renault, 2011). Table 3 presents the most recent examples of electrochemical biosensors developed for BOD measurement and for the determination of specific (groups of) analytes. To modify cell resistance and sensitivity towards toxic compounds, microorganisms may be genetically modified. Buonasera *et al.* recently combined on a single biosensing platform amperometric and optical modes of transduction as well as several genetically modified algal strains harbouring various degrees of sensitivity and resistance towards pesticides. The system allowed detecting different subclasses of pesticides in the 0.1 to 10 nM range (Buonasera *et al.*, 2010). To enhance selectivity, genetical modification of the cells is also possible by fusing a natural regulatory circuit existing in the microorganism with a promotorless gene encoding for an easily measurable protein expressed only when the analyte(s) is present. The most common gene used for electrochemical detection is *lacZ* encoding β-galactosidase Activation of β-galactosidase is generally followed through the increase of its enzymatic activity using *p*-aminophenyl β-D-galactopyranoside (PAPG) as substrate. PAPG is transformed into *p*-aminophenol oxidized at the amperometric electrode. Tag et al. (Tag et al., 2007) proposed another method of detection using lactose as deputy substrate.

Potentiometric and conductometric biosensors have been also developed for the determination of specific pollutants. For example, a biosensor based on *P. aeruginosa* JI104 immobilized on a chloride ions-selective solid-state electrode has been reported for trichloroethylene detection in waters. More recently, Hnaien *et al.* (Hnaien *et al.*, 2011) proposed a fast, sensitive and miniaturized whole cell conductometric biosensor for the determination of the same pollutant. The biosensor assembly was prepared by immobilizing *P. putida* F1 bacteria at the surface of gold microelectrodes through a three dimensional alkanethiol self-assembly monolayer/arbon nanotubes architecture functionalised with *Pseudomonas* antibodies. pH electrodes have also been widely used to detect H^+ ions produced through enzymatic reactions (Kumar et al., 2008).

2.2.3.2 Optical biosensors

Optical biosensors rely on the modulation of cell optical properties (UV-Visible absorption and biochemiluminescence, reflectance, fluorescence) following interaction with compounds present in the sample. Most of the optical biosensors proposed are based on bioluminescence or fluorescence detection. The so-called "light-off" systems measure a decrease in the celullar light emission following exposure to the pollutant(s). They are mainly used for water toxicity assessment (§2.3.5). The detection of specific analytes or groups of analytes is rather performed using "light-on" type biosensors, where the interaction causes an increase of light signal proportional to the analyte concentration. "Light-on" microorganisms are produced naturally or via genetic engineering. The most frequently used genes are *lux* gene encoding for luminescent luciferase and *gfp* gene encoding for fluorescent GFP (green fluorescent protein). A variety of well-characterized promoters is available for genetic manipulations and has been used to construct new organisms able to sense specifically different classes of pollutants, including metals (copper, mercury, lead, cadmium, arsenic ...), hydrocarbons and organic solvents or pesticides. Many examples have been reported in the literature (Daunert et al., 2000; Lei et al., 2006). Naturally emitting bacteria have been also used for BOD determination (Lin et al., 2006; Sakaguchi et al., 2007).

Target	Microorganism	Transduction	Detection limit	Reference
BOD				
	Saccharomyces cerevisiae	Amperometry	6.6 mg L^{-1}	Nakamura et al., 2010
	Microbial consortium (BODSEED)	Amperometry	< 5 mg L^{-1}	L. Liu et al., 2010
	Escherichia coli DH5a	Potentiometry	1mg L^{-1}	Chiappini et al., 2010
	Photobacterium phosphoreum IFO 13896	Luminescence	1mg L^{-1}	Sakaguchi et al., 2007
	B. licheniformis, D. maris, M. marinus	Optical fibre	0.2 mg L^{-1}	Lin et al., 2006
Phenolic compounds				
Phenol	*Pseudomonas putida DSM 50026*	Amperometry	500 μM	Timur et al.
p-nitrophenol	*Pseudomonas* sp.	Amperometry	< 10 μM	Timur et al.
Organophosphorous pesticides				
Paraoxon, parathion	Modified *P. putida* JS444	Amperometry	0.001 μM	Lei et al., 2005
Fenithron, EPN			0.005 μM	Lei et al., 2007
Heavy metals				
Cu	*S. cerevisiae* 19.3C/ *CUP1::lacZ*	Amperometry	0.1 μM	Tag et al., 2007
	S. cerevisiae SEY6210/ *CUP1::lacZ*		33 μM	
Endocrine disrupting compounds				
	S. cerevisiae Y190 *medER ::lacZ*	Amperometry	-	Ino et al., 2009
Antibiotics				
Cephalosporins	*P. aeruginosa* MTCC 647	Potentiometry	100 μM	Kumar et al., 2008
Organic solvents				
Benzene	*P.putida* L2	Amperometry	10 μM	Lanyon et al., 2006
Trichlorethylene	*P. aeruginosa* JI104	Potentiometry	0.22 μM	Han et al., 2002
	P. putida F1	Conductometry	0.07 μM	Hnaien et al., 2011

Table 3. Some recent examples of cell based biosensors for the detection of specific (groups of) pollutants

2.3 Application of biosensors to the assessment of aquatic toxicity

Most of biosensors developed for toxicity assessment exploit toxic effects of the pollutants, including enzyme inhibition, such as AchE inhibition by neurotoxic compounds, interaction with a specific receptor (androgenicity, estrogenicity), interaction with and damage of DNA or RNA (genotoxicity). The detection of some biomarkers of toxicity may be also used.

2.3.1 Enzyme biosensors

A major contribution of enzyme biosensors to ecotoxicological studies concerns aquatic neurotoxicity assessment. This latter may be due to organophosphate and carbamate pesticides, heavy metals or detergents that inhibit esterase enzymes. Neurotoxicity biosensors proposed in the literature are mostly based on two enzymes belonging to this family, acethylcholinesterase (AchE) and butylcholinesterase. Many works and two reviews have been published on this type of biosensor (Jaffrezic-Renault, 2001; S. Liu et al., 2008). Current developments aim to improve enzymatic activity, either by genetic modification (Bucur et al., 2006) or by a better immobilisation on the transducer. The use of new materials based on gold, silver or iron nanoparticles (Du et al. 2008; Gan et al., 2010; Gong et al. 2009; Shulga et al., 2007; W. Zhao et al., 2009) or on carbon nanotubes (Viswanathan et al., 2009) also allows significant increase of the sensitivity of electrochemical and optical biosensors. A portable system using a potentiometric transduction and AchE as bioreceptor has been recently validated on different samples of water (Hildebrandt et al., 2008). Cortina et al. (Cortina et al., 2008) proposed an enzyme-based array that used three AchE enzymes: the wild type and two different genetically modified enzymes. Multianalyte devices combining informations from several different types of enzymes have been also proposed. For example, Soldatkin et al. recently developped an amperometric multibiosensor using the inhibition of acetylcholinesterase, butyryl- cholinesterase, urease, glucose oxidase, and three-enzyme system (invertase, mutarotase, glucose oxidase) for water toxicity assessment (Soldatkin et al., 2009).

2.3.2 Estrogen receptor-based biosensors

Endocrine disruptors (EDCs) are chemical substances that cause hormonal imbalances and impair endocrine or nervous systems. Some of these compounds affect the synthesis of endogenous hormones or that of their receptors. Others are structurally similar to estrogens and bind to their receptors, leading to their inactivation or to abnormal behaviours. Many molecules, such as synthetic hormones or chemical substances such as phthalates, surfactants, PCBs, alkylphenols, parabens, PAHs, dioxins and some pesticides, are EDCs. To date, 320 priority substances suspected to disrupt endocrine system have been identified by the European Community. Some of them (nonylphenol, di-2-ethylhexylphthalate and polybrominated diphenyl ethers) have been included in the list of priority substances of the Water Framework Directive. A review has addressed the use of biosensors for environmental EDCs monitoring (Rodriguez-Mozaz et al. 2004).

Toxicity biosensors rely on EDCs binding on estrogen receptors immobilised on the surface of a transducer. The estrogen receptor of human origin (ER-α) is the most often used. Different transduction modes such as fluorescence, cyclic voltammetry, SPR, electrophoretic mobility, have been proposed. Portable systems, mainly based on SPR detection have also been developed (Habauzit et al., 2007). Recently, a biosensor containing carbon nanotubes functionalised with the α-type human estrogen receptor and using a FET as transducer has been reported. The response time was extremely rapid (2 min) (Sanchez-Acevedo et al., 2009). Another biosensor, using impedance as transduction mode, was fabricated by immobilizing ER-α in a supported bilayer lipid membrane modified with Au nanoparticles. (Xia et al., 2010) The results indicated that the biosensor was able to detect the natural estrogen 17β-estradiol with an acceptable linear correlation ranging from 5 to 150 ng/L and a detection limit of 1 ng/L. The biosensor could also detect bisphenol A and 4-nonylphenol. Im et al. (Im et al., 2010) propose to bind ER-α receptor covalently on a gold electrode for impedimetric detection of 17β-estradiol. The detection limit was 1 μM.

Toxicity mechanism	Pollutants	Microorganisms	Transduction	Reference
Inhibition of AChE activity	Cd, Zn	*C. vulgaris*	Conductometry	Chouteau et al., 2005
Inhibition of AP activity	OPs	*C. vulgaris*	Conductometry	Chouteau et al., 2005
	Cd	*C. vulgaris*	Conductometry	Guedri et al., 2008
	Cd, Zn	*C. vulgaris*	Amperometry	Chong et al., 2008
Inhibition of respiratory activity	Antibiotics	*E.coli* JM 105	Amperometry	Mann & Mikkelsen, 2008
	Hg, Cu, Zn, Ni, phenolic compounds	*E.coli*	Amperometry	H. Wang et al., 2008
	KCN, As$_2$O$_3$, Hg^{2+}	*E.coli* DHα	Amperometry	C. Liu et al., 2009
Inhibition of photosynthetic activity	Atrazine, DCMU Formaldehyde	*C. vulgaris*	Amperometry	Shitanda et al., 2009
		C. vulgaris / *P. subcapitata*/ *C.reinhardtii*	Amperometry	Tatsuma et al., 2009
Inhibition of luminol peroxidase activity	Pb^{2+}, Hg^{2+}, Cu^{2+}	*Vibrio fischeri*	Luminescence	Komaitis et al., 2010
Genotoxicity	Nalidixic acid, mitomycin C, H$_2$O$_2$	*E.coli* RFM443 *nrdA :: luxCDABE*	Luminescence (induction)	Hwang et al., 2008
	Mitomycin C, nalidixic acid , MNNG, 4-NQQ	*E. coli* RFM443 with *recA, NrdA, dinI, sbmC, recN, sulA or alkA* promoters and *luxCDABE* reporter	Luminescence (induction)	Ahn et al., 2009
	Nalidixic acid IQ	*E. coli* RFM443 *sulA ::phoA* *S. typhimurium* TA1535 *umuC::lacZ*	Amperometry (induction of AP)	Ben Yoav et al., 2009
	Mitomycin C, ethidium bromide,H$_2$O$_2$, toluene, pyrene, benzo[a]pyrene, MMS	*Acinetobacter baylyi* ADP1 *recA::luxCDABE*	Luminescence (induction)	Song et al., 2009
	Mitomycin, H$_2$O$_2$	*E. coli* RFM443 *grpE::luxCDABE* and *recA::luxCDABE*	Luminescence (induction)	Eltzov et al., 2009
	Mitomicyn C, pentachlorophenol, H$_2$O$_2$	*E. coli* K12 *recA ::luxCDABE and ColD::luxCDABE*	Luminescence (induction)	Kotova et al., 2010

Toxicity mechanism	Pollutants	Microorganisms	Transduction	Reference
Protein damage	Phenol	*E. coli DnakA::lacZ* *E. coli grpE::lacZ*	Amperometry (induction of β-galactosidase)	Popovtzer et al., 2006
Membrane damage	Phenol	*E. coli fabA::lacZ*	Amperometry (induction of β-galactosidase)	Popovtzer et al., 2006
Heat shock	Mitomicyn C, pentachlorophenol, H_2O_2	*E. coli grpE: :luxCDABE* *E. coli plbpA::luxCDABE*	Luminescence (induction)	Kotova et al., 2010
Oxidative stress	Cd^{2+}, Cu^{2+}, Pb^{2+}, Zn^{2+} H_2O_2, menadione, selenite, arsenite, triphenyltin naphthalene	*E. coli* DH5α *pRSET::roGFP2*	Fluorescence (induction)	Arias-Barreiro et al., 2010
	Mitomicyn C, pentachlorophenol, H_2O_2	*E.coli K12 katG ::luxCDABE* and *SoxS::luxCDABE*	Luminescence (induction)	Kotova et al., 2010
	Paraquats and derivatives , H_2O_2	Various strains and promoters with *luxCDABE* reporter	Luminescence (induction)	J.Y. Lee et al., 2007

AchE: acetylcholinesterase, AP: alkaline phosphatase; DCMU: 3-(3,4-dichlorophenyl)-1,1-diethylurea, IQ : 2-amino-3-methylimidazo[4,5-f]quinoline, MNNG: 1-methyl-1-nitroso-N-methylguanidine, MMS: Methyl methanesulfonate , 4-NQQ: 4-nitroquinoline N-oxide

Table 4. Some recent examples of toxicity cell-based biosensors.

2.3.3 Immunosensors

As seen in section 2.2.2, a large number of applications of immunosensors relate to the determination of pollutants or groups of target pollutants. It is also possible to exploit them for the detection of substances, called biomarkers, that are produced by an organism following exposure to specific pollutants. Vitellogenin, for example, is a phospholipo-serum glycoprotein secreted in large quantities by fish exposed to endocrine disruptors. Its presence is suitable for identifying oestrogenotoxic effects of natural or anthropogenic substances. Vitellogenin may be detected using electrochemical, optical or piezoelectric biosensors based on carp anti-vitellogenin antibody (Bulukin et al., 2007).

2.3.4 DNA biosensors

DNA structure is extremely sensitive to the influence of environmental pollutants such as heavy metals, polycyclic aromatic compounds and PCBs. These substances possess high affinity for DNA, at the origin of mutagenic and carcinogenic effects. Biosensors, measuring the interactions between these substances and single or double strand DNA molecules immobilised on a transducer, have been developed and used for water genotoxicity assessment. Electrochemical transduction is the most commonly used (Nowicka et al., 2010). The compounds bound to DNA are detected, either directly when electroactive species are involved, or through the modification of DNA electrochemical signal. Toxicity biosensors based on the detection of DNA bases oxidation (mainly guanine, but also guanosine and adenosine) or on the degradation of the strands using an electrochemical probe, have been developed and applied to the analysis of water samples containing different types of genotoxic aquatic contaminants (metals, pesticides, PCBs, aromatic amines ...). Some of these biosensors were favorably compared to commercial genotoxic assays.

Other types biosensors using either optical (SPR, fluorescence) or mechanical transduction have been also proposed (Palchetti & Mascini, 2008).

2.3.5 Biosensors based on whole cells

Bacteria, yeasts, algae and fish cells have been also used for the development of toxicity biosensors (Baronian, 2004; Daunert et al., 2000; Hansen & Sorensen, 2001; Lagarde & Jaffrezic-Renault, 2011; Lei et al., 2006; Girotti et al., 2008; Van der Meer et al., 2010; Woutersen et al., 2010). The biosensor response may be due to a change in cell metabolism (inhibition of enzyme activity, respiration or photosynthesis), cell alteration, death, or change in the expression of certain genes (modified organisms).

Many examples of electrochemical and optical biosensors proposed for toxicity assessment may be found in the different reviews cited above. The most recent ones are given in Table 4. Optical biosensors are mainly based on luminescent modified bacteria, using typically the *recA, uvrA, NrdA* promoters for DNA damage detection, the *grpE* and *dnaKp* promoters for protein damage detection, and the *fab A* promoter for cell membrane damage (Woutersen et al., 2010).

3. Conclusion

Despite the large number of works carried out on the field of biosensors for water analysis, and although they have many benefits, very few systems have so far been marketed, unlike bioassays. Most commercial biosensors are versatile and suitable for applications in various fields such as environment, biological analysis or medical (Rodriguez-Mozaz et al., 2005).

Significant efforts still have to be done to obtain selective, robust, rapid and sensitive tools usable in the field. The main limitation of the proposed systems come from the biological elements. Current developments include enhancement of their sensitivity, selectivity and their stability by genetic engineering (Girotti et al., 2008; Campas et al., 2009; Conroy et al., 2009). Recent progress in this area as well as in data numerisation, transmission and processing allows now the construction of arrays of microorganisms or of enzymes arranged on a single detection platform for the determination of several parameters at the same time. In parallel, the development of new biomimetic receptors such as that molecularly imprinted polymers (MIP) or aptamers (synthetic oligonucleotides) is expanding to overcome the fragility of natural bioreceptors (Wang et al., 2007; Guan et al., 2008). Methods allowing a more efficient immobilisation of the bioreceptor will also have to be developed to improve the robustness and sensitivity of biosensors. The exploration of new materials, including gold nanoparticles, carbon nanotubes or quantum dots is an extremely promising route to achieve this goal.

Essential progress has also been made in recent years in the miniaturization of transducers (nanoelectrodes, nanowaveguides, BioMEMS) and will contribute to reduce significantly the amount of biological entity required, but also to improve integration of the systems in labs on chips (Ligler, 2009; Wei et al., 2009).

4. References

Ahn, J.-M.; Hwang, E.T.; Youn, C.-H.; Banub, D.L.; Kim B.C.; Niazi, J.H. & M.B. Gu (2009). Prediction and classification of the modes of genotoxic actions using bacterial biosensors specific for DNA damages. *Biosensors & Bioelectronics*, Vol.25, No.4, pp. 767–772, ISSN 0956-5663

Arduini, F.; Amine, A.; Moscone, D. & Palleschi G. (2010) Biosensors based on cholinesterase inhibition for insecticides, nerve agents and aflatoxin B1 detection (review). *Microchimica Acta*, Vol.170, No.3-4, pp.193–214, ISSN 0026-3672

Allan, I.J.; Vrana B.; Greenwood, R.; Mills G.A.; Roig B. & Gonzalez, C. (2006). A "toolbox" for biological and chemical monitoring requirements for the European Union's Water Framework Directive. *Talanta*, Vol.69, No.2, pp.302–322, ISSN: 0039-9140

Arias-Barreiro, C.R.; Okazaki, K.; Koutsaftis, A.;. Inayat-Hussain, S.H.; Tani, A.; Katsuhara, M.; Kimbara, K. & Mor, I.C. (2010). A Bacterial Biosensor for Oxidative Stress Using the Constitutively Expressed Redox-Sensitive Protein roGFP2. *Sensors*, Vol.10, No.7, pp. 6290-6306, ISSN 1424-8220

Badihi-Mossberg, M.; Buchner, V. & Rishpon, J. (2007). Electrochemical biosensors for pollutants in the environment. *Electroanalysis*, Vol. 19, No.19-20, pp. 2015-2028, ISSN 1521-4109

Bakker E. & Pretsch, E. (2008). Nanoscale potentiometry. *Trends in Analytical Chemistry*, Vol.27, No.7 , pp.612-618, ISSN: 0165-9936

Banik R.M.; Mayank; Prakash, R. & Upadhyay, S.N. (2008). Microbial biosensor based on whole cell of *Pseudomonas sp.* for online measurement of *p*-nitrophenol. *Sensors Actuators B*, Vol.131, No.1, pp.295-300, ISSN: 0925-4005

Barcelo, D. & Hansen, P.-D. (Eds). (2009). *Handbook of environmental chemistry. Volume 5J: Biosensors for environmental monitoring of aquatic systems*, Springer, Berlin, Heidelberg, ISBN 978-3-540-00278-9

Baronian, K.H.R. (2004) The use of yeast and moulds as sensing elements in biosensors. *Biosensors & Bioelectronics*, Vol.19, No.9, pp. 953-962, ISSN 0956-5663

Ben-Yoav H.; Biran, A; Pedahzur, R; Belkin S.; Buchinger, S.; Reifferscheid, G. & Shacham-Diamand, Y. (2009). A whole cell electrochemical biosensor for water genotoxicity bio-detection. *Electrochimica Acta* 54:6113-6118

Bratov A.; Abramova N. & Ipatov, A. (2010). Recent trends in potentiometric sensor arrays-A review. *Analytica Chimica Acta*, Vol.678, No.2, pp.149-159, ISSN 0003-2670

Bucur, B.; Fournier, D.; Danet, A. & Marty, J.-L. (2006). Biosensors based on highly sensitive acetylcholinesterases for enhanced carbamate insecticides detection. *Analytica Chimica Acta*, Vol.562, No1, pp. 115-121, ISSN 0003-2670

Bulukin, E.; Meucci, V.; Minunni, M.; Pretti, C.; Intorre, L.; Soldani, G. & Mascini, M. (2007). An optical immunosensor for rapid vitellogenin detection in plasma from carp (Cyprinus carpio). *Talanta*, Vol.72, No.2, pp. 785-790, ISSN: 0039-9140

Buonasera K.; Pezzotti, G.; Scognamigli, V.; Tibuzzi, A. & Giardi, M.T. (2010). New platform of biosensors for prescreening of pesticide residues to support laboratory analyses. *Journal of Agricultural Food and Chemistry*, Vol.58, No.10, pp.5982–5990, ISSN 0021-8561

Campas, M.; Prieto-Simon, B. & Marty, J.L. (2009). A review of the use of genetically engineered enzymes in electrochemical biosensors. *Seminars Cell & Developmental Biology*, Vol.20, No.1, pp. 3-9, ISSN: 1084-9521

Carralero, V.; Mena, M.L.; Gonzalez-Cortes, A.; Yanez-Sedeno, P. & Pingarron, J.M. (2006). Development of a high analytical performance-tyrosinase biosensor based on a composite graphite–Teflon electrode modified with gold nanoparticles. *Biosensors & Bioelectronics*, Vol.22, No.5, pp.730–736, ISSN 0956-5663

Chen, H.; Mousty, C.; Cosnier, S.; Silveira, C.M.; Moura, J.J.G. & Almeida, M.G. (2007) Highly sensitive nitrite biosensor based on the electrical wiring of nitrite reductase by [ZnCr-AQS] LDH. *Electrochemical Communications*, Vol.9, No.9, pp. 2240-2245, ISSN: 1388-2481

Chen, J. & Jin, Y. (2010). Sensitive phenol determination based on co-modifying tyrosinase and palygorskite on glassy carbon electrode. *Microchimica Acta*, Vol.169, No.3-4, pp.249–254, ISSN 0026-3672

Chen, L.; Zeng, G.; Zhang, Y.; Tang, L.; Huang, D.; Liu, C.; Pang, Y. & Luo, J. (2010). Trace detection of picloram using an electrochemical immunosensor based on three-dimensional gold nanoclusters. *Analytical Biochemistry*, Vol. 407, No.2 172–179, ISSN 0003-2697

Chiappini, S.A.; Kormes, D.J., Bonetto, M.C.; Sacco, N. & Cortona, E. (2010). A new microbial biosensor for organic water pollution based on measurement of carbon dioxide production. *Sensors and Actuators B*, Vol.148, No.1, pp.103-109, ISSN 0925-4005

Chong, K.F.; Loh K.P.; Ang, K. & Ting, Y.P. (2008). Whole cell environmental biosensor on diamond. *Analyst*, Vol.133, No.6, pp.739–743, ISSN 0003-2654

Chouteau, C.; Dzyadevych, S.; Durrieu ,C. & Chovelon, J.M. (2005). A bi-enzymatic whole cell conductometric biosensor for heavy metal ions and pesticides detection in water samples. *Biosensors & Bioelectronics*, Vol.21, No.2, pp.273–281, ISSN 0956-5663

Conroy, P.J.; Hearty, S.; Leonard, P. & O'Kennedy, R.J. (2009) Antibody production, design and use for biosensor-based applications. *Seminars in Cell & Developmental Biology*, Vol.20, No.1, pp. 10-26, ISSN: 1084-9521

Cortina, M.; del Valle, M. & Marty J.-L. (2008) Electronic Tongue Using an Enzyme Inhibition Biosensor Array for the Resolution of Pesticide Mixtures, *Electroanalysis*, Vol.20, No. 1, pp.54-60, ISSN 1521-4109

D'Souza, S.F. (2001). Immobilization and stabilization of biomaterials for biosensor applications. *Applied Biochemistry and Biotechnology*, Vol.96, No.1-3, pp. 225-238, ISSN 0273-2289

Daunert, S.; Barrett, G.; Feliciano, J.S.; Shetty, R.S.; Shrestha, S. & Smith-Spencer, W. (2000). Genetically engineered whole-cell sensing systems: coupling biological recognition with reporter genes. *Chemical Reviews*, Vol.100, No.7, pp. 2705-2738, ISSN 0009-2665

De Albuquerque, Y. D. T. & Ferreira, L. F. (2007). Amperometric biosensing of carbamate and organophosphate pesticides utilizing screen-printed tyrosinase-modified electrodes. *Analytica Chimica Acta*, Vol.596, No.210–221, ISSN: 0003-2670

Deo, R. P.; Wang, J.; Block, I.; Mulchandani, A.; Joshi, K.A.; Trojanowicz, M.; Scholz, F.; Chen, W. & Lin Y. (2005). Determination of organophosphate pesticides at a carbon nanotube/organophosphorus hydrolase electrochemical biosensor, *Analytica Chimica Acta*, Vol.530, No.2, pp.185–189, ISSN: 0003-2670

Du, D.; Huang X.; Cai, J. & Zhang, A. (2007). Amperometric detection of triazophos pesticide using acetylcholinesterase biosensor based on multiwall carbon nanotube–chitosan matrix, *Sensors and Actuators B*, Vol.127, No.2, pp. 531–535, ISSN: 0925-4005

Du, D.; Chen, S.; Cai, J. & Zhang, A. (2008). Electrochemical pesticide sensitivity test using acetylcholinesterase biosensor based on colloidal gold nanoparticle modified sol-gel interface. *Talanta*, Vol.74, No.4, pp. 766-772, ISSN: 0039-9140

Du, D.; Wang, J.; Smith, J. N.; Timchalk, C. & Lin, Y. 2009. Biomonitoring of Organophosphorus Agent Exposure by Reactivation of Cholinesterase Enzyme Based on Carbon Nanotube-Enhanced Flow-Injection Amperometric Detection. *Analytical Chemistry*, Vol.81, No.22, pp.9314–9320, ISSN: 0003-2700

Du; D.; Chen, W.; Zhang, W.; Liu, D.; Li, H.; Lin Y. (2010). Covalent coupling of organophosphorus hydrolase loaded quantum dots to carbon nanotube/Au nanocomposite for enhanced detection of methyl parathion, *Biosensors & Bioelectronics* , Vol.25, No.6, pp. 1370–1375, ISSN 0956-5663

Eguilaz, M.; Moreno-Guzman, M.; Campuzano, S.; Gonzalez-Cortes, A.; Yanez-Sedeno, P. & Pingarron, J.M. (2010). An electrochemical immunosensor for testosterone using functionalized magnetic beads and screen-printed carbon electrodes. *Biosensors & Bioelectronics*, Vol.26, No.2, pp.517–522, ISSN 0956-5663

Eltzov, E.; Marks, R.S.; Voost, S.; Wullings, B.A. & Heringa M.B. (2009) Flow-through real time bacterial biosensor for toxic compounds in water. *Sensors & Actuators B*, Vol.142, No.1, pp. 11–18, ISSN: 0925-4005

Evtugyn, G.A.; Eremin, S.A.; Shaljamova, R.P.; Ismagilova, A.R. & Bidnikov, H.C. (2006). Amperometric immunosensor for nonylphenol determination based on peroxidase indicating reaction. *Biosensensors & Bioelectronics*, Vol.22, No 1, pp. 56-62, ISSN 0956-5663

Fan, X.; White, I.M.; Shopova, S.I.; Zhu, H.; Suter, J.D. & Sun, Y. (2008). Sensitive optical biosensors for unlabeled targets: a review, *Analytica Chimica Acta*, Vol.620, No.1-2, pp. 8-26, ISSN 0003-2670

Faridbod, F.; Ganjali, M.R.; Dinarvand, R. & Norouzi, P. (2008). Developments in the field of conducting and non-conducting polymer based potentiometric membrane sensors for ions over the past decade. *Sensors*, Vol.8, No.8, pp.2331-2412, ISSN 1424-8220

Farre, M.; Brix, R. & Barcelo, D. (2005).Screening water for pollutants using biological techniques under European Union funding during the last 10 years. *Trends in Analytical Chemistry*, Vol.24, No.6, pp. 532-545, ISSN: 0165-9936

Farre, M.; Martínez; E.; Ramon J.; Navarro A.; Radjenovic, J.; Mauriz, E.; Lechuga, L.; Marco, M.P. & Barcelo, D. (2007). Part per trillion determination of atrazine in natural water samples by a surface plasmon resonance immunosensor. *Analytical & Bioanalytical Chemistry*, Vol.388, No1, pp.207–214

Gan, N.; Yang, X.; Xie, D.; Wu, Y. & Wen, W. (2010). A Disposable Organophosphorus Pesticides Enzyme Biosensor Based on Magnetic Composite Nano-Particles Modified Screen Printed Carbon Electrode, *Sensors*, Vol.10, No.1, pp.625-638, ISSN 1424-8220

Gilbert, L.; Browning, S.; Jenkins, A.T.A. & Hart, J.P. (2010). Studies towards an amperometric phosphate ion biosensor for urine and water analysis. *Microchimica Acta*, Vol.170, No.3-4, pp.331–336, ISSN 0026-3672

Girotti, S.; Ferri, E.N.; Fumo, M.G. & Maiolini, E. (2008). Monitoring of environmental pollutants by bioluminescent bacteria. *Analytica Chimica Acta*, Vol.608, No.1, pp. 2-29, ISSN 0003-2670

Glazier, S.A.; Campbell, E.R. & Campbell, W.H. (1998). Construction and characterization of nitrate reductase-based amperometric electrode and nitrate assay of fertilizers and drinking water. *Analytical Chemistry*, Vol.70, No.8, pp. 1511-1515, ISSN: 0003-2700

Gong J.; Wang, L. & Zhang L. (2009) Electrochemical biosensing of methyl parathion pesticide based on acetylcholinesterase immobilized onto Au–polypyrrole interlaced network-like nanocomposite. *Biosensors & Bioelectronics*, Vol.24, No.7, pp. 2285–2288, ISSN 0956-5663

Goure, J.-P. & Blum, L. (2009). Biosensors and chemical sensors based upon guided optics. In: *Chemical and biological microsensors: Applications in Fluid Media*, J. Fouletier & P. Fabry (Eds), John Wiley & Sons, ISBN: 978-1-84821-142-1

Grieshaber, D.; MacKenzie, R.; Vörös, J. & Reimhult, E. (2008). Electrochemical biosensors – Sensor principles and architectures, *Sensors*, Vol.8, No.3, pp. 1400-1458, ISSN 1424-8220

Guan, G.; Liu, B.; Wang, Z. & Zhang, Z. (2008) Imprinting of molecular recognition sites on nanostructures and its applications in chemosensors. *Sensors*, Vol.8, No.12, pp. 8291-8320, ISSN 1424-8220

Guedri, H.; Durrieu, C. (2008). A self-assembled monolayers based conductometric algal whole cell biosensor for water monitoring. *Microchimica Acta*, Vol.163, No.3-4, pp.179–184, ISSN 0026-3672

Habauzit, D.; Chopineau, J. & Roig, B. (2007) SPR-based biosensors: a tool for biodetection of hormonal compounds. *Analytical & Bioanalytical Chemistry*, Vol.387, No.4, pp. 1215-1223, ISSN 1618-2642

Han, T.S.; Sasaki, S.; Yano, K.; Ikebukuro, K.; Atsushi, K.; Nagamune, T. & Karube, I. (2002). Flow injection microbial trichloroethylene sensor. *Talanta*, Vol.57, No.2, pp.271–276, ISSN: 0039-9140

Hansen, L.H. & Sorensen, S.J. (2001). The use of whole-cell biosensors to detect and quantify compounds or conditions affecting biological systems. *Microbial Ecology*, Vol.42, No.4, pp. 483-494 , ISSN 0095-3628

Hassen, K.M. & Thundat, T. (2005) Microcantilever biosensors. *Methods*, Vol.37, No.1, pp. 57-64, ISSN: 1046-2023

Hervas Perez, J.P.; Sanchez-Paniagua Lopez M., Lopez-Cabarcos, E. & Lopez-Ruiz, B. (2006) Amperometric tyrosinase biosensor based on polyacrylamide microgels. *Biosensors & Bioelectronics*, Vol.22, No.3, pp.429–439, ISSN 0956-5663

Hildebrandt, A.; Jordi Ribas, Bragos, R.; Marty, J.-L.; Tresanchez, M. & Lacorte, S. (2008). Development of a portable biosensor for screening neurotoxic agents in water samples. *Talanta*, Vol.75, No.5, pp.1208–1213, ISSN: 0039-9140

Hleli, S.; Martelet, C.; Abdelghani, A.; Bessueille, F.; Errachid, A.; Samitier, J.; Burais, N. & Jaffrezic-Renault, N. (2006) Atrazine analysis using an impedimetric immunosensor based on mixed biotinylated self-assembled monolayer. *Sensors Actuators B*, Vol.113, No.2, pp. 711-717, ISSN: 0925-4005

Hnaien, M.; Lagarde, F.; Bausells, J.; Errachid, A.; Jaffrezic-Renault N. (2011). A new bacterial biosensor for trichloroethylene detection based on a three dimensional carbon nanotubes bioarchitecture. *Analytical & Bioanalytical Chemistry*, special issue "Microorganisms in Analysis", in press, DOI 10.1007/s00216-010-4336-x, ISSN 1618-2642

Hoa, X.D.; Kirk, A.G. & Tabrizian, M. (2007) Towards integrated and sensitive surface plasmon resonance biosensors: a review of recent progress. *Biosensors & Bioelectronics*, Vol.23, No.2, pp. 151-160, ISSN 0956-5663

Hwang, E.T.; Ahn, J.-M.; Kim, B.C. & Gu, M.B. (2008) Construction of a *nrdA::luxCDABE* Fusion and Its Use in *Escherichia coli* as a DNA Damage Biosensor. *Sensors*, Vol.8, No.2, pp.1297-1307, ISSN 1424-8220

Im, J.-E.; Han, J.-A.; Kim, B.K.; Han, J.H.; Park, T.S.; Hwang, S.; Cho, S.I.; Lee , W.-Y. & Kim, Y.-R. (2010). Electrochemical detection of estrogen hormone by immobilized estrogen receptor on Au electrode. *Surface & Coatings Technology*, Vol.205, Sup.1, pp.S275–S278, ISSN 0257-8972

Ino, K.; Kitagawa, Y.; Watanabe, T.; Shiku, H.; Koide, M.; Itayama, T.; Yasukawa, T. & Matsue, T. (2009). Detection of hormone active chemicals using genetically engineered yeast cells and microfluidic devices with interdigitated array electrodes. *Electrophoresis*,Vol.30, No.19, pp.3406-3412, ISSN 0173-0835

Jaffrezic-Renault, N. & Dzyadevych, S.V. (2008) Conductometric microbiosensors for environmental monitoring. *Sensors*, Vol.8, No.4, pp. 2569-2588, ISSN 1424-8220

Jaffrezic-Renault, N. (2001). New trends in biosensors for organophosphorous pesticides. *Sensors*, Vol.1, No.2, pp. 60-74, , ISSN 1424-8220

Jang, E.; Son, K. J.; Kimb, B. & Koh, W.-G. (2010). Phenol biosensor based on hydrogel microarrays entrapping tyrosinase and quantum dots. *Analyst*, Vol.135, No.11, pp.2871–2878, ISSN 0003-2654

Joshi, K.A.; Prouza; Kum, M.M.; Wang, J.; Tang, J.; Haddon, R.; Chen, W. & Mulchandani,A. (2006) V-Type Nerve Agent Detection Using a Carbon Nanotube-Based Amperometric Enzyme Electrode, *Analytical Chemistry*, Vol.78, No.1, pp.331-336, ISSN: 0003-2700

Karnati, C.; Du, H.; Ji, H.-F.; Xu, X.; Lvov, Y.; Mulchandani, A.; Mulchandani, P. & Chen,W. (2007). Organophosphorus hydrolase multilayer modified microcantilevers for organophosphorus detection. *Biosensors & Bioelectronics*, Vol.22, No.11, pp. 2636–2642, ISSN 0956-5663

Katz, E. & Willner, I. (2003). Probing biomolecular interactions at conductive and semiconductive surfaces by impedance spectroscopy: Routes to impedimetric immunosensors, DNA-sensors, and enzyme biosensors.*Electroanalysis*, Vol.15, No.11, pp. 913-947, ISSN 1040-0397

Khadro, B.; Namour, P.; Bessueille, F.; Léonard, D. & Jaffrezic-Renault, N. (2009). Validation of a conductometric bienzyme biosensor for the detection of proteins as marker of organic matter in river samples. *Journal of Environmental Sciences*, Vol.21, No.4, pp. 545-551, ISSN 1001-0742

Kim, S. J.; Gobi, K V, Iwasaka, H., H. Tanaka & N. Miura (2007). Novel miniature SPR immunosensor equipped with all-in-one multi-microchannel sensor chip for detecting low-molecular-weight analytes. *Biosensors & Bioelectronics*, Vol. 23, No.5, pp. 701–707, ISSN 0956-5663

Kochana, J.; Gala, A.; Parczewski, A. & Adamski, J. (2008). Titania sol-gel-derived tyrosinase-based amperometric biosensor for determination of phenolic compounds in water samples. Examination of interference effects. *Analytical & Bioanalytical Chemistry*, Vol.391, No.4, pp. 1275-1281, ISSN 1618-2642

Komaitis, E.; Vasiliou, E.; Kremmydas, G.; Georgakopoulos, D.G. & Georgiou, C. (2010). Development of a Fully Automated Flow Injection Analyzer Implementing Bioluminescent Biosensors for Water Toxicity Assessment . *Sensors*, Vol.10, No.8, pp.7089-7098, ISSN 1424-8220

Kong, L.; Huang, S.; Yue, Z.; Peng, B.; Li, M. & Zhang, J. (2009). Sensitive mediator-free tyrosinase biosensor for the determination of 2,4-dichlorophenol. *Microchimica Acta*, Vol.165, No.1-2, pp.203–209, ISSN 0026-3672

Kotova, V.Y.; Manukhov, I. V. & Zavilgelskii G. B. (2010). Lux_Biosensors for Detection of SOS_Response, Heat Shock, and Oxidative Stress. *Applied Biochemistry and Microbiology*,Vol. 46, No. 8, pp. 781–788, ISSN 0175-7598

Kumar, S.; Kundu, S.; Pakshirajan, K. & Dasu, V.V. (2008). Cephalosporins determination with a novel microbial biosensor based on permeabilized *Pseudomonas aeruginosa* whole cells. *Applied Biochemistry & Biotechnology*, Vol.151, No.2-3, pp.653–664, ISSN 0175-7598

Kwan, R. C. H.; Leung, H. F.; Hon, P.Y.T.; Barford J. P. & Renneberg, R. (2005) A screen-printed biosensor using pyruvate oxidase for rapid determination of phosphate in synthetic wastewater *Applied Microbiology and Biotechnology*, Vol. 66, No.4, pp. 377–383, ISSN 0175-7598

Lagarde, F.; Jaffrezic-Renault N. (2011) Cell-based electrochemical biosensors for water quality assessment. *Analytical & Bioanalytical Chemistry*, special issue "Microorganisms in Analysis", in press, DOI 10.1007/s00216-011-4816-7, ISSN 1618-2642

Lanyon, Y.H.; Tothill, I.E. & Mascini, M. (2006). An amperometric bacterial biosensor based on gold screen-printed electrodes for the detection of benzene. *Analytical Letters*, Vol.39, No. 7-8, pp.669-1681, ISSN 0003-2719

Lee, C.-S.; Kim, S.K. & Kim, M. (2009). Ion-sensitive field-effect transistor for biological sensing, *Sensors*, Vol.9, No.9, pp.7111-7131, ISSN 1424-8220

Lee , J.H; Youn, C.H.; Kim, B.C. & Gu, M.B. (2007). An oxidative stress-specific bacterial cell array chip for toxicity analysis. *Biosensors & Bioelectronics*, Vol. 22, No.9-10, pp.2223–2229, ISSN 0956-5663

Lee, J.H.; Park, J.Y.; Min, K.; Cha, H.J.; Choi, S. S. & Yoo, Y.J. (2010). A novel organophosphorus hydrolase-based biosensor using mesoporous carbons and carbon black for the detection of organophosphate nerve agents, *Biosensors and Bioelectronics*, Vol.25 , No.7, pp.1566–1570, ISSN 0956-5663

Lee, Y.-J.; Lyu, Y.-K.; Choi, H. N. & Lee W.-Y. (2007). Amperometric Tyrosinase Biosensor Based on Carbon Nanotube – Titania – Nafion Composite Film. *Electroanalysis*, Vol.19, No.10, pp.1048-1054, ISSN 1521-4109

Lei ,Y.; Mulchandani, P.; Wang, J.; Chen, W. & Mulchandani, A. (2005). A highly sensitive and selective amperometric microbial biosensor for direct determination of *p*-nitrophenyl-substituted organophosphate nerve agents. *Environmental Science & Technology*, Vol.39, No.22, pp.8853-8857, ISSN 0013-936X

Lei, Y.; Chen, W. & Mulchandani, A. (2006) Microbial biosensors. *Analytica Chimica Acta*, Vol.568, No.1-2, pp. 200-210, ISSN 0003-2670

Lei, Y.; Mulchandani, P.; Chen, W. & Mulchandani, A. (2007). Biosensor for direct determination of fenitrothion and EPN using recombinant *Pseudomonas putida* JS444 with surface-expressed organophosphorous hydrolase. 2. Modified carbon paste electrode. *Applied Biochemistry & Biotechnology*, Vol.136, No.3, pp.243-250, ISSN 0273-2289

Ligler, F.S. (2009) Perspective on optical biosensors and integrated sensor systems. *Analytical Chemistry*, Vol.81, No. 2, pp. 519-526, ISSN: 0003-2700

Lin, L.; Xiao, L.-L.; Huang, S.; Zhao, L.; Cui, J.-S.; Wang, X.-H. & Chen X. (2006). Novel BOD optical fiber biosensor based on co-immobilized microorganisms in ormosils matrix. *Biosensors & Bioelectronics* , Vol.21, No.9, pp.1703–1709, ISSN 0956-5663

Liu, C.; Suna, T.; Xub, X. & Dong S (2009). Direct toxicity assessment of toxic chemicals with electrochemical method. *Analytica Chimica Acta*, Vol.641, No.1-2, pp.59-63, ISSN 0003-2670

Liu, G. & Lin, Y. (2006). Biosensor Based on Self-Assembling Acetylcholinesterase on carbon Nanotubes for Flow Injection/Amperometric Detection of Organophosphate Pesticides and Nerve Agents. *Analytical Chemistry*, Vol.78, No.3, pp. 835-843, ISSN: 0003-2700

Liu, L.; Shang, L.; Liu, C.; Liu, C.; Zhang, B. & Dong, S. (2010). A new mediator method for BOD measurement under non-deaerated condition. *Talanta*, Vol.81, No.4-5, pp.1170-1175, ISSN: 0039-9140

Liu, S.; Yuan, L.; Yue, X.; Zheng, Z. & Tang, Z. (2008) Recent advances in nanosensors for organophosphate pesticide detection. *Advanced Powder Technology*, Vol.19, No.5, pp. 419-441, ISSN 0921-8831

Long, F.; He, M.;. Shi, H.C & Zhu, A.N. (2008). Development of evanescent wave all-fiber immunosensor for environmental water analysis. *Biosensors & Bioelectronics*, Vol. 23, No.7, pp.952–958, ISSN 0956-5663

Luckarift, H. R.; Balasubramanian, S.; Paliwal, S.; Johnson, G.R. & Simonian, A.L. (2007). Enzyme-encapsulated silica monolayers for rapid functionalization of a gold surface, *Colloids and Surfaces B: Biointerfaces* , Vol.58, No.1, pp. 28–33, ISSN 0927-7765

Luong, J.H.T.; Maleb, K.B. & Glennon, J.D. (2009). Boron-doped diamond electrode: synthesis, characterization, functionalization and analytical applications. *Analyst* , Vol.134, No.10, pp.1965-1979, ISSN 0003-2654

Luppa, P.B.; Sokoll, L.J. & Chan, D.W. (2001) Immunosensors – Principles and applications to clinical chemistry. *Clinica Chimica Acta*, Vol.314, No.1-2, pp. 1-26, ISSN0009-8981

Mann, T.S. & Mikkelsen, S.R. (2008). Antibiotic susceptibility testing at a screen-printed carbon electrode array. *Analytical Chemistry*, Vol.80, No.3, pp.843-848, ISSN 1572-6657

Marchesini, G.R.; Meulenberg, E.; Haasnoot, W. & Irth, H. Biosensor immunoassays for the detection of bisphenol A. (2005). *Analytica Chimica Acta*, Vol.528, No.1, p. 37-45, ISSN 0003-2670

Mauriz, E.; Calle, A.; Lechuga, L.M.; Quintana, J.; Montaya, A. & Manclus, J.J. (2006). Real-time detection of chlorpyrifos at part per trillion levels ground, surface and drinking water samples by a portable surface plasmon resonance immunosensor. *Analytica Chimica Acta.*, Vol.561, No.1-2, pp. 40-47, ISSN 0003-2670

Mauriz, E.; Calle, A.; Manclus, J.J.; Montoya, A.; Hildebrandt, A.; Barcelo, D. & Lechuga, L.M. (2007).Optical immunosensor for fast and sensitive detection of DDT and related compounds in river water samples. *Biosensors & Bioelectronics*,Vol.22,No.7, pp. 1410-1418, ISSN 0956-5663

Mita, D.G.; Attanasio, A.; Arduini, F.; Diano, N.; Grano, V.; Bencivenga, U.; Rossi, S.; Amine, A. & Moscone, D. (2007) Enzymatic determination of BPA by means of tyrosinase immobilized on different carbon carriers. *Biosensors & Bioelectronics*, Vol.23, No.1, pp.60-65, ISSN 0956-5663

Mitchell, J. 2010. Small Molecule Immunosensing Using Surface Plasmon Resonance. *Sensors*, Vol.10, No.8, pp.7323-7346, ISSN 1424-8220

Mousty, C.; Cosnier, S.; Shan, D. & Mu, S.L. (2001) Trienzymatic biosensor for the determination of inorganic phosphate. *Analytica Chimica Acta*, Vol.443, No.1, pp. 1-8, ISSN 0003-2670

Nakamura, H.; Suzuki, K.; Ishikuro, H.; Kinoshita, S.; Koizumi, R.; Okuma, S.; Gotoh, M. & Karube, I. (2007). A new BOD estimation method employing a double-mediator system by ferricyanide and menadione using the eukaryote *Saccharomyces cerevisiae*. *Talanta*, Vol.72, No.1, pp.210-216, ISSN: 0039-9140

Nakamura, H. (2010). Recent organic pollution and its biosensing methods. *Analytical Methods*, Vol.2, No.5, pp.430–444, ISSN: 1759-9660

Neufeld T, Biran D, Popovtzer R, Erez T, Ron EZ, Rishpon J (2006) Genetically engineered pfabA pfabR bacteria: an electrochemical whole cell biosensor for detection of water toxicity. Anal Chem 78:4952-4956

Notsu, H.; Tatsuma, T. & Fujishima, A. (2002) Tyrosinase-modified boron doped diamond electrodes for the determination of phenol derivatives. *Journal of Electroanalytical Chemistry*, Vol.523, No.1, pp. 86-92, ISSN 1572-6657

Nowicka, A.M.; Kowalczyk, A.; Stojek, Z. & Hepel, M. (2010) Nanogravimetric and voltammetric DNA-hybridization biosensors for studies of DNA damage by common toxicants and pollutants. *Biophysical Chemistry*, Vol.146, No.1, pp. 42-53, ISSN 0301-4622.

Palchetti, I. & Mascini, M. (2008). Nucleic acid biosensors for environmental pollution monitoring. *Analyst*, Vol.133, No.7, pp. 846-854, ISSN 0003-2654

Perez Lopez, B. & Merkoci, A. (2009) Improvement of the electrochemical detection of catechol by the use of a carbon nanotube based biosensor. *Analyst*, Vol.134, No.1, pp.60–64, ISSN 0003-2654

Poghossian, A.; Ingebrandt, S.; Offenhäusser, A. & Schöning, M.J. (2009). Field-effect devices for detecting cellular signals. Seminars in Cell Developmental Biology, Vol.20, No.1, pp:41–48, ISSN 1084-9521

Ponomareva, O. N., Arlyapov, V.A.; Alferov, V.A. & Reshetilov, A.N. (2011). Microbial Biosensors for Detection of Biological Oxygen Demand (a Review), *Applied Biochemistry and Microbiology,*Vol.47, No.1, pp.1–11, ISSN 0003-6838

Popovtzer R, Neufeld T, Ron EZ, Rishpon J, Shacham-Diamand Y (2006) Electrochemical detection of biological reactions using a novel nano-bio-chip array. *Sensors Actuators B* 19:664-672

Prodromidis, M.I. (2010). Impedimetric immunosensors—A review. *Electrochimica Acta*, Vol.55, pp. 4227–4233

Quan, D.; Nagarale, R.K. & Shin, W. (2010). A Nitrite Biosensor Based on Coimmobilization of Nitrite Reductase and Viologen-Modified Polysiloxane on Glassy Carbon Electrode, *Electroanalysis*, Vol.22, No. 20, pp.2389-2398, ISSN 0956-5663

Rahman, M.A.; Shiddiky, M.J.A.; Park, J.-S. & Shim, Y.-B. (2007). An impedimetric immunosensor for the label-free detection of bisphenol A. *Biosensors & Bioelectronics*, Vol.22, No.11, pp.2464–2470, ISSN 0956-5663

Ramanathan, M. & Simonian A.L. (2007) Array biosensor based on enzyme kinetics monitoring by fluorescence spectroscopy: Application for neurotoxins detection, *Biosensors and Bioelectronics*, Vol.22, No.12, pp.3001–3007, ISSN 0956-5663

Rodriguez-Mozaz, S.; Marco, M.P.; Lopez de Alda, M. & Barcelo, D. (2004) Biosensors for environmental monitoring of endocrine disruptors: a review article. *Analytical &. Bioanalytical Chemistry*, Vol.378, No.3, pp. 588-598, ISSN 1618-2642

Rodriguez-Mozaz, S.; Lopez de Alda, M.; Marco, M.P. & Barcelo, D. (2005) Biosensors for environmental monitoring. A global perspective. *Talanta*, Vol.65, No.2, pp. 291-297, ISSN: 0039-9140

Rodriguez-Mozaz, S.; Lopez de Alda, M.; Marco, M.P. & Barcelo, D. (2007) Advantages and limitations of on-line solid phase extraction coupled to liquid chromatography-mass spectrometry technologies versus biosensors for monitoring of emerging contaminants in water. *Journal of Chromatography A*, Vol.1152, No.1-2, pp. 97-115, ISSN 0021-9673

Rogers, K.R. (2006). Recent advances in biosensor techniques for environmental monitoring, *Analytica Chimica Acta*, Vol.568, No.1-2, pp. 222-231, ISSN 0003-2670

Sakaguchi, T.; Morioka, Y.; Yamasaki, M.; Iwanaga, J.; Beppu, K.; Maedac, H.; Morita, Y. & Tamiya E. (2007). Rapid and onsite BOD sensing system using luminous bacterial cells-immobilized chip. *Biosensors & Bioelectronics*, Vol.22, No.7, pp.1345–1350

Sanchez-Acevedo, Z.C.; Riu, J. & Rius, F.X. (2009) Fast picomolar selective detection of bisphenol A in water using a carbon nanotube field effect transistor functionalized with estrogen receptor-α. *Biosensensors & Bioelectronics*, Vol.24, No9, pp. 2842-2846, ISSN 0956-5663

Scarano, S.; Mascini, M.; Turner, A.P.F. & Minunni, M. (2010). Surface plasmon resonance imaging for affinity-based biosensors, *Biosensors & Bioelectronics*,Vol.25, No5, pp. 957-966, ISSN 0956-5663

Schöning, M.J. & Poghossian, A. (2006). Bio FEDs (Field-Effect Devices): State-of-the art and new directions. *Electroanalysis*, Vol.18, No.19-20, pp. 1893-1900, ISSN 0956-5663

Sepulveda, B.; Sanchez del Rio, J.; Moreno, M.; Blanco, F.J.; Mayora, K.;Dominguez, C. & Lechuga, L.M. (2006) Optical biosensor microsystems based on the integration of highly sensitive Mach-Zehnder interferometer devices. *Journal of Optics A*, Vol.8, No7, pp. S561-S566, ISSN 1464-4258

Shitanda, I.; Takamatsu, S.; Watanabe, K. & Itagaki, M. (2009). Amperometric screen-printed algal biosensor with flow injection analysis system for detection of environmental toxic compounds. *Electrochimica Acta*, Vol. 54, No.21, pp.4933–4936, ISSN 0013-4686

Shulga, O. & Kirchhoff, J.R. (2007). An acetylcholinesterase enzyme electrode stabilized by an electrodeposited gold nanoparticle layer, *Electrochemistry Communications*, Vol.9, No.5, pp.935–940, ISSN: 1388-2481

Sohail M. & Adeloju, S.B. (2009) Fabrication of Redox-Mediator Supported Potentiometric Nitrate Biosensor with Nitrate Reductase, *Electroanalysis*, Vol.21, No.12, pp.1411–1418, ISSN 0956-5663

Soldatkin, O.O.; Pavluchenko, O.S.; Kukla, O.L., Kucherenko, I.S., Peshkova, V.M.; Arkhypova, V.M.; Dzyadevych, S.V. , Soldatkin, A.P. & El'skaya, A.V. (2009). Application of enzyme multibiosensor for toxicity analysis of real water samples of different origin. *Biopolymers and cell*. Vol. 25, No. 3, pp.204-209, ISSN 1993-6842

Song, Y .; Li, G., Thorton, S.F., Thompson, I.P.; Banwart, S.A.; Lerner, D.N. & Huang W. (2009). Optimization of Bacterial Whole Cell Bioreporters for Toxicity Assay of

Environmental Samples *Environmental Science & Technology*, Vol.43, No.20 pp.7931–7938, ISSN 0013-936X

Tag K.; Riedel, K.; Bauer H.J.; Hanke, G.; Baronian, K.H.R. & Kunze, G. (2007). Amperometric detection of Cu^{2+} by yeast biosensors using flow injection analysis (FIA). *Sensors Actuators B*, Vol.122, No.2, pp.403–409, ISSN: 0925-4005

Tatsuma, T.; Yoshida, Y.; Shitanda, I. & Notsu, H. (2009). Algal biosensor array on a single electrode. *Analyst*, Vol.134, No.2, pp.223–225, ISSN 0003-2654

Teles, F.R.R. & Fonseca, L.P. (2008). Applications of polymers for biomolecule immobilization in electrochemical biosensors. *Materials Science and Engineering C*, Vol.28, No.8, pp. 1530-1543, ISSN 0928-4931

Timur, S.; Anik U.; Odaci D. & Gorton L. (2007). Development of a microbial biosensor based on carbon nanotube (CNT) modified electrodes. *Electrochemical Communications*, Vol.9, No.7, pp.1810-1815, ISSN: 1388-2481

Tschmelak, J.; Kumpf, M.; Kappel, N.; Proll, G. & Gauglitz, G. (2006). Total internal reflectance fluorescence (TIRF) biosensor for environmental monitoring of testosterone with commercially available immunochemistry: antibody characterization, assay development and real sample measurements. *Talanta*, Vol.69, pp. 343-350, ISSN: 0039-9140

Tschmelak, J.; Proll, G.; Riedt, J.; Kaiser, J.; Kraemmer, P.; Barzaga, L.; Wilkinson, J.S.; Hua, P.; Hole, J.P.; Nudd, R.; Jackson, M.; Abuknesha, R.; Barcelo, D.; Rodriguez-Mozaz, S.; Lopez de Alda, M.J.; Sacher, F.; Stien, J.; Slobodnik, J.; Oswald, P.; Kozmenko, H.; Korenkova, E.; Tothova, L.; Krascsenits, Z.& Gauglitz, G. (2005). Automated water analyser computer supported system (AWACSS).PartII: Intelligent, remote-controlled, costeffective,on-line, water-monitoring measurement system. *Biosensors & Bioelectronics*, Vol.20, No.8, pp. 1509-1519, ISSN 0956-5663

Tymecki, L; Glab, S. & Koncki, R. (2006). Miniaturized planar ion-selective electrodes fabricated by means of thick-film technology, *Sensors*, Vol.6, No.6, pp.390-396, ISSN 1424-8220

Van der Meer, J.R. & Belkin, S. (2010). Where microbiology meets microengineering: design and applications of reporter bacteria. *Nature Reviews Microbiology*, Vol.8, No.7, pp.511-522, ISSN: 17401526

Vidal, J.C.; Bonel, L. & Castillo, J. R. (2008) A Modulated Tyrosinase Enzyme-Based Biosensor for Application to the Detection of Dichlorvos and Atrazine Pesticides. *Electroanalysis*, Vol. 20, No.8, pp.865 – 873, ISSN 1521-4109

Viswanathan, S.; Radecka, H. & Radecki, J. (2009) Electrochemical biosensor for pesticides based on acetylcholinesterase immobilized on polyaniline deposited on vertically assembled carbon nanotubes wrapped with ssDNA. *Biosensors & Bioelectronics*, Vol.24, No9, pp. 2772-2777, ISSN 0956-5663

Wang H.; Wang X.J.; Zhao J.F. & Chen, L. (2008). Toxicity assessment of heavy metals and organic compounds using CellSense biosensor with *E.coli*. *Chinese Chemical Letters*, Vol.19, No.2, pp.211-214, ISSN 1001-8417

Wang, S.; Tan, Y.; Zhao, D., Liu G. (2008) Amperometric tyrosinase biosensor based on Fe3O4 nanoparticles–chitosan nanocomposite. *Biosensors & Bioelectronics*, Vol.23, No.12, pp.1781–1787, ISSN 0956-5663

Wang, L.; Ran, Q.; Tian, Y.; Ye S.; Xu, J.; Xian, Y.; Peng, R. & Jin, L. (2010). Covalent grafting tyrosinase and its application in phenolic compounds detection. *Microchimica Acta*, Vol.171, No(3-4), pp.217–223, ISSN 0026-3672

Wang, Z.; Wilkop, T.; Xu, D.; Dong, Y.; Ma, G. & Cheng, Q. (2007) Surface plasmon resonance imaging for affinity analysis of aptamer- protein interactions with PDMS microfluidic chips. *Analytical & Bioanalytical Chemistry*, Vol.389, No.3, pp. 819-825, ISSN 1618-2642

Wei, D.; Bailey, M.J.A.; Andrew, P. & Ryhanen, T. (2009) Electrochemical biosensors at the nanoscale. *Lab on Chip*, Vol.9, No.15, pp. 2123-2131, ISSN 1473-0197

Wilmer, M.; Trau, D.; Rennenberg, R. & Spener, F. (1997). Amperometric immunosensor for the detection of 2,4-dichlorophenoxyacetic acid (2,4-D) in water. *Analytical Letters*, Vol.30, No.3 pp. 515-525, ISSN 0003-2719

Woutersen, M.; Belkin S., Brouwer, B.; van Wezel, A.P. & Heringa, M. B. (2010). Are luminescent bacteria suitable for online detection and monitoring of toxic compounds in drinking water and its sources? *Analytical & Bioanalytical Chemistry* DOI 10.1007/s00216-010-4372-6, ISSN 1618-2650

Xia, W; Li, Y.; Wan, Y., Chen, T.; Wei, J.; Lin, Y. & Xu, S. (2010). Electrochemical biosensor for estrogenic substance using lipid bilayers modified by Au nanoparticles. *Biosensors and Bioelectronics*, Vol. 25, No.10, pp.2253–2258, ISSN 0956-5663

Xu, Z.; Chen, X. & Dong, S. (2006). Electrochemical biosensors based on advanced bioimmobilization matrices. *Trends in Analytical Chemistry*, Vol.25, No.9, pp. 899-908

Xuejiang, W.; Dzyadevych, S.V.; Chovelon, J.M.; Jaffrezic-Renault, N.; Ling, C.; Siqing, X. & Jianfu, Z. (2006) Conductometric nitrate biosensor based on methyl viologen/Nafion®/nitrate reductase interdigitated electrodes. *Talanta*, Vol.69, No.2, pp. 450-455, ISSN: 0039-9140

Yildiz, H.B.; Castillo, J.; Guschin, D.A., Toppare, L. & Schuhmann W. (2007). Phenol biosensor based on electrochemically controlled integration of tyrosinase in a redox polymer, *Microchimica Acta*, Vol.159, No.1-2, pp. 27–34, ISSN 0026-3672

Yuan, C.-J.; Wang, C.-L.; Wu, T. Y.; Hwang, K.-C.; Chao, W.-C. (2011). Fabrication of a carbon fiber paper as the electrode and its application toward developing a sensitive unmediated amperometric biosensor. *Biosensors & Bioelectronics*, Vol.26, No.6, pp. 2858–2863, ISSN 0956-5663

Zhang, X.; Guo, Q. & Cui, D. (2009). Recent advances in nanotechnology applied to biosensors. *Sensors*, Vol.9, No.2, pp. 1033-1053, ISSN 1424-8220

Zhang, Z.; Jaffrezic-Renault, N.; Bessueille, F.; Léonard, D.; Xia, S.; Wang, X.; Chen, L. & Zhao, J. (2008) Development of a conductometric phosphate biosensor based on tri-layer maltose phosphorylase composite films. *Analytica Chimica Acta*, Vol.615, No.1, pp. 73-79, ISSN 0003-2670

Zhang, Z.; Xia, S.; Léonard, D.; Jaffrezic-Renault, N., Zhang, J.; Bessueille, F.; Goepfert, Y.; Wang, X.; Chen, L.; Zhu, Z.; Zhao, J.; Almeida, M.G. & Silveira, C.M. (2009) A novel nitrite biosensor based on conductometric electrode modified with cytochrome c nitrite reductase composite membrane. *Biosensors & Bioelectronics*, Vol.24, No6, pp. 1574-1579, ISSN 0956-5663

Zhao, J.; Zhi, J.; Zhou Y. & Yan, W. (2009). A Tyrosinase Biosensor Based on ZnO Nanorod Clusters/Nanocrystalline Diamond Electrodes for Biosensing of Phenolic Compounds. *Analytical Sciences*, Vol. 25, No.9, pp.1083-1088, ISSN 10830910-6340

Zhao, W.; Ge, P. -Y., Xu, J.-J. & Chen, H.-Y. (2009) Selective Detection of Hypertoxic Organophosphates Pesticides via PDMS Composite based Acetylcholinesterase-

Inhibition Biosensor, *Environmental Science & Technology*, Vol.43, No.17, pp.6724–6729, ISSN 0013-936X

Zhou, Y. & Zhi, J. (2006). Development of an amperometric biosensor based on covalent immobilization of tyrosinase on a boron-doped diamond electrode. *Electrochemistry Communications*, Vol.8, No.12, pp.1811–1816, ISSN: 1388-2481

Zourob, M.; Simonian, A.; Wild, J.; Mohr, S.; Fan, X.; Abdulhalime, I. & Goddard, N. J. (2007) Optical leaky waveguide biosensors for the detection of organophosphorus pesticides, *Analyst*, Vol.132, No.2, pp. 114–120, ISSN 0003-2654

Part 2

Materials Design and Developments

7

New Materials for Biosensor Construction

Man Singh[1], R. K. Kale[1] and Sunita Singh[2]
[1]School of Chemical Sciences, Central University of Gujarat,
[2]Department of Biochemistry, Shivaji College, University of Delhi, New Delhi.
India

1. Introduction

Currently materiel development and characterization have attained a central position in science and technology to meet out conventional and unconventional challenges in exponentially growing demands of civil society. Science, technology and society now are closely correlated to serve mutually where materials sever as most fundamental interface and initiator for dedicated and non-dedicated applications.

Materials are most significant entities which constitute body of all objective things may be living or non-living being on the basis of structural stability. The structural components which develop binding forces are critical components of the materials. The materials are of two basic categories such as non-molecular and materials molecular. Of course, under non-molecular materials (NMM), the composites, blends, alloys, eutectic mixtures, acoustic materials constitute most industrially useful sensors mainly for non-biocompatible applications. However, some of them serve as a most biocompatible material such as polymer blends as contact less, stainless steel as bone substitute to normalize a function of bone when it fractures. The NMM serve as basic materials for communication, building materials, machine materials, computer ware materials and many others.

2. Constitutional biosensing property genesis

The molecular materials due to structural identity do sever as most excellent biosensors because of certain structural changes which are trapped in form of certain impulse in electronegativity, chemical potential, electrical signal, change in shared electron pair, free energy, entropy, internal energy change, enthalpy change, friccohesity change. In general, such changes are referred to as physicochemical properties (PCP). When the PCP predicts or conveys certain signal or senses as an outcome of biological, biochemical, biophysical, biotechnological, bioengineering phenomenon and similar others then the PCP act as physicochemical indicators (PCI). The PCI also encompasses certain signals created due to oxidation and reduction process and hydrophilic and hydrophobic interactions along with ionic responses.

In this context a new materials to be prepared and act as biosensor must have definite sites in molecules which can either undergo change or indicate some shear stress and shear strain to be detected as a signal to analyze a nature and type of phenomenon being executed at the operational site. Several molecules such as DNA, RNA, proteins, enzymes, hormone etc. do have giant size with manifolds responding sites which are covered under a new concept

IMMFT (intramolecular multiple force theory) reported elsewhere [1]. For example, egg phosphatidylcholine (EPC) with multiple force points behaves as an emulsifier type sensor to detect hydrophilic interactions while the cetylpyrridiunium bromide (CPB) with cetyl carbon chain with uniform covalent bonding force (UCBF) due to sp^3 orbital hybridization behaves as surfactant type sensor to depict hydrophobic interactions. Thus the EPC is a biosensor for scaling viscous property through intermolecular forces whiles the CPB as a cohesive forces scaling through hydrophobic mode [1, 2].

2.1 Fundamental energetics

Fundamentally a molecular mechanism of biosensor is thermodynamic in nature as each and every structural change is accompanied with free energy and entropic change. According to first law of thermodynamic the energy neither destroyed nor created but transformed from one form to another. It is depicted as q = ΔE +PdV, q is enthalpy change, ΔE is internal molecular change and PdV is physical work done in energy change process as the P is pressure and dV, a volume change. Also the structural changes are depicted with simple thermodynamic change for energy transformation equation 1.

$$dG_{mix} = \left(\frac{\partial G}{\partial p}\right)_{T,n1,n2} dP + \left(\frac{\partial G}{\partial T}\right)_{p,n1,n2} dT + \left(\frac{\partial G}{\partial n_1}\right)_{T,P,n2} dn_1 + \left(\frac{\partial G}{\partial n_2}\right)_{T,P,n1} dn_2 \tag{1}$$

It analyzes a change on mixing n1 and n2 moles at dP = 0, dT = 0 and depicted as equation 2.

$$dG_{mix} = \left(\frac{\partial G}{\partial n_1}\right)_{T,P,n2} dn_1 + \left(\frac{\partial G}{\partial n_2}\right)_{T,P,n1} dn_2 \tag{2}$$

Equations 1 and 2 define chemical potential response as base for biosensor with no physical changes. The internal energy also matters a lot in biosensor q = ΔE + PdV. Thus the molecular constitutions fundamentally illustrate a fundamental science and mechanism of biosensors without which it is impossible to develop any molecular device for biosensing to be used efficiently. Thus for understanding a theme, theory and mechanism of sensor, a broad view is an essential step and knowledge before initiating and designing application of biosensors. The molecular identity is must during biosensing action mechanism. Of course, the biosensors do have potential and befitted in current trends of nanotechnology and green sciences in analytical and materials sciences. Technically, there is a lot more hidden potential within molecular world if it is unearthed and proliferated in right direction and suitable applications. The biosensor could not only open new gateways of ample new opportunities but could also constitute exponential synergies to synchronize molecular potential to a level of nuclear sciences for positive works. Currently, molecular sciences have gathered enough momentum for wider applications in chemicals, medical, biochemical, biotechnological, biophysical, biotransformational, bioelectronics biomembrane, osmolarities, electrical conducting charges along interesting and powerful potential of bioconductance in field of fluid sciences (1, 2). Quarts materials with pi-mesons and similar other subatomic particles could also simulate biosensor sciences. In layman language, the sensor word reveal or do indicate some dynamic activities associated with molecular materials. Of course, the non-molecular world could equally be benefitted in applied sciences in space research, telecommunications, electronics etc. Here, the focus is on biosensor which could generate some formative signal of particular activities which is readable or measurable with accurate and reproducible analytical equipment. The biosensors are molecular materials with definite molecular constituents organized in a specific electronic configuration and geometries to have integrated impact like benzene when dissolved in water no thermodynamic properties water are changed. In contrary to this, when ethanol is dissolved in water the thermodynamics is changed and a resultant aqueous ethanol mixture exists with totally new sensing properties. The benzene sp^2 hybridization is intact when is brought in contact of water but the sp^3 hybridization of the ethanol quickly senses and responds a presence of the water environment. It is very true in case of alcohol intake where the later quickly bind with body water and tightens the muscle and body becomes stiff on alcohol intake.

2.2 Biothermidynamics support

This is a real mechanism of biosensor. For example, DNA, RNA, proteins etc. with definite biocompatible molecular constituents do perform critical role in body, and also DNA finger printing is a great biosensor. Thus there are certain molecular accessories which initiate energy ridges and anchor interacting tools to be nourished as a phenomenon for furtherance or amplification of specific signal. Thus the highly biocompatible molecules like proteins, drugs, biopolymers etc. act as biosensors where such molecules undergo several mechanism like oxidation and reduction; hydrophilic and hydrophobic; interstitial placements, shifting shared electron pair to more electronegative atom within molecules. The abovementioned interacting molecular accessories (IMA) as critically potential molecular nut bolt (MNB) do induce and exert molecular interacting engineering (MIE) which are responsible for molecules to behave as biosensor.

The IMA and MNB together create paces and phases with different electric or chemical potentials to develop working and biosensing thermodynamic spontaneity (BTS) (2) to develop interacting vacancy in search of equilibrium due to $\Delta G \neq 0$. The BTS is tool to equilibrate $\Delta G \neq 0$ into $\Delta G = 0$ with a thrust to homogenize liquid mixture, may be in context of chemical potential $\mu = (dG/dn)_{P,T}$. The fundamental function μ generates IMA followed by MIE under constant P (pressure) and T (temperature). Thus for biosensor efficiency and mechanism the MIE and BTS are unavoidable tools, in fact, these are the fundamental molecular accessories to generate required properties of any individual molecule to act as biosensor. Therefore, the materials which are to be constructed to work as biosensor must have capability and capacity to initiate and cause the BTS. Incorporating abovementioned factors specifically the IMA and MNB to have effective MIE with excellent BTS, lead to simulate changes and integrate certain signal to visualize measurable or sensible indicators. Such generation, construction and formulation of MIE lead to certain sciences like surface and bulk reorientation in molecular energies (3).

For broad elaboration of developing effective biosensors few molecular activities or initiative dents could be noted as essential accessories for successful biosensing mechanism. These are as collision and kinetic activities, electrostatic, solvent activation, pH change, photosensitization, temperature induction, activity coefficient, solubility product, complex product, ionic exchange, solvent saturation point, colloidal formation, critical micelle concentration. Any of the above mentioned activities either independently or in combination of others could integrate in terms of structural potential which could complete a biosensing processes and products then the foundation of biosensor is excelled in practical uses for their effective and reliable validity industrial as well academic applications. Also the reproducibility of any individual molecules to act as biosensor is based on the molecular reorientation, optimization, energetics, friccohesity, interaction engineering, localized molecular potential, molecular potential distribution for efficient working and reducing energy barriers of individual processes with advanced binding or lock and key model based kinetics.

3. Electrostatic science of biosensor

Broadly speaking, the biosensor, in general, specifies a change in chemical systems as environmental systems where if temperature goes up then surrounding too respond accordingly. For example, global warming is due to solar radiation absorption which raises temperature of the globe due to solar radiation absorptions which are electromagnetic radiation (EMR). The EMR interacts with polar gases HCl, N_2O, SO_X, CO_X ($x=1$, 2, 3) and water vapor (H_2O) which are noted green house gases and sensor for global warming where whole ecosystems influenced. Thus environmental biosensor too play critical role for sustainment of ecosystem. For example, during weather change, the people get cold or some other disease that is due to temperature variation as the human body responds temperature changes and dissipate and dimensionalized the energy level enhanced due to rise in temperature. This theory is based on $\Delta G \neq 0$ where the BTS theory equilibrates $\Delta G \neq 0$ into $\Delta G = 0$ to homogenize body equilibrium.

Fundamentally, the biosensor is not an industrial phenomenon but a way out of chemical thermodynamics where an additional energy of molecules accumulated on specific molecular constituents is utilized in reorientation to get optimization in prevailing environment in the body or other living beings. For example, when Hg (mercury) thermometer is kept in $10°C$ water bath, the Hg column remains near bottom but when the temperature goes to $50°C$, the temperature goes to high level. In this process, a well defined thermodynamics work is done explained with $q = \Delta E + P\Delta V$ equation. The physical work w done in Hg process is $w = P\Delta V$ where the p is pressure and v is change volume. The $v = \pi r^2 h$, r column radii and h is its height, hence work done $w = p \pi r^2 h$. Notably, the h is either higher or lower not because of a pressure change but due nature of Hg which on heating expands and runs up which is an excellent sensor of temperature control, and could be extended to similar other systems.

Previously so many direct physical devices were used to make change and to check the temperature points. Similarly, the low pressure in specific area of globe cause storm to come which is very much fitted in $\Delta G \neq 0$ as a strong BTS is developed to equilibrate $\Delta G \neq 0$ into $\Delta G = 0$ with no pressure change. So biosensor is basically used for different uses. The bioscience human body is associated with specific molecules to perform several functions to do the work or dock specific molecule or ions or cells to normalize the body functioning. For example, certain medium when given do reduce body temperature which is an indication of their effect on the body. So biosensors are basically used for different used. Soap or detergent is another biosensor to remove dirt particle from textile fiber similar in $\Delta G \neq 0$ to $\Delta G = 0$ manner. So several highly useful products are in use in society where biosensor do play constructive role. For example nutrients, essential oils, salts macronutrients which automatically indicate either surplus or deficiency in body. The biosensor only predict and hint about so many products and properties of molecular like conductance, viscosity, surface tension, chemical potential, Gibbs free energy and devised according as optical sensors, electrical sensors, temperature sensor, electronic, volume or pressure of gases in blood, sound sensor based on acoustic biomolecules, DNA sensor, protein sensor and others.

4. Conformations backup to biosensor

Current thrust is to construct or synthesize materials which could be used as safer and sure efficient biomolecular sensors (EBS). The EBS mechanism is conceptually works on structural reorientations and conformations (SRC) of molecular constituents which are highly compatible to human activities in body. Sure and pure study of all possible SRC are studied in detail and ensured that no unwanted structural conformation is developed in newly developed biosensor. As the structural configurations do develop binding and interacting capacities with different molecules in body, for example, the phytic acid (Fig.1) intake binds with calcium ions and cause deficiency of calcium. The phytic acid is in hulls of nuts, seeds, grains but cooking food slightly reduces it so a sprouted food with skin is good behave as excellent materials. The phytic acid is a chelator of minerals Ca^{2+}, Mg^{2+}, Fe^{3+}, Zn^{2+} and causes mineral deficiencies in people that result into osteoporosis. The osteoporosis is a bone disease accessible to increased risk of fracture due to lower bone mineral density (BMD). It aqueous mixtures cause unique IMF with drastic change due to much structural interactions depicted with Friccohesity data.

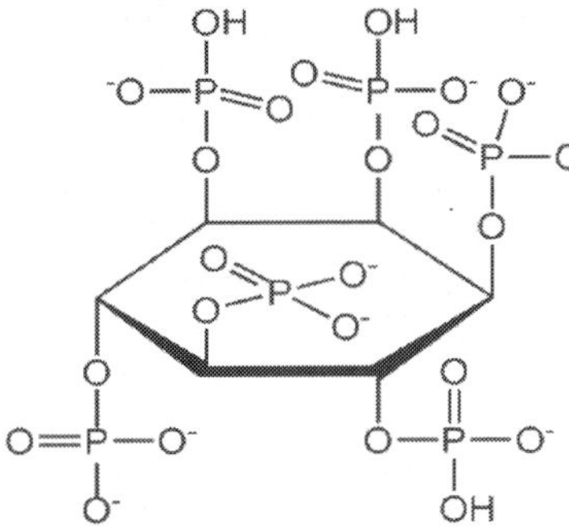

Fig. 1. Phytic acid

Thus the few new molecules are being developed in reference as construction of new materials. Hunger is another sensor to take food, a thirsty another sensor to maintain the water intake in body. Ambition is another sensor to work hard and curiosity too to know more and more, sleeping is another sensor to provide rest to muscles.

5. Novel science of biosensor materials

Biosensor science is a pure thermodynamics process because when plastic touched with electric connection and the electronic wire does and allow bulb to light. The metallic wires coated with plastic due to electronic and non-electric release respectively. Hence ideal materials can't work as biosensors in body but get deposited like cholesterols. There are ample opportunities in body as several biomolecules are gigantic heteromolceucles with several functional groups which could act as ideal biosensor if studies are devised and conducted accordingly. Each molecule does develop specific Friccohenics to diversify the forces to operate molecular forces of biosensor. A hidden potential of biosensor and operational biophysics must be fabricated in scientific manner with befitting biophysical MIE with BTS for biosensor molecular models (BMM) to integrate and exploit or harness the friccohesity model from fundamental molecular unrest and optimization to strengthen potential structural reorientation. In this context few newly developed biosensors (NDB) like poly-N-vinylpyrrolidone oximo-L-valyl-siliconate (POVS) [4], melamine-formaldehyde-polyvinylpyrrolidone (MFP) supramolecules [5] and 2, 4, 6 Tridiethymalonate-Triazine

(TDEMTA) and 2, 4, 6 Hexadiethylmalonate-Triazine (HDEMTA) dendrimers [6] could be of great significance in field of biosensors. Especially, the dendrimers are with unique architectural structures and widely suitable to be as biosensor due to functional end groups in forms of hyper-branched tree like macromolecules.

6. Fundamental molecular engineering

Modern research trends are of innovative supramolecular and nanoparticle-based systems for novel phenomena and applications in different fields. The NDB integrates grey areas of molecular research approaches in the fields of supramolecular, dendrimer, biosensor and smart biomolecular chemistry of hydrophilic and hydrophobic quanta dots within a molecule (7). The NDB could also enhance understanding a technology of liquid crystal devices (LCD), liquid electronic devices (LED) insulators etc. The NDB are most efficient and work based on $v = 0, 1, 2, n$. are vibrational quantum numbers and total molecular energy (E_{mol}) which is constituted of electronic, vibrational, rotational, nuclear and translational components with many degree of freedoms that lead to either $\Delta G \neq 0$ or $\Delta G = 0$ as $E_{mol} = E_{electronic} + E_{vibrational} + E_{rotational} + E_{nuclear} + E_{translational}$

The $E_{electronic}$ is electronic or a potential energy surface at equilibrium geometry and energies of these components vary with the oscillations. The rheological and IMA, MNB, MIE with TBS could be structurally and quantitatively analyzed with Survismeter [8] which works on $E_{electronic}$, $E_{vibrational}$, $E_{rotational}$, $E_{nuclear}$ and $E_{translational}$ conceptual framework. Such systems have been widely studied (4-6). Intramolecular multiple force theory [IMMFT] is proposed to explain molecular interactions of olive oil-water- egg-phosphatidylcholine (EPC) mixtures with a possible correlation of surface and bulk reorientations with microstructures depicted with SEM. Frictional and cohesive forces as Friccohesity have been noted as driving forces to assert for validity of the IMMFT model and its link with SEM [2].

7. Detailed view of biosensor mechanism

Molecular activities of biosensors are initials and signatures for origin of scientific simulations and frameworks for academic as well as new industrial upcoming. How do the molecules maintain identity under different polarity and electrostatics for sequential structural reorientations? It becomes more important in case of biomolecules such as EPC which being weakly polar involved in biosensor signal as emulsifying agent (9-11). For example, a detection of bound water at moon is an input and incentive to intensify search and research of undiscovered molecular world (12). In year 1953, Stanley and Miller experiment was a pioneer model to signify molecular signatures and several experiments and functions have strengthened a concept of molecular science (13) in search of newer mixtures and properties (14). The biomolecular sciences were studied by many scientists (15) for peculiar structural reorientation optimized to facilitate interactions (16). Since 17th to 19th centuries, several workers intensified efforts to elucidate hidden biomolecular combinations in different polarity (17). The friccohesity of such molecular materials do have excellent control due to their fluid dynamic and collision pattern in medium based on the Brownian motions. The friccohesity incorporated cohesive and frictional forces together with mansingh equation depicted below in equation 3.

$$\sigma = \sigma_0 \left[\left(\frac{t}{t_0} \pm \frac{B}{t} \right) \left(\frac{n}{n_0} \pm 0.0012(1-\rho) \right) \right] \qquad (3)$$

For example, Van der Waals emphasizes on conducting and transporting properties along binding forces (18). Debye Huckle theory and Lennard Jones potential distinguished a basic difference in potential and kinetics of biosensor dispersion and motions (19). Theoretically Schrödinger and Born Oppenheimer Approximation (BOA) focused nuclear charge contribution based on quantum chemistry support. A molecular potential before mixing is zero but on mixing is not zero due to interactions (20). For example, oil and water (21) are not much soluble due to weaker interactions (22) but the BOA conceptually explains interactions extended to simple organic or inorganic complexes (23). In such situations the forces inside within a molecule are confined to a centre of control. Onsager and Debye-Huckle, explained a contribution of electrostatics poles of either single ions Na^+, Ca^+, NH_4^+ or Zwitterions like amino acids as alignments of solvents were fitted with biosensor electrostatics (24). For biomolecule such as olive oil does not hold any charge so theories like Debye Huckle could not be fitted for its biosensor activities. Debye could not offer adequate solutions for macromolecules such as proteins where partially ionic peptide bonds were embedded in folding and electrostatic poles were not clearly exposed to solvents (25).

8. Interdisciplinary alignments of biosensor concept and applications

Thus the medium reorientation is effective for embedded electrostatics poles to unfolds the proteins and similar others. Tanford conducted substantial studies on interactions useful for biochemists and biothermodynamists (26). Ludvig Lorenz refined such interactions and approaches of molecular forces, especially of weaker electrolytes as surfactants and mildly partial such as olive oil where forces are confined, redistributed in a local arrangements. This made a better understanding of organic mixtures in a wider way to study oil-water muslins for industrial purposes. For centuries, the molecular framework has been a fascination to scientists for molecular design, polarity, electronic configurations and others. Since origin of life, it has been a never ending process and the scientists, chemists, biochemists, biotechnologists continued their pursuit for further search of either developing new mixtures in laboratories or extracting out of natural sources animal or plants (27). Further, Vander Waals, Lennard and others worked on primitive part of molecular sciences and realized a lot more potential and science hidden which has been furthered various new scientific theories, options, surfaces, intra-surfaces like nanotechnology. Thermodynamics was tried to retrieve hidden molecular energetics such as entropic and free energy changes (28). The studies of biosensor are continued and scientists remarkably contributed to further elaborate and signify their medical potential for industrial uses. From 18th to 19th centuries, a shift from bioionic to biomolecular approaches was noted on a pattern of big debate on existence of atomic theory put forward by Nellie Bohr with evidence and an existence of atomic theory was accepted. Thermodynamic mixtures are easy to explain but the oil and water mixtures with zwitterions as additives need classical support in favor of molecular origin of forces responsible for biosensor sciences. Though several intensely diversified conflicts of biosensor sciences were noted but a molecular force theory (MFT) was unanimously accepted which is still continued.

9. Intramolecular multiple force theory (IMMFT)

Since late 20th to 21th century, trends to develop supra or giant molecules in laboratories either based on metallic ions such as transitional and lanthanides metals (29) or certain molecular rings as core or centermost part or then after branching as an extension for molecular shape and sizes developed for industrial uses and relevance (30). These sciences revolved around MFT, and originated several diversified molecular forces theories (DMFT) to understand vivid molecular fascinations using biothermodynamics, chemical kinetics, electrochemistry which thinly peeped into inner sides of molecular interactions. The DMFT was put forward and new phenomenon such as friccohesity was emerged to find out a relevance of materials to biosensor such as polymers, drugs, cosmetics, sol gel, electrolytes, solvents, pesticides, disinfectants and others.

It is aimed for opening new gates of knowledge about for industrialization of molecular concept in interest of society. For example, olive oil-water-bio-surfactants mixtures were never designed and studied, despite huge industrial potential (31). Fundamentally, biosensors involve bond disruption, reorientation, breaking, bond angle and bond energy to correlate and reveal mechanism for generating signal such as liquid crystals. It could be in a most conceptual manner such as intramolecular multiple force theory (IMMFT) to intuitively explain a sensing device that could excellently interpret surface and bulk phase dynamics responsible for microstructure designing depicted with SEM. In history of biosensor the IMMFT is a novel step forward and approaches for effective use. For example, around phosphate atom of EPC, three different alkyl chains with O atoms are fitted which cause different force centers within EPC (Figs.2 and 3). It sensed diffused hydrophilic response useful for emulsification or homogenous dispersion in bulk water phase and not moves to surface contrary to surfactants sensor such CTAB, CPC and CPB which accumulate at surface (32).

10. Experimental support

Experimental verifications could be made with surfactants and L-α-phosphatidylcholine (source egg yolk) by mixing with olive oil in 10^{-5} mol kg^{-1} with 2-10 mm kg^{-1} BS. Their densities, viscosity and surface tension with Survismeter measured of biophysical significance under sterilized condition. The EPC behaved as excellent emulsifier only because of many forces centers operational in developing interactions at separate points within a molecule. Hence it could not pushed to interface but remained suspended in hydrophilic water while the CPC, CPB and CTAB reduced almost 43% ST as they tend to surface and saturate the same by reducing surface energy or tension due to integrated hydrophobic CBF but the EPC does not have integrated CBF along the alkyl chain because of the O atoms are in chains. Hence comparatively the EPC behave as best emulsifier whiles the CPC, CPB and CTAB as best cationic surfactants. This observation is also supported by a SEM pattern, with stronger intermolecular forces with CTAB and olive oil which produced a higher compatibility and density (Fig. 2).

Surface tension

The surface tensions of BS-olive oil-water are lower than of the water by 1.5 mN m^{-1} except the EPC, with slightly lower values. With OOW, the higher surface forces than of the water are attributed to the stronger hydration spheres with the CPC, CPB and CTAB. The γ^0 are as EPC > CTAB > CPC > CPB (Table 1). The γ^0 data inferred a role of cohesive and adhesive forces due to activities in addition to electrostatic and Newtonian forces. The higher γ^0 value with the EPC exerted a stronger IMF with stronger hydrophobic-hydrophobic interactions of longer alkyl chain. The IMMFT model explained an action mechanism of the EPC due to multiple force centers such as two alkyl chain, oxygen atoms, PO_4^{-3} group and quaternary nitrogen atoms.

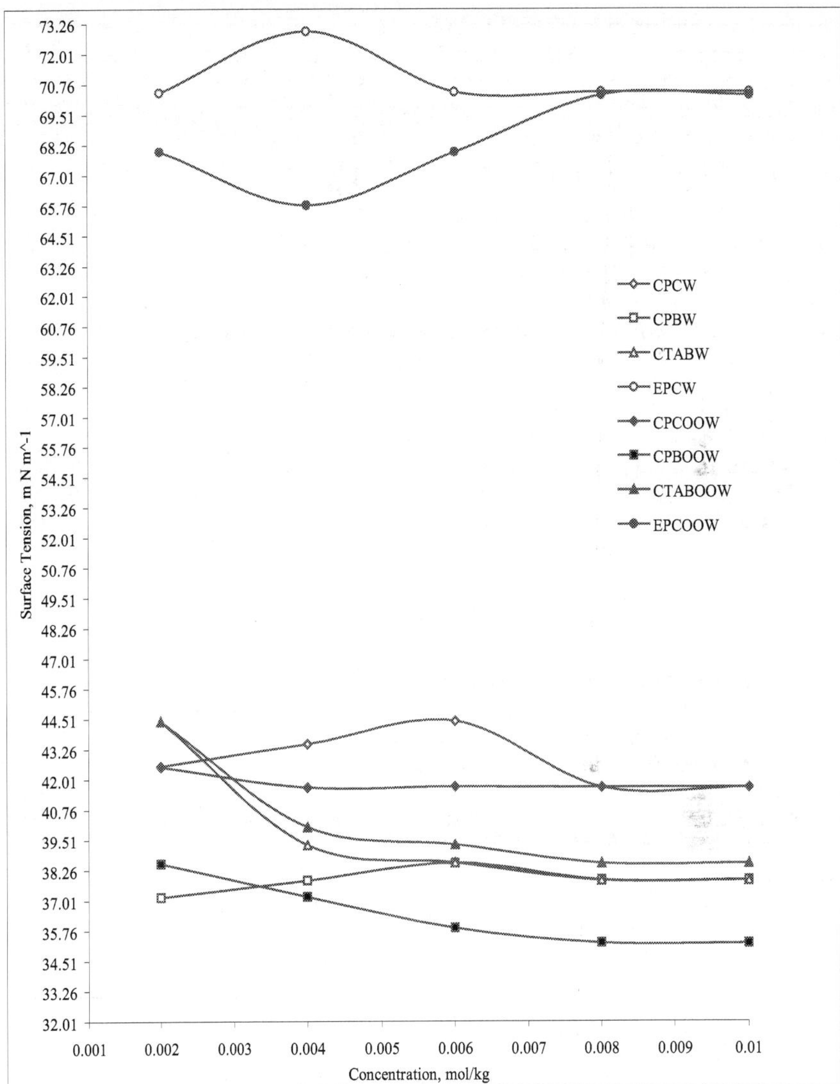

Fig. 2. A big gap in surface tension values of EPC and surfactants as physicochemical indicator where EPC and surfactants act as excellent biosensor in biofluids, bioemulsions, biocolloids and others. The biosensing mechanism is executed through stronger cohesive forces in case of the EPC while the weaker frictional forces with the surfactants. This defines a role of Friccohesity in successful and effective biosensors for based on fluid. The EPC works through hydrophilic and the surfactants through hydrophobic interactions based on IMMFT model. dynamics.

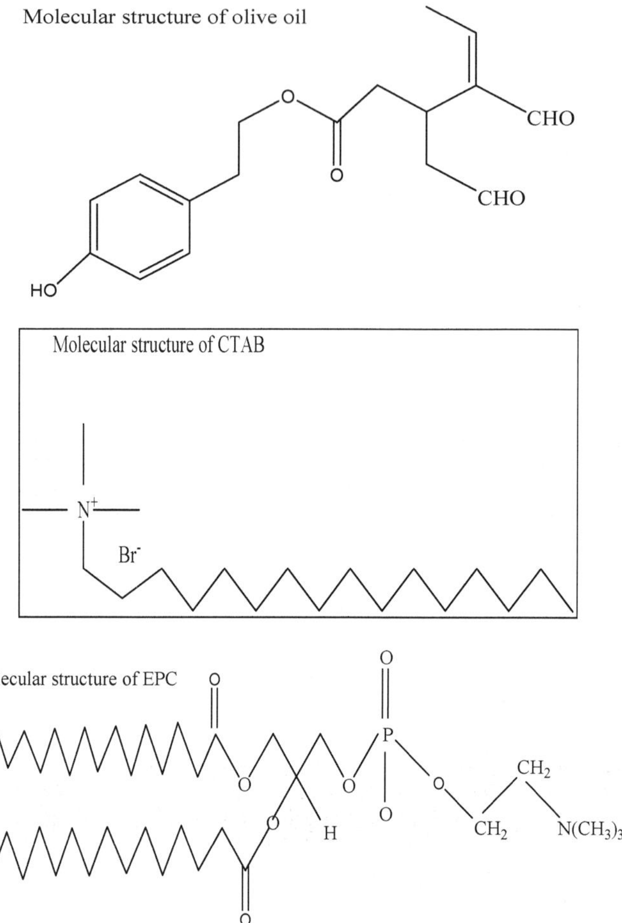

Fig. 3. Molecular structure of olive oil, CPC, CPB, CTAB and EPC.

The PO_4^{-3} group weakly disrupted the water, and highly asymmetricity in the EPC structure initiated the stronger cohesive forces. The higher entropic changes due to multiple force centers for interactions caused stronger hydrophobic-hydrophobic interactions. The IMMFT predicted that the EPC molecules are not able to disrupt the hydrogen bonded water and the multiple forces of electrostatical points EPC exert higher tension.

11. Hydrophobic and hydrophobic phenomenal support

For example, from 8 mmkg⁻¹ BS, the γ^0 is linear with no further change in surface forces, due to a complete kind of reorientation of water along little monomer formation (Fig. 3) which further increased with increasing concentration. The 8 mmkg⁻¹ BS, initiated micelles formation, and is a critical micelle concentration point (Figs 2 and 3). The SEM illustrated dispersed structures of biosurfactants with higher intermolecular forces between them in aqueous olive oil mixtures (Fig.2). The IMMFT model excellently explained geometry of microstructures as a function of frictional and cohesive forces at multiple points, especially with the EPC The molecular dispersion have maximum surface forces but an olive oil brought them together causing intermolecular forces which reduced an exposed area with reduction surface tension as EPC > CTAB > CPC > CPB. Their comparative study illustrated higher surface tension and lowest dispersion with EPC (Fig. 2). The stronger intermolecular forces between the EPC-EPC with higher cohesive and adhesive forces in olive oil-water are responsible for such a behavior. The SEM illustrated comparatively uniform dispersion with the EPC with weaker Van der Waals and London dispersion forces and is attributed to IMMFT. Usually oil develops colloidal solution with diffuse interface forces while smaller biosurfactants are well defined hydrated units depicted by SEM. The biosurfactants breakdown the water structure while olive oil is not able to do so but it shifts the bulk structure of water around itself with a cage formation. The dispersion of the olive oil in water and of CPC, CPB, CTAB and EPC molecules with olive oil-water mixture is depicted by the SEM as CTAB > CPC > CPB > EPC > OOW (Fig. 2). Here, the SEM studies illustrate that the olive oil structure in water does not get disrupted due to stronger intermolecular forces but it remains in its original structure. The thread like structure inferred the C backbone of olive oil (Fig. 2). The EPC has caused least disruption into n-fragmentations due to stronger solute-solute interactions, results higher surface forces (Fig. 2). The CTAB though weakened the frictional forces but also caused some integrated or packed patches or the smaller globules which is dewetting depicted with SEM picture (Fig. 4). It describes the oil-water-surfactants mode.

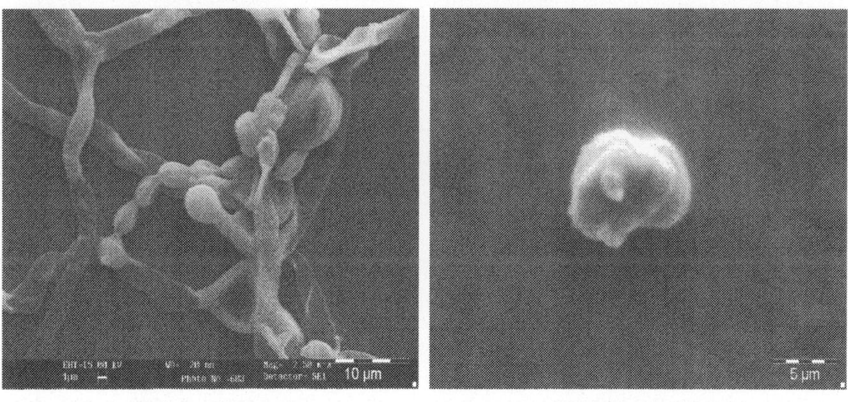

[Olive oil-water]   [CTAB with OOW]

Fig. 4. SEM pictures infer excellent biosensor activities of the biofluids.

12. Physicochemical significance

The McLachlan theory predicts that van der Waals attractions in media are weaker than in vacuum based on like dissolves like where the different types of atoms interact more weakly than identical types of atoms. In contrast to combinatorial rules or Slater-Kirkwood model that illustrated classical force fields which were supported by Jacob Israelachvili with

Intermolecular and surface forces model. The IMMFT model is advantageous by incorporating distribution of forces intramolecularly for better signal. The model of multiple force fields could be used for unfolding of protein structures, for example, energies of hydrogen bonds in protein engineering. The molecular biotics, fashionable configurations of atoms within spatial framework of covalent bonds, for example proteins, amino acid with intradisciplinary molecular structures like organic, semi-organic, complex, supramolecular prototypes activated chip. These perform several functions where Schrödinger quantum mechanics and wave mechanism of energy distribution occurs. The theology of scientific up gradations on ionic to molecular coordination have now at center stage not because of nanotechnology but because of molecular potential to resolve various complicated issues of the matters. In this context, the IMMFT model is most suitable and a step forwards for resolving better understanding of the giants and supamolecular structures.

13. SEM correlation

Conclusively, a correlation between physicochemical properties and SEM microstructures is noted in case of the EPC, a very common ingredient of the food digestion process. A fundamental difference in interacting behaviors in EPC and other surfactants CPC, CPB and CTAB is of surface activities. With EPC due to IMMFT the EPC does not pushed to surface and could not reduce surface tension as compared to others which reduce about 43 %. Since they saturate the surface due to stronger CBF but EPC is missing CBF. So EPC is a best emulsifier while they CPC, CVPB and CTAB are best surfactant. With EPC three alkyl chain surround phosphate with steric and induction effects due to CBF and each alkyl has double bond with sp^2 configuration, also contribute to emulsification with negligible surface excess concentration. Hence IMMFT is excellent model to structurally explain molecular forces responsible for emulsification action of the EPC as it has not largely reduced surface tension by not ending towards interface. Contrary to EPC, the CPC, CPB and CTAB reduced surface tension more than 43% with higher surface excess concentration and comparatively stronger hydrophobic interaction than those of the hydrophilic. The EPC is as excellent emulsifier only because of the many forces centers operation in developing interactions within single EPC molecule and is not pushed to interface but remained suspended in hydrophilic. Hence comparatively the EPC behave as best emulsifier whiles the others as best cationic surfactants. The slopes values with EPC are lower than those of CPC, CPB and CTAB because the EPC is emulsified or dispersed homogenously in bulk water phase. So the EPC does not move towards surface with increment in its concentration contrary to CTAB, CPC and CPB which tend towards surface and get accumulated there on that seriously affect the surface tension. Higher concentration accumulation on surface need higher work to be done and hence it strongly weakens tension. Thus the slopes with CPC, CPB and CTAB concentrations the higher reduction in surface tension is required as compared to the EPC. Hence the IMMFT explained the maximum dissolution with EPC

14. Novel molecules

Few new and novel molecules such as 2,4,6 tridiethymalonate-triazine (2,4,6 TDEMTA) 1st (G$_1$) and 2,4,6 hexadiethylmalonate-triazine (2,4,6 HDEMTA) 2nd (G$_2$) tier dendrimers as efficient biosensors were developed [synthetic] and found with exceptionally high biosensing activities due to octopus like geometry with $7.12 \times 10^{145} k$ J mol^{-1}K^{-1} frequency factor range estimated with Arrhenius equation. Their higher surface area with significant void spaces has many channels and cavities with an ability to trap foreign material in medical sciences, drug delivery systems, in biomedical, biophysical and biochemical fields. Their biosensing activities A′ and activation energy E′ calculated from $[\eta]$ vs $1/T$ with $[\eta] = \log A′-(E′/2.303\ RT)$ Arrhenius equation log. E′/2.303 are slope and R gas constant $(8.314\ J\ mol^{-1}K^{-1})$.

14.1 Limiting viscosity η^0

Their biosensing activities A′ and activation energy E′ may monitor branching and π conjugation where both the factors contribute to modulate intermolecular forces. The η^0 values define state of electrostatic forces for shape and hydrodynamic structure a most prominent requirement for biosensor. The G_2 structure with larger π conjugation numbers develop hydrogen bonding among branches with higher molar volume. Intrinsic viscosity B is associated to the shape and size and is very low whose value indicates heteromolecular forces with medium and rotational motions. As the viscosity is an arrangement of intermolecular forces to get oriented to interact through certain activation energy. In general, the dendrimers show higher viscosities than of the water because higher activation energy E′ and longer time is required for their reorientation. For example, activation energy of G_1 is 854.39 kJ $mol^{-1}K^{-1}$ which is because unexpectedly very high rotations and reorientations. These forces increase many times for G_2 with greater rotational and electronic rearrangement due to greater entropy. The B values for G_2 are slightly higher than those of the G_1 with a similar interacting dynamics where more branching develops higher hydrodynamic volume. Several theories such as Mark-Houwink-Sakurada Equation structurally illustrate action mechanism of biosensors in context of degree polymerization as 6 and 2 times, respectively.

15. Binding volume (v_m) impacts

Boundary is to be determined that where from a restriction on binding volume becomes effective when the biosensor undergoes activities. So far no concrete studies are cited to define a critical point drawing a line for restriction to be effective or not, and the solvents also contribute. For example, the water is a poor solvent for macromolecules (2), and such data become highly relevant for G_2 type biosensors where chain length is 2 times higher than of G_1 type biosensor. The v_m values for G_2 are higher that prove stronger biosensing action. Newly developed biosensors must be stable at workable temperatures. In this context their temperature stability is studied with thermal gravimetric analysis (TGA) as their disintegration ($\partial m/\partial T$, g/Kelvin or g K^{-1}) during biosensing activities would defeat the purpose to use them as biosensor.

16. Thermal stability of new materials

This technique predicts a weight loss (g min^{-1}) on thermal decomposition or disintegration ($\partial m/\partial T$) that infers their thermal stability where compositional or thermally induced transitions do not disrupt original biosensor structure on a pattern of the catalyst. Their thermogramme must be a straight line with time with respect to variable, may be temperature of the process or the pressure or the polarity of the medium. The G_1 and G_2 both have marked slight decomposition from 80 to 120°C due to loss of weakly bound water with @ 2.4 g min^{-1}. The 2nd small zone noted around 250-260°C, which confirmed higher temperature stability of the targeted biosensors. At 280-360°C a sharp weight loss with 22.5

g min^{-1} was observed. The TDEMTA is slightly complicated decomposed with several rates. The dendrimer shows 22% weight loss @ 2 g min^{-1} and 19.63% weight loss @ 1.2 g min^{-1} at 620°C and 800°C respectively. Hence the structure of dendrimer comparatively resists temperature change. The thermogramme shows least loss at around 80-100°C with almost a stable curve. The dendrimer consists 3 polar zones with lone pair of electron on nitrogen; also the π conjugation develops polarity responsible for binding of the polar molecules such as water and ethanol. At 250°C, a sharp weight loss @ 5.8 g min^{-1} with molecular reorientation infers weakening of hydrogen-bonded structure. The critical changes due to thermal energy were observed after 320°C, may be, due to breaking of branches but at about 420 and 450°C. The G_2 (2, 4, 6 HDEMTA) shows 11 thermal decomposition zones with different rates with sequence of physical transformation and breaking of bonds. The G_2 shows comparatively higher thermal stability than those of G_1 with weight loss @ 0.05 g min^{-1} from 50 to 80°C. At 340.82, 382.45°C, the 3rd and 4th reactive zones infer elimination of weakly bonded and hydrogen bonded molecules with 39.6 % weight loss @ 2.5 and 4.0 g min^{-1} as it is highly branched with a web like structure. The G_2 with 12 terminal branches radiated from a central hub and the zones at around 450 and 560°C demonstrate breaking of branches with decomposition of 62.05% weight loss. Physicochemical, spectroscopic and thermal decomposition studies inferred similarity in structures of 1st and 2nd tier dendrimers except degree of polymerization 2 and 6 times, respectively. It infers that larger branching that inhibits the IMMFT linkage for biosensing linkages due to more stability. The G_1 depicted below.

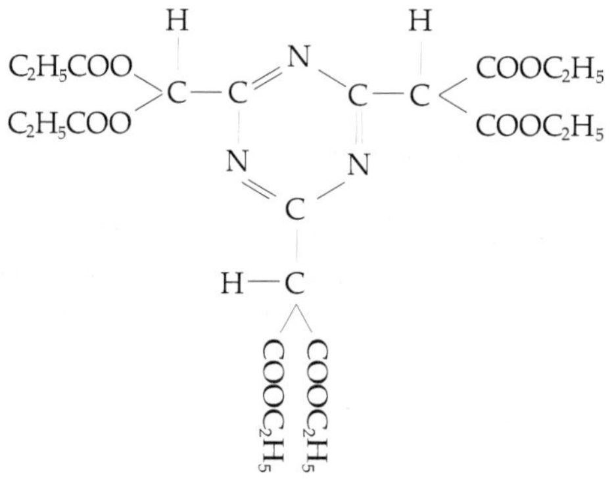

Fig. 5. 2,4,6 TDEMTA G_1, a white solid washed with water dried at NTP stored in P2O5 filled desiccator

16.1 IMMFT

Figs. 5 and 6 depicts most effective and befitting model fitted in IMMFT concept for working as excellent biosensors due to multifunctional force centers that monitor biosensing activities and interacting energies. Each force centre has individual electrostatic force (ESF) confined and aligned in A most distributive manner based on Boltzmann energy distribution concept. The ESF is defined with densities. Similarly the biosensor develop cumulative intermolecular force (IMF) based on ESF and also due to covalent bonding forces (CBF). The CBF deals with viscosities. For example, the G_1 and G_2 densities and viscosities are measured with Antaan Paar densitometer and Survismeter (Calibration no.

0607582/1.01/C-0395, NPL, Govt. of India) respectively at NTP. Their data are given in Tables 1 and 2. Further excellent newly developed conceptual parameter named as friccohesity could serve as a most useful data to illustrate dynamic movements of the dendrimers during operational course

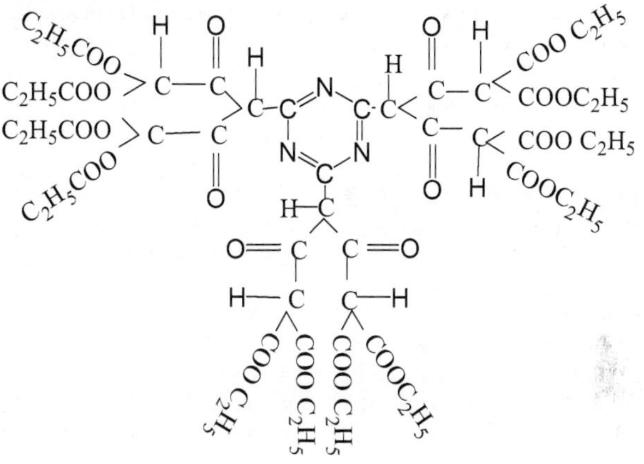

Fig. 6. The 2,4,6 HDEMTA G_2 tier dendrimers, white precipitate filtered, washed with warm water (Singh et al, Synthetic Commun 38, 2857, 2008)

16.2 Data support

dendrimers	Temperature (K)	ρ^0	S_d	S_d^*
G_1	298.15	0.9967	0.1948	-12.9449
G_1	303.15	0.9914	0.1238	-6.5065
G_2	298.15	0.99684	0.1947	-12.9449
G_2	303.15	0.9915	0.1238	-6.5065

Table 1. Limiting densities $\rho^0/10^3$kg m^{-3}, 1st and 2nd degree slopes, $S_d/10^3$kg^2 m^{-3}mol^{-1} and $S_d^*/10^3$kg^4 m^{-3}mol^{-3}, on regression densities against c%.

dendrimers	Temperature (K)	B	D	D'
G_1	298.15	45.3607	- 13623.30	947527.20
G_1	303.15	- 6.4835	2357.29	-170666.91
G_2	298.15	45.3318	-13596.20	945149.85
G_2 dendrimer	303.15	-6.34206	2306.44	-165958.33

Table 2. Intrinsic viscosity B/0.1 kg mol^{-1}, slopes D/(0.1 kg mol^{-1})^2 and D'/(0.1 kg mol^{-1})^4.

17. Novel biocompatibility for new biosensor materials development

Historically, polyvinylpyrrolidone (PVP) in 2nd world war, was used as artificial blood plasma as anticoagulant to compensate shortage of the blood due to infinite war causalities.

Since the PVP molecule is sizable and was used to develop melamine-formaldehyde-polyvinylpyrrolidone (MFP) resin with excellent biosensor activities and porous structure to perform several activities such immobilization of the bacterial growth, adsorption of toxic metal, alkaloids, narcotic drug, protein unfolding, biomacromolecular engineering and others. Its microstructure obtained with scattering electron microscopy in depicted in Fig. 7 which is highly porous with sizable void spaces. Dispersive and adsorptive activities of the MFP biosensor were studied by blending it with glycerol, a biocompatible molecule and an impact is also depicted in Fig. 7. The glycerol is trapped and equally distributed as per Boltzmann distribution law where the friccohesity (equation 3) resolves the thermodynamic assistance to distribution as its motions are developed due to concentration gradients.

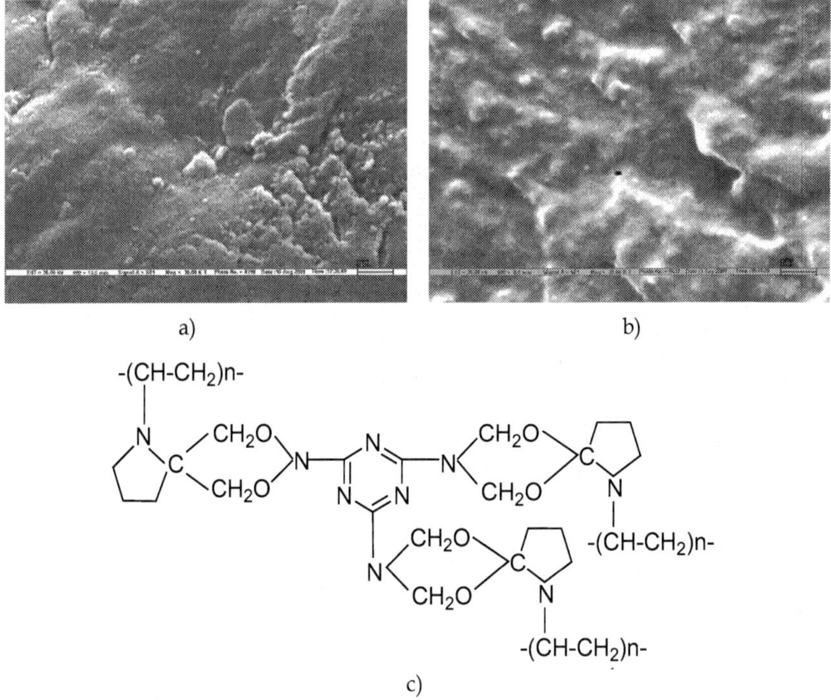

Fig. 7. a) MFP; b) MFP + glycerol; c) Molecular structure of MFP, core is belongs to melamine bridged via CH_2O to PVP units (Singh et al, J. Appl. Polym. Sci. 114, 1870, 2009)

18. Biocompatibility essential condition

The biocompatibility of newly developed molecules to use them as biosensor is essential condition hence few molecules illustrated in Figs. 8a and b are developed. It could be achieved when the highly biocompatible molecules are used in their preparation. The 2,4,6 triglcerate triazine, Tri (2, 4, 6) glycerate triazine (TTGTA)biosensor which have been prepared with glycerol a most biocompatible molecule and widely used in biochemical, biophysical, and essential part of the oils and fats.

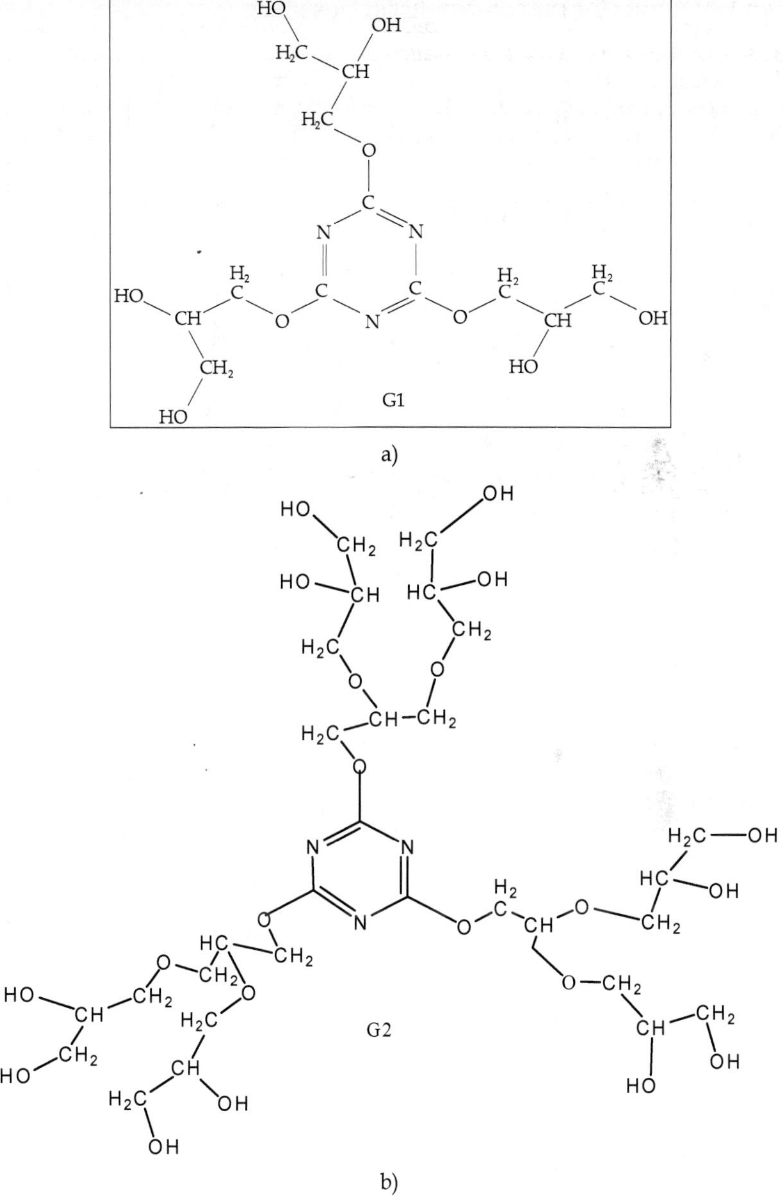

Fig. 8. a) 2, 4, 6 triglcerate triazine (TGTA); b) Tri (2, 4, 6) glycerate triazine (TTGTA) biosensor dendrimer with anticancer mechanism and drug carrier Branching act as tentacles to rupture cancerous cells and cutoff food supply from normal cells

19. Supramolecular biosensor or metallic biosensor

Silicon is well known and established biocompatible metal and has been used in several surgery, implants and artificial catheter preparation. Thus Poly-N-vinylpyrrolidone oximo-L-Valyl-Siliconate, a supramolecular biosensor, was prepared using silicic acid as core unit with the PVP and amino acid. The structure is depicted in Fig. 9. It has wider intramolecular voids to bioremediate toxic metals such as Hg, Cd and intrafacial interactions.

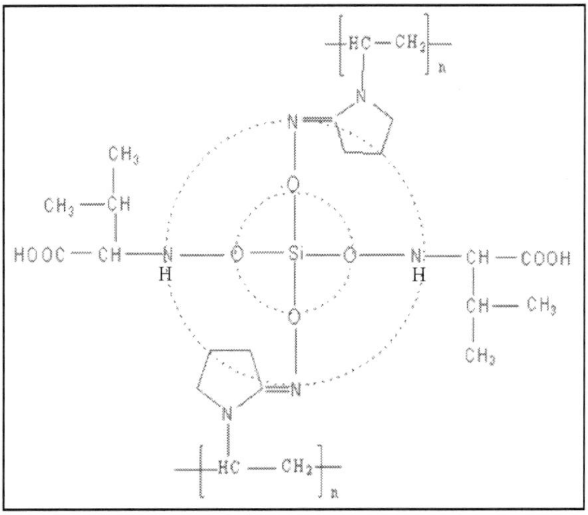

Fig. 9. Poly-N-vinyl pyrrolidone oximo-L-Valyl-Siliconate (Singh et al, Bull. Korean Chem. Soc. 31, 1869, 2010)

20. Tentropy

In general, the dendrimers do have effective tentacles which have effective movements and entropy induced changes. Fundamentally any effective change that cause a dent in particular dimension with effective friccohesity incorporating cohesive and frictional forces associated with dynamic motions is noted as tentropy. This an in significant feature of the biosensors that lead the message to viable mode and to amply so that a minor change is measurable and noticed in case of some critical turns. Since friccohesity is a turn force theory (TFT) or dual force theory (DFT) hence such changes are most desired to be measured in case of the biosensor sciences. The tentropy is most desired parameter with molecules such as TTGTA.

21. Biosensor science

Truly the biosensor is a most fabulous science and closely associated with thermodynamics and biothermodynamics. Its friccohesity could offer a most relevant and legitimate information about the signal sensing which is reported in terms of the chemical change without disrupting original structure of the biosensor except moderate and reversible reorientation and motions to suit the signal to be transported and reached to a measuring devices. Thus the electrochemical signal noted as force is depicted with equation of the net force theory where the IMMFT is really immensely related to inner and outer correlation of the surface and bulk phase structure changes of the biosensors. Such forces could trap with equation 4 which newly formulated to study weaker and molecular forces in mixtures of different polarity and thermodynamic changes.

$$F_{nimf} = F_w - F_{pm} = \frac{1}{4\pi\varepsilon_0}\left[\frac{q_w^- q_w^+}{r_w^2} - \frac{q_{pm}^- q_{pm}^+}{r_{pm}^2}\right] \qquad (4)$$

The F_{nimf} is net intermolecular force, F_w and F_{pm} as water and polar molecular forces, q denotes their respective electrostatic poles with r as their distances, the ε_0 medium permittivity. A net intermolecular force is responsible to generate the signal where electronic charge conduction moves over atoms due to atomic combinations. In general, the molecules such as oils also develop electrostatic potential ($-e^2/r$) and kinetic energy ($p^2/2m$) in contact of the $H^{\delta+}$-$O^{2\delta-}$-$H^{\delta+}$, a polar water molecule. It generates weaker Vander Waals and Lennard Jones Potential. The molecular forces, F along several motions, orientations with definite cohesive forces which are highly useful for acoustics type biosensors and useful for structural protein unfolding as an excellent model of molecular motions. In general, the newly developed biosensors have been found in a nanorange depicted in Fig. 10.

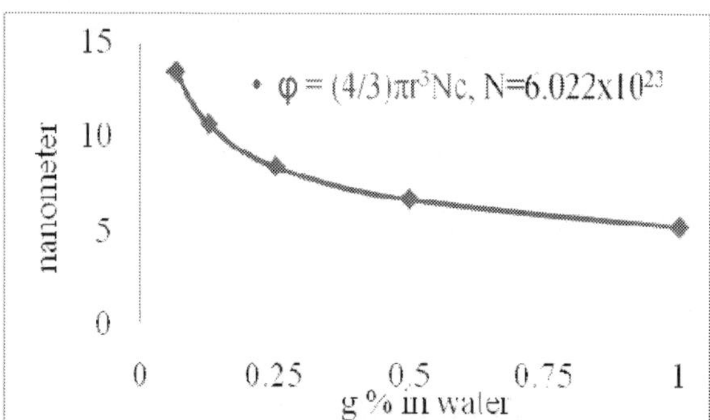

Fig. 10. In general, molecular size of the newly prepared biosensor

The φ is volume fraction, N_c or N Avogadro number, r radii of the molecule.

22. Induced electronegativity

When polar molecule is mixed with partially polar molecules the electronegativity is induced that enhanced higher wetting as compared non-induced electronegativity. The equation 7 with electrostatic potential ($-e^2/r$) of the water and kinetic energy ($p^2/2m$) of organic molecules caused stronger interaction with stronger CBF, Van der Waals and Lennard Jones Potential. A hydrophobicity of organic molecules could be noted that could also cause several motions, orientations and bond stretching with shear stress and strain. Pauling described interaction due to electronegativities as in equation given below.

$$\chi_A - \chi_B = \frac{1}{eV^{1/2}}\sqrt{E_d(AB) - \left[\frac{E_d(AA) + E_d(BB)}{2}\right]}$$

The E_d is dissociation energy, the χ_A and χ_B are electronegativities of A-B and A-A and B-B type molecules such water and p-cresol with a mixing with rise in temperature. Oscosurvismeter could also be noted as an effective instrument to deal with such molecules as it tracks osmotic pressure, conductance, surface tension and viscosity together. Also econoburette for simple titrations could be noted as indicators or sensor for indicating pollutants in water after doing titrations with this.

23. Sustainable tech for physicochemical characterization of biosensor: Borosil Mansingh Survismeter

Architectural molecules are the most sensitive and befitting to act as ideal biosensor with 100% activities which are altogether different as compared to ideal molecule with 0% activities. Thermodynamically $\Delta G_{biosensor} < 0 < \Delta G_{ideal}$ relationship exits with $\Delta S_{biosensor} > 0 \neq \Delta S_{ideal}$. The ΔG and ΔS stand for Gibbs free energy and entropy respectively. In general ideal systems are away from thermodynamic systems due to an absence of interactions with them that generate and design interacting and reacting molecular materials analyzed with green analytical device noted as Borosil mansingh survismeter. It is most sensitive to elucidate structural and tentropic changes associated with systems. The biosensor are covered under interacting and reacting molecular materials (IRMM) developed as academically and industrially functional materials (AIFM) and currently cover wider industrial and academic horizons of research and application. Two categories of materials transparently stand namely (a) thermodynamically, optically, electrically, chemically interacting molecular materials and (b) thermodynamically, electrically, optically, chemically reacting molecular materials. Both interaction and reacting molecular materials (IRMM) require accurate and précised sustainable trusted analytical devices (STAD) for trapping, capturing, modernizing, manipulating, manifolding molecular potential for developing new and novel materials. Differentiating and integrating profile (DAIP) of interacting, reacting and allied properties (IRAP) of molecular materials under material sciences need STAD for prerequisite experimental analysis (PREA) for efficient quality control, quality formulations and assurance, testing, calibration and similar others of industrially materials with certain physicochemical indicators (PCI) with interesting physicochemics and molionic materials.

In this context, surface tension, interfacial tension wetting coefficient, viscosity and Friccohesity as useful PCI offer ample opportunities in checking potential of biosensor including cosmetics, petroleum & oils, polymer and textile, paper pulp, soap & detergent, liquid soaps, drug designing, molecular weight determinations, water binding, holding, molecular aggregations, polymerization, dissociations, tiers are regularly and routinely analyzed. Frequent laboratory experiments for such purposes requires manifolds resources, time, skilled manpower, infrastructure and other accessories are urgently used where individual devices are failed, discarded redundant, beyond reach, troublesome, tiring, accessible to accidents and cost intensive. Borosil Mansingh Survismeter, being a green science STAD measures abovementioned PCI together with 98% saving of experimental resources, infrastructure and skilled manpower, for Interacting and Reacting Molecular Materials. It lays down a foundation of new fundamental chemistry noted as Friccochemistry and most suitable for giants, supramolecular, dendrimers and similar others. The friccochemistry defines a fundamental science responsible and operation for creating vacancies of PCI and IMMFT simulations, combinations and synergies. For example, the thermodynamics and simulations of biphasic systems do cause a practical potential for extractions, salting out or in phenomenon for recovery of target drug, proteins or other molecules. Novel sciences are urgently required for safer and sustainable technique to detect extra elements present in biosensor molecules as the elements such as halides (chloride, bromide and iodide), nitrogen and sulphur do induce viable sensing electrostatic forces for enabling strength to the functional groups. For such identifications Safer Technique for Sustainable Sodium Extract Preparation for Extra Elements Detection listed as Nonbreakable Sodium Ignition Apparatus (NOSIA) was developed and used for wider experimental determinations.

It becomes a milestone in exponentiation and boosting up science and research on biosensors to open new gateways and explore newer opportunities. Especially chemical interaction and interfacing (CII) with several interdisciplinary for learning and understanding complex through laboratory experiments. The surface tension and viscosity measurements need cost intensive Tensiometer and Brookfield viscometer which he could not afford. Work on biosensors associated with protein unfolding dynamics needs accurate and précised surface tension and viscosity data. For few experiments looking for some novel feature of fluid dynamics but suddenly experimental failed to produce reliable results and idea was revised and experimented several times but there was no outcome despite best efforts and economics. The failure led to a novel concept of dealing surface dynamics and viscous features of BSA protein aqueous solutions. There was a historic moment when a failure diverted attention to another hidden science of highest potential of measuring surface tension and viscosity together. Borosil Mansingh Survismeter Singapore Patent no. 126089, New Green Analytical Tech Laboratory Equipment is easily affordable within ordinary experimental conditions. Borosil Mansingh Survismeter is new and novel breakthrough in laboratory instrument with wider analytical potential for quality control and formulations in material applied sciences. It has unique cutting edge and salient facilities and feature based on n-in-one based on "On and Off" or 0 and 1" circuitry functional loops.

- Fundamentally, it measures surface tension, interfacial tension, wetting coefficient, viscosity and Friccohesity together of aqueous, non-aqueous, aprotic dipolar, polar, protic polar and non-polar solvents and mixtures.
- The absolute and relative parameters are measured with 95.5% CV (confidence variance).
- It works on theory of R4M4 of materials and methods with highly précised and accurate experimental results along 100% inhibition of polluting discharges in experimental determinations.
- The parameters measured with it are highly significant for quality analysis of pharmaceuticals, biochemicals, cosmetics, agrochemicals, food and beverages, petroleum and oils, polymer and proteins, sol gels, soaps and detergents, inks, colloids, emulsion technology, cutting oils, lubricating, viscous, moderately viscous and highly viscous materials.
- It is especially useful for characterizing liquid mixtures of biopolymers, spuramolecular chemistry, biotechnological processes and molecular interacting engineering of the biomolecular devices, tracking interacting molecular forces, water binding capacities and structural changes during processes.
- It is an asset for volatile, moderately and highly volatile liquids and mixtures, volatile organic compounds, flammable liquids, carcinogenic materials as samples are completely jacketed for surface tension determinations.

Environmental friendly:

- Currently, environmental and user safeties have been in focus and urgent needs, so steps are being initiated to implement innovative ideas in experimental devices to cop up and to save recurring in laboratory practices.
- The Survismeter is a most recent practical solution to apply multifaceted, multipurpose, multidimensional, fast track and most inspiring science.
- The Survismeter is a most defective analytical tool for industries that excellently enrich a spirit of research initiatives and to commercialize innovative ideas to benefit the society at large.
- Historically, Inventor Industry Interaction (I_3) has been on cards for time immemorial to resolve the analytical problems in material sciences.
- Since its inception the students and scientists are fascinated to apply it in various fields of applied sciences like cosmetics, sol gels, drug designing, pesticides, insecticides, syrups, coatings etc.
- In general, syllabi curricula of graduate and postgraduate classes of chemistry/chemical technology, pharmacy, physical sciences, do have experimental provisions for measuring surface tension, interfacial tension, molecular surface areas, wetting coefficient and viscosity.
- It saves electricity, water, manpower, chemicals, laboratory infrastructure, glass materials, glassblowing gases LPG, oxygen and others by 97% along 98% reduction in dissipation of heat contents.
- The equipment works on principle of pressure gradients generated with help of "Cutting Off and On Devices" in form of continuum and non-continuum models of fluid dynamics.
- The survismeter is most excellent model for study of Liquid-Liquid Interfaces (LLI) of two immiscible solvents. In general, homogenous liquid mixtures of components i and j do follow $dG_{ij} = 0$ where both i and j are fully dissolved and attain equilibrium with zero value of the Gibbs free energy change (dG_{ij}).

Asset for Physical, chemical biophysical, pharmaceutical, biotechnological sciences laboratories, for measuring surface tension, surface excess concentration, molecular surface area, interfacial tension and viscosities of samples with single unit. Since centuries, these parameters are being measured individually with much experimental resources and laboratory infrastructures.

- The Survismeter has reduced resources including support materials like glass, blowing gases, manpower, laboratory space, electricity, water, chemicals, by 97%. A statistical analysis of the resources being used for measuring surface tension, interfacial tension and viscosity in 219 Indian Universities was made and noted about 98% saving if Survismeter is used in place of individual apparatuses.
- Apart from Universities and colleges, characterization of materials for coating, polishing, cosmetics, VOC, thinners, ethers, acetone, hexane, ethanol, benzene, CCl_4, printing, calendaring, dry cleaning, textile cleaning-drying, interior decorations, fumigation, sprayants.
- Generally nonideal systems of surfactants industrially act as emulsifier, defoaming agents, spreaders, special detergents, household and industrial detergents, liquid detergent, industrial application-textile, agriculture, pulp and paper, etc, scouring agent, emulsion polymerization, dispersant.

- It is safe in handling, sample loading with no hazards and no discharge of polluting fumes/materials. It occupies minimum laboratory infrastructure.

The Survismeter determines Gibbs adsorption isotherm (surface area Γ, mol m^{-2}), wetting coefficient of phase forming liquids with high accuracy & precision, and is multipurpose and fascinating. The surface and interfacial tensions are beneficial to surfactants, waxes, inks, soaps-detergents, cosmetics, pharmaceutics, oils-petroleum, sol-gels, emulsions polymers, solvents, supercritical solvent, lubricants, textile and others. The equation given below explains wetting coefficient of biosensor.

$$\frac{\eta h^2}{t} = \left(\frac{\eta h^2}{t}\right)^0 + A\gamma$$

The mansingh equation given under also defines Brownian and Boltzmann energy distribution of biosensors.

$$\sigma = \sigma_0 \left[\left(\frac{t}{t_0} \pm \frac{B}{t}\right)\left(\frac{n}{n_0} \pm 0.0012(1-\rho)\right) \right]$$

The B/t and 0.0012(1-ρ) range from 10^{-7} to 10^{-6}, and are omitted then equation becomes as

$$\sigma = \sigma_0 \left[\left(\frac{t}{t_0}\right)\left(\frac{n}{n_0}\right) \right]$$

$$\text{or } \sigma = \sigma_0 \left[\left(\frac{tn}{t_0 n_0}\right) \right]$$

$$\text{or } \sigma = \frac{\sigma_0}{t_0 n_0}\left[(tn) \right]$$

Putting $\frac{\sigma_0}{t_0 n_0} = M_c$ then $\sigma = M_c\left[(tn)\right]$, the M_c is Mansingh constant which illustrate activity coefficient of the biosensor under defined physicochemical constants. The mansingh constant is a vital tool of friccochemistry applications, efficient and reliable characterizations and categorization of the biosensors.

24. References

[1] Buhleier, E. Wehner W., Vogtle F., Synthesis, 155 (1978).

[2] Tomalia D. A. Starbrush TM/cascade dendrimer: fundamental building blocks for new nanoscopic chemistry set. Aldrichim. Acta. 26, 91 (1993).

[3] Lee, J.J.; Ford, W.T.; Moore, J. A.; Li, Y., Macromolecules 1994, V27, 4632-4634. Tomalia, D. A.; Naylor, A. M.; Goddard III, W. A. Angew. Chem., Int. Ed.Engl. 29, 138 (1990).

[4] Cook, Renert, T. J.; Sunlick, K.S. J. Am. Chem. Soc. 1986.

[5] Tomalia, D.A., Polymer Journal, V12, No.1, 117 (1985).

[6] Peppas N.A. Star Polymer and dendrimers: Prospects of their use in drug delivery and Pharmaceutical applications Controlled Release Soc. News. 12, 12 (1995).

[7] Man Singh, Hema Chand, K.C. Gupta 'Chemistry & Biodiversity, 2 (6) 809 (2005).

[8] Flory P. J. and Leutner F. S., J. Polymer Sci., 3, 880 (1948); 5, 267 (1950).

[9] Einstein A., "Investigations on the theory of the Brownian Movement," chap. III, Dover, New York (1956).

[10] Tomito B, J Appl. Polym. Sci, Vol. 15, 2347(1977).

[11] Braun, D & Gunther P, Angew. Chem. 128 (1984) 1.

[12] Freemntle, M., C& EN, November 1, 27 (1999)

[13] Lee, J.J.; Ford, W.T.; Moore, J. A.; Li, Y., Macromoluclues V27, 4632-4634 (1994).

[14] Wells, M and Crooks, R.M., J Am. Chem. Soc. V16, 3988 (1996).

[15] Weaton, R. E, J Phys Chem, A, 105, 1656 (2001).

[16] Bhyrappa, P,'Young, J.K.; Moore, J.S.; Sunlick, K.S., Am. Chem. Soc. V118, 5708 (1996).

[17] Tomito B, J Appl Polym Sci. 15, 2347 (1997).

[18] Vogtle, F., and Fisher M, Angew. Chem. Int. Ed. V38, 884 (1999).

[19] Nagasaki, T., Kimura, O., Ukom, M., Arimori, S., Hamachi, I., and [1] Shinkai, S., J.Chem. Soc. Pekin Trans. 75 (1994).

[20] J.M.J. Frechet, C.J. Hawker, I. Gitsov, J.W.Leon, J.W.J Macromol. Sci. Puer Appl.Chem. A33, 1399 (1996).

[21] Elecrochimica Acta, 1977, Vol.22, pp. 1183-1187, Pergamon Press. Printed in Great Britain.

[22] Man Singh, J. Appl. Polym. Sci. 92, 3437 (2004).

[23] Man Singh, Ind. J. Chem. 43A, 1696 (2004).

[24] Janca J Ed Static exclusion liquid chromatography of polymer (mercel Decker, Inc, New York), 1984.

[25] Mohammad, A; Wahab, M; Azhar Ali; and Mohammad A Mottaleb; Bull. Korean Chem. Soc. Vol.23, No. 7, 959 (2002).

[26] Man Singh and Sanjay Kumar, J. Appl. Polym. Sci. 93, 47 (2004).

[27] Man Singh, J. Instr. Exp. Techn. 48(2) 270 (2005).

[28] Man Singh and Ajay Kumar, J. Sol. Chem., 2006, 35(4) 567-582.

[29] Man Singh, J. Chem. Thermodyn, 39, 240-246, 2006.

[30] Man Singh and Vinod Kumar, Intern. J. Thermodyn. 10 (3) 121-133, 2007.

[31] Man Singh and Hideki Matsuoka, Surf. Rev. Lett., 16 (4), 509-608 (2009).

[32] Man Singh, Neeti Chaudhary, R. K. Kale and H. S. Verma, Arabian. J. Chem. In press.

8

Silicon and Silicon-related Surfaces for Biosensor Applications

Ahmed Arafat and Muhammad Daous
King Abdulaziz University
Saudi Arabia

1. Introduction

Biosensing systems, such as enzyme, immunosensors, and DNA microarrays, are widely used in the field of medical care and medicine manufacturing [Spochiger-Kuller, 1998]. Recent developments in these devices require a high performance integrated micro-multi-biosensing system, which can be employed for the recognition of an individual biomolecule and the analysis of bioreactions at the single molecular level. Constructing a highly sensitive biosensing system, precise fabrication of the electrode parts for molecular recognition is of significant importance. In these context, organic monolayers have self-assemble ability onto surfaces [Ulman, 1991]. Monolayer-modified electrode is suitable as the template for ordered immobilization of biomolecules. On the other hand, it is preferable that the detection system can detect the signal immediately, with a high sensitivity.

Formation of covalently-bound organic monolayers has been particularly developed in the last two decades. The main benefit of organic monolayers is to add functionality to inorganic surface via the adaptable tailoring of surface properties. These monolayers keep the bulk features of the material (electrical, optical, magnetic, mechanical and structural), while their surface properties (wetting, passivation, bioresistance, biochemical affinity, etc...) can be tuned through a nanometer-sized grafting.

This chapter provides substantial information on modification of silicon and silicon-related surfaces by organic monolayers to get the reader acquainted with the different techniques employed in tailoring the surface properties towards biosensing capability.

2. Silicon and silicon related surfaces

2.1 Silicon surfaces

Silicon was discovered by Berzelius in 1824 and isolated as amorphous brown powder. Crystalline silicon was first prepared in 1854 as a grey material with metallic luster. Normally, silicon is prepared by reduction of silica, using different reducing agents. Silicon has a crystal structure similar to diamond, with Si-Si bond length of 2.3 Å [Cotton & Wilkinson, 1999].

Cleavage of a silicon crystal results in a large variety of surfaces. Several investigations on these surfaces have been carried out under ultra high vacuum (UHV) conditions [Hamers & Wang, 1996]. The surfaces are characterized by their Miller indices, which refer the plane thorough which the crystal was originally cleaved.

2.2 Surface orientations of silicon

The most common surface orientations of commercially available silicon are Si(100) and Si(111), *See Fig. 1.*

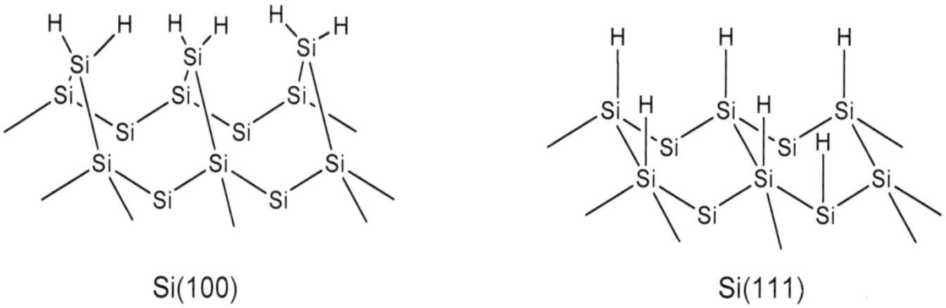

Si(100)   Si(111)

Fig. 1. Hydrogen-terminated Si(100) and Si(111) surfaces.

Upon exposure to air both become rapidly coated with a self-limiting, thin native oxide (SiO_2) that can be removed thermally under UHV conditions or chemically by immersion in aqueous fluoride-containing solutions [Buriak, 2002 & Leftwich, et al., 2008 & Sieval, et al., 2000a & Wayner & Wolkow, 2002]. Typically, Si(100) wafers are treated with 2.5% HF to yield dihydride-terminated Si(100) surfaces that are on the nanometer scale still rough. In contrast, Si(111) yields atomically flat terraces with monohydride-termination, *(See Fig. 2)* because during etching in argon-saturated 40% NH_4F solution the initially rough Si(111) surface will spontaneously smoothen as a result of the differences in reactivities of different crystal faces [Allongue, et al., 2000]. Both hydrogen-terminated Si surfaces are sufficiently stable that they can be handled in air for short periods of time (tens of seconds), allowing wet-chemical modification routes like the formation of organic monolayers. Because the lattice constant of thermally grown silicon dioxide (SiO_2) matches best with the crystal plane of Si(100), for electronic devices that use the oxide as an electrical insulator Si(100) is the most used crystal orientation, since these results in the lowest concentration of defects at the SiO_2-Si interface. However, due to its atomic flatness and nearly defect-free hydrogen-termination, Si(111) is the best substrate for new hybrid organic monolayer-silicon devices [Hamers & Wang, 1996].

2.3 Silicon-related materials

A special attention has given in recent years to the formation of organic monolayers on other surfaces than silicon, namely, silicon-rich silicon nitride (Si_xN_4, 3.5 < x < 4.5) and silicon carbide (SiC). The exceptional mechanical and chemical robustness of Si_xN_4 and SiC make these substrates attractive for applications where harsh conditions and/or when prolonged exposure are applied. Moreover, Si_xN_4 and SiC have properties that differ from those of materials commonly used for the formation of organic monolayers (gold, glass, or silicon), and the chemistry presented herein thus provides the scientist or engineer with more choices in the selection of a suitable substrate.

Stoichiometric silicon nitride (Si_3N_4) can form robust insulating coatings, but this material can develop a very high surface stress that negatively affects its mechanical properties. In comparison, silicon-rich silicon nitride displays very low residual stress, and can form homogeneous coatings by chemical vapor deposition (CVD) [Andersen, et al., 2005]. The composition of the material can be controlled by tuning the proportions of the compounds used as precursors in the CVD process (usually NH_3/SiH_2Cl_2). This material is indeed used commonly, for example, for the coating of micro-fabricated membranes or microelectromechanical systems (MEMS).

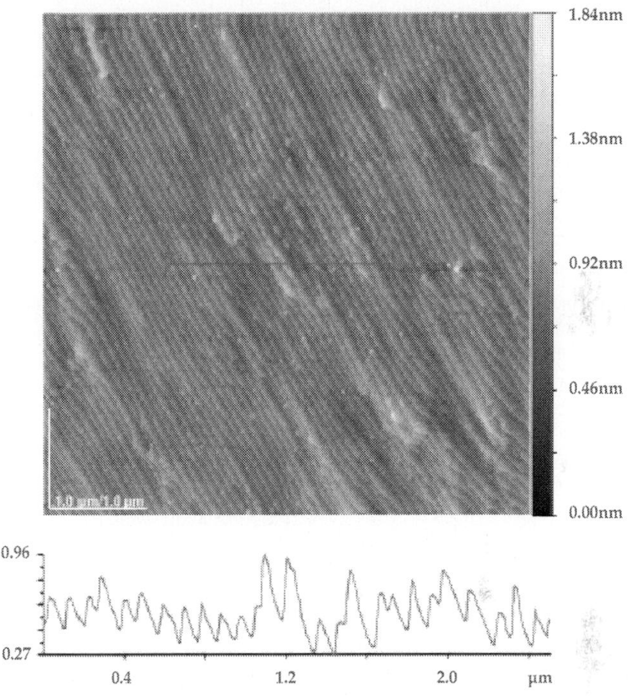

Fig. 2. H-terminated Si(111) surface, and section analysis along a line across the surface.

Silicon carbide had long been perceived as a potential replacement for silicon in electronic applications, but it was only in 1989, with the founding of CREE Inc., that SiC LED's (light emitting diodes) and high quality wafers became widely available due to the breakthrough of "step-controlled epitaxy"[Saddow & Agarwal, 2004]. Since then, the quality and availability of SiC materials have been steadily improved, and applications in high-power electronics and sensors are increasingly investigated. Application of Si_xN_4 and SiC can even be enhanced if effective surface modification techniques are becoming available.

3. Organic monolayers on solid substrates

Organic monolayers are layers that are precisely one organic molecule thick, and which are attached in a dense packing on a solid substrate. The attachment can be weak or strong, and can rely on either physical adsorption (e.g. electrostatic interactions) or chemical adsorptions (formation of chemical bonds). From the time when the revolutionary work on organic monolayers on gold [Nuzzo & Allara, 1983], glass [Maoz & Sagiv, 1984] and oxidized aluminum [Allara & Nuzzo, 1985], the field of organic monolayers has grown tremendously, and nowadays organic monolayers on numerous metals, oxides and semiconductors have been reported in literature. With these extremely thin organic films (typical thickness ca. 2-5 nm) the surface properties of the underlying substrate can be precisely controlled, and therefore organic monolayers find rapidly increasing application in many fields of interest, including surface hydrophilicity and lubrication, surface passivation, chemical and biological sensing, and molecular electronics [Love, et al., 2005a & b & Onclin, et al., 2005a & b].

Organic monolayers of alkylthiols on gold and alkylsilanes on oxidized surfaces are obviously the most extensively studied systems (Fig. 3a and b) [Love, et al., 2005a & Onclin, et al., 2005b & Ulman, 1996]. Due to the high affinity of the thiol group for the gold surface, the self-assembly of alkylthiol monolayers on gold is a highly flexible process, which is clearly displayed by the wide variety of functional and rather complex monolayers that have been prepared [Love, et al., 2005a]. In addition, the semi-covalent nature of the Au-S bond allows diffusion of already absorbed chains along the surface, and as a result well-ordered and nearly defect-free monolayers can be obtained in a simple and reproducible manner [Love, et al., 2005a & Ulman, 1996]. However, the semi-covalent Au-S bond is also the shortcoming of these monolayers, because its limited strength provides alkylthiol monolayers with only moderate thermal and chemical stability. This stability, both thermally and chemically, is significantly increased by the use of a covalent C-Si-O linkage to an oxide, as results from the attachment of alkylsilanes onto oxide surfaces [Onclin, et al., 2005b].

The increased stability comes at a price, however, as the preparation of alkylsilane monolayers on oxidized surfaces is highly dependent on the reaction conditions, and therefore considerably less simple and reproducible than achievable for alkylthiols on gold. In addition, while organosilane-derived monolayers can be prepared with a wide variety of functional moieties, their long-term applicability remains less than ideal since the interfacial Si–O bonds are susceptible to hydrolysis.

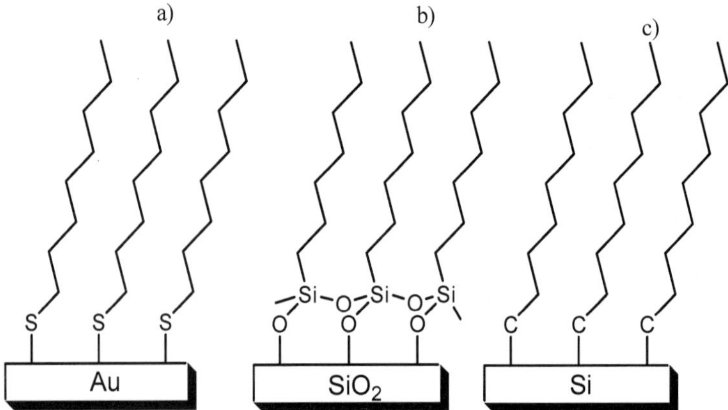

Fig. 3. Some examples of organic monolayers: (a) alkylthiols on gold, (b) alkylsilanes onto glass, and (c) 1-alkenes on oxide-free silicon.

4. Monolayers on oxide-free, hydrogen-terminated silicon surfaces

Due to the ongoing down-sizing of semiconductor devices, there is a significant interest in the surface modification of silicon. In this perspective, organic monolayers directly bound to oxide-free, hydrogen-terminated silicon are interesting candidates as they can easily be implemented in existing technology for the fabrication of silicon-based micro and nano-structured devices (Fig. 3c). The direct covalent linkage (Si–C bond) to the silicon surface provides a well-defined organic monolayer-silicon interface, and the nonpolar character of this strong bond make these monolayers thermally and chemically very robust [Linford, et al., 1995 & Sung, et al., 1997]. Moreover, because an intervening SiO_2 layer is essentially absent, direct electronic coupling between any organic functionality and the silicon substrate is possible, which provides an opportunity to enhance the device performance compared to SiO_2-covered electronic devices [Aswal, et al., 2006 & Cahen, et al., 2005 & Hiremath, et al., 2008 & Salomon, et al., 2006 & Vilan, et al., 2010]. Furthermore, using a semiconductor instead of a metal as a substrate/electrode has the advantage that – depending on the desired electronic properties of the final device – semiconductors with different doping levels and doping types can be used [Boukherroub, 2005 & Cahen, et al., 2005 & Salomon, et al., 2006 & Salomon, et al., 2007]. As a result organic monolayers on oxide-free silicon have great potential in the field of biosensors, molecular electronics and photovoltaic devices [Har-Lavan, et al., 2009 & Maldonado, et al., 2008].

Since the first reports of Chidsey and Linford [Linford & Chidsey, 1993 & Linford, et al., 1995], numerous new methods have been reported, and nowadays organic monolayers on oxide-free, hydrogen-terminated silicon can be prepared under a variety of conditions with both 1-alkenes and 1-alkynes. Over the last ten years several reviews about this topic have appeared in literature [Boukherroub, 2005 & Buriak, 2002 & Shirahata, et al., 2005]. Although initially harsh conditions (neat 1-alkenes or 1-alkynes with radical initiators and heat) [Sieval, et al., 1998] were required for the modification of planar silicon surfaces, the last decade displays a trend towards milder reaction conditions. In 1999 Sieval et al. [Sieval, et al., 1999] already showed that instead of neat 1-alkenes also dilute solutions of 1-alkenes can be used for monolayer formation on H-Si(100) under thermal conditions. Subsequently, Cicero et al. [Cicero, et al., 2000] demonstrated monolayer assembly on H-Si(111) by UV illumination at room temperature, and Stewart and Buriak reported visible light-promoted modification of porous silicon with 1-alkenes and 1-alkynes [Stewart & Buriak, 2001 & 1998]. Not much later, it was shown by Sun et al. [Sun, et al., 2005 & Sun, et al., 2004] on planar silicon surfaces visible light can initiate monolayer formation, even in dilute solutions.

Nowadays, it is widely accepted that monolayer formation occurs via a radical-chain mechanism on the surface (Fig. 4), even during mild visible light-induced monolayer assembly at room temperature [Eves, et al., 2004]. However, the exact initiation mechanism of the radical chain reaction, especially under these mild reaction conditions, is not yet completely understood. Radical initiators [Linford, et al., 1995] and UV light [Effenberger, et al., 1998] are capable of breaking the H-Si bond homolytically, which yields silicon radicals (silicon dangling bonds) that can act as a starting point for the radical chain propagation (Fig. 4, route 1). In contrast, using thermal conditions [Sieval, et al., 2001 & Sieval, et al., 2000b] or visible light at room temperature [Eves, et al., 2004 & Sun, et al., 2005 & Sun, et al., 2004], insufficient energy for homolytic cleavage of the strong H-Si bond is available. Nevertheless, as evidenced by scanning tunnelling microscopy (STM) (*see Fig. 5*) monolayer formation still occurs via island growth [De Smet, et al., 2005 & Eves, et al., 2004 & Mischki, et al., 2009]. This implies that propagation of the radical chain reaction still proceeds, but a different initiation mechanism must be active under mild reaction conditions.

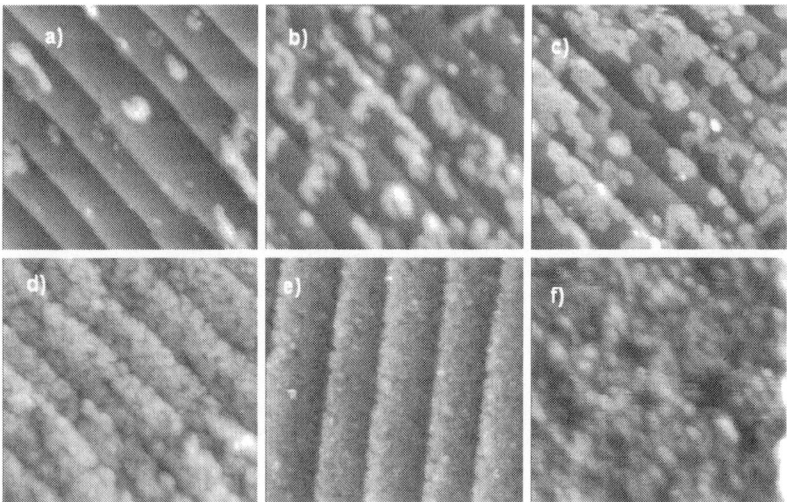

Fig. 4. The radical chain mechanisms for modification of H-terminated silicon surface with 1-alkenes (1) with radical initiators or UV irradiation and (2) with thermal activation or visible-light irradiation.

Fig. 5. STM images of hexadecyl monolayers on Si(111) surface taken at different time imtervals, a) 3 min., b) 15 min., c) 30 min, d) 2 h, e) 15 h and f) 24 h of irradiation with 447 nm [Eves, et al., 2004].

Inspired by the visible light-induced monolayer formation at room temperature, Sun et al. [Sun, et al., 2005] proposed an initiation mechanism based on photo-excited electron-hole pairs (excitons) near the silicon surface. These electron-hole pairs are susceptible to nucleophilic attack by a 1-alkene or 1-alkyne resulting in the formation of a Si–C bond and a carbon radical at the β-position +1 ion (Fig. 3, route 2). This radical can then abstract a hydrogen atom from an adjacent H-Si site and leaves highly-reactive silicon radical at the surface. A new incoming alkene or alkyne molecule can react with this silyl radical and in this way propagate the radical chain reaction at the H-Si surface. However, we note that although the increasingly milder reaction conditions that were shown to work with 1-alkenes will extend the range of functional groups that can be attached directly onto Si [De Smet, et al., 2003], at the same time the quality and thus the stability of these organic monolayers is decreased with respect to those obtained under more harsh attachment conditions [Sun, et al., 2005].

5. Crucial issues for organic monolayers on silicon

For all potential applications the stability of the monolayer and its oxide-free monolayer silicon interface are the crucial issues. Both depend, in principle, on the exclusion of water and oxygen from the monolayer-silicon interface. If water and oxygen can get to the interface via some defects in the monolayer, they will react with the many remaining H-Si sites (45-50% of the H-Si sites remain after completion of an alkyl monolayer) [Cicero, et al., 2000 & Linford, et al., 1995 & Sieval, et al., 2001 & Sieval, et al., 2000c & Wallart, et al., 2005 & Yuan, et al., 2003] and some small oxide patches will be formed. These trace amounts of oxide facilitate hydrolysis-based degradation of the monolayer via an excavation mechanism, and introduce electrically active interface states that change the electronic properties of the underlying Si atoms drastically. Thus, the primary role of the organic monolayer is to provide a hydrophobic environment that is not readily penetrated by water and oxygen molecules, and therefore the densest possible packing of the monolayer is desirable. As monolayer formation occurs via a meandering radical chain reaction on the silicon surface, and because diffusion of already absorbed chains to improve the ordering – as observed for alkyl-thiol monolayers on gold [Ulman, 1996] – cannot take place due to the strong covalent Si–C bond, steric hindrance of the covalently bound chains prevents insertion of new chains. Consequently, filling the last pinholes in the monolayer is hard and thus organic monolayers on oxide-free silicon are in general less ordered and almost never completely defect free. As a result the oxide-free monolayer-silicon interface, generally, has a limited long-term stability [Faber, et al., 2005 & Seitz, et al., 2006]. Furthermore, because many functional groups (including -OH, -CHO, $-NH_2$, -Br, -SH) are reactive towards a H-Si surface [Asanuma, et al., 2005 & Boukherroub, et al., 2000 & Faucheux, et al., 2006], preparation of ω-functionalized monolayers on H-Si is considerably more difficult than, for instance, with alkylthiols on gold. Here, the use of protected precursors, which do not react with the H-Si surface and after completion can be deprotected to yield the desired functional monolayer, could offer an outcome [Böcking, Till, et al., 2007 & Fabre & Hauquier, 2006 & Sieval, et al., 2001 & Strother, et al., 2000]. However, often quite harsh deprotection conditions are required that consequently affect the quality of the monolayer-substrate interface. As mentioned above, also the use of milder reaction conditions could be helpful. A nice example is the carboxylic acid (-COOH) functionality, which binds to the H-Si surface at elevated temperatures [Linford & Chidsey, 1993], whereas under mild photochemical reaction conditions carboxylic acid-terminated monolayers with only small to negligible indications of upside-down attachment were reported [Perring, et al., 2005]. Nevertheless, hydrogen bonding causes acid bilayer formation, which makes these monolayers hard to clean while for further functionalization an additional activation step via carboxylic anhydrides or N-hydroxysuccinimide (NHS) chemistry is still needed [Fabre & Hauquier, 2006 & Fabre, et al., 2008 & Hauquier, et al., 2008 & Strother, et al., 2000]. In addition, we note that the last years some interesting ω-functionalized monolayers are prepared, which showed no signs of upside-down attachment, are easy to clean, and allow further functionalization in a single step [Böcking, T., et al., 2006 & Ciampi, et al., 2007 & Li, et al., 2010 & Ng, et al., 2009 & Scheres, et al., 2010 & Yang, M., et al., 2008]. Finally, in view of the broad range of available patterning techniques [Garcia, et al., 2006 & Woodson & Liu, 2007] it is somewhat remarkable that thus far, only a limited number of patterning routes for organic monolayers on oxide-free silicon has been reported. In particular, because monolayer formation on H-Si can be initiated with UV or visible light, mainly photolithographic procedures were applied [Voicu, et al., 2004 & Wojtyk, et al., 2001 & Yin, et al., 2004]. In addition, micro-contact printing (µCP) – a fast and simple patterning technique, which is frequently used for alkylthiols on gold and alkylsilanes on oxide surfaces [Xia & Whitesides, 1998] – is currently not feasible with 1-alkenes and 1-alkynes directly on H-Si, due to the extended reaction times required for monolayer formation and related difficulties to remain a oxide-free monolayer-silicon interface. Only recently a number of elegant soft lithographic [Jun, et al., 2002 & Mizuno & Buriak, 2008 & Perring, et al., 2007] and scanning probe [Niederhauser, et al., 2001 & Niederhauser, et al., 2002 & Yang, L., et al., 2005 & Yang, M., et al., 2009] methods for patterning of organic monolayer on oxide-free silicon were published.

6. Organic monolayers on the surfaces of silicon-rich materials

Alkene-based monolayers were also formed on flat Si_xN_4 [Arafat, et al., 2007 & Arafat, et al., 2004], (Fig. 6) and 6H-SiC and polycrystalline 3C-SiC [Rosso, et al., 2008b] using thermal conditions close to those used for the surface modification of silicon. Good quality monolayers were obtained with several simple alkenes (e.g. water contact angles up to 107 ° for hexadecene-derived monolayers on both SiC and Si_xN_4, Table 1). The UV-induced formation of monolayers of semi-carbazide on H-terminated Si_5N_4 surfaces prepared under UHV conditions was also reported by Coffinier *et al. [Coffinier, et al., 2007]*. In addition, methyl- and ester-terminated monolayers were also formed on $Si_{3.9}N_4$ [Arafat, et al., 2004] and 3C-SiC substrates [Rosso, et al., 2008b & Rosso, et al., 2009], using wet etching with HF and UV irradiation in the presence of alkenes, under ambient conditions of temperature and pressure. Semi-carbazides and esters can be easily converted to amine [Coffinier, et al., 2007] and acid groups [Rosso, et al., 2009], respectively, which can serve for further attachment of biomolecules or biorepelling molecules and polymers [Asanuma, et al., 2006 & Coffinier, et al., 2005 & Love, et al., 2005a].

The advantage of alkene-based monolayers is their stability, mainly due to the absence of a silicon oxide layer, and the presence of stable and non-polar Si-C bonds, in the case of Si [Linford & Chidsey, 1993] and Si_xN_4, [Arafat, et al., 2007] [Rosso, et al., 2008b] and stable C-O-C bonds in the case of SiC surfaces [Rosso, et al., 2008a]. Stability measurements revealed the outstanding stability of thermally produced 1-hexadecene monolayers on Si_xN_4 substrates, in acidic or basic conditions at 60 °C, with changes in contact angles of less than 5° after 4 h of such a treatment [Arafat, et al., 2007]. On SiC substrates, a good stability is also obtained after 4 h treatments in 2M HCl at 90 °C and at pH 11 at 60 °C, with resulting water contact angle values of 106° and 96°, respectively (coming from 108/109° for the original alkyl monolayer) [Rosso, et al., 2009]. Even after 1 h in 2.5% HF solution, under which Si-O-Si bonds dissolve rapidly, 1-hexadecene monolayers on SiC still displayed water contact angles of 99°, which indicates the presence of a stable hydrophobic coating.

Compound (concentration)	$\theta \pm 1°$
$CH_2=CH-C_{20}H_{41}$ (0.4 M)	102
$CH_2=CH-C_{16}H_{33}$ (Neat)	107
$CH_2=CH-C_{16}H_{33}$ (0.4 M)	104
$CH_2=CH-C_{14}H_{29}$ (Neat)	107
$CH_2=CH-C_{14}H_{29}$ (0.4 M)	106
$CH_2=CH-C_{12}H_{25}$ (0.4 M)	105
$CH_2=CH-C_{10}H_{21}$ (0.4 M)	106
$CH\equiv C-C_{16}H_{33}$ (0.4 M)	104
$CH\equiv C-C_{14}H_{33}$ (0.4 M)	103
$CH_2=CH-(CH_2)_8CO_2CH_2CF_3$ (0.4 M)	85

Table 1. Water contact angle $\theta°$ of different monolayers on $Si_{3.9}N_4$ surface.

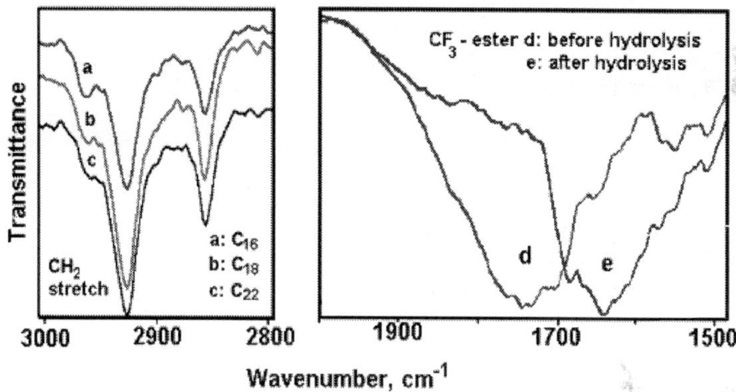

Fig. 6. Infrared reflection absorption Spectroscopy (IRRAS) data of modified Si_3N_x. (left) CH_2 vibrations after reaction of Si_3N_x with different 1-alkenes. (right) C=O vibrations after reaction of Si_3N_x with $CH_2=CH-(CH_2)_8CO_2CH_2CF_3$, before (d) and after (e) hydrolysis with 0.25 M potassium *tert*-butoxide [Arafat, et al., 2004].

Hydrogen-free diamond surfaces were also reacted under UHV conditions [Buriak, 2001 & Hovis, et al., 2000 & Knickerbocker, et al., 2003] with alkenes via a [2+2] cycloaddition or Diels-Alder mechanism. Hydrogen-terminated diamond surfaces could be functionalized with alkenes under UV irradiation [Knickerbocker, et al., 2003 & Strother, et al., 2002]. The reaction forms new carbon-carbon bonds, ensuring a robust grafting of monolayers. The same modification can be applied to amorphous carbon [Ababou-Girard, et al., 2007 & Ababou-Girard, et al., 2006]. Functional molecules, including DNA could be grafted in this way to diamond surfaces [Knickerbocker, et al., 2003] and the hybridization with complementary DNA strands could be monitored on the surfaces. In this case, the surface reactivity differs from that of silicon: the reaction initiation on diamond and amorphous carbon surfaces is due to their negative electron affinity [Nichols, et al., 2005] Upon irradiation with sub-band gap wavelengths, electrons are ejected from the surface into the surrounding alkenes, causing the formation of charged reactive species in the liquid close to the diamond surface [Wang, et al., 2007].

7. Biodetection based on organic monolayers on surfaces

Numerous examples of biodetection using organic monolayers have been described: for example, those based on fluorescence [Cattaruzza, et al., 2006], Raman scattering [Yonzon, et al., 2004], ellipsometry [Arwin, 2000 & 2001], infra-red [Liao, et al., 2006] or simply [Masuda, et al., 2005] have used the specificity of monolayer-functionalized surfaces. The geometry of the sensor also has a remarkable importance for the realization of the devices. Besides classical transmission and reflection modes, the need for compact sensors has favored the development of systems based on optic fibers and waveguide materials to transport light from the source to the binding area and to the detector [Sharma & Gupta, 2007]. Miniaturized microfluidics sensors, for example, can use silicon oxide or silicon nitride as waveguide and immobilization platform. In particular, Si_xN_4 is widely used, for example, as waveguide material in refractometric [Karymov, et al., 1995] or fluorescence [Anderson, et al., 2008] detection.

Among all optical techniques, one type of surface-based detection technique, surface plasmon resonance (SPR), has had magnificent success in the last 20 years [Phillips, K.S. & Cheng, 2007 & Phillips, M.M., et al., 2010]. Surface plasmon resonance (SPR) has been successful so far for many reasons include the easy cleaning and the infinite reuse of the sensor. Thiol based monolayers are normally grafted to gold or silver surfaces allow specific detection of DNA [Buhl, et al., 2007] or organic pollutants [Farré, et al., 2007 & Mauriz, et al., 2006] and pathogens [Chah & Zare, 2008] with coupling of DNA and antibodies. A significant improvement has been achieved with the formation of more multifaceted sensing architectures using the assembly of nanoparticles onto the gold surface of the SPR: the signal amplification caused by the coupling of the SPR signal with the localized surface plasmons of the metal or semiconductor nanoparticles (Fig. 7) has resulted in a 1000-fold increase in sensitivity, allowing the detection of picomolar concentrations, and approaches the performances of classical fluorescence based detection method of DNA hybridization [Hutter & Pileni, 2003].

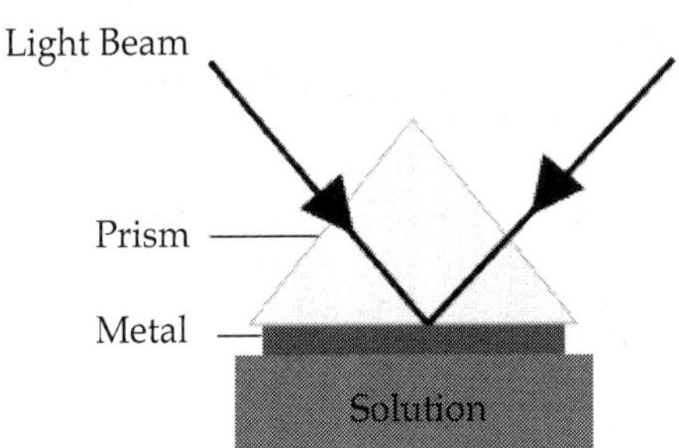

Fig. 7. SPR configuration for SPR detection: the light beam is directed towards the backside of a metal (gold or silver), by means of a prism. The front side of the metal is in contact with the solution to analyze. Adsorption of biomolecules on the metal changes the refractive index of the surface and decreases the intensity of the reflected beam.

The fluorescence of metallic or semiconducting nanoparticles has found their applications in biosensing and imaging [Bruchez Jr, et al., 1998 & Michalet, et al., 2005]. Similarly to SPR, the vibrating free electrons oscillations in metal nanoparticles can be together with environmental factors result in variations in optical properties [Xu, et al., 2008]. DNA hybridization on surfaces modified with organic monolayers can cause the subsequent immobilization of functionalized nanoparticles (sandwich assay) [Bailey, et al., 2003]; their presence on the glass sensor can then be monitored directly, or after an amplification step, often carried out by reduction of the metal salt to increase the size of nanoparticles [Yang, N., et al., 2007].

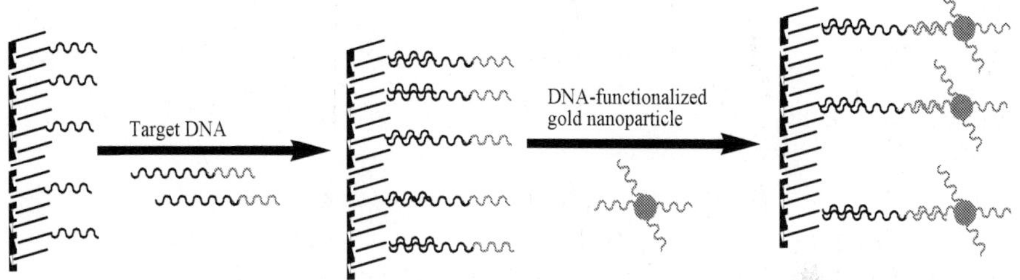

Fig. 8. Simplified SPR detection: nanoparticle-enhanced SPR.

In a similar way, the aggregation of metal nanoparticles functionalized with thiol-modified DNA single strands in the presence of the complementary DNA will cause a change in their melting behavior that can be observed by a change in their optical absorption, fluorescence or simply the visual aspect of the nanoparticles (Fig. 8) [Elghanian, et al., 1997]. Additionally, DNA, variations on this method have also enabled the detection of single amino acids or metal ions [Han, et al., 2006].

An additional use of monolayer-functionalized metal nanoparticles involves the coupling of their localized plasmon resonance (LSP) with fluorescence [Tam, et al., 2007] or chemiluminescence: the attachment of fluorescent molecules to metallic nanoparticles via organic monolayers can result in a dramatic increase in detection sensitivity, making single molecule detection and imaging possible (Fig. 9).

Another recent interesting sensing technique based on functionalized solid surfaces involves the use of photonic crystals, a type of mesoscopic structure formed by the periodical arrangement of nanosized objects. The resulting periodical variation of refractive index between bulk and void in the crystal causes the appearance of a photonic bang-gap: typically this causes sharp peaks in the transmission spectrum of the material. The position of these peaks is highly dependent on the refractive index in the voids of the crystal, allowing for an optical monitoring of adsorption processes at this location. Even the binding of small molecules on the highly developed surface of the crystals can cause a significant frequency shift in the resonant photonic crystal mode.

Several studies have investigated the use of surface-modified photonic crystals for biosensing applications; in particular, silicon can be used as a material for the fabrication of photonic crystals sensors, when combined with surface modification by hydrosilylation reactions with alkenes. Indeed, alkene-based monolayers can be readily formed onto the surface of porous silicon structures and the attachment and subsequent surface binding events can be monitored by the shift of the crystal optical band-gap [Alvarez, et al., 2009]. Such a silicon photonic crystal was developed to monitor the protease activity of biological samples [Kilian, et al., 2007], by immobilizing the protein angiotensine on the walls of the crystal: the degradation of the protein caused by the presence of a protease induces a shift in the photonic band gap of the crystal.

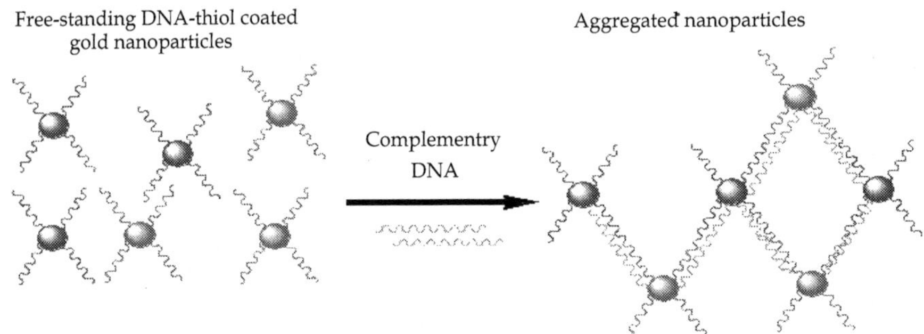

Fig. 9. Two types of DNA-functionalized gold nanoparticles, which carry different parts of the sequence to detect, aggregate when the target DNA is introduced, causing a shift in optical properties and melting behavior.

Besides optical techniques, electrical measurements have been intensively exploited for the development of biosensors. Metal and semiconductor surfaces coated with organic monolayers can be characterized by e.g. resistance, impedance or capacitance measurement [Chaki & Vijayamohanan, 2002]. The wealth of combinations of measurement modes and functionalized monolayers creates a large variety of sensor designs. Some examples of sensors include resistivity measurements to detect DNA hybridization [Hianik, et al., 2003], capacitance measurements to monitor the adsorption of nanoparticle-functionalized antibodies or the voltammetric detection of copper (II) ions on a gold electrode functionalized with a thiolderived sequence of chelating amino-acids (Gly-Gly-His, Fig. 10). Impedance measurements were also used to detect antibody–antigen interaction, by coating of silicon nitride surfaces with antibodies with an alkylsilane linker. The adsorption of rabbit immunoglobulin causes a change in the capacitance of the layer, which is used to monitor the interaction.

An important class of electrical biosensors is constituted by enzymatic electrodes: enzymes are immobilized onto metal electrodes and usually kept in presence of a red-ox mediator. In the presence of the analyte to detect (e.g. glucose [Alexander & Rechnitz, 2000], gluconic acid [Campuzano, et al., 2011], phenolic compounds [Liu, et al., 2006], the enzyme is converting the mediator into the other red-ox form, which is then detected at the electrode. Thiols have been extensively used to immobilize enzymes onto gold electrodes, often using acid-terminated monolayers and the subsequent formation of a strong amide bond with free amines of the enzyme. For enzyme electrodes, numerous studies describe the use of artificial bilayer membranes deposited onto electrodes: the two-dimensional liquid environment of lipid bilayers not only stabilizes the enzyme on the surface but also allows enough conformational freedom for the enzyme to function like in a natural membrane. Although artificial membranes are out of the scope of this review, it is important to highlight the positive role of covalent organic monolayers in the stabilization if these membranes. Compact linear alkyl monolayers, for instance, have been shown to stabilize bilayers membranes sitting on top of them [Zhang, et al., 2000]. In other studies, the integration of some thiol compound within the bilayers also stabilizes the formation of synthetic membranes 303 or vesicles; such glycolipid-containing vesicles attached to gold electrodes, upon binding with concanavalin A, can decrease the typical reduction current of a solution of Fe(CN) (Fig. 11).

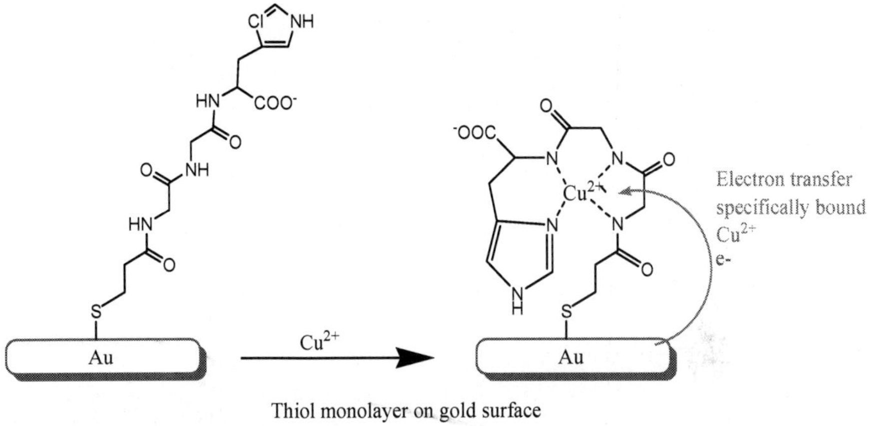

Thiol monolayer on gold surface

Fig. 10. Specific binding of copper (II) ions onto gold electrode using a thiol-based monolayer of oligopeptide (Gly-Gly-His), the detection is carried out by voltammetry.

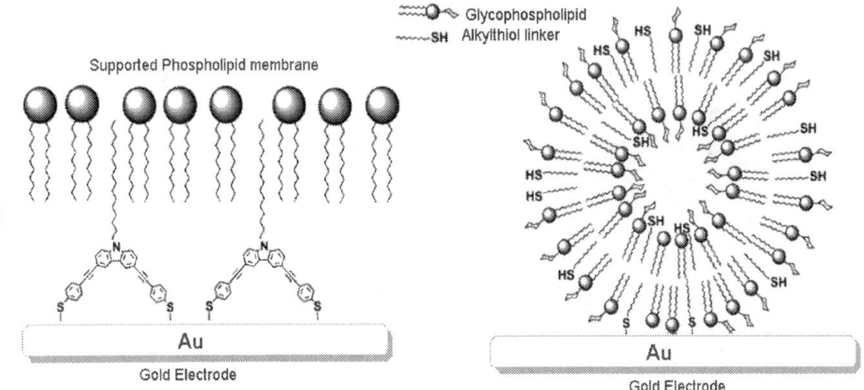

Fig. 11. Two examples of biomimetic membrane: a) bulky alkyl thiol linker to anchor flat phospholipid membrane and b) mixed vesicle prepared from glycophospholipids (for the specific binding of concanavalin A) and linear alkylthiol linkers.

The formation of covalent organic monolayers is an optimal way to control the surface of inorganic materials. The carefully designed bulk characteristics of a conducting, transparent or nanostructured inorganic device can be maintained even if surface properties need to be adapted to changing environments or applications. In this respect, they allow a supplementary freedom in the design of nano- and microdevices. However, monolayers not only serve material science in this secondary function; they are nowadays at the center of new fields of research, such as nanopatterning or biocompatible surfaces.

So far, we have presented the general methods for the formation of covalent organic monolayers onto inorganic surfaces and a number of relevant applications where these nanometer-thick structures play a central role. Despite the wealth of applications, the fundamental aspects of monolayer formation are still undergoing intensive investigations; a basic internet query with ISI Web of Science on the topic "organic monolayers" shows an ever-increasing number of hits since 1990, which reaches almost 400 articles for the year 2008. Considering the current overwhelming trend in favor of nanotechnological research, which has brought an intense light upon a world dominated by interfacial effects, the interest in organic monolayers is likely to persist for a long time.

8. References

Ababou-Girard, S., Sabbah, H., Fahre, B., Zellama, K., Solal, F. & Godet, C., (2007). Covalent grafting of organic layers on sputtered amorphous carbon: Surface preparation and coverage density. *Journal of Physical Chemistry* C.Vol 111, No. 7, pp 3099-3108, 19327447 (ISSN).

Ababou-Girard, S., Solal, F., Fabre, B., Alibart, F. & Godet, C., (2006). Covalent grafting of organic molecular chains on amorphous carbon surfaces. *Journal of Non-Crystalline Solids*.Vol 352, No. 9-20 SPEC. ISS., pp 2011-2014, 00223093 (ISSN).

Alexander, P.W. & Rechnitz, G.A., (2000). Enzyme inhibition assays with an amperometric glucose biosensor based on a thiolate self-assembled monolayer. *Electroanalysis*.Vol 12, No. 5, pp 343-350, 10400397 (ISSN).

Allara, D.L. & Nuzzo, R.G., (1985). Spontaneously organized molecular assemblies. 1. Formation, dynamics, and physical properties of n-alkanoic acids adsorbed from solution on an oxidized aluminum surface. *LANGMUIR*.Vol 1, No. 1, pp 45-52, 07437463 (ISSN).

Allongue, P., Henry De Villeneuve, C., Morin, S., Boukherroub, R. & Wayner, D.D.M., (2000). Preparation of flat H-Si(111) surfaces in 40% NH4F revisited. *Electrochimica Acta*.Vol 45, No. 28, pp 4591-4598, 00134686 (ISSN).

Alvarez, S.D., Derfus, A.M., Schwartz, M.P., Bhatia, S.N. & Sailor, M.J., (2009). The compatibility of hepatocytes with chemically modified porous silicon with reference to in vitro biosensors. *Biomaterials*.Vol 30, No. 1, pp 26-34, 01429612 (ISSN).

Andersen, K.N., Svendsen, W.E., Stimpel-Lindner, T., Sulima, T. & Baumgärtner, H., (2005). Annealing and deposition effects of the chemical composition of silicon-rich nitride. *Applied Surface Science*.Vol 243, No. 1-4, pp 401-408, 01694332 (ISSN).

Anderson, A.S., Dattelbaum, A.M., Montano, G.A., Price, D.N., Schmidt, J.G., Martinez, J.S., Grace, W.K., Grace, K.M. & Swanson, B.I., (2008). Functional PEG-modified thin films for biological detection. *Langmuir*.Vol 24, No. 5, pp 2240-2247, 07437463 (ISSN).

Arafat, A., Giesbers, M., Rosso, M., Sudhölter, E.J.R., Schroën, K., White, R.G., Li, Y., Linford, M.R. & Han, Z., (2007). Covalent biofunctionalization of silicon nitride surfaces. *Langmuir*.Vol 23, No. 11, pp 6233-6244, 07437463 (ISSN).

Arafat, A., Schroën, K., De Smet, L.C.P.M., Sudhölter, E.J.R. & Zuilhof, H., (2004). Tailor-made functionalization of silicon nitride surfaces. *Journal of the American Chemical Society*.Vol 126, No. 28, pp 8600-8601, 00027863 (ISSN).

Arwin, H., (2000). Ellipsometry on thin organic layers of biological interest: Characterization and applications. *Thin Solid Films*.Vol 377-378, No. pp 48-56, 00406090 (ISSN).

Arwin, H., (2001). Spectroscopic ellipsometry for characterization and monitoring of organic layers. *Physica Status Solidi (A) Applied Research*.Vol 188, No. 4, pp 1331-1338, 00318965 (ISSN).

Asanuma, H., Lopinski, G.P. & Yu, H.Z., (2005). Kinetic control of the photochemical reactivity of hydrogen-terminated silicon with bifunctional molecules. *Langmuir*. Vol 21, No. 11, pp 5013-5018, 07437463 (ISSN).

Asanuma, H., Noguchi, H., Uosaki, K. & Yu, H.Z., (2006). Structure and reactivity of alkoxycarbonyl (ester)-terminated monolayers on silicon: Sum frequency generation spectroscopy. *Journal of Physical Chemistry B*.Vol 110, No. 10, pp 4892-4899, 15206106 (ISSN).

Aswal, D.K., Lenfant, S., Guerin, D., Yakhmi, J.V. & Vuillaume, D., (2006). Self assembled monolayers on silicon for molecular electronics. *Analytica Chimica Acta*.Vol 568, No. 1-2, pp 84-108, 00032670 (ISSN).

Bailey, R.C., Nam, J.M., Mirkin, C.A. & Hupp, J.T., (2003). Real-Time Multicolor DNA Detection with Chemoresponsive Diffraction Gratings and Nanoparticle Probes. *Journal of the American Chemical Society*.Vol 125, No. 44, pp 13541-13547, 00027863 (ISSN).

Böcking, T., Kilian, K.A., Gaus, K. & Gooding, J.J., (2006). Single-step DNA immobilization on antifouling self-assembled monolayers covalently bound to silicon (111). *Langmuir*.Vol 22, No. 8, pp 3494-3496, 07437463 (ISSN).

Böcking, T., Salomon, A., Cahen, D. & Gooding, J.J., (2007). Thiol-Terminated Monolayers on Oxide-Free Si: Assembly of Semiconductor−Alkyl−S−Metal Junctions. *Langmuir*.Vol 23, No. 6, pp 3236-3241, 0743-7463.

Boukherroub, R., (2005). Chemical reactivity of hydrogen-terminated crystalline silicon surfaces. *Current Opinion in Solid State and Materials Science*.Vol 9, No. 1-2, pp 66-72, 13590286 (ISSN).

Boukherroub, R., Morin, S., Sharpe, P. & Wayner, D.D.M., (2000). Insights into the formation mechanisms of Si-OR monolayers from the thermal reactions of alcohols and aldehydes with SidlD-H. *Langmuir*.Vol 16, No. 19, pp 7429-7434, 07437463 (ISSN).

Bruchez Jr, M., Moronne, M., Gin, P., Weiss, S. & Alivisatos, A.P., (1998). Semiconductor nanocrystals as fluorescent biological labels. *Science*.Vol 281, No. 5385, pp 2013-2016, 00368075 (ISSN).

Buhl, A., Metzger, J.H., Heegaard, N.H.H., Von Landenberg, P., Fleck, M. & Luppa, P.B., (2007). Novel biosensor-based analytic device for the detection of anti-double-stranded DNA antibodies. *Clinical Chemistry*.Vol 53, No. 2, pp 334-341, 00099147 (ISSN).

Buriak, J.M., (2001). Diamond surfaces: Just big organic molecules? *Angewandte Chemie - International Edition*.Vol 40, No. 3, pp 532-534, 14337851 (ISSN).

Buriak, J.M., (2002). Organometallic chemistry on silicon and germanium surfaces. *Chemical Reviews*.Vol 102, No. 5, pp 1271-1308, 00092665 (ISSN).

Cahen, D., Naaman, R. & Vager, Z., (2005). The cooperative molecular field effect. *Advanced Functional Materials*.Vol 15, No. 10, pp 1571-1578, 1616301X (ISSN).

Campuzano, S., Kuralay, F., Lobo-Castañón, M.J., Bartošík, M., Vyavahare, K., Paleček, E., Haake, D.A. & Wang, J., (2011). Ternary monolayers as DNA recognition interfaces for direct and sensitive electrochemical detection in untreated clinical samples. *Biosensors and Bioelectronics*.Vol 26, No. 8, pp 3577-3583, 09565663 (ISSN).

Cattaruzza, F., Cricenti, A., Flamini, A., Girasole, M., Longo, G., Prosperi, T., Andreano, G., Cellai, L. & Chirivino, E., (2006). Controlled loading of oligodeoxyribonucleotide monolayers onto unoxidized crystalline silicon; fluorescence-based determination of the surface coverage and of the hybridization efficiency; parallel imaging of the process by Atomic Force Microscopy. *Nucleic Acids Research*.Vol 34, No. 4, 03051048 (ISSN).

Chah, S. & Zare, R.N., (2008). Surface plasmon resonance study of vesicle rupture by virus-mimetic attack. *Physical Chemistry Chemical Physics*.Vol 10, No. 22, pp 3203-3208, 14639076 (ISSN).

Chaki, N.K. & Vijayamohanan, K., (2002). Self-assembled monolayers as a tunable platform for biosensor applications. *Biosensors and Bioelectronics*.Vol 17, No. 1-2, pp 1-12, 09565663 (ISSN).

Ciampi, S., Böcking, T., Kilian, K.A., James, M., Harper, J.B. & Gooding, J.J., (2007). Functionalization of acetylene-terminated monolayers on Si(100) surfaces: A click chemistry approach. *Langmuir*.Vol 23, No. 18, pp 9320-9329, 07437463 (ISSN).

Cicero, R.L., Linford, M.R. & Chidsey, C.E.D., (2000). Photoreactivity of unsaturated compounds with hydrogen-terminated silicon(111). *Langnuir*.Vol 16, No. 13, pp 5688-5695, 07437463 (ISSN).

Coffinier, Y., Boukherroub, R., Wallart, X., Nys, J.P., Durand, J.O., Stiévenard, D. & Grandidier, B., (2007). Covalent functionalization of silicon nitride surfaces by semicarbazide group. *Surface Science*.Vol 601, No. 23, pp 5492-5498, 00396028 (ISSN).

Coffinier, Y., Olivier, C., Perzyna, A., Grandidier, B., Wallart, X., Durand, J.O., Melnyk, O. & StiÃ©venard, D., (2005). Semicarbazide-functionalized Si(111) surfaces for the site-specific immobilization of peptides. *Langmuir*.Vol 21, No. 4, pp 1489,

Cotton, F.A. & Wilkinson, G.,*Advanced Inorganic Chemistry* 6th (Eddition). John Wiley & Sons, ISBN-13: 9788126513383, New York, NY, USA.

De Smet, L.C.P.M., Pukin, A.V., Sun, Q.Y., Eves, B.J., Lopinski, G.P., Visser, G.M., Zuilhof, H. & Sudhölter, E.J.R., (2005). Visible-light attachment of SiC linked functionalized organic monolayers on silicon surfaces. *Applied Surface Science*.Vol 252, No. 1 SPEC. ISS., pp 24-30, 01694332 (ISSN).

de Smet, L.C.P.M., Stork, G.A., Hurenkamp, G.H.F., Sun, Q.-Y., Topal, H., Vronen, P.J.E., Sieval, A.B., Wright, A., Visser, G.M., Zuilhof, H. & Sudhölter, E.J.R., (2003). Covalently Attached Saccharides on Silicon Surfaces. *J. Am. Chem. Soc*.Vol 125, No. 46, pp 13916-13917,

Effenberger, F., Götz, G., Bidlingmaier, B. & Wezstein, M., (1998). Photoactivated preparation and patterning of self-assembled monolayers with 1-alkenes and aldehydes on silicon hydride surfaces. *Angewandte Chemie - International Edition*.Vol 37, No. 18, pp 2462-2464, 14337851 (ISSN).

Elghanian, R., Storhoff, J.J., Mucic, R.C., Letsinger, R.L. & Mirkin, C.A., (1997). Selective colorimetric detection of polynucleotides based on the distance-dependent optical properties of gold nanoparticles. *Science*.Vol 277, No. 5329, pp 1078-1081, 00368075 (ISSN).

Eves, B.J., Sun, Q.Y., Lopinski, G.P. & Zuilhof, H., (2004). Photochemical attachment of organic monolayers onto H-terminated Si(111): Radical chain propagation observed via STM studies. *Journal of the American Chemical Society*.Vol 126, No. 44, pp 14318-14319, 00027863 (ISSN).

Faber, E.J., De Smet, L.C.P.M., Olthuis, W., Zuilhof, H., Sudhölter, E.J.R., Bergveld, P. & Van Den Berg, A., (2005). Si-C linked organic monolayers on crystalline silicon surfaces as alternative gate insulators. *ChemPhysChem*.Vol 6, No. 10, pp 2153-2166, 14394235 (ISSN).

Fabre, B. & Hauquier, F., (2006). Single-Component and Mixed Ferrocene-Terminated Alkyl Monolayers Covalently Bound to Si(111) Surfaces. *The Journal of Physical Chemistry B*.Vol 110, No. 13, pp 6848-6855, 1520-6106.

Fabre, B., Hauquier, F., Herrier, C., Pastorin, G., Wu, W., Bianco, A., Prato, M., Hapiot, P., Zigah, D., Prasciolu, M. & Vaccari, L., (2008). Covalent Assembly and Micropatterning of Functionalized Multiwalled Carbon Nanotubes to Monolayer-Modified Si(111) Surfaces. *Langmuir*.Vol 24, No. 13, pp 6595-6602, 0743-7463.

Farré, M., Martínez, E., Ramón, J., Navarro, A., Radjenovic, J., Mauriz, E., Lechuga, L., Marco, M.P. & Barceló, D., (2007). Part per trillion determination of atrazine in natural water samples by a surface plasmon resonance immunosensor. *Analytical and Bioanalytical Chemistry*.Vol 388, No. 1, pp 207-214, 16182642 (ISSN).

Faucheux, A., Gouget-Laemmel, A.C., Henry De Villeneuve, C., Boukherroub, R., Ozanam, F., Allongue, P. & Chazalviel, J.N., (2006). Well-defined carboxyl-terminated alkyl monolayers grafted onto H-Si(111): Packing density from a combined AFM and quantitative IR study. *Langmuir*.Vol 22, No. 1, pp 153-162, 07437463 (ISSN).

Garcia, R., Martinez, R.V. & Martinez, J., (2006). Nano-chemistry and scanning probe nanolithographies. *Chemical Society Reviews*.Vol 35, No. 1, pp 29-38, 03060012 (ISSN).

Hamers, R.J. & Wang, Y., (1996). Atomically-Resolved Studies of the Chemistry and Bonding at Silicon Surfaces. *Chemical Reviews*.Vol 96, No. 4, (June 1996), pp 1261-1290, 0009-2665.

Han, M.S., Lytton-Jean, A.K.R., Oh, B.K., Heo, J. & Mirkin, C.A., (2006). Colorimetric screening of DNA-binding molecules with gold nanoparticle probes. *Angewandte Chemie - International Edition*.Vol 45, No. 11, pp 1807-1810, 14337851 (ISSN).

Har-Lavan, R., Ron, I., Thieblemont, F. & Cahen, D., (2009). Toward metal-organic insulator-semiconductor solar cells, based on molecular monolayer self-assembly on n-Si. *Applied Physics Letters*.Vol 94, No. 4, 00036951 (ISSN).

Hauquier, F., Ghilane, J., Fabre, B. & Hapiot, P., (2008). Conducting ferrocene monolayers on nonconducting surface. *Journal of the American Chemical Society*.Vol 130, No. 9, pp 2748-2749, 00027863 (ISSN).

Hianik, T., Vitovic, P., Humenik, D., Andreev, S.Y., Oretskaya, T.S., Hall, E.A.H. & Vadgama, P., (2003). Hybridization of DNA at the surface of phospholipid monolayers. Effect of orientation of oligonucleotide chains. *Bioelectrochemistry*.Vol 59, No. 1-2, pp 35-40, 15675394 (ISSN).

Hiremath, R.K., Rabinal, M.K., Mulimani, B.G. & Khazi, I.M., (2008). Molecularly controlled metal-semiconductor junctions on silicon surface: A dipole effect. *Langmuir*.Vol 24, No. 19, pp 11300-11306, 07437463 (ISSN).

Hovis, J.S., Coulter, S.K., Hamers, R.J., D'Evelyn, M.P., Russell Jr, J.N. & Butler, J.E., (2000). Cycloaddition chemistry at surfaces: Reaction of alkenes with the diamond(001)-2 x 1 surface [11]. *Journal of the American Chemical Society*.Vol 122, No. 4, pp 732-733, 00027863 (ISSN).

Hutter, E. & Pileni, M.P., (2003). Detection of DNA hybridization by gold nanoparticle enhanced transmission surface plasmon resonance spectroscopy. *Journal of Physical Chemistry B*.Vol 107, No. 27, pp 6497-6499, 15206106 (ISSN).

Jun, Y., Le, D. & Zhu, X.Y., (2002). Microcontact printing directly on the silicon surface. *Langmuir*.Vol 18, No. 9, pp 3415-3417, 07437463 (ISSN).

Karymov, M.A., Kruchinin, A.A., Tarantov, Y.A., Balova, I.A., Remisova, L.A. & Vlasov, Y.G., (1995). Fixation of DNA directly on optical waveguide surfaces for molecular probe biosensor development. *Sensors and Actuators: B. Chemical*.Vol 29, No. 1-3, pp 324-327, 09254005 (ISSN).

Kilian, K.A., Böcking, T., Gaus, K., Gal, M. & Gooding, J.J., (2007). Peptide-Modified Optical Filters for Detecting Protease Activity. *ACS Nano*.Vol 1, No. 4, pp 355-361, 1936-0851.

Knickerbocker, T., Strother, T., Schwartz, M.P., Russell Jr, J.N., Butler, J., Smith, L.M. & Hamers, R.J., (2003). DNA-modified diamond surfaces. *Langmuir*.Vol 19, No. 6, pp 1938-1942, 07437463 (ISSN).

Leftwich, T.R., Madachik, M.R. & Teplyakov, A.V., (2008). Dehydrative cyclocondensation reactions on hydrogen-terminated Si(100) and Si(111): An ex situ tool for the modification of semiconductor surfaces. *Journal of the American Chemical Society*.Vol 130, No. 48, pp 16216-16223, 00027863 (ISSN).

Li, Y.H., Wang, D. & Buriak, J.M., (2010). Molecular layer deposition of thiol - ene multilayers on semiconductor surfaces. *Langmuir*. Vol 26, No. 2, pp 1232-1238, 07437463 (ISSN).

Liao, W., Wei, F., Liu, D., Qian, M.X., Yuan, G. & Zhao, X.S., (2006). FTIR-ATR detection of proteins and small molecules through DNA conjugation. *Sensors and Actuators, B: Chemical*.Vol 114, No. 1, pp 445-450, 09254005 (ISSN).

Linford, M.R. & Chidsey, C.E.D., (1993). Alkyl monolayers covalently bonded to silicon surfaces. *Journal of the American Chemical Society*.Vol 115, No. 26, pp 12631-12632, 00027863 (ISSN).

Linford, M.R., Fenter, P., Eisenberger, P.M. & Chidsey, C.E.D., (1995). Alkyl monolayers on silicon prepared from 1-alkenes and hydrogen-terminated silicon. *Journal of the American Chemical Society*.Vol 117, No. 11, pp 3145-3155, 00027863 (ISSN).

Liu, G., Quynh, T.N., Chow, E., Böcking, T., Hibbert, D.B. & Gooding, J.J., (2006). Study of factors affecting the performance of voltammetric copper sensors based on Gly-Gly-His modified glassy carbon and gold electrodes. *Electroanalysis*.Vol 18, No. 12, pp 1141-1151, 10400397 (ISSN).

Love, J.C., Estroff, L.A., Kriebel, J.K., Nuzzo, R.G. & Whitesides, G.M., (2005a). Self-assembled monolayers of thiolates on metals as a form of nanotechnology. *Chemical Reviews*.Vol 105, No. 4, pp 1103-1169, 00092665 (ISSN).

Love, J.C., Estroff, L.A., Kriebel, J.K., Nuzzo, R.G. & Whitesides, G.M., (2005b). Self-assembled monolayers of thiolates on metals as a form of nanotechnology. *Chemical Reviews*.Vol 105, No. 4, pp 1103,

Maldonado, S., Knapp, D. & Lewis, N.S., (2008). Near-ideal photodiodes from sintered gold nanoparticle films on methyl-terminated Si(111) surfaces. *Journal of the American Chemical Society*.Vol 130, No. 11, pp 3300-3301, 00027863 (ISSN).

Maoz, R. & Sagiv, J., (1984). On the formation and structure of self-assembling monolayers. I. A comparative atr-wettability study of Langmuir-Blodgett and adsorbed films on flat substrates and glass microbeads. *Journal of Colloid and Interface Science*.Vol 100, No. 2, pp 465-496, 00219797 (ISSN).

Masuda, T., Yamaguchi, A., Hayashida, M., Asari-Oi, F., Matsuo, S. & Misawa, H., (2005). Visualization of DNA hybridization on gold thin film by utilizing the resistance effect of DNA monolayer. *Sensors and Actuators, B: Chemical*.Vol 105, No. 2, pp 556-561, 09254005 (ISSN).

Mauriz, E., Calle, A., Montoya, A. & Lechuga, L.M., (2006). Determination of environmental organic pollutants with a portable optical immunosensor. *Talanta*.Vol 69, No. 2 SPEC. ISS., pp 359-364, 00399140 (ISSN).

Michalet, X., Pinaud, F.F., Bentolila, L.A., Tsay, J.M., Doose, S., Li, J.J., Sundaresan, G., Wu, A.M., Gambhir, S.S. & Weiss, S., (2005). Quantum dots for live cells, in vivo imaging, and diagnostics. *Science*.Vol 307, No. 5709, pp 538-544, 00368075 (ISSN).

Mischki, T.K., Lopinski, G.P. & Wayner, D.D.M., (2009). Evidence for initiation of thermal reactions of alkenes with hydrogen- terminated silicon by surface-catalyzed thermal decomposition of the reactant. *LANGMUIR*.Vol 25, No. 10, pp 5626-5630, 07437463 (ISSN).

Mizuno, H. & Buriak, J.M., (2008). Catalytic stamp lithography for sub-100 nm patterning of organic monolayers. *Journal of the American Chemical Society*.Vol 130, No. 52, pp 17656-17657, 00027863 (ISSN).

Ng, A., Ciampi, S., James, M., Harper, J.B. & Gooding, J.J., (2009). Comparing the reactivity of alkynes and alkenes on silicon 100 surfaces. *Langmuir*.Vol 25, No. 24, pp 13934-13941, 07437463 (ISSN).

Nichols, B.M., Butler, J.E., Russell Jr, J.N. & Hamers, R.J., (2005). Photochemical functionalization of hydrogen-terminated diamond surfaces: A structural and mechanistic study. *Journal of Physical Chemistry B*.Vol 109, No. 44, pp 20938-20947, 15206106 (ISSN).

Niederhauser, T.L., Jiang, G., Lua, Y.Y., Dorff, M.J., Woolley, A.T., Asplund, M.C., Berges, D.A. & Linford, M.R., (2001). A new method of preparing monolayers on silicon and patterning silicon surfaces by scribing in the presence of reactive species. *Langmuir*.Vol 17, No. 19, pp 5889-5900, 07437463 (ISSN).

Niederhauser, T.L., Lua, Y.Y., Jiang, G., Davis, S.D., Matheson, R., Hess, D.A., Mowat, I.A. & Linford, M.R., (2002). Arrays of chemomechanically patterned patches of homogeneous and mixed monolayers of 1-alkenes and alcohols on single silicon surfaces. *Angewandte Chemie - International Edition*.Vol 41, No. 13, pp 2353-2356, 14337851 (ISSN).

Nuzzo, R.G. & Allara, D.L., (1983). Adsorption of bifunctional organic disulfides on gold surfaces. *Journal of the American Chemical Society*.Vol 105, No. 13, pp 4481-4483, 00027863 (ISSN).

Onclin, S., Ravoo, B.J. & Reinhoudt, D.N., (2005a). *Angew. Chem. Int. Ed*.Vol 105, No. pp 1103-1169,

Onclin, S., Ravoo, B.J. & Reinhoudt, D.N., (2005b). Engineering silicon oxide surfaces using self-assembled monolayers. *Angewandte Chemie - International Edition*.Vol 44, No. 39, pp 6282-6304, 14337851 (ISSN).

Perring, M., Dutta, S., Arafat, S., Mitchell, M., Kenis, P.J.A. & Bowden, N.B., (2005). Simple methods for the direct assembly, functionalization, and patterning of acid-terminated monolayers on Si(111). *Langmuir*.Vol 21, No. 23, pp 10537-10544, 07437463 (ISSN).

Perring, M., Mitchell, M., Kenis, P.J.A. & Bowden, N.B., (2007). Patterning by Etching at the Nanoscale (PENs) on Si(111) through the controlled etching of PDMS. *Chemistry of Materials*.Vol 19, No. 11, pp 2903-2909, 08974756 (ISSN).

Phillips, K.S. & Cheng, Q., (2007). Recent advances in surface plasmon resonance based techniques for bioanalysis. *Analytical and Bioanalytical Chemistry*.Vol 387, No. 5, pp 1831-1840, 16182642 (ISSN).

Phillips, M.M., Case, R.J., Rimmer, C.A., Sander, L.C., Sharpless, K.E., Wise, S.A. & Yen, J.H., (2010). Determination of organic acids in Vaccinium berry standard reference materials. *Analytical and Bioanalytical Chemistry*.Vol 398, No. 1, pp 425-434, 16182642 (ISSN).

Rosso, M., Arafat, A., SchroÃ«n, K., Giesbers, M., Roper, C.S., Maboudian, R. & Zuilhof, H., (2008a). Covalent attachment of organic monolayers to silicon carbide surfaces. *Langmuir*.Vol 24, No. 8, pp 4007-4012,

Rosso, M., Arafat, A., Schroën, K., Giesbers, M., Roper, C.S., Maboudian, R. & Zuilhof, H., (2008b). Covalent attachment of organic monolayers to silicon carbide surfaces. *Langmuir*.Vol 24, No. 8, pp 4007-4012, 07437463 (ISSN).

Rosso, M., Giesbers, M., Arafat, A., Schroën, K. & Zuilhof, H., (2009). Covalently attached organic monolayers on SiC and SixN 4 surfaces: Formation using UV light at room temperature. *Langmuir*.Vol 25, No. 4, pp 2172-2180, 07437463 (ISSN).

Saddow, S.E. & Agarwal, A.,*Advances in Silicon Carbide: Processing and Applications* (Eddition). Biomed Central Med, ISBN 1580537405-xiv

Salomon, A., Böcking, T., Gooding, J.J. & Cahen, D., (2006). How important Is the interfacial chemical bond for electron transport through alkyl chain monolayers? *Nano Letters*.Vol 6, No. 12, pp 2873-2876, 15306984 (ISSN).

Salomon, A., Boecking, T., Seitz, O., Markus, T., Amy, F., Chan, C., Zhao, W., Cahen, D. & Kahn, A., (2007). What is the barrier for tunneling through alkyl monolayers? Results from n- And p-Si-Alkyl/Hg junctions. *Advanced Materials*.Vol 19, No. 3, pp 445-450, 09359648 (ISSN).

Scheres, L., Maat, J.T., Giesbers, M. & Zuilhof, H., (2010). Microcontact printing onto oxide-free silicon via highly reactive acid fluoride-functionalized monolayers. *Small*.Vol 6, No. 5, pp 642-650, 16136810 (ISSN).

Seitz, O., Böcking, T., Salomon, A., Gooding, J.J. & Cahen, D., (2006). Importance of monolayer quality for interpreting current transport through organic molecules: Alkyls on oxide-free Si. *Langmuir*.Vol 22, No. 16, pp 6915-6922, 07437463 (ISSN).

Sharma, A.K. & Gupta, B.D., (2007). On the performance of different bimetallic combinations in surface plasmon resonance based fiber optic sensors. *Journal of Applied Physics*. Vol 101, No. 9, 00218979 (ISSN).

Shirahata, N., Hozumi, A. & Yonezawa, T., (2005). Monolayer-derivative functionalization of non-oxidized silicon surfaces. *Chemical Record*.Vol 5, No. 3, pp 145-159, 15278999 (ISSN).

Sieval, A.B., Demirel, A.L., Nissink, J.W.M., Linford, M.R., Van Der Maas, J.H., De Jeu, W.H., Zuilhof, H. & Sudhölter, E.J.R., (1998). Highly stable Si-C linked functionalized monolayers on the silicon (100) surface. *Langmuir*.Vol 14, No. 7, pp 1759-1768, 07437463 (ISSN).

Sieval, A.B., Linke, R., Heij, G., Meijer, G., Zuilhof, H. & Sudhölter, E.J.R., (2001). Amino-terminated organic monolayers on hydrogen-terminated silicon surfaces. *Langmuir*. Vol 17, No. 24, pp 7554-7559, 07437463 (ISSN).

Sieval, A.B., Linke, R., Zuilhof, H. & Sudhölter, E.J.R., (2000a). High-quality alkyl monolayers on silicon surfaces. *Advanced Materials*. Vol 12, No. 19, pp 1457-1460, 09359648 (ISSN).

Sieval, A.B., Opitz, R., Maas, H.P.A., Schoeman, M.G., Meijer, G., Vergeldt, F.J., Zuilhof, H. & Sudhölter, E.J.R., (2000b). Monolayers of 1-alkynes on the H-terminated Si(100) surface. *Langmuir*. Vol 16, No. 26, pp 10359-10368, 07437463 (ISSN).

Sieval, A.B., Van Den Hout, B., Zuilhof, H. & Sudhölter, E.J.R., (2000c). Molecular modeling of alkyl monolayers on the Si(111) surface. *Langmuir*.Vol 16, No. 7, pp 2987-2990, 07437463 (ISSN).

Sieval, A.B., Vleeming, V., Zuilhof, H. & Sudhölter, E.J.R., (1999). Improved method for the preparation of organic monolayers of 1-alkenes on hydrogen-terminated silicon surfaces. *Langmuir*.Vol 15, No. 23, pp 8288-8291, 07437463 (ISSN).

Spochiger-Kuller, U.E.,*Chemical Sensors and Biosensors for Medical and biological Applications* (Eddition). Wiley-VCH, ISBN 3-527-28855-4, Washington, DC.

Stewart, M.P. & Buriak, J.M., (2001). Exciton-mediated hydrosilylation on photoluminescent nanocrystalline silicon. *Journal of the American Chemical Society*.Vol 123, No. 32, pp 7821-7830, 00027863 (ISSN).

Stewart, M.P. & Buriak, J.M., (1998). Photopatterned hydrosilylation on porous silicon. *Angewandte Chemie - International Edition*.Vol 37, No. 23, pp 3257-3260, 14337851 (ISSN).

Strother, T., Cai, W., Zhao, X., Hamers, R.J. & Smith, L.M., (2000). Synthesis and characterization of DNA-modified silicon (111) surfaces. *Journal of the American Chemical Society*.Vol 122, No. 6, pp 1205-1209, 00027863 (ISSN).

Strother, T., Knickerbocker, T., Russell Jr, J.N., Butler, J.E., Smith, L.M. & Hamers, R.J., (2002). Photochemical functionalization of diamond films. *Langmuir*.Vol 18, No. 4, pp 968-971, 07437463 (ISSN).

Sun, Q.Y., De Smet, L.C.P.M., Van Lagen, B., Giesbers, M., Thüne, P.C., Van Engelenburg, J., De Wolf, F.A., Zuilhof, H. & Sudhölter, E.J.R., (2005). Covalently attached monolayers on crystalline hydrogen-terminated silicon: Extremely mild attachment by visible light. *Journal of the American Chemical Society*.Vol 127, No. 8, pp 2514-2523, 00027863 (ISSN).

Sun, Q.Y., De Smet, L.C.P.M., Van Lagen, B., Wright, A., Zuilhof, H. & Sudhölter, E.J.R., (2004). Covalently attached monolayers on hydrogen-terminated Si(100): Extremely mild attachment by visible light. *Angewandte Chemie - International Edition*.Vol 43, No. 11, pp 1352-1355, 14337851 (ISSN).

Sung, M.M., Kluth, G.J., Yauw, O.W. & Maboudian, R., (1997). Thermal behavior of alkyl monolayers on silicon surfaces. *LANGMUIR*.Vol 13, No. 23, pp 6164-6167, 07437463 (ISSN).

Tam, F., Goodrich, G.P., Johnson, B.R. & Halas, N.J., (2007). Plasmonic enhancement of molecular fluorescence. *Nano Letters*.Vol 7, No. 2, pp 496-501, 15306984 (ISSN).

Ulman, A., (1996). Formation and structure of self-assembled monolayers. *Chemical Reviews*.Vol 96, No. 4, pp 1533-1554, 00092665 (ISSN).

Ulman, A., *An Introduction to Ultrathin Organic Films*. (Eddition). Academic Press, ISBN 0-12-708230-1, Boston, MA, USA.

Vilan, A., Yaffe, O., Biller, A., Salomon, A., Kahn, A. & Cahen, D., (2010). Molecules on Si: Electronics with chemistry. *Advanced Materials*.Vol 22, No. 2, pp 140-159, 09359648 (ISSN).

Voicu, R., Boukherroub, R., Bartzoka, V., Ward, T., Wojtyk, J.T.C. & Wayner, D.D.M., (2004). Formation, characterization, and chemistry of undecanoic acid-terminated silicon surfaces: Patterning and immobilization of DNA. *Langmuir*.Vol 20, No. 26, pp 11713-11720, 07437463 (ISSN).

Wallart, X., De Villeneuve, C.H. & Allongue, P., (2005). Truly quantitative XPS characterization of organic monolayers on silicon: Study of alkyl and alkoxy monolayers on H-Si(111). *Journal of the American Chemical Society*.Vol 127, No. 21, pp 7871-7878, 00027863 (ISSN).

Wang, X., Colavita, P.E., Metz, K.M., Butler, J.E. & Hamers, R.J., (2007). Direct photopatterning and SEM imaging of molecular monolayers on diamond surfaces: Mechanistic insights into UV-initiated molecular grafting. *Langmuir*.Vol 23, No. 23, pp 11623-11630, 07437463 (ISSN).

Wayner, D.D.M. & Wolkow, R.A., (2002). Organic modification of hydrogen terminated silicon surfaces. *Journal of the Chemical Society, Perkin Transactions 2*.Vol 2, No. 1, pp 23-34, 14701820 (ISSN).

Wojtyk, J.T.C., Tomietto, M., Boukherroub, R. & Wayner, D.D.M., (2001). "Reagentless" micropatterning of organics on silicon surfaces: Control of hydrophobic/hydrophilic domains [19]. *Journal of the American Chemical Society*.Vol 123, No. 7, pp 1535-1536, 00027863 (ISSN).

Woodson, M. & Liu, J., (2007). Functional nanostructures from surface chemistry patterning. *Physical Chemistry Chemical Physics*.Vol 9, No. 2, pp 207-225, 14639076 (ISSN).

Xia, Y. & Whitesides, G.M., (1998). Soft lithography. *Angewandte Chemie - International Edition*.Vol 37, No. 5, pp 550-575, 14337851 (ISSN).

Xu, X.H., Feng, Y., Li, B.Q. & Chu, J.R., (2008). Integration of displacement sensor into bulk PZT thick film actuator for MEMS deformable mirror. *Sensors and Actuators, A: Physical*.Vol 147, No. 1, pp 242-247, 09244247 (ISSN).

Yang, L., Lua, Y.Y., Lee, M.V. & Linford, M.R., (2005). Chemomechanical functionalization and patterning of silicon. *Accounts of Chemical Research*.Vol 38, No. 12, pp 933-942, 00014842 (ISSN).

Yang, M., Teeuwen, R.L.M., Giesbers, M., Baggerman, J., Arafat, A., De Wolf, F.A., Van Hest, J.C.M. & Zuilhof, H., (2008). One-step photochemical attachment of NHS-terminated monolayers onto silicon surfaces and subsequent functionalization. *Langmuir*.Vol 24, No. 15, pp 7931-7938, 07437463 (ISSN).

Yang, M., Wouters, D., Giesbers, M., Schubert, U.S. & Zuilhof, H., (2009). Local probe oxidation of self-assembled monolayers on hydrogen-terminated silicon. *ACS Nano*.Vol 3, No. 10, pp 2887-2900, 19360851 (ISSN).

Yang, N., Su, X., Tjong, V. & Knoll, W., (2007). Evaluation of two- and three-dimensional streptavidin binding platforms for surface plasmon resonance spectroscopy studies of DNA hybridization and protein-DNA binding. *Biosensors and Bioelectronics*.Vol 22, No. 11, pp 2700-2706, 09565663 (ISSN).

Yin, H.B., Brown, T., Wilkinson, J.S., Eason, R.W. & Melvin, T., (2004). Submicron patterning of DNA oligonucleotides on silicon. *Nucleic Acids Research*.Vol 32, No. 14, 13624962 (ISSN).

Yonzon, C.R., Haynes, C.L., Zhang, X., Walsh Jr, J.T. & Van Duyne, R.P., (2004). A Glucose Biosensor Based on Surface-Enhanced Raman Scattering: Improved Partition Layer, Temporal Stability, Reversibility, and Resistance to Serum Protein Interference. *Analytical Chemistry*.Vol 76, No. 1, pp 78-85, 00032700 (ISSN).

Yuan, S.L., Cai, Z.T. & Jiang, Y.S., (2003). Molecular simulation study of alkyl monolayers on the Si(111) surface. *New Journal of Chemistry*.Vol 27, No. 3, pp 626-633, 11440546 (ISSN).

Zhang, L., Longo, M.L. & Stroeve, P., (2000). Mobile phospholipid bilayers supported on a polyion/alkylthiol layer pair. *Langmuir*.Vol 16, No. 11, pp 5093-5099, 07437463 (ISSN).

9

Polymer Based Biosensors for Pathogen Diagnostics

Katrine Kiilerich-Pedersen, Claus Riber Poulsen, Johannes Daprà,
Nikolaj Ormstrup Christiansen and Noemi Rozlosnik
Department of Micro- and Nanotechnology, Technical University of Denmark
Denmark

1. Introduction

Over the past three decades researchers have witnessed an enormous amount of activity in the area of biosensors. The major processes involved in any biosensor system are analyte recognition, signal transduction, and readout. Due to their specificity, speed, portability, and low cost, biosensors offer exciting opportunities for numerous decentralized clinical applications – point of care systems.

The ongoing trend in biomedicine is to go smaller. For almost a decade, the buzz word has been nano, and the analytical micro devices are now appearing in the clinic. The progress within microfluidic technologies has enabled miniaturization of biomedical systems and biosensors. The down-scaling has several advantages: refined control of fluidics, low sample consumption, applicability to point of care, and low cost.

Point of care is an emerging field within medical diagnostics and disease monitoring, and eventually disease control. Employing specially designed micro systems, a patient can be monitored continuously at bed side, and save precious time on commuting between home, doctor and hospital. The technological advancements in the biosensor technology within recent years have accelerated the R&D in point of care devices.

Cost benefit is always an important factor in development of novel medical devices. To reduce the expenses of biosensors, the use and cleanroom processing of noble metals should be kept at a minimum. Therefore, we predict a shift in the usage of gold and platinum to degradable polymer materials.

This chapter will look further into the advantages and applications of all-polymer microfluidic devices for biomedical diagnostics and compare with traditional systems. In many biosensor appliitions, only one analyte is of interest, and preferentially it should be isolated from an inhomogeneous patient sample. Section 2 provides the reader with an overview of the different novel microfluidic separation techniques in polymeric devices. Conductive polymers are the focus of section 3. They have many excellent properties and in fact, they can compete with gold in many applications. The focus of section 4 is sensitivity and specificity of biosensors. High sensitivity and specificity is crucial and can be achieved by functionalization with different molecules. The section will primarily center around the use of aptamers which is favourable above antibodies. Different detection methods are applied in biosensors, some of the promising techniques will be summarized in section 5. Finally, section 6 gives an overview of the current status in biosensor development while focusing on ongoing research.

2. Novel microfluidic separation techniques for sample preparation

The progress in microfabrication and lab-on-a-chip technologies is a major field for development of new approaches to bioanalytics and cell biology. Microfluidics has proven successful for cell and particle handling, and the interest in microdevices for separation of particles or cells has increased significantly (Giddings (1993); Nolan & Sklar (1998); Toner & Irimia (2005)).

Biological samples comprise a heterogeneous population of cells or particles, which is inconvenient for many biomedical applications, where the objective of study is often just one species. For example, the isolation of CD4+ T-lymphocytes from whole blood is essential to diagnose human immunodeficiency virus (HIV) (Kuntaegowdanahalli et al. (2009)), the isolation of leukocytes is important in drug screening assays, and the isolation of specific micro particles from blood plasma is critical for our understanding of inflammatory diseases. Thus, separation of cells or particles has a wide range of applications within different areas of medicine such as diagnostics, therapeutics, drug discovery, and personalized medicine (Gossett et al. (2010)).

Flow cytometry has remained the preferred method for cell sorting by many biologists because the technique is well established and has both high sensitivity and high throughput. Recently, fluorescence based sorting of cells and particles has also been implemented in microfluidic devices.

The microfluidic separation techniques are broadly classified as being either passive or active, depending on the operating principles (Table 1). Active separation of particles requires an external force (i.e. electrical power, mechanical pressure or magnetic force), whereas passive separation techniques rely on channel geometry and inherent hydrodynamic forces for functionality (i.e. pillars, pressure field gradient or hydrodynamic force). The following section will introduce a couple of novel separation principles with application in biomedical sensors. For further reading on continuous separation of particles, see review papers by Lenshof and Laurell (2010), Gossett et al. (2010), and Bhagat et al. (2010).

	Method	Mechanism
Active	Acoustophoresis	Acoustic waves
	Optical tweezers	Optical
	Dielectrophoresis	Electric field
Passive	Obstacles	Laminar flow
	Induced lift	Inertial force

Table 1. Active and passive separation technique with application in biomedical sensors

2.1 Active separation techniques
2.1.1 Acoustophoresis

Acoustophoresis is the separation of particles using high intensity sound waves. In a microfluidic system, particles with an induced acoustic standing wave will experience a force towards a node or anti node dependent on their physical properties (Lenshof & Laurell

(2010)). If two particles suspended in a fluid have opposite acoustic contrast, a separation will occur gathering one at node and the other at anti-node. Generally, rigid particles will have negative phase and move toward the node, whereas air bubbles and lipid vesicles gather at the anti-node (Lenshof & Laurell (2010)). After separation, the properties of the laminar flow in the microfluidic channel ensure that particles remain at their position in the channel, hence they can be collected separately with a flow splitter.

Both particles with opposite and similar acoustic contrast can be separated using this technique. The size of particles will influence the time scale. Large particles experience a higher force than small than the smaller ones, and thus gather at the node faster than the small particles. Peterson et al. (2007) described a microfluidic system with three inlets (Fig. 1), where a sample composed of different sized particles was introduced at the sides of a microfluidic channel with a sheath fluid in the middle to keep particles in close proximity to channel walls. The system is designed such that an ultrasonic transducer induces a force on the particles, which forces them towards the middle of the channel. Since the larger particles experience a higher force than small particles, they large particles immediately gather at the center of the channel. Particles are thus allocated proportional to their size. Making use of a flow splitter, particles are separated according to their size. Applying this technique, Peterson et al. (2007) demonstrated separation of a mixture of different sized particles.

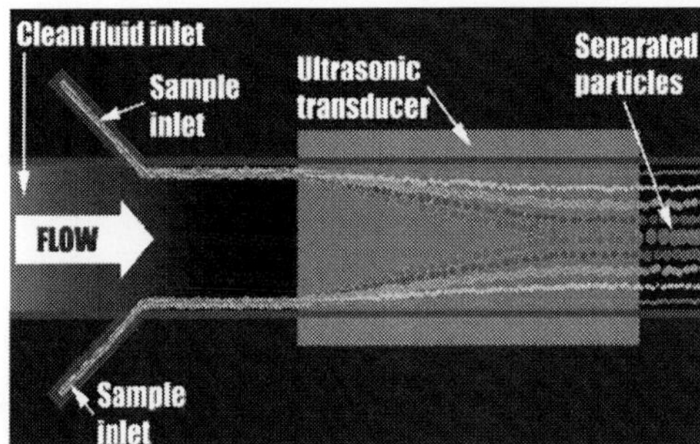

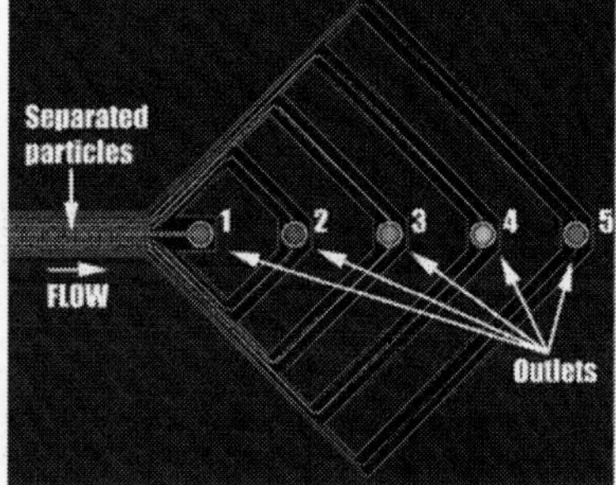

Fig. 1. Acoustophoresis. (a) Particles entering main channel from two side inlets. Particles are positioned near channel walls because clean sheat fluid is introduced at a third inlet. The flow of particles is controlled by the acoustic waves, which are introduced by an ultrasonic transducer. After this point, the particles distribute proportional to size. (b) Flow splitters are used for separation of different sized particles. Nine fractions of the flow can be gathered at five outlets (Adapted from Peterson et al. (2007)).

2.1.2 Ion depletion

Ion depletion is a microfluidic technique for separation and concentration of proteins. As the name indicates, the method is based on ion transfer in a nanofluidic channel (approximately 50 nm in depth). Counter-ions will migrate from the Debye layer through the nanochannel to a higher extent than co-ions, so that a net transfer of counter-ions is transferred from the anodic side to the cathodic side. Thus, the concentration of counter-ions decreases on the anodic side and an increase is achieved on the cathodic side. If a protein in solution is part of the co-ion population, this protein will be trapped in a plug on either side of the ion depletion region, and is hence separated from the bulk solution. The principle of ion depletion is illustrated on Fig. 2 and 3 (Wang et al. (2005)).

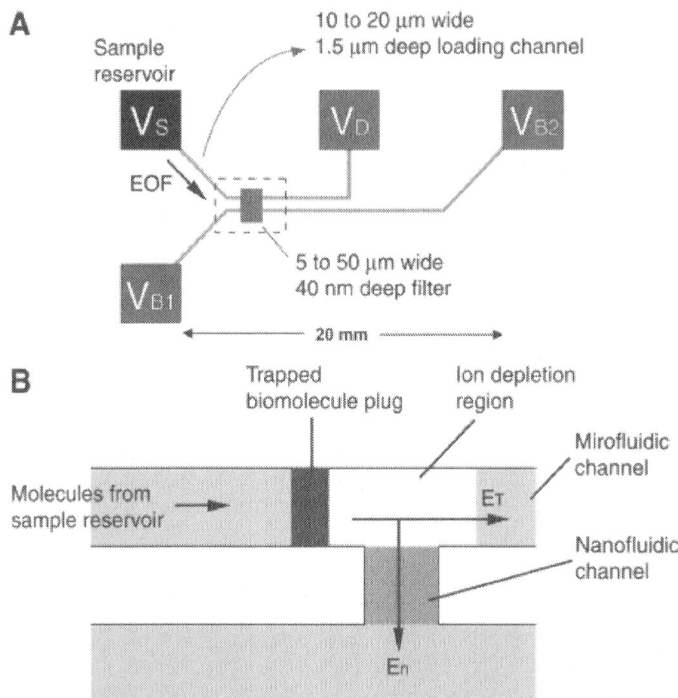

Fig. 2. Nanofluidic protein concentrating device by ion depletion: (A) Layout of the device. (B) Schematic diagram showing the concentration mechanism. Once proper voltages are applied, the trapping region and depletion region will be formed as indicated. The ET specifies the electrical field applied across the ion depletion region, while the En specifies the cross nanofilter electrical field (Adapted from Wang et al., (2005)).

2.2 Passive separation techniques
2.2.1 Obstacles
Obstacles arranged in microfluidic channels are commonly applied for preventing particles from entering certain areas or used to manipulate the flow of fluid in a microchannel. Deterministic lateral displacement is a method for size separation of particles or cells, accomplished by placing posts asymmetrically in a microchannel (Fig. 4) and thus forcing particles of different sizes to follow different flow paths.

2.2.2 Spiral microchannels
Separation of particles in a spiral microchannel was described by Kuntaegowdanahalli and colleagues (2009) (see Fig. 5).
It is a passive separation technique based on the centrifugal force. Centrifugal based techniques have been demonstrated using flows in curvilinear microchannels (Gregoratto et al. (2007); Seo et al. (2007)). In general, the flow of fluid through a curvilinear channel experiences a centrifugal acceleration, directed radially outward. The channel geometry

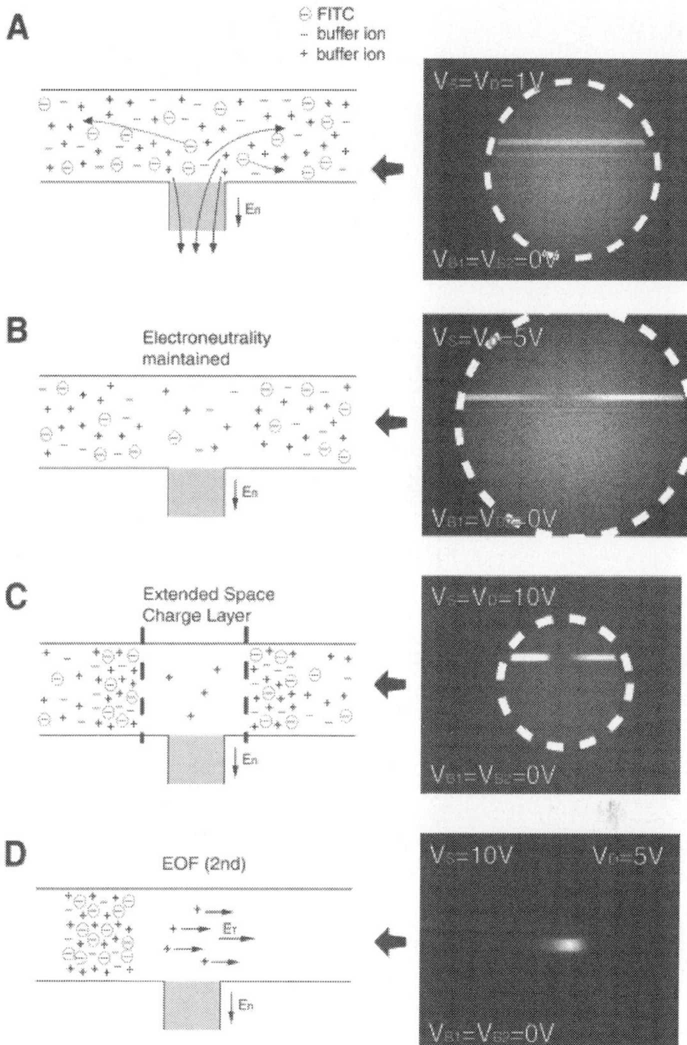

Fig. 3. Mechanism of preconcentration in the nanofilter device (A) No concentration polarization is observed when a small electrical field (En) is applied across the nanofilter. (B) As the En increases, the transport of ions becomes diffusion-limited and generates the ion depletion zone. However, the region maintains its electroneutrality. (C) Once a strong field (En) is applied, the nanochannel will develop an induced space charge layer, where electroneutrality is no longer maintained. (D) By applying an additional field (ET) along the microfluidic channel in the anodic side (from VS to VD), a nonlinear electrokinetic flow (called electroosmosis of the second kind) is induced, which results in fast accumulation of biomolecules in front of the induced space charge layer. (Adapted from Wang et al. (2005))

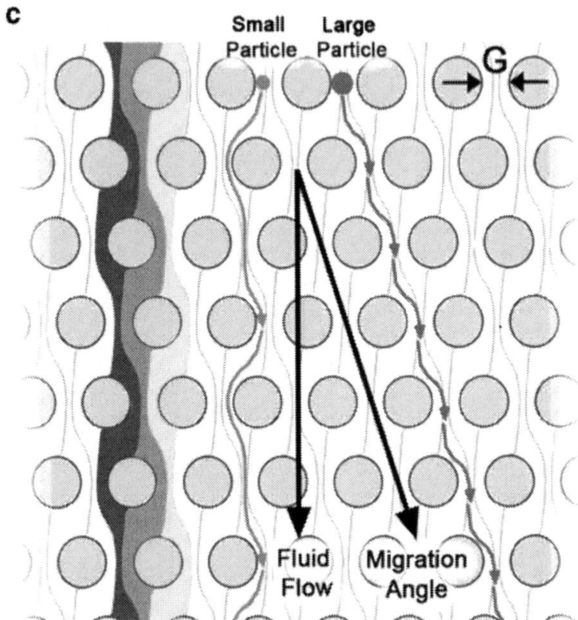

Fig. 4. Deterministic lateral displacement (Adapted from Gossett et al., (2010)).

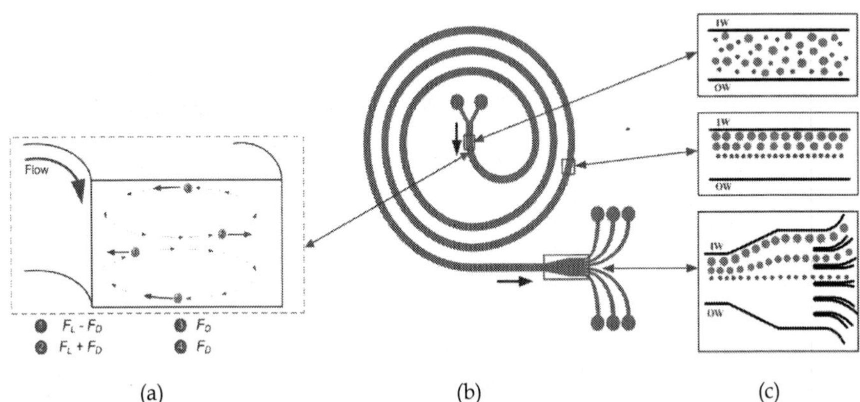

Fig. 5. Spiral microschannels. (a) Neutral buoyant particles suspended in a medium in a spiral shaped channel experience forces and drag. Resultantly, particles redistribute within the microchannel. (b) Schematic representation of spiral channel for particle separation. (c) Different sized particles equilibrate at different positions in microchannel, and are collected at different outlets. (Adapted from Kuntaegowdanahalli et al. (2009) and Bhagat et al. (2008)).

gives rise to vortices, which are exploited for separation of different sized particles. Particles in the center of the channel will experience a drag away from the center, whereas particles in the proximity of the channel walls experience repulsion from the walls. Consequently, particles align at four equilibrium positions in the channel and different sized particles can thus be collected at different outlets (Bhagat et al. (2008); Di Carlo et al. (2007)).

3. Conductive polymers for sensing

Modern biosensors for medical diagnostics must be specific, quick, and producible at reasonable cost. A major cost factor is the electrode material - often a noble metal - demanding extensive production steps in cleanroom facilities. To cut down on these expenses there is a trend to utilize conductive polymers for sensing. This section will give an introduction to advantages of conductive polymers compared to noble metals, and guide through the considerations associated with selecting an appropriate polymer material for biosensor applications.

3.1 Polymers or metals?

The application of polymers as supporting materials in microfluidic systems is well established, however the electronic sensing units in most chips are fabricated from metallic conductors such as platinum or gold.

Biocompatibility, high sensitivity and specificity are a demand in modern medical biosensors. Biocompatibility is required because some biological applications involve living cells, bacteria or virus. High specificity and sensitivity is essential for detecting highly diluted analytes in biological samples, because the samples contain a cocktail of similar components, which can influence a measurement. All of these requirements can be fulfilled by the metal electrode materials such as solid platinum or gold (Prodromidis & Karayannis, 2002). Though, a major disadvantage of the noble metals is the high cost, which is continuously increasing.

Conjugated polymers are an alternative to the traditional electrode materials. The electronic structure of these compounds gives them properties similar to inorganic semiconductors. In 1977, Shirakawa et al. discovered that doping polyacetylene with halogens increased the conductivity by up to four orders of magnitude. The following research on this topic by Shirakawa, MacDiarmid and Heeger was awarded with the Nobel prize in chemistry in 2000.

Over the years, electronically conductive polymers have been proposed for many applications (Jagur-Grodzinski, 2002; Olson et al., 2010) - from biomedical sensors to nanowire integration in photovoltaic cells or printable RFID antennae - yet only few have made it to the market. Among those are electrochromic coatings for windows, antistatic coatings, organic light emitting diodes (OLEDs), corrosion protection for metals or surface finish for printed circuit boards (Groenendaal et al., 2000; Gustafsson et al., 1994; Wessling, 2001).

The usage of conductive polymer electrode in biosensor application is rising. The immediate advantage of conductive polymer electrodes is the much lower cost of the raw materials and the inexpensive production steps. Certain polymers offer high biocompatibility and options for modifying the properties by varying side groups. This can be useful for probe immobilization, which is a crucial procedure in biosensors. Conductive polymers allow a broad range of chemical modifications for covalent attachment of enzymes, antibodies, DNA or other bioprobes (Sarma et al., 2009; Teles & Fonseca, 2008).

In summary, replacing metals with polymers as electrode material does not only limit the cost on the materials themselves, but also allows for the inexpensive mass production by modern ink-jet printing methods (Loffredo et al., 2009; Mabrook et al., 2006) or agarose stamping (Hansen et al., 2007).

3.2 Polymer selection

As mentioned in section 3.1, biocompatibility is a very important factor in selecting an appropriate polymer. Biocompatibility is mainly influenced by the intrinsic toxicity of a material but also by hydrophilicity. Many conjugated polymers suffer from degradation because of irreversible oxidation processes, or they lose their conductive properties over time. A constant and reliable signal is crucial for sensor devices, and accordingly the polymer should be stable over a certain period of time.

In order to provide a good signal to noise ratio in electrochemical measurements, a low ohmic resistance (i. e. high conductivity) is preferred. Currently, these requirements are met by few polymers on the market.

3.2.1 Polypyrrole

The physical properties of Polypyrrole (PPy, figure 6(a)) makes it suitable for biosensor applications. PPy has high decomposition temperature (180–237 °C), glass transition temperature (T_g , 160–170 °C), and relatively high conductivity of up to 3 S·cm^{-1} (Biswas & Roy (1994)). Besides, PPy has a good environmental stability and different facile processing methods (Wang et al. (2001)).

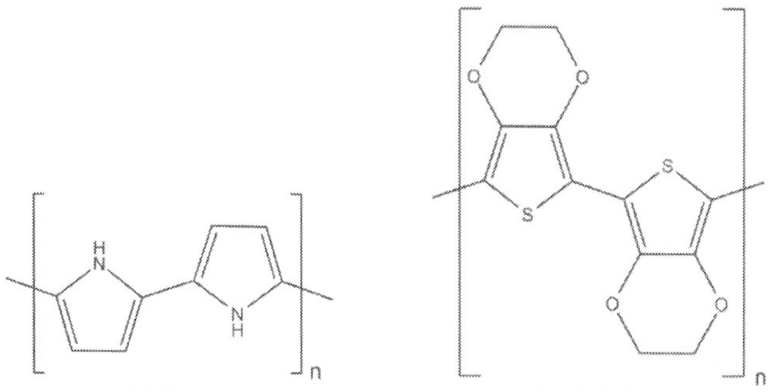

(a) Ppy   (b) PEDOT

Fig. 6. Monomer units of (a) polypyrrole (PPy) and (b) poly(3,4-ethylenedioxythiophene) (PEDOT).

In 2005, Dubois et al. developed a PPy based biosensor for label-free detection of peanut agglutinin. The lactosyl probe unit was immobilized on a biotinylated PPy film via avidin bridges. Their findings demonstrated that the bioprobe could be immobilized directly on the functionalized electrode surface, facilitating label-free detection by electrochemical methods. There are different strategies to functionalize the electrode surface, and another approach was described by Campbell et al. (1999). They incorporated human erythrocytes into the

PPy matrix, and upon capture and binding of Anti-Rhesus (D) antibody, a resistance change could be detected. Other techniques will be discussed in section 4.

3.2.2 PEDOT

Improved properties compared to PPy were found for poly(3,4-ethylenedioxythiophene) or PEDOT. It is either chemically or electrically polymerized from the commercially available monomer 3,4-ethylenedioxythiophene. As can be seen in figure 6(b), it has some structural similarities with PPy. PEDOT has exceptional high conductivity (up to 600 S cm−1), high environmental stability and is biocompatible and transparent for visible light. The most common dopants used for PEDOT are poly(styrene sulfonate) (PSS) and tosylate (Rozlosnik (2009)).

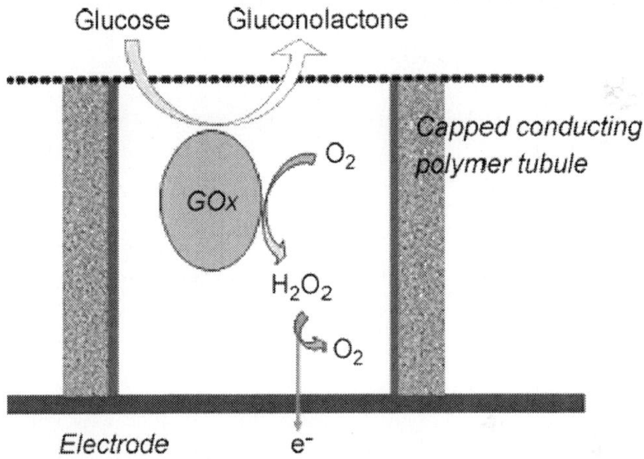

Fig. 7. Glucose oxidase is imprisoned inside a PEDOT microtube covered with a non-conductive polymer (Park et al., 2008).

The work by Balamurugan & Chen (2007) and Vasantha & Chen (2006) show the high potential and superior qualities of PEDOT, and this conductive polymer has been employed in a number of biosensor microdevices.

An interesting study was presented by Kumar et al. (2006). A biosensor was developed to determine the concentration of the important mammalian neurotransmitter, dopamine via an electrochemical process. Since the concentration of ascorbic acid is around a thousand times higher than dopamine in a biological sample, and the two analytes have similar electrochemical potentials, the challenge was to measure the concentration of dopamine in presence of ascorbic acid. Kumar et al. (2006) employed glassy carbon electrodes coated with PEDOT, and their findings demonstrated significant peak separation and improved anti-fouling properties compared to the more common electrode material glassy carbon, making PEDOT a good candidate for further applications in this field.

Glucose detection for blood sugar monitoring of diabetes patients is a huge and growing market for disposable biosensors. The established commercial systems make use of metal electrodes (typically Pt) coated with a gel containing the enzyme glucose oxidase, and the effectively measured agent is thus the oxidation product, hydrogen peroxide (H_2O_2). In contrast to the direct oxidation of dopamine on the electrodes in the example above, this indirect detection of glucose is more complicated. Considering the current market price of platinum of about 41 €/g (http://platinumprice.org), replacing the electrode material with a low cost polymer such as PEDOT seems sensible. Park et al. (2008) imprisoned glucose oxidase in hollow PEDOT micro-tubules on an indium-tin-oxide (ITO) glass surface (figure 7). In this configuration, the enzymes are surrounded by the electrode, and therefore their activity is not constrained by immobilization on a surface or incorporation into a polymer. Although the performance of this biosensor cannot meet the requirements of a classic system, it can be refined by increasing the enzyme density or improving the conductivity.

Many biosensors for pathogen detection are based on antibodies as probes, and deliver an indirect signal. These immunosensors require a fluorescently tagged second antibody, which reacts with occupied immobilized antibodies in a so-called sandwich assay.

A different approach was tested by Kim et al. (2010), who worked on the development of a point of care system for prostate specific antigen/α1-antichymotropsin (PSA-ACT) complex detection. This cancer marker is associated with prostate tumors and important for preoperative diagnosis and screening. Instead of using the conventional optical methods, they constructed an organic electrochemical transistor (OECT) based on PEDOT. The antigen was captured by immobilized antibodies on the conductive polymer. For signal enhancement, a secondary antibody with a covalently tethered gold nanoparticle was used. The system provided a detection limit as low as 1 pg mL−1 and is thus sensitive enough for reliable PSA-ACT analysis.

3.2.3 PEDOT derivatives

A field effect transistor (FET) based biosensor was demonstrated by Xie et al. (2009). The working principle is fundamentally different, considering it uses conductive polymer nanowires, which were electropolymerized between two gold electrodes. For minimizing the distance between polymer and binding event it was necessary to couple the probe (an aptamer, see also section 4.3) directly to the electrode material. Normal PEDOT offers no possibility for covalent bonding of other molecules, so a derivative bearing a carboxylic acid group was used. With this functional group the oligonucleotide for thrombin detection was attached with a simple 1-ethyl-3-(3-dimethylaminopropyl)carbodiimide/N-hydroxysuccinimide procedure (see section 3.3, (EDC/NHS)). Thrombin binds specifically to aptamers and becomes immobilized on the surface. The positively charged protein influences the transistor, so that the current flow changes. This type of biosensor has a broad dynamic range covering the physiologically interesting thrombin concentration range from a few nanomoles to several hundred nanomoles.

Other PEDOT derivatives have also been investigated (Akoudad & Roncali, 2000; Ali et al., 2007; Daugaard et al., 2008). The structural formulas of the most commonly used monomers are shown in figure 8; PEDOT-OH is more hydrophilic than normal PEDOT, and the azide modified PEDOT-N3 polymerizes slowly and has decreased conductivity. The only commercially available monomer is (2,3-dihydrothieno[3,4-b][1,4]dioxin-2-yl)methanol (commonly known as hydroxymethyl-EDOT or EDOT-OH) (8(a)), and it can be used as a basis for further modifications.

3.3 Coupling methods

There are different techniques for immobilization of biomolecules (e.g. DNA) on an electrode surface. The most popular methods are formation of a biotin-streptavidin complex, formation of different covalent bonds like esters or amides, or click chemistry.

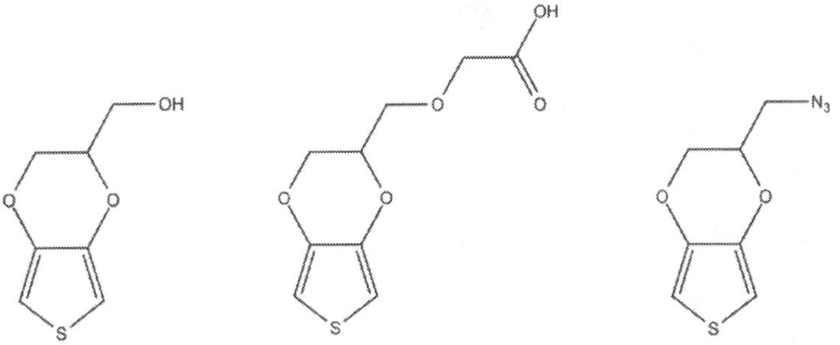

(a) EDOT-OH   (b) EDOT-COOH   (c) EDOT-N3

Fig. 8. Different derivatives of 3,4-ethylenedioxythiophene (Ali et al., 2007; Daugaard et al., 2008).

3.3.1 Biotin-streptavidin complex

Streptavidin is a protein consisting of four identical subunits, each of which has an extremely high affinity for biotin. A biotinylated surface can be coated with streptavidin so it offers reactive sites for fixation of likewise biotin tagged (bio)molecules. The biotin-streptavidin interaction is one of the strongest non-covalent bonds in nature and it is very specific. Moreover, the system is easy to handle and very biocompatible.

Despite the many advantages of streptavidin, a major drawback is the instability at low or high pH values, and high temperature. For some detection methods the rather thick protein layer between electrode and probe can substantially decrease the sensitivity of the sensor.

3.3.2 Covalent bonding

Different activation methods have been used for a long time in chemistry, which requires the availability of certain functional groups on the surface. The activation of a carboxylic acid group with 1-ethyl-3-(3-dimethylaminopropyl)carbodiimide (EDC) and N-hydroxysuccinimide (NHS) is often applied for amide-bond formation under mild conditions and can be used for binding molecules bearing free amino groups (Balamurugan et al., 2008; Xie et al., 2009).

For hydroxyl functionalized polymers and target molecules, a technique from DNA synthesis can be employed. The alcohol groups are activated with phosphoramidites to form a phosphoester, which then reacts with another hydroxyl moiety and links the target molecules covalently to the surface (Pirrung, 2002).

3.3.3 Click chemistry

A very elegant approach for probe immobilisation is the usage of so called "click-chemistry". In the Cu-catalysed Huisgen-type 1,3-cycloaddition suggested by Daugaard et al. (2008) an azide reacts in high yield with an alkyne to form a five-membered heterocycle. This bond is very stable and also the precursors have advantages such as stability toward hydrolysis and dimerization or ease of introduction (Kolb et al. 2001).

However, the azide functionalization of PEDOT downgraded its conductive properties significantly and remaining Cu catalyst could influence biological systems.

4. Electrode functionalization

Functionalization of electrodes is essential for achieving high sensitivity and specificity of electrochemical biosensors. This section provides an overview of the current trend in electrochemical sensors for medical diagnostics.

4.1 Recognition of pathogens

Point of care diagnostic devices present a viable option for rapid and sensitive detection and analysis of pathogens. Biosensors can play an important role in the early diagnosis of acute viral disease and confine the spread of virulent disease outbreaks. Biosensors can also play an important role in early detection and diagnosis of cancer and autoimmune disorders based on specific biochemical markers.

As discussed in section 2, separation and isolation of large quantities of a specific analyte would be preferable for many medical applications.

Patient samples comprise of a heterogeneous population of particles and cells, hence challenging the isolation of a single species in a high background concentration. For this reason, biosensors must be very specific and sensitive, allowing precise detection of very small quantities.

4.2 Antibodies

Many techniques for preparing functional biological surfaces for studies of cells, viruses or disease markers have been described in the literature. Refer to 3.3 for an overview of different coupling methods.

Immunoglobulins (IgG) are large Y-shaped proteins produced by the immune system, and are most abundant in blood plasma. Two identical antigen binding sites are formed from several loops of the polypeptide chain. These loops allow many chemical groups to close in on a ligand and link to it with many weak (non-covalent) reversible bonds. An antibody-antigen bond is highly specific because of the molecular structure of the protein.

Antibodies are the most common recognition molecule in biosensors. It is a naturally occuring protein and can only be produced in a host against immunogenic substances, giving rise to batch variation and a limited target range. For research purposes, monoclonal or polyclonal antibodies can be applied as recognition molecule. Typically, monoclonal antibodies will ensure a higher specificity than polyclonal antibodies.

In medical sensor applications, functional orientation of the antibodies on the surface is crucial to ensure high sensitivity and specificity. It can be achieved by immobilizing the proteins on a supporting layer of protein A.

4.3 Aptamers

For many years, antibodies have been applied for surface functionalization in biosensors, ensuring specificity and sensitivity of sensors. Artificial nucleic acid ligands - known as aptamers - can cover the same field of application as antibodies. In the recent years, the use of aptamers has increased (Han et al. (2010); Syed & Pervaiz (2010)), and they are in many ways superior to antibodies, as it will be disscused in this section.

4.3.1 The properties of aptamers

Aptamers are oligonucleotides with a typical length of forty to eighty basepairs, and were discovered in the 1980's as naturally occurring regulation elements in prokaryotic cells, and they showed high affinity for viral and cellular proteins.

In 1990, Tuerk & Gold developed a convenient process for in vitro aptamer production, the so-called systematic evolution of ligands by exponential enrichment (SELEX, see in section 4.3.2).

Aptamers are in many regards better than antibodies as summarized in table 2. The affinity for the target molecules of aptamers is similar to antibodies, and in some cases even higher compared to antibodies. The specificity is also higher for aptamers, as they can distinguish between targets of the same family, like it was shown for the molecules caffeine and theophylline (Zimmermann et al., 2000). Selection and production of the nucleic acid ligands can be done in vitro, and once the correct sequence has been determined, the oligonucleotide can be synthesized in an automated chemical procedure. The range for possible target molecules is very wide and - in comparison with the mentioned biomolecules - comprises all kinds of smaller ions, organic compounds and even whole cells. Contrary to antibodies, aptamers can be selected against toxic compounds.

Due to the chemical synthesis, there is no significant batch variation and it allows for easy chemical modification, like attachment of certain end groups for surface immobilization.

Reversible thermal denaturation makes aptamers potentially recyclable and their very high stability promises long self life (Lee et al., 2008).

4.3.2 The SELEX process

Aptamer production is accomplished in the SELEX (Systematic Evolution of Ligands by Exponential Enrichment) process (see figure 9). A pool of single stranded oligonucleotides with a random section of about 25 to 70 basepairs (the library) is incubated with the target molecule. Some nucleic acid strands will interact with the target molecule and form strong non-covalent bonds. Target-DNA-complexes are partitioned from unbound DNA. After dissolution of the complex, the selected oligonucleotides are amplified in a standard PCR process. DNA strands are separated and the whole procedure is repeated for up to 20 times in order to select the best fitting sequences.

If RNA is used, a transcription step must be inserted before and after PCR. In order to increase specificity for the target molecule and exclude unspecific binding, counter selection steps can be employed. In those selection rounds no target is used and DNA strands with affinity to the support and container material are removed from the pool

4.3.3 Biosensor applications

In order to eliminate systematic problems with sandwich assays, the development of label free biosensors is an interesting topic. Xiao et al. (2005) modified a thrombin specific aptamer with a thiol group for immobilization on a gold surface. The strand was partially hybridised with a not fully complementary strand bearing a methylene blue (MB) tag.

In presence of thrombin, the strands separated and the MB redox tag was approximated to the Au surface (figure 10(a)).

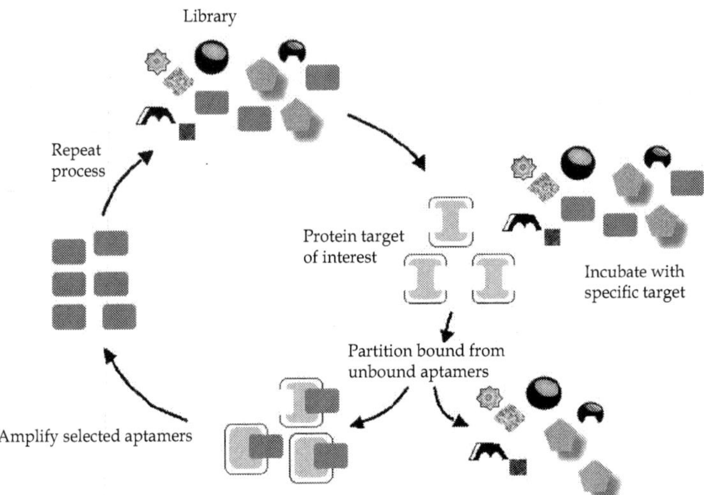

Fig. 9. The SELEX process for use with a DNA library (Nimjee et al. (2006)).

Similar signal-on detectors were developed by Baker et al. (2006) and Lai et al. (2006). The single stranded aptamer had the ability to hybridize with itself and form three loops upon target binding (see figure 10(b)). The conformational change brings the MB tag in proximity to the gold surface and allows for an electrical measurement. Both systems could be regenerated to a high degree, and thus are potentially reusable. Baker's system could detect cocaine concentrations as low as 500 µM in biological fluids even in the presence of contaminants.

So et al. (2005) attached thrombin binding aptamer to a single walled carbon nanotube (SWNT) which connected two electrodes.

Binding the charged protein induced an electrostatic gate potential and changed the source-drain current. The field effect transistor (FET) biosensor was able to detect thrombin in a concentration range of 10 – 100 nM.

	Aptamers	Antibodies
Affinity	Low nM – pM	Low nM – pM
Specificity	High	High
Production	In vitro chemical process	In vivo biological process
Target range	Wide: ions – whole cells	Narrow: immunogenic compounds
Batch to batch variation	Little or no	Significant
Chemical modification	Easy and straightforward	Limited
Thermal denaturation	Reversible	Irreversible
Shelf life	Unlimited	Limited

Table 2. Differences between aptamers and antibodies. Advantages are emphasised (Lee et al., 2008

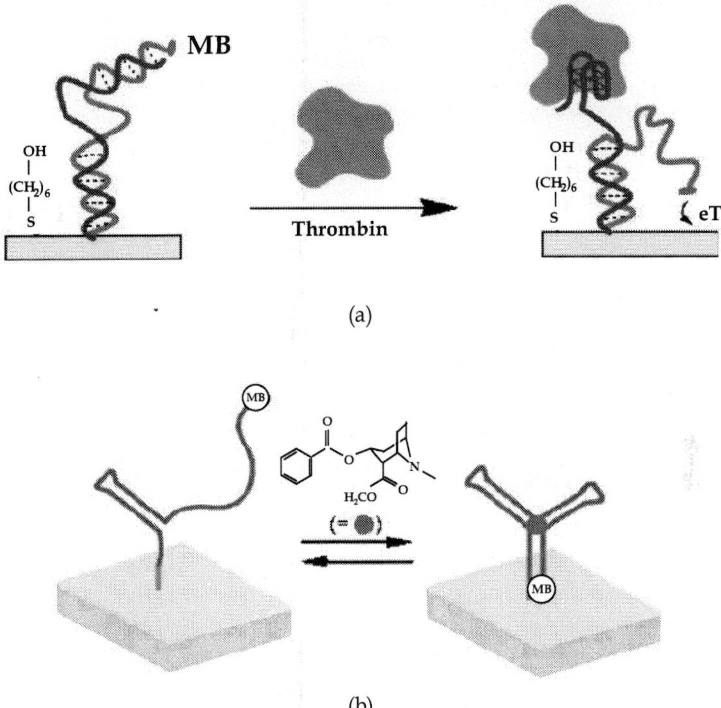

Fig. 10. Aptamers with a methylene blue redox tag for thrombin (a) and cocaine detection (b). The binding event induces a conformational change in the aptamer and brings the redox active tag closer to the gold surface Baker et al. (2006); Lai et al. (2006). An analogue sensor was described earlier in section 3.2. Xie et al. (2009) used carboxylic acid modified PEDOT nanowires instead of SWNTs as FET.

5. Electrical detection methods

The electrical detection has traditionally received the major share of the attention in biosensor development. Such devices produce a simple, inexpensive and yet accurate and sensitive platform for patient diagnosis. The name electrochemical biosensor is applied to a molecular sensing device which intimately couples a biological recognition element to an electrode transducer. The purpose of the transducer is to convert the biological recognition event into a useful electrical signal.

Electrochemical systems are extremely sensitive to the processes that take place on the surfaces of the electrodes, and in this sense the electrodes are direct transducers in biomedical applications. Several types of electrochemical methods are used in biosensors; the two most common ones are the amperometry and impedance spectroscopy (EIS) (Lazcka et al. (2007)). Recently, all-polymer field effect transistors for biosensing have been introduced (Lee et al. (2010)).

5.1 Amperometry

Amperometry is a method of electrochemical analysis in which the signal of interest is a current that is linearly dependent upon the concentration of the analyte. As certain chemical species are oxidized or reduced (redox reactions) at the electrodes, electrons are transferred from the analyte to the working electrode or to the analyte from the electrode. The direction of flow of electrons depends upon the properties of the analyte and can be controlled by the electric potential applied to the working electrode.

Amperometric biosensors operate by applying a constant potential and monitoring the current associated with the reduction or oxidation of an electroactive species involved in the recognition process. The amperometric biosensor is attractive because of its high sensitivity and wide linear range.

5.2 Conductivity and impedance spectroscopy

Electrochemical impedance spectroscopy (EIS) combines analyses of both the resistive and capacitive properties of materials, based on the perturbation of the system by a small-amplitude sinusoidal AC signal. The impedance of the system can be scanned over a wide range of AC signal frequencies. The amplitude of the current, potential signals, and the resulting phase difference between voltage and current dictates the system impedance Therefore, the impedance signal is dependent on the nature of the system under study.

Equivalent circuit models fitted to the impedance curves are useful tools for characterizing the system. Although this methodology is widely accepted because of ease of use, extreme care must be taken to ensure that the equivalent circuit obtained makes physical sense. An advantage of EIS compared to amperometry is that redox labels are no longer necessary, which simplifies the sensor preparation.

5.3 Organic field effect transistors

Organic field effect transistors (Organic FETs) have a potential being the active matrix for many electronic devices, including biosensors for biological material. An organic field-effect transistor consists of a source and drain electrode, an organic semiconductor (which is in this case a conductive polymer), a gate dielectric, and a gate electrode. A number of different studies have demonstrated conductance-based sensors employing a molecular receptor layer immobilized on the surface of a semiconductor device. The receptor molecules provide the means to achieve highly selective sensing because they can be engineered to have much higher binding affinities with the desired target molecules than the other species in the analyte solution (see section 4). Although the organic FET is a promising candidate for biosensor applications, optimization of the device structure and operating conditions is still required.

6. Outlook

In recent years, fascinating developments of a wide range of commercial applications have occurred. Elegant research on new sensing concepts has opened the door to a wide variety of microsystem based biosensors for clinical applications. Such devices are extremely useful for delivering diagnostic information in a fast, simple, and low cost fashion, and are thus uniquely qualified for meeting the demands of point of care systems, e.g. for cancer screening. The high sensitivity of the modern biosensors should facilitate early detection and treatment of diseases, and lead to increased patient survival rates.

In the future, one of the main challenges is to bring the new biosensor techniques to bed side for use by non-laboratory personnel without compromising accuracy and reliability. The internal calibration and reference is also a major requirement, and provoke researchers to reshape the existing methods. From a clinical point of view, the in vivo biosensors that are biocompatible and can remain in the body for weeks or months will also be a demand. Special attention should be given to non-specific adsorption issues that commonly control the detection limits of electrochemical bioaffinity assays. The stability of biosensors remains an important issue in the fabrication and use of these devices for many application areas.

By measuring abnormalities within few minutes, disposable cartridges containing electrode strips and simple sample processing could offer early and fast screening of diseases in a point of care setting.

It has become apparent that the field of polymer biosensors has reached a new level of maturity. In the near future it is highly likely that pathogen detection will undoubtedly benefit from the integration of biosensors into all-polymer microdevices, and thus in some regards revolutionize the medical diagnostics.

7. Acknowledgements

This work was supported by the Danish Research Counsil for Technology and Production Sciences and the Technical University of Denmark.

8. References

Akoudad, S. & Roncali, J. (2000). Modification of the electrochemical and electronic properties of electrogenerated poly(3,4-ethylene-dioxythiophene) by hydroxymethyl and oligo(oxyethylene) substituents, Electrochemistry Communications 2(1): 72.

Ali, E. M., Kantchev, E. A. B., Yu, H.-h. & Ying, J. Y. (2007). Conductivity shift of polyethylenedioxythiophenes in aqueous solutions from side-chain charge perturbation, Macromolecules 40(17): 6025–6027.

Baker, B. R., Lai, R. Y., Wood, M. S., Doctor, E. H., Heeger, A. J. & Plaxco, K. W. (2006). An electronic, aptamer-based small-molecule sensor for the rapid, label-free detection of cocaine in adulterated samples and biological fluids., J. Am. Chem. Soc. 128(10): 3138–9.

Balamurugan, A. & Chen, S.-M. (2007). Poly(3,4-ethylenedioxythiophene-co-(5-amino-2-naphthalenesulfonic acid)) (pedot-pans) film modified glassy carbon electrode for selective detection of dopamine in the presence of ascorbic acid and uric acid., Anal. Chim. Acta 596(1): 92–8.

Balamurugan, S., Obubuafo, A., Soper, S. A. & Spivak, D. A. (2008). Surface immobilization methods for aptamer diagnostic applications., Anal Bioanal Chem 390(4): 1009–21.

Bhagat, A. A. S., Bow, H., Hou, H. W., Tan, S. J., Han, J. & Lim, C. T. (2010). Microfluidics for cell separation, Medical & Biological Engineering & Computing 48(10, Sp. Iss. SI): 999–1014.

Bhagat, A. A. S., Kuntaegowdanahalli, S. S. & Papautsky, I. (2008). Continuous particle separation in spiral microchannels using dean flows and differential migration, Lab on a Chip 8(11): 1906–1914.

Biswas, M. & Roy, A. (1994). Thermal, stability, morphological, and conductivity characteristics of polypyrrole prepared in aqueous medium, Journal of Applied Polymer Science 51(9): 1575.

Campbell, T. E., Hodgson, A. J. & Wallace, G. G. (1999). Incorporation of erythrocytes into polypyrrole to form the basis of a biosensor to screen for rhesus (d) blood groups and rhesus (d) antibodies, Electroanalysis 11(4): 215.

Daugaard, A. E., Hvilsted, S., Hansen, T. S. & Larsen, N. B. (2008). Conductive polymer functionalization by click chemistry, Macromolecules 41(12): 4321.

Di Carlo, D., Irimia, D., Tompkins, R. & Toner, M. (2007). Continuous inertial focusing, ordering, and separation of particles in microchannels, PNAS 104(48): 18892–18897.

Dubois, M.-P., Gondran, C., Renaudet, O., Dumy, P., Driguez, H., Fort, S. & Cosnier, S. (2005). Electrochemical detection of arachis hypogaea (peanut) agglutinin binding to monovalent and clustered lactosyl motifs immobilized on a polypyrrole film., Chem. Commun. (Camb.) (34): 4318–20.

Giddings, J. (1993). Field-flow fractionation: analysis of macromolecular, colloidal, and particulate materials, Science 260(5113): 1456–465.

Gossett, D. R., Weaver, W. M., Mach, A. J., Hur, S. C., Tse, H. T. K., Lee, W., Amini, H. & Di Carlo, D. (2010). Label-free cell separation and sorting in microfluidic systems, Analytical and Bioanalytical Chemistry 397(8): 3249–3267.

Gregoratto, I., McNeil, C. & Reeks, M. (2007). Micro-devices for rapid continuous separation of suspensions for use in micro-total-analysis-systems , SPIE .

Groenendaal, L., Jonas, F., Freitag, D., Pielartzik, H. & Reynolds, J. R. (2000). Poly(3,4-ethylenedioxythiophene) and its derivatives: Past, present, and future, Advanced Materials 12(7): 481.

Gustafsson, J., Liedberg, B. & Inganäs, O. (1994). In situ spectroscopic investigations of electrochromism and ion transport in a poly (3,4-ethylenedioxythiophene) electrode in a solid state electrochemical cell, Solid State Ionics 69(2): 145.

Han, K., Liang, Z. & Zhou, N. (2010). Design Strategies for Aptamer-Based Biosensors, Sensors 10(5): 4541–4557.

Hansen, T., West, K., Hassager, O. & Larsen, N. (2007). Direct fast patterning of conductive polymers using agarose stamping, Advanced Materials 19(20): 3261.

Jagur-Grodzinski, J. (2002).Electronically conductive polymers, Polymers for Advanced Technologies 13(9): 615.

Kim, D.-J., Lee, N.-E., Park, J.-S., Park, I.-J., Kim, J.-G. & Cho, H. J. (2010). Organic electrochemical transistor based immunosensor for prostate specific antigen (psa) detection using gold nanoparticles for signal amplification., Biosens Bioelectron 25(11): 2477–82.

Kolb, Hartmuth C., Finn, M. G., and Sharpless, K. Barry (2001): Click Chemistry: Diverse Chemical Function from a Few Good Reactions, Angewandte Chemie International Edition 40(11):2004

Kumar, S. S., Mathiyarasu, J., Phani, K. L. N. & Yegnaraman, V. (2006). Simultaneous determination of dopamine and ascorbic acid on poly (3,4-ethylenedioxythiophene) modified glassy carbon electrode, Journal of Solid State Electrochemistry 10(11): 905.

Kuntaegowdanahalli, S. S., Bhagat, A. A. S., Kumar, G. & Papautsky, I. (2009). Inertial microfluidics for continuous particle separation in spiral microchannels, Lab on a Chip 9(20): 2973–2980.

Lai, R. Y., Seferos, D. S., Heeger, A. J., Bazan, G. C. & Plaxco, K. W. (2006). Comparison of the signaling and stability of electrochemical dna sensors fabricated from 6- or 11-carbon self-assembled monolayers., Langmuir 22(25): 10796–800.

Lazcka, O., Campo, F. J. D. & Muñoz, F. X. (2007). Pathogen detection: A perspective of traditional methods and biosensors, Biosensors and Bioelectronics 22(7): 1205 – 1217.

Lee, J.-O., So, H.-M., Jeon, E.-K., Chang, H., Won, K. & Kim, Y. H. (2008). Aptamers as molecular recognition elements for electrical nanobiosensors., Anal Bioanal Chem 390(4): 1023–32.

Lee, M.-W., Lee, M.-Y., Choi, J.-C., Park, J.-S. & Song, C.-K. (2010). Fine patterning of glycerol-doped pedot:pss on hydrophobic pvp dielectric with ink jet for source and drain electrode of otfts, Organic Electronics 11(5): 854 – 859.

Lenshof, A. & Laurell, T. (2010). Continuous separation of cells and particles in microfluidic systems, Chemical Society Reviews 39(3): 1203–1217.

Loffredo, F., Mauro, A. D. G. D., Burrasca, G., La Ferrara, V., Quercia, L., Massera, E., Di Francia, G. & Sala, D. D. (2009). Ink-jet printing technique in polymer/carbon black sensing device fabrication, Sensors and Actuators B: Chemical 143(1): 421.

Mabrook, M., Pearson, C. & Petty, M. (2006). Inkjet-printed polypyrrole thin films for vapour sensing, Sensors and Actuators B: Chemical 115(1): 547.

Nimjee, S. M., Rusconi, C. P. & Sullenger, B. A. (2006). Aptamers to proteins, The aptamer handbook: functional oligonucleotides and their applications .

Nolan, J. & Sklar, L. (1998). The emergence of flow cytometry for sensitive, real-time measurements of molecular interactions., Nat. Biotechnol. 16(7): 633–8.

Olson, T. S., Pylypenko, S., Atanassov, P., Asazawa, K., Yamada, K. & Tanaka, H. (2010). Anion-exchange membrane fuel cells: Dual-site mechanism of oxygen reduction reaction in alkaline media on cobalt–polypyrrole electrocatalysts, The Journal of Physical Chemistry C 114(11): 5049.

Park, J., Kim, H. K. & Son, Y. (2008). Glucose biosensor constructed from capped conducting microtubules of pedot, Sensors and Actuators B: Chemical 133(1): 244.

Peterson, F., Åberg, L., Swård-Nilsson, A. & Laurell, T. (2007). Free Flow Acoustophoresis: Microfluidic-Based Mode of Particle and Cell Separatio, Anal. Chem. 79(14): 5117–5123.

Pirrung, M. C. (2002). How to make a dna chip., Angew. Chem. Int. Ed. Engl. 41(8): 1276–89. Prodromidis, M. & Karayannis, M. (2002). Enzyme based amperometric biosensors for food analysis, Electroanalysis 14(4): 241.

Rozlosnik, N. (2009). New directions in medical biosensors employing poly(3,4-ethylenedioxy thiophene) derivative-based electrodes., Analytical and Bioanalytical Chemistry 395(3): 637–45.

Sarma, A. K., Vatsyayan, P., Goswami, P. & Minteer, S. D. (2009). Recent advances in material science for developing enzyme electrodes., Biosens Bioelectron 24(8): 2313–22.

Seo, J., Lean, M. & Kole, A. (2007). Membrane-free micro?ltration by asymmetric inertial migration, Appl. Phys. Lett. 91(033901).

Shirakawa, H., Louis, E. J., MacDiarmid, A. G., Chiang, C. K. & Heeger, A. J. (1977). Synthesis of electrically conducting organic polymers: halogen derivatives of polyacetylene, (CH) x , Journal of the Chemical Society, Chemical Communications (16): 578.

So, H.-M., Won, K., Kim, Y. H., Kim, B.-K., Ryu, B. H., Na, P. S., Kim, H. & Lee, J.-O. (2005). Single-walled carbon nanotube biosensors using aptamers as molecular recognition elements., J. Am. Chem. Soc. 127(34): 11906–7.

Syed, M. A. & Pervaiz, S. (2010). Advances in Aptamers, Oligonucleotides 20(5): 215–224.

Teles, F. & Fonseca, L. (2008). Applications of polymers for biomolecule immobilization in electrochemical biosensors, Materials Science and Engineering: C 28(8): 1530.

Toner, M. & Irimia, D. (2005). Blood on a Chip, Annu. Rev. Biomed Eng. pp. 77–103.

Tuerk, C. & Gold, L. (1990). Systematic evolution of ligands by exponential enrichment: Rna ligands to bacteriophage t4 dna polymerase, Science 249(4968): 505.

Vasantha, V. & Chen, S. (2006). Electrocatalysis and simultaneous detection of dopamine and ascorbic acid using poly(3,4-ethylenedioxy)thiophene film modified electrodes, Journal of Electroanalytical Chemistry 592(1): 77.

Wang, L.-X., Li, X.-G. & Yang, Y.-L. (2001). Preparation, properties and applications of polypyrroles, Reactive and Functional Polymers 47(2): 125–139.

Wang, Y., Stevens, A. & Han, J. (2005). Million-fold preconcentration of proteins and peptides by nanofluidic filter, Analytical Chemistry 77(14): 4293–4299.

Wessling, B. (2001). From conductive polymers to organic metals, Chemical Innovation(USA) 31(1): 34–40.

Xiao, Y., Piorek, B. D., Plaxco, K. W. & Heeger, A. J. (2005). A reagentless signal-on architecture for electronic, aptamer-based sensors via target-induced strand displacement., J. Am. Chem. Soc. 127(51): 17990–1.

Xie, H., Luo, S.-C. & Yu, H.-H. (2009). Electric-field-assisted growth of functionalized poly(3,4-ethylenedioxythiophene) nanowires for label-free protein detection., Small 5(22): 2611–7.

Zimmermann, G. R., Wick, C. L., Shields, T. P., Jenison, R. D. & Pardi, A. (2000). Molecular interactions and metal binding in the theophylline-binding core of an rna aptamer, RNA-A Publication of the RNA Society 6(5): 659–667.

10

Hybrid Film Biosensor for Phenolic Compounds Detection

Joanna Cabaj and Jadwiga Sołoducho*
*Faculty of Chemistry, Department of Medicinal Chemistry and Microbiology, Wrocław
University of Technology, Wrocław
Poland*

1. Introduction

Phenolic compounds are industrial chemicals widely used in the manufacture of products. Most of them are generated artificially and are found in the wastewaters from chemical plants, exhaust gases from incinerators, the side stream smoke from cigarettes. They are easily adsorbed in humans, regardless of their form. High levels of phenols have been shown to have detrimental effects on animal health, and some phenolic compounds are reportedly carcinogenic (N. Li et al., 2005) and allergenic (Haghighi et al., 2003), and due to their toxic effects, their determination and removal in the environment are of great importance.

Biosensors can make ideal sensing systems to monitor the effects of pollution on the environment, due to their biological base, ability to operate in complex matrices, short response time and small size. The determination of phenol and its derivative compounds is of the environmental greatness, since these species are released into the environment by a large number of industries, *e.g.* the manufacture of plastics, dyes, drugs, antioxidants and waste waters from pulp and paper production. This group of biosensors is of great interest because of their application in food and pharmaceutical industry.

Among enzymes, laccases and tyrosinases (Duran et al., 2002) or horse-radish peroxidase (Freire et al., 2001) as well as polyphenol oxidase are groups of enzymes that catalyze the transformation of a large number of phenolic compounds. The mechanism for tyrosinase, laccase and peroxidase in the electrochemical biosensors are different. Enzyme molecules are re-reduced by phenolic compounds after they were oxidized by oxygen (for tyrosinase and laccase) or hydrogen peroxide (for peroxidase) on the surface to the electrode. The tyrosinase biosensors are applicable to the monitoring of phenolic compounds with at least one free *ortho*-position. On the other hand, the laccase biosensor can detect phenolic compounds with free *para-* and *metha-* position with a complicated catalytic cycle. Horseradish peroxidase (HrP) based biosensors are most sensitive for a great number of phenolic compounds since phenols can be act as electron donors for peroxidase (Yang et al., 2006).

As seen, phenolooxidases have wide substrate specificity and a great potential for the determination of phenolic compounds. Furthermore, fungal laccases catalyze demethylation reactions an important and initial step of the biodegradation process of the lignin polymer chain, and subsequently decompose the lignin macromolecule by splitting aromatic rings and C–C bonds in the phenolic substructures (Freire et al., 2001).

In enzymatic devices, efforts have been concentrated on the control over enzyme activity, which is highly dependent on the interface between the nanocomposite and the enzyme. Such control has led to immobilization techniques suitable for anchoring the enzyme close to electrode with preservation of biological activity. In these type of devices, where preservation of the enzyme activity at the nanocomposite/enzyme interface is the key for designing efficient electrode, charge transfer between enzyme and electrode should be fast and reversible. This charge transfer may be also optimized with some mediating particles being used in conjunction with the biological molecules at the electrode surface. Conducting polymers are often used as support materials for amperometric biosensors (Kuwahara et al., 2005). The interlaced polymer is expected to facilitate the electron transfer as well enhancing the sensor sensitivity.

It is essential for the sensitivity of the system that the recognition units have optimized surface density, good accessibility, long-term stability and minimized non-specific interactions with compounds other than the analyte. Such model molecular assemblies can be prepared by Langmuir-Blodgett (LB) and Langmuir-Schaefer (LS) techniques in which we have to the moment successful experience (Cabaj et al., 2010a; Cabaj et al., 2010b), layer-by-layer (LbL) or by employing self-assembly monolayers (SAMs) as well as electrodeposition. Construction of novel phenol detecting biosensor is challenge for new technologies and the key problem is modification of electrode by enzyme using thin film preparation methods.

2. Hybrid thin film fabrication

Proteins are usually adsorbed on the surface of a solid substrate in a disordered orientation and are apt to suffer from conformational change, which renders the functional sites inaccessible to interacting molecules and substrates for enzymatic reactions (Kumada et al., 2007; Nakanishi et al., 2001). The extent of the conformational change of proteins in their adsorbed state differs with the type of protein, the kinds/states of the surface including hydrophobicity/hydrophilicity, and with the environmental conditions, such as temperature, pH, and ionic strength of the solution.

Most immobilization methods developed thus far involve modification or coating of the surface with appropriate substances to change the surface property and/or provide functional groups for the binding of protein (Zhu & Snyder, 2003). On the other hand, immobilization of proteins on a bare surface with no modification utilizes specific interactions between the protein and the surface, which includes immobilization on the Au surface *via* thiol (sulfhydryl) groups and necessitates using an affinity peptide that is specific to the particular surface. Conceptual schemes for the various immobilization methods are summarized in Table 1.

2.1 Non-covalent immobilization
2.1.1 Physical adsorption
Extensive work in area of biosensors is mainly connected with immobilization of protein in miniaturized structures, which may also contain hybrid materials for enhancing sensitivity and selectivity (Siqueira et al., 2010). Many biosensing devices integrate biomolecules with metal nanoparticles (Debabov 2004), carbon nanotubes, solid matrices (S.N. Kim et al., 2007)

and polymers (Hammond 2004). Proteins can be immobilized in a variety of structures as in transmembrane pores, and phospholipids Langmuir-Blodgett/Langmuir-Schaefer films, layer-by-layer (LbL) structures (Cabaj et al., 2010a; Cabaj et al., 2010b). For electrical detection, in particular, mediators or conducting matrices may be used to enhance conduction.

Immobilization method	Surface for immobilization	Capturing mechanism	*Ref.*
Physical adsorption	Polystyrene, nitrocellulose, LB/LS, LbL films	Physical adsorption	Cabaj et al., 2010a; Cabaj et al., 2010b; Nakanishi et al., 2008
Immobilization using hydrogel	Glass plate	Entrapment in gel	Nakanishi et al., 2008
Immobilization using coiled coil interaction	Octadecyltrichlorosilane (OTS) coated surface covalently bound with artificial polypeptide containing Leu zipper	Coiled coil association of a heterodimeric Leu zipper pair	Nakanishi et al., 2008
Immobilization on Au surface	Au surface	Chemisorption of SH-groups on Au	Nakanishi et al., 2008
Glutatione/S-Transferase-mediated immobilization	Polystyrene coated with protein followed by covalently coupling with glutathione	Affinity between glutathione and S-Transferase	Nakanishi et al., 2008
Silane coupling method	Glass modified with bifunctional silane coupling reagents containing aldehyde	Schiff's base linkage between aldehyde and amino groups	Nakanishi et al., 2008
Immobilization using polystyrene (PS)-affinity peptide	Hydrophilic polystyrene	Affinity of the affinity peptide to PS surface	Nakanishi et al., 2008

Table 1. Various immobilization methods

Direct adsorption of biomolecules on a solid surface may also lead to the rapid, simple immobilization. For example, a gold surface can directly adsorb biomolecules *via* strong electrostatic interaction and thiol-based chemical binding under ambient conditions, and a SiO_2 surface can directly adsorb biomolecules mainly by van der Waals force and hydrogen bonding. The atomically flat sapphire (Al_2O_3) substrate with a hydrophilic surface may be used for hydrophilic adsorption (Yoshida et al., 1998).

However, such a direct adsorption of biomolecules on the surface of metals or metal oxides is usually non-specific and the orientation of the adsorbed structures remains largely random (D.C. Kim et al., 2009). As direct adsorption may result in significant conformational change in the native structure of proteins and severe loss of biological activity.

2.1.2 Layer-by-layer method

The technique of production of sensing structures in the form of thin films has to comply with three requirements: molecular recognition, signal transduction and signal detection by the abiotic component. That is the key problem in construction of miniaturized sensing devices. Nanofilms produced with LbL and LB methods may give a satisfactory results in controlling of layer architecture.

In the LbL method, solid supports are immersed into aqueous solutions of materials to immobilize. Alternating layers of positively and negatively charged materials are adsorbed in a sequence, which may lead to multilayers. The procedure of the film preparation is depicted in Figure 1.

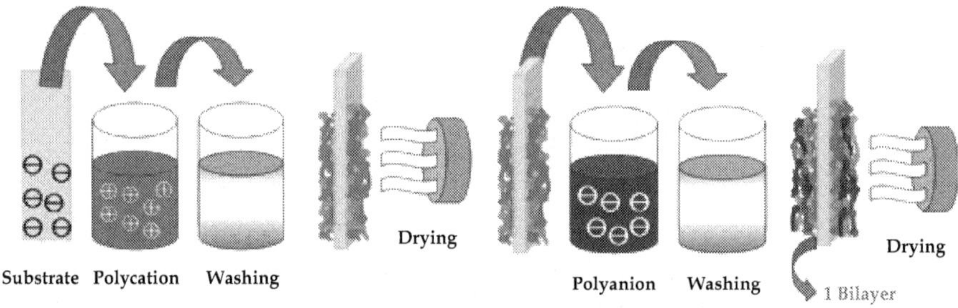

Fig. 1. Schematic procedure for fabrication of LbL films. The substrate is alternately dipped into solutions containing polycations and polyanions to produce bilayers (Siqueira et al., 2010)

The amount of material adsorbed on the support and the surface topography depend on various parameters, including concentration of material to be adsorbed, pH, ionic strength. The obtaining films are thermally stable and resist to washing in aqueous solutions. In order to sensors fabrication, the LbL layers have been prepared of proteins, nucleic acids and polysaccharides (Siqueira et al., 2010; Liu et al., 2008).

Deposition of a large number of layers can be performed by using a mechanical robot (Portnov et al., 2006) or a simple fluidic system. The main drawbacks of LbL deposition are sensitivity of the receptor containing multilayer to variations of ionic strength, generally leading to destabilization of an assembly and an influence of the drying process on the multilayer structure (Lourenco et al., 2007). LbL deposition was used *i.e.* to incorporate Glucose oxidase (GOx) into a multilayer of poly(styrene sulfonic acid)/polypyrrole (PSS/PPY) on the surface of *in-situ* polymerized PPY (Ram et al., 2000).

2.1.3 Langmuir–Blodgett technique

On the contrary to LbL procedure, the LB film-forming materials are obligatory insoluble in water. LB films are obtained by transferring a Langmuir layer from air/water interface onto the solid support *via* vertical dipping into subphase. Multilayers can be deposited by repeating the solid immersion/withdrawal steps.

Langmuir–Blodgett technique was widely used at the end of 1980s–beginning of 1990s as a relatively simple technology to get highly ordered organic films with artificial or natural receptors. However the system assembled by external forces does not lead to highly stable and defect free films. Therefore this technique was used mostly in combination with other immobilization procedures. Indeed, phospholipids matrices have been used to host biocatalyst molecules that are then co-deposited onto the solid supports as LB films (Cabaj et al., 2010a; Cabaj et al., 2010b; Girart-Ergot et al., 2005) , as shown in Figure 2.

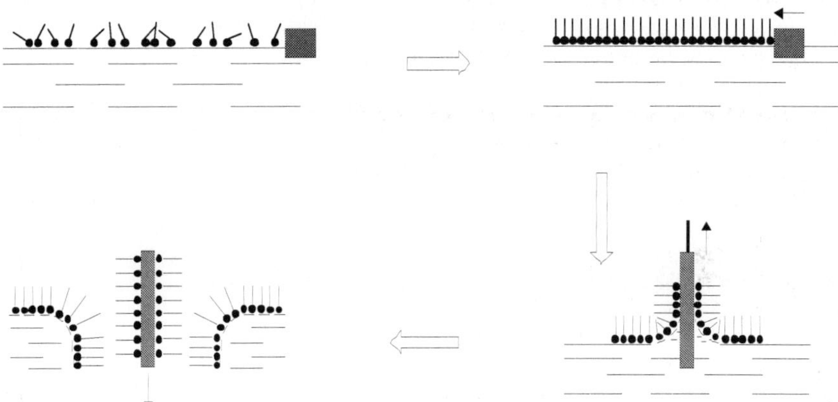

Fig. 2. Schematic path of LB technique

Hydrophobic derivatives of (poly(3- dodecylthiophene), poly(3-hexylthiophene)) were used in a mixture of stearic acid and galactose oxidase (Sharma et al., 2004), laccase, tyrosinase or GOx (J.R. Li et al., 1989; Singhal et al., 2002) to be deposited by Langmuir–Blodgett technique. During the last years the application of Langmuir–Blodgett approach was almost completely replaced by the technologies of self-assembly based on layer-by-layer deposition or on binding of thiols to some metals, modification by silanization and subsequent covalent immobilization.

As proteins are not ideal amphiphilic molecules, the techniques need to be adapted, either by chemical methods (*e.g.*, derivatization methods (Riccio et al., 1996) or varying the subphase composition (Erokhin et al., 1995) or by applying some mimetic systems of biological membranes, due to preserve their native structure and function in monolayer. But the LB technique has been found to be a suitable approach for the development of protein nanostructured matrixes. This technique has been utilized for production to sensitive elements based on protein molecules (Cabaj et al., 2010a; Cabaj et al., 2010b), and more recently, for fabrication of protein-based templates to be employed in nanocrystallography (Pechkova & Nicolini 2004). Generally, this method not only does not destroy the protein structure and function, but also provides new useful properties, such as protein thermal and temporal stability and film anisotropy (Sivozhelezov et al., 2009). The proteins in solutions start to denaturate at 60-70°C where as the secondary structures of proteins in the LB film is slightly affected only till 200°C (Sivozhelezov et al., 2009). This property is not possible to find in chaotically oriented layers. Furthermore, due to this procedure it is possible to control the order of the protein structures, as follows from improved properties of protein crystals and improved helical content of proteins when LB technique is applied.

The quality of protein-monolayer formation at the air-water interface is related to the degree of preservation of the native properties of all proteins. The magnitude of the electrostatic forces maintaining the protein structure is comparable with that of the surface tension. Proteins tend to form stable monolayers at the air-water interface because of their mixture of hydrophilic and lipophilic groups. Often spreading species such as proteins at the air-water interface can effect the conformation of the molecule such as causing unfolding. For example insulin or ovalbumin unfold completely whereas myoglobin and cytochrome C are only partially unfolded (Birdi 1999). This again is thought to be a function of the ratio of polar to non-polar amino acids residues. Highly polar proteins such as xanthine oxidase do not form stable monolayers (Erokhin et al., 1995). In all these circumstances the convenient matrix may be required. For instance, according to Girart-Ergot et al. (2005) enzyme bioactivity in mixed lipid LB films is preserved due to the lipid molecular assembly protects the enzyme, positioning the polypeptide moiety in such a way as to allow the recognition and signal events. In fact, phospholipids have been used as protecting agents for several types of material, not only for membrane cell proteins (Caseli et al., 2002) but also for polysaccharides (Pavinatto et al., 2007) and synthetic polymers (Caseli et al., 2001).
The lipid fraction of biological membranes is mainly composed of phospholipids with different chain lengths and ionic character. Therefore, phospholipids are widely used as mimetic systems in studies involving the cross resistance to drugs.

2.1.4 Langmuir–Schaefer technique

Langmuir-Schaefer (Fig. 3) as well as Langmuir-Blodgett techniques represent a "classical" tool for biofilm engineering, recently improved and extended to a wide variety of proteins (Nicolini 1997). It is also very useful in deposition other than protein rigid layers.
The method implies horizontal touching with the substrate of the preformed monolayer. For protein layers it is not necessary to use a grid as a separator because the layer is rather soft, but is important to compensate the charge of the molecules at the water surface before deposition (Nicolini 1996).
Moreover, it is seemed that LS methodology in case of deposition of tyrosinase is more effective than vertical process (Cabaj et al., 2010b).

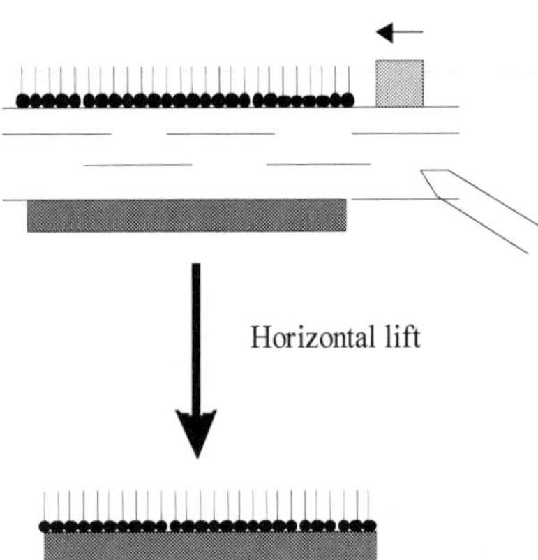

Fig. 3. Langmuir-Schaefer technique

2.2 Covalent immobilization

The covalent immobilization of functional, biological molecules onto a defined and also conductive surface provides the basis for sophisticated biomolecular architectures with numerous applications for *in vitro* studies on the behavior of biological structures such as proteins and cells, for implant or for biosensor devices. In recent years, much work has been devoted to the development of new techniques for protein immobilization on different substrates.

Among the chemical immobilization methods, the following techniques can be distinguished:

- covalent attachment to a water-insoluble matrix,
- cross-linking with the use of a multifunctional and low molecular weight reagent,
- *co*-cross-linking with other neutral substances, for instance proteins.

Cross-linking can be described as a technique, in which an enzyme is linked to the supporting polymeric material by the mean of chemical bonding. The particular method utilizes bifunctional reagents, *i.e.* glutaraldehyde. The main disadvantages of this technique include: enzyme damage, limited diffusion of the substrate, poor mechanical strength (Eggins 2002).

Covalent coupling with the electrogenerated polymer can be defined as an immobilization method which employs formation of bond between a functional group of the enzyme and the support polymeric matrix . The coupling of molecules to the polymer surface is usually carried out by water-soluble carbodiimide or N-hydroxysuccinimide, or the mixture of both reagents. However, the presence of additional chemicals in such a solution may lead to partial denaturation of the enzyme as well as damages within the polymer structure. Among others, nucleophilic groups of the enzyme amino acids, which do not play a major role in the catalytic mechanism, seem to be the best targets. The most common nucleophilic groups used for coupling are: NH_2, COOH, OH, SH, C_6H_4OH and imidazole. One of the biggest advantages of this method is that the enzyme will not be released from the support when performing the catalysis (Klis et al., 2007).

Analyte	Polymer/monomer/ way of polymerization	Receptor	Type of transducing	Ref.
Boronic acid	Poly(aniline boronic acid)/electrochemical polymerization	Glycoproteins	Photometric	Anzenbacher et al., 2004
Na^+, K^+, NH_4^+	Overoxidized polypyrrole	Valinomycin	Amperometric	Izaoumen et al., 2005
Cyclodextrin	Poly(3-methylthiophene)/ electrochemical polymerization	Dopamine neurotransmitters	Amperometric	Bouchta et al., 2005
Atrazine	Polypyrrole/ electrochemical	Tyrosinase	Amperometric	Gerard et al., 2002
Phenolic compounds	LB film of bis(thiophene)carbazole	Laccase	Photometric	Cabaj et al., 2008

Table 2. Selected examples of sensors with receptors covalently immobilized to conducting polymers

Chemical bonding of receptors by covalent reaction between functional groups of receptor and *i.e.* polymer provides the strongest immobilization. Various derivatives of pyrrole and thiophene modified by ionophore units were reported (Table 2). These derivatives were electropolymerized mostly in organic solvents. Low solubility of modified monomers in aqueous media and often low conductivity of resulting polymer can be crucial for biomolecule immobilization. Suitable derivatives are pyrrole and thiophene bearing carboxyl function at the β-position (Rahman et al., 2003; G. Li et al, 2000; Cha et al., 2003; Freitas et al., 2005; Kong et al., 2003; Kuwahara et al., 2005; Peng et al., 2005) or *N*-substituted pyrrole (Culvo-Munoz et al., 2005).

Besides PPY, such polymers as polythionine are found to be convenient for the immobilization due to the presence of free amino functions in the polymer structure and reversible and stable behavior of the polymer film in biological compatible media (Ferreira et al., 2006). A sensor based on tyrosinase linked to electroactive poly(dicarbazole) *via* carboxy group was presented in Cosnier et al. (2001). Various strategies to preparation of polymer with synthetic (polyalkyl ether, crown ether, pyridyl-based ligands) and biological receptors are reviewed in McQuade et al., 2000.

3. Methods of detection

The choice method of detection is based mainly on the type of material employed for producing the sensing units and on the analyte to be detected. Electrochemical methods normally require electroactivity either of the film itself or induced by the analytes. Using electrochemistry may be advantages because distinct experimental procedures can be used, including amperometric measurements and cyclic voltammetry (Siqueira et al., 2010).

Electrical impedance spectroscopy, in turn, offers advantage in being in principle applicable to any type of molecule immobilized in the sensing structure as well as to any type of analyte. In this technique, used in characterization of dielectric properties of materials, the interaction between external field with the dipole moments of the dielectric material is measured as a function of frequency (Neto et al., 2003).

The optical methods are generally more restrictive with regard to the number of system for which they can be used. They include UV-vis absorption or fluorescence spectroscopies, where the former may be used when the analyte or a mediator absorbs in the UV-vis region. If fluorescence measurement are used, fluorescent chromophores must be incorporated into the system, and there are some important advantages: simple optical monitoring may suffice for detection, thus making it possible to produce low-cost sensors. Furthermore, sensing based on fluorescence is usually highly efficient.

Detecting of phenolic compounds is mostly electrochemical process (S. Yang et al., 2006; Shan et al., 2003; Serra et al., 2003), rarely UV absorption (Cabaj et al., 2010a; Cabaj et al., 2010b).

4. Imaging of surface structure

The morphology of the deposited multilayered structures is characterized at nanometer level by AFM (atomic force microscopy) that is a tool with different possibilities and limits. Multilayered structures are possible to form because the technique gives the possibility to form multilayered architectures controlled at molecular level.

AFM can image biological samples under aqueous conditions with high resolution in three dimensions without the use of any probes. AFM has been successfully used to image isolated phase separated bilayers and peptide–lipid domains in supported bilayers. Also monolayers containing glycosphingolipids and cholesterol have been imaged (Yuan & Jhnston 2000) as well as phenoloxidases or glucose oxidase mixed (with linoleic acid, phospholipids) LB/LS films (Kuwahara et al., 2005).

The phenoloxidases (laccase, tyrosinase) and glucose oxidase hetero layers were visualized by contact mode AFM (Fig. 4). The enzyme molecules were fairly well deposited onto solid substrate. Immobilized phenoloxidases as well as glucose oxidase (Sołoducho & Cabaj 2010) were observed as an aggregated pattern in solid-like state with keeping their characteristic random cloud-like or island structure. The heterogeneous films roughness was found relatively high (especially in case of tyrosinase film) for an LB film, which indeed shows that the enzymes were transferred. The roughness of linoleic acid - laccase film was measured as 7.17 nm (similar results was found for film of lipase (Baron et al., 2005)), when the roughness value of tyrosinase film was found as 19 nm. To compare, the roughness of glucose oxidase LB film has been measured as 0.38 nm (Singhal et al., 2003) or 1.9 for LS films. These obtained values were attributed to the immobilization process of comparatively large molecule aggregates of enzymes (laccase, tyrosinase, glucose oxidase) incorporated to LB/LS films. This leads to conclude that there is sometime formation of an agglomerate of enzymes rather than an organized monolayer at the air/water interface. The AFM results showed that the effect could be also associated with changes in the enzyme conformations A monolayer rearrangement, such as two-dimensional formation or hindered molecular orientation, might take place during the phase transition behaviour resulted in the molecular aggregates on the protein layer.

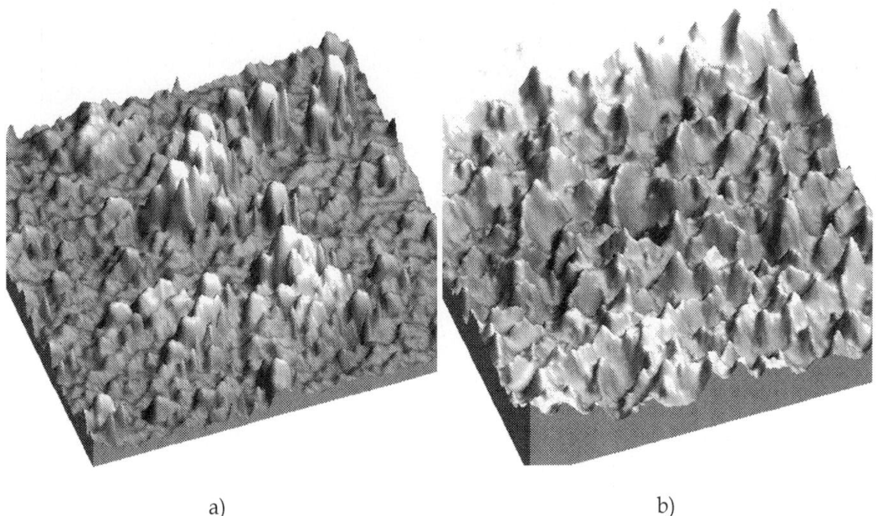

a)   b)

Fig. 4. AFM topography images of a) linoleic acid – laccase LB film, b) phospholipids – glucose oxidase LS film (Sołoducho & Cabaj 2010)

5. Biomolecules for phenolic compounds detection

Biosensors based on the coupling of a biological entity with a suitable transducer offer an effective route for detection of phenolic compounds. For phenolic compounds determination biosensors modified with tyrosinase, peroxidase, laccase and polyphenol oxidase have been reported. Electrodes modified with these enzymes have the advantage that the detection of phenolic compounds can be carried out between -0.2 and 0.05 V *versus* SCE and the interface is minimized. The mechanism for tyrosinase, laccase and peroxidase in the electrochemical biosensors is different (S. Yang et al., 2006).

5.1 Horseradish peroxidase

Peroxidases are enzymes of the class EC 1 .1 1. defined as oxidoreductases using hydroperoxides as electron acceptor. It has been found that most of the peroxidases: plant peroxidases, cytochrome c peroxidase, chloroperoxidase, lactoperoxidase, etc. are heme-proteins with a common catalytic cycle (Everse et al., 1991). Horseradish peroxidase (HrP) has been most thoroughly studied and frequently used to exemplify the peroxidase reaction cycle:

$$\text{Native peroxidase}(Fe^{3+}) + H_2O_2 \rightarrow \text{Compound - I} + H_2O \qquad\qquad (1a)$$
$$\text{Compound - I} + AH_2 \rightarrow \text{Compound - II} + AH^* \qquad\qquad (1b)$$
$$\text{Compound - II} + AH_2 \rightarrow \text{Native peroxidase}(Fe^{3+}) + AH^* + H_2O \qquad (1c)(\text{Ruzgas et al., 1996})$$

The first reaction (1a) involves a two-electron oxidation of the ferriheme prosthetic group of the native peroxidase by H_2O_2 (or organic hydroperoxides). This reaction results in the formation of an intermediate, compound-I, consisting of oxyferryl iron ($Fe^{4+}=O$) and a porphyrin π cation radical. In the next reaction (1b), compound-I loses one oxidizing equivalent upon one-electron reduction by the first electron donor AH_2 and forms compound-II (oxidation state +4). The later in turn accepts an additional electron from the second donor molecule AH_2 in the third step (1c), whereby the enzyme is returned to its native resting state, ferriperoxidase.

The oxidation products formed during the peroxidase reaction depend on the nature of the substrate. Electron donors such as aromatic amines and phenolic compounds are oxidized to free radicals, AH^* (reactions (1b) and (1c)) (Ruzgas et al., 1996). The following discussion about the bioelectrochemistry of peroxidases is directed for a more detailed presentation of the electrochemical reactions of peroxidase providing the basis for amperometric peroxidase-modified electrodes. Horseradish peroxidase based biosensors are most sensitive for a great number of phenolic compounds since phenols can be act as electron donors for peroxidase (S. Yang et al., 2006).

5.2 Laccases

Laccase (EC 1.10.3.2) is defined as a blue oxidase capable of oxidizing phenols and aromatic amines by reducing molecular oxygen to water by a multicopper system. Laccases belong to a large group of the multicopper enzymes, which includes among others ascorbic acid oxidase and ceruloplasmin (Hublik & Schinner 2000).

Laccase demonstrates a broad substrate specificity (Vianello et al., 2007). It catalyses the oxidation of such diverse compounds as: *o, p*-diphenols, aminophenols, polyphenols, polyamines, lignin, some inorganic ions, aryl diamines, benzenthiols, phenothiazines (Vianello et al., 2007). The substrate for laccase is also molecular oxygen, hence the enzyme plays a role of terminal electron acceptor in a four electron process in which water is the final product. Moreover, this blue oxidase is known to demethylate lignin and methoxyphenol acids (Vianello et al., 2007).

Typical laccase reaction (Fig. 5) leads to conversion of the phenolic substrate into aryloxyradical by the mean of a one-electron oxidation process. During the second stage of the oxidation the active species can be converted into a quinone. In the next step, both the quinone and the free radical product undergo a non-enzymatic coupling reactions leading to the polymerization resulting in dimers, oligomers and polymers (Duran et al., 2002; Riva 2006). Products of the oxidative coupling reactions result from either C-O and C-C coupling of the phenolic reactants or N-N and C-N coupling of the aromatic amines. The particular reaction is known as the detoxification of phenolic contaminants (Hublik & Schinner 2000).

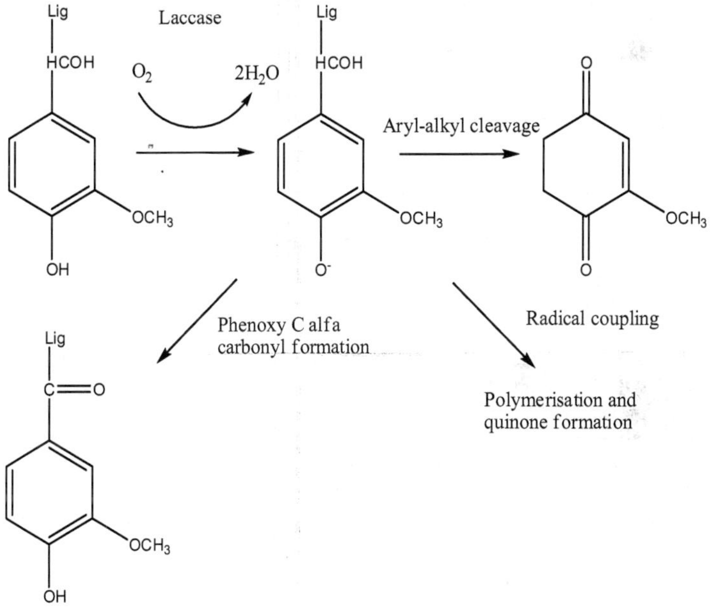

Fig. 5. Laccase-catalysed oxidation of phenolic groups (according to Archibald et al. (1997))

5.3 Tyrosinases

Tyrosinase (monophenol monooxygenase, EC 1.14.18.1) is a binuclear copper containing metalloprotein which catalyses, in the presence of molecular oxygen, two different reactions: (1) the transformation of -mono-phenols into o-diphenols (monophenolase activity) and (2) the oxidation of o-diphenols to o-quinones (diphenolase activity, Fig. 6).

This enzyme, found in prokaryotes as well as in eukaryotes, is involved in the formation of pigments such as melanins and other polyphenolic compounds. Each of the two copper ions is bounded by three conserved histidines residues (Lerch 1988).

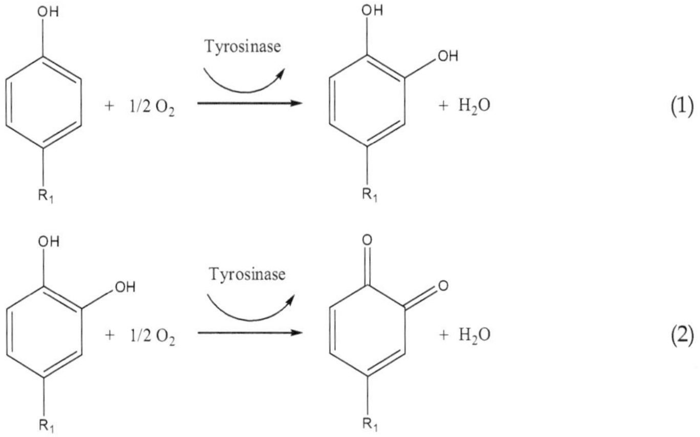

$$\text{(phenol)} + \tfrac{1}{2}\,O_2 \xrightarrow{\text{Tyrosinase}} \text{(catechol)} + H_2O \qquad (1)$$

$$\text{(catechol)} + \tfrac{1}{2}\,O_2 \xrightarrow{\text{Tyrosinase}} \text{(quinone)} + H_2O \qquad (2)$$

Fig. 6. Tyrosinase-catalysed oxidation of phenolic compounds

6. Biosensors for phenolic compounds

Electrochemical biosensors are rather cheap, simple to fabricate, and reusable. They have high stability and sensitivity. This kind of sensors can potentially be used for other species with the necessary modifications. Many phenolic compounds are successfully detected using electrochemical sensors as most sensors are oxidized at readily accessible potentials.

6.1 Horseradish peroxidase electrodes for phenolic compounds detection

Many different methods such as covalent immobilization (Imabayashi et al., 2001), sol-gel derived matrix (Rosatto et al., 2002), recently LbL assembly was employed for modification of electrodes (S. Yang et al., 2006). And still the combination of oxidoreductases and amperometric electrodes is the most commonly studied biosensor concept (Fig. 7).

Imabayaschi et al. (2001) reported also the HrP biosensor constructed by enzyme covalently immobilized on the mercaptonic acid self-assembled monolayer (SAM) on the gold electrode. The most simple electrode for the detection of peroxide consists of a layer of peroxidase molecules adsorbed onto the electrode surface. If the electrode is placed into a sample and poised at a potential more negative than 0.6 V *vs.* SCE then a proportionality between the registered reduction current and the peroxide concentration is observed. This phenomenon was observed for horseradish peroxidase adsorbed on carbon black, graphite, carbon fibers, gold, and platinum electrodes (Ruzgas et al., 1996). The electrode current is due to an electrochemical reduction of compound-I and –II (Fig. 8).

The response of the peroxidase biosensors to phenolic compounds is based on the double displacement or "ping-pong" mechanism in which two substrates, H_2O_2 and the electron-donating phenolic compounds are involved. At the electrode surface, peroxidase molecules are oxidized by H_2O_2 followed of its reduction by phenolic compounds. In the last reaction, the phenolic compounds are mainly converted into quinones or free radical products, which are electroactives and can be electrochemically reduced on the electrode surface. The reduction current is proportional to the phenolic compounds concentration in the solution, as long as the H_2O_2 concentration is not limiting (S. Yang et al., 2006).

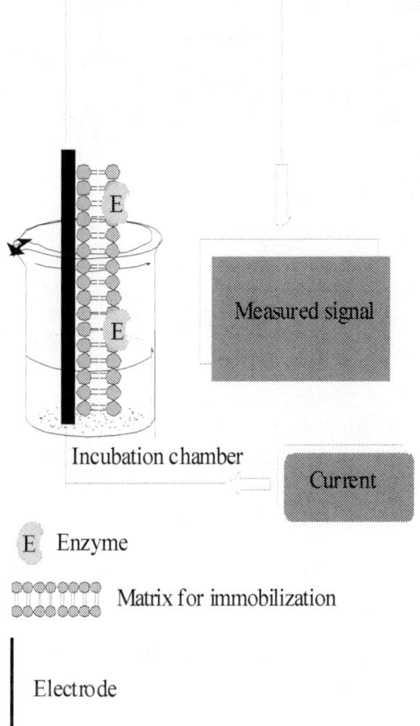

Fig. 7. Simplified biosensor system for detection of phenolic compounds

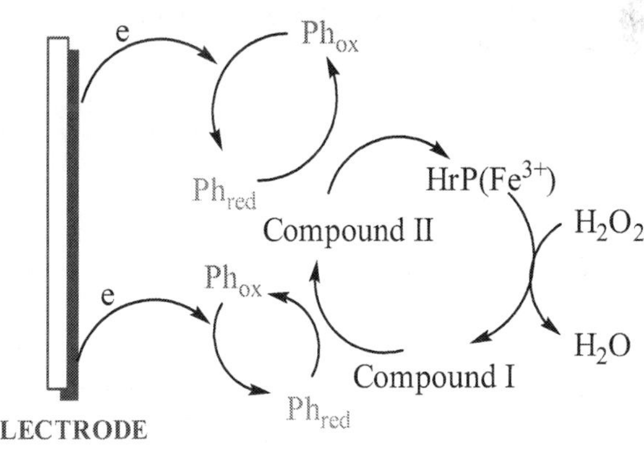

Fig. 8. Scheme of the reactions occurring at the surface of the electrode modified with horseradish peroxidase; Ph_{ox} and Ph_{red} are the oxidized and reduced forms of the phenolic compounds, respectively

Oxidation of phenolic compounds by horseradish peroxidase in solution can be enhanced by nitrogen ligands such as ammonium ions as wel as aromatic amines (*i.e. o*-dianisidine, benzidine, *o*-aminophenol, *p*-aminophenol, *p*-cresol, aniline) (Ruzgas et al., 1996). These observations gave the possibility to correlate the increase of the reduction current of peroxidase electrode with the concentration of aromatic amines.

The monitoring of the enzyme reaction is accomplished by the electrode reduction of the phenoxy radicals formed, the current being proportional to the concentration of phenolic compounds as long as the H_2O_2 concentration is not limiting. Therefore, an excess of H_2O_2 should be added to the working solution in order for the biosensor to be able to respond to the phenolic compounds (Serra et al., 2001). However, it is well known that the presence of a high concentration of H_2O_2 causes inhibition of the activity of peroxidase (Scheller et al., 1997). Moreover, H_2O_2 is unstable in solution. These facts cause some difficulties in the practical use of the peroxidase biosensor for the detection of phenolic compounds. In order to improve this type of detection, there are developed composite multienzyme systems.

Serra et al. (2001) reported the sensing system for phenolic compounds where horseradish peroxidase is mixed with glucose oxidase (GOx). In this biosensor, GOx was responsible for generating *in situ* H_2O_2 needed for the enzyme reaction with the phenolic compounds. For the sensor design, matrices of graphite and Teflon were selected. The enzymatic electrodes were constructed by simple physical inclusion of the enzymes (HrP, GOx) into the bulk of graphite-Teflon pellet with no covalent attachments.

Serra et al. described also the three enzyme system with HrP, GOx and tyrosinase to monitor possibly large number of phenolic compounds (Serra et al., 2001).

6.2 Laccase and tyrosinase electrodes for phenolic compounds detection

Laccase and tyrosinase are both copper containing oxidases catalyzing the oxidation of phenolic compounds in the presence of oxygen. In these reactions oxygen is reduced directly to water on the surface of enzymatic electrode (Fig. 9).

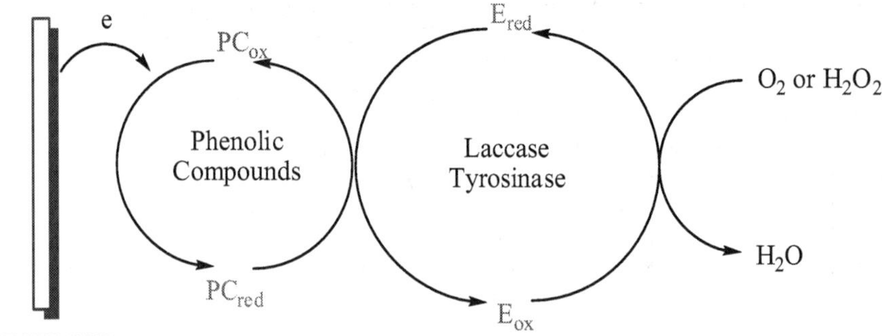

Fig. 9. Mechanism of the reactions on the laccase/tyrosinase biosensor. PC: phenolic compound; E: enzyme; *red* and *ox* are the reduced and oxidized forms

For the design of biosensor different methods of enzyme immobilization were employed. They include the modification of solid graphite (Ortega et al., 1994), incorporation of enzyme into carbon paste, immobilization on surface of different membranes (Yarpolov et al., 1995), Langmuir-Blodgett hybrid films (Cabaj et al., 2010a; Cabaj et al., 2008; Sołoducho & Cabaj 2010). The most sensitive biosensors are based on tyrosinases (Psrellada et al., 1997), however, in order to low stability of this class of enzymes, these devices usually present short life-times (Vianello et al., 2004). Alternatively to tyrosinases – laccases are often used. Laccase is one of the best candidate for use in analytical systems for determination of chlorinated and polyphenolic compounds (Freire et al., 2002), mainly due to its great stability (Gianfreda et al., 1999), wide range of detectable phenolic compounds (Freire et al., 2002) and it is not susceptible to product inhibition. Laccase can act over the phenolic compounds that are not reactive with the other mentioned enzymes (Freire et al., 2002). In addition, one of the most important point in using laccase is its sensitivity for phenolic compounds, such as chloroguaiacols and guaiacols, which are considered very toxic (Freire et al., 2002). The enzymatic oxidation of phenolic compounds and anilines by laccases generate radicals that react with each other to form dimers, oligomers or polymers covalently coupled by C–C, C–O and C–N bonds. In the case of substituted compounds, the reaction can be accompanied by partial demethylations and dehalogenations (Claus 2004).

However, an exhaustive overview in the basic aspects of immoblilization of laccase and tyrosinase has been reported. Whereas, to retain enzyme's specific biological function, their immobilization on solid matrix is a key factor in preparing biosensors. So far several immobilization strategies have been commonly used to immobilize small molecules onto appropriately functionalized glass slides, including covalent immobilization with Staudingeer ligation (Kohn et al., 2003). The immobilization methods for laccase or tyrosinase such as physical adsorption (Cabaj et al., 2010a; Cabaj et al., 2010b), covalent attachment (Vianello eta al., 2004), incorporation within carbon paste (Duran et al., 2002), immobilization in polymer films (Timur et al., 2004), entrapment in some sol–gel matrices (Duran et al., 2002) have been also reported in the literature. Vianello et al. (2004) presented a high-sensitivity flow biosensor based on a monomolecular layer of laccase immobilized on a gold support. This biosensor detects phenols in the low micromolar range, *i.e.* below European Community limits (Vianello et al., 2004).

Laccase belongs to the restricted groups of redox enzymes that show efficient direct heterogeneous electron transfer at electrodes (Ghindilis et al., 1997). When laccase was adsorbed on graphite electrode, bioelectrocatalytic reduction of oxygen occurs and is observed as a reduction current caused by direct (mediatorless) electron transfer from the electrode to the immobilized laccase and then further to molecular oxygen in solution. Oxygen is reduced to water in a four-electron transfer mechanism. In this mechanism the electron donor (substrate) penetrates the active site of the enzyme where it is oxidized in a single electron oxidation step often producing electrochemically active compound (possibly a radical) that it turn can be re-reduced at the electrode surface in a mediated electron transfer step. This creates an electron-shuttle process between the electrode and the laccase providing the basis for reduction of molecular oxygen by mediated electron transfer at the enzymatic electrode (Haghighi et al., 2003).

In particular, several biosensors based on tyrosinase were developed for determination of phenols (Mai Anh et al., 2002; Dzyadevych eta al., 2002). The tyrosinase was immobilized on an electrode's surface as a thin film or in a membrane on a Clark oxygen electrode (Macholan 1990), chemically bonded to a solid graphite electrode (Cosnier & Innocent 1993) or controlled-pore glass (Zachariah & Mottola 1989) and using electropolymerization of an amphilic pyrrole derivative–enzyme mixture (Cosnier & Popescu 1996). Tyrosinase was also adsorbed on the surface of phospholipids Langmuir-Schaefer film (Cabaj et al., 2010b).

Tyrosinase-based electrochemical biosensors suffer from low enzyme stability and significant inhibition of the enzyme by reaction products; both these factors deteriorate electrode characteristics in phenol determination. In the case of HrP, the limitation is the necessity of hydrogen peroxide presence to complete the biocatalytic cycle.

Nevertheless, tyrosinase- and laccase-based amperometric biosensors have proved to be very useful for the determination of phenols and substituted phenols at low levels (Ghindilis et al., 1997).

7. Modification of electrodes by conducting polymers

Conducting polymers have found increased applications in various industries. Some of the main classes of conducting polymers that are available for various applications include polyacetylene, polyaniline (PANI), polypyrrole (PPY), polythiophene (PTH), polyethylenedioxythiophene (PEDOT), poly(paraphenylene), poly(paraphenylenevinylene), polyfluorene, polycarbazole, and polyindole (PI), etc. Conducting polymers exhibit intrinsic conductivity when the conjugated backbone of the polymer is oxidized or reduced (Bredas & Street 1985). Apart from its conductivity, the change of electronic band in the conducting polymer affects the optical properties in the UV-visible and near IR region. The changes in conductivity and optical properties make them candidates for use as optical sensors or element able to modify enzymatic electrodes.

Synthetic and biological receptors can be used to manipulate the sensitivity of a conducting polymer for different analytes (Adhikari & Majumdar 2004; Ahuja et al., 2007). Some conducting polymers that have been modified with various receptors are listed in Table 3. To immobilize the receptor, it is bonded to the polymer matrix through covalent or noncovalent interaction. Physical adsorption (Lopez et al., 2006), the Langmuir-Blodgett technique (Cabaj et al., 2010ab), layer-by layer deposition technique (Ram et al., 2000), and mechanical embedding method (J. Kan et al., 2004) are used to bind the receptor to the matrix through noncovalent bonding. Gerard et al. (2002) have discussed the advantages and limitations of these techniques.

8. Future of biosensing

Despite the rapid progress in biosensor development, *i.e.* clinical applications of biosensors are still rare, with glucose monitor as an exception. Sensors, biosensors have a number of disadvantages compared to classical chemical monitoring methods, however, they ensue a number of requirements of current and emerging environmental pollution or medical monitoring that chemical methods fail to address. Ongoing developments in material technology, computer technology, and microelectronics are expected to help to omit many of these problems. It is expected that progress in the development of tools and strategies to identify, record, store, and transmit parameter data will help in expanding the scope of the use of sensors on a broader scale (Blasco & Pico 2009).

The proceeding researches have to be carried out in all the building blocks of biosensors, which include transducers, recognition molecules, immobilization strategies, as well as transduction mechanisms. Particular attention should be also given to the control of molecular architectures afforded by thin film forming methods (LB/LS, LbL, electrodeposition), especially when biomolecules could be combined with layers of polymers or metallic nanoparticles (Au). Relevantly, the success of LB, LbL films for

Analyte	Receptor	Conducting polymer	Type of immobilization	Type of transduction	Ref.
H_2O_2	Horseradish peroxidase	PANI/ polyethylene terephthalate	Physical adsorption	Optical	Caramori & Fernandes 2004; Borole et al., 2005; Fernandes et al., 2005
H_2O_2	Horseradish peroxidase	PEDOT/PSS (poly(styrene-sulfonate))	Physical adsorption	Amperometric	Asberg & Inganas 2003
Hydro-quinone	Laccase	Poly-*o*-phenylenediamine	Physical adsorption	Amperometric	Pałys et al., 2007
Catechins	Laccase	Poly (tertthiophene)	Covalent immobilization	Amperometric	Rahman et al., 2008
Phenol	Tyrosinase	PEDOT	Physical adsorption	Amperometric	Vedrine et al., 2003
Catechol	Tyrosinase	Poly(dicarbazole)	Physical adsorption	Amperometric	Cosnier et al., 2001
Catechol	Tyrosinase	Poly(1,8-diaminocarbazole)	Covalent immobilization	Amperometric	Skompska et al., 2007

Table 3. Examples of conducting polymer-based biosensors for phenol compounds detection

biosensing can be attributed to the fact that entrained water remains activity of biomolecules. When extremely high sensitivity is required, the optical methods (luminescence) are favorable. Also important is the possibility of integrating sensing molecules with silicon-based technology, a path to accomplish low-cost biosensors with scalable production (Siqueira et al., 2010).

New devices based on microelectronics and related (bio)-micro-electro-mechanical systems and (bio)-nano-electro-mechanical systems are expected to provide technological solutions (Bezbaruah & Kalita 2010). Miniaturized sensing devices, microfluidic delivery systems, and multiple sensors on one chip are needed. High reliability, potential for mass production, low cost of production, and low energy consumption are also expected and some progress has already been achieved in these areas (Farre et al., 2007).

A wireless sensor network comprising spatially-distributed sensors or biosensors to monitor environmental conditions will contribute enormously towards continuous environmental monitoring especially in environments that are currently difficult to monitor such as coastal areas and open seas (Farre et al., 2009). Blasco and Pico (2009) expect that such a network can provide appropriate feedback during characterization or remediation of contaminated sites. The laboratory-on-a-chip is another concept that is going to impact future sensor technology. These chips involve microfabrication to achieve miniaturization and/or minimization of components of the analytical processes. It has been suggested that nanoscale and ultra-miniaturized sensors could dominate the production lines in the next generation of biotechnology-based industries (Farre et al., 2007).

9. Acknowledgments

Authors are gratefully acknowledged for financial support of the Polish National Centre of Progress of Explorations Grant no. NR05-0017-10/2010 and Wrocław University of Technology.

10. References

Adhikari, B.; Majumdar, S. (2004) Polymers in sensor applications. *Prog. Polym. Sci.*, 29(7), pp. 699–766.

Ahuja, T.; Mir, I. A.; Kumar, and Rajesh, D. (2007) Biomolecular immobilization on conducting polymers for biosensing applications. *Biomaterials*, 28(5), pp. 791–895

Anzenbacher, J.; Jursikova, K.; Aldakov, D.; Marquez, M.; Pohl, R. (2004) Materials Chemistry Approach to the Anion-Sensor Design. *Tetrahedron*, 60, pp. 11163-11168.

Archibald, F. S.; Bourbonnais, R.; Jurasek, L.; Paice, M. G.; Reid, I. D. (1997) Kraft pulp bleaching and delignification by Trametes versicolor. *J Biotechnol.*, 53, pp. 215–36.

Asberg, P.; Inganas, O. (2003) Biomolecular and Organic Electronics. *Biosens. Bioelectron.*, 19, pp. 199-207.

Baron, A. M.; Lubambo, A. F.; Lima, V. M. G.; de Camargo, P. C.; Mitchell, D. A.; Krieger, N. (2005). Atomic Force Microscopy: a useful tool for evaluating aggregation of lipases. *Microsc. Microanal.*, 11, pp. 74-77.

Birdi, K. S.; *Self-assembly monolayer structures of lipids and macromolecules at interfaces*, Kluwer Academic Press, Dordrecht, 1999.

Borole, D. D.; Kapadi, U. R.; Mahulikar, P. P. and Hundiwale, D. G. (2005). Glucose oxidase electrodes of a terpolymer poly(aniline-co-*o*-anisidine-co-*o*-toluidine) as biosensors. *Eur. Polym. J.*, 41(9), pp. 2183-2188.

Bouchta, D.; Izaoumen, N.; Zejli, H.; Kaoutit, M. E.; Temsamani, K. R. (2005) A novel electrochemical synthesis of poly-3-methylthiophene-γ-cyclodextrin film: Application for the analysis of chlorpromazine and some neurotransmitters. *Biosens. Bioelectron.*, 20, pp. 2228-2235.

Blasco, C. and Pico, Y. (2009) Prospects for combining chemical and biological methods for integrated environmental assessment. *Trends Anal. Chem.*, 28(6), pp. 745–757.

Bezbaruah, A. N. and Kalita, H. *Treatment of Micropollutants in Water and Wastewater* J. Virkutyte, V. Jegatheesan and R. S. Varma (Eds). IWA Publishing, London, 2010.

Bredas, J. L. and Street, G. B. (1985) Polarons, bipolarons, and solitons in conducting polymers. *Acc. Chem. Res.*, 18(10), pp. 309–315

Cabaj, J.; Idzik, K.; Sołoducho, J.; Chyla, A.; Bryjak, J.; Doskocz, J. (2008) Well-ordered thin films as practical components of biosensors. *Thin Solid Films*, 516, pp. 1171-1174.

Cabaj, J.; Sołoducho, J.; Nowakowska-Oleksy, A. (2010) Langmuir–Blodgett film based biosensor for estimation of phenol derivatives. *Sens. Actuat. B*, 143, pp. 508-515;

Cabaj, J.; Sołoducho, J.; Świst, A. (2010) Langmuir–Schaefer films of tyrosinase— Characteristic. *Sesn. Actuat. B*, 150, pp. 505-512.

Calvo-Munoz, M. L.; Bile, B. E.-A.; Billon, M.; Bidan, G. (2005) Electrochemical study by a redox probe of the chemical post-functionalization of N-substituted polypyrrole films: Application of a new approach to immobilization of biotinylated molecules. *J. Electroanal. Chem.*, 578, pp. 301-313.

Caramori, S. S. and Fernandes, K. F. (2004) Covalent immobilization of horseradish peroxidase onto poly(ethylene terephthalate)–poly(aniline) composite. *Process Biochem.*, 39(7), pp. 883–888.

Caseli, L. ; Zaniquelli, M. E. D. ; Furriel, R. P. M. ; Leone, F. A. (2002) Enzymatic activity of alkaline phosphatase adsorbed on dimyristoylphosphatidic acid Langmuir-Blodgett films. *Colloid Surf. B-Biointefaces*, 25, pp. 119-128.

Caseli, L.; Nobre, T. M.; Silva, D. A. K.; Loh, W.; Zaniquelli, M. E. D. (2001) Flexibility of the triblock copolymers modulating their penetration and expulsion mechanism in Langmuir monolayers of dihexadecyl phosphoric acid. *Colloid Surf. B-Biointerfaces*, 22, pp. 309-321.

Cha, J.; Han, J.I.; Choi, Y.; Yoon, D.S.; Oh, K.W.; Lim, G. (2003) DNA hybridization electrochemical sensor using conducting polymer. *Biosens. Bioelectron.*, 18, pp. 1241-1247.

Claus, H. (2004) Laccases: Structure, Reactions, Distribution. *Micron*, 35, pp. 93-96.

Cosnier, S. and Innocent, C. (1993) A new strategy for the construction of a tyrosinase-based amperometric phenol and o-diphenol sensor. *Bioelectrochem. Bioenerg.*, 31, pp. 147-160.

Cosnier, S. and Catalin Popescu, I. (1996) Poly(amphiphilic pyrrole)-tyrosinase-peroxidase electrode for amplified flow injection-amperometric detection of phenol. *Anal. Chim. Acta*, 319, pp. 145-151.

Cosnier, S.; Szunerits, S.; Marks, R. S.; Lellouche, J. P.; Perie, K. (2001) Mediated electrochemical detection of catechol by tyrosinase-based poly(dicarbazole) electrodes. *J. Biochem. Biophys. Methods*, 50, pp. 65-77.

Debabov, V.G. (2004) Bacterial and Archaeal S-Layers as a Subject of Nanobiotechnology. *Mol. Biol.* 38, pp. 482-493.

Durán, N.; Rosa, M. A.; D'Annibale, A.; Gianfreda, L.(2002), Applications of laccases and tyrosinases (phenoloxidases) immobilized on different supports: a review. *Enzyme and Microbial Technology*, 31, pp. 907–931.

Dzyadevych, S. V.; Mai Anh, T.; Soldatkin, A. P.; Duc Chien, N.; Jaffrezic-Renault, N. and Chovelon, J.-M. (2002) Development of enzyme biosensor based on pH-sensitive field-effect transistors for detection of phenolic compounds. *Bioelectrochemistry*, 55, pp. 79-81

Eggins, B. R. *Chemical Sensors and Biosensors*, Chichester, John Wiley & Sons, 2002.

Erokhin, V.; Facci, P.; Nicolini, C. (1995) Two-dimensional order and protein thermal stability: high temperature preservation of structure and function. *Biosens. Bioelectron.*, 10, pp. 25-34.

Everse, J.; Everse, K. E. and Grisham, M. B. (Eds.), *Peroxidases in Chemistry and Biology*, CRC Press, Boca Raton, 1991.

Farre, M.; Kantiani, L. and Barcelo, D. (2007) Advances in immunochemical technologies for analysis of organic pollutants in environment. *Trends Anal. Chem.*, 26(11), pp. 1100–1112.

Farre, M.; Kantiani, L.; Perez, S. and Barcelo, D. (2009) Sensors and biosensors in support of EU Directives. *Trends Anal. Chem.,* 28(2), pp. 170–185.

Fernandes, K. F.; Lima, C. S.; Lopes, F. M. and Collins, C. H. (2005) Hydrogen peroxide detection system consisting of chemically immobilised peroxidase and spectrometer. *Process Biochem.,* 40(11),pp. 3441–3445.

Ferreira, V.; Tenreiro, A.; Abrantes, L. M. (2006) Electrochemical, microgravimetric and AFM studies of polythionine films Application as new support for the immobilisation of nucleotides. *Sens. Actuat. B,* 119, pp. 632-641.

Freire, R. S.; Duran, N.; Kubota, L. T. (2001) Effects of fungal laccase immobilization procedures for the development of a biosensor for phenol compounds.*Talanta,* 54, pp. 681–686.

Freire R. S.; Dura´n, N.; Kubota, L. T. (2002) Development of a laccasebased flow injection electrochemical biosensor for the determination of phenolic compounds and its application for monitoring remediation of Kraft E1 paper mill effluent. *Anal. Chim. Acta,* 463, pp. 229-238.

Freitas, J. M.; Leonor Duarte, M.; Darbre, T.; Abrantes, L. M. (2005) Electrochemical synthesis of a novel conducting polymer: the poly[(3-pyrrolyl)-octanoic acid]. *Synth. Met.,* 155, 549-554.

Gerard, M.; Chaubey, A.; and Malhotra, B. D. (2002) Application of conducting polymers to biosensors. *Biosens. Bioelectron.,* 17, pp. 345–359.

Ghindilis, A. L.; Krishnan, R.; Atanasov, P.; Wilkins, E. (1997) Flow-through amperometric immunosensor: Fast 'sandwich' scheme immunoassay. *Biosens. Bioelectron.,* 12, pp. 415-423.

Gianfreda, L.; Xu, F. and Bollag, J. M. (1999) Laccases: a useful group of oxidoreductive enzymes. *Bioremed. J.,* 3, pp. 1-25.

Girart-Ergot, A. P.; Godoy, S.; Blum, L. J. (2005) Enzyme association with lipidic Langmuir-Blodgett films: Interests and applications in nanobioscience. *Adv. Colloid Interface Sci,* 116, pp. 205-225.

Haghighi, B.; Gorton, L.; Ruzgas, T.; Jönsson L.J. (2003) Characterization of graphite electrodes modified with laccase from Trametes versicolor and their use for bioelectrochemical monitoring of phenolic compounds in flow injection analysis, *Analytica Chimica Acta,* 487, pp. 3–14.

Hammond, P. T. (2004) Form and Function in Multilayer Assembly: New Applications at the Nanoscale. *Adv. Mater.,* 16, pp. 1271-1293.

Hublik, G.; Schinner, F. (2000) Characterization and immobilization of the laccase from *Pleurotus ostreatus* and its use for the continuous limination of phenolic pollutant. *Enzyme Microb. Technol.,* 27, 330–336.

Imabayashi, S.; Kong, Y.; Watanabe, M. (2001) Amperometric biosensor for polyphenol peroxidase immobilized on gold electrodes. *Electroanal.,* 13(5), pp. 408-412.

Izaoumen, N.; Bouchta, D.; Zejli, H.; Kaoutit, M. E.; Stalcup, A. M.; Temsamani, K. R. (2005) Electrosynthesis and analytical performances of functionalized poly (pyrrole/β-cyclodextrin) films. *Talanta,* 66, pp. 111-117.

Kan, J.; Pan, X. and Chen, C. (2004) Polyaniline–uricase biosensor prepared with template process. *Biosens. Bioelectron.,* 19(12), pp. 1635–1640.

Kim, D. C.; Jang, A-R.; Kang, D. J. (2009) Biological functionality of active enzyme structures immobilized on various solid surfaces. *Cur. Appl. Phys.,* 9, pp. 1454-1458.

Kim, S. N. ; Rusling, J. F.; Papadimitrakopoulos, F. (2007) Carbon nanotubes for electronic and electrochemical detection of biomolecules. *Adv. Mater*, 19, pp. 3214-3228.

Klis, A.; Maicka, E.; Michota, A.; Bukowska, J.; Sek, S.; Rogalski, J.; Bilewicz, R. (2007) Electroreduction of Laccase Covalently Bound to Organothiol Monolayers on Gold Electrodes. *Electrochimica Acta*, 52, pp. 5591–5598.

Kohn, M.; Wacker, R.; Peters, C.; Schroeder, H.; Soulere, L.; Breinbauer, R.; Niemeyer, R.; Waldmann, C.M.H. (2003) Staudinger ligation: a new immobilization strategy for the preparation of small-molecule arrays. *Angew. Chem. Int.* 42, pp. 5830-5834.

Kong, Y. T.; Boopathi, M.; Shim, Y. B. (2003) Direct electrochemistry of horseradish peroxidase bonded on a conducting polymer modified glassy carbon electrode. *Biosens. Bioelectron.*, 19, 227-232.

Kumada, Y.; Zhao, C. ; Imanaka, H.; Imamura, K. and Nakanishi, K. (2007) Protein interaction analysis using an affinity peptide tag and hydrophilic polystyrene plate. *J. Biotechnol.*, 128, pp. 354-361.

Kuwahara, T.; Oshima, K.; Shimomura, M.; Miyauchi, S. (2005) Immobilization of glucose oxidase and electron-mediating groups on the film of 3-methylthiophene/thiophene-3-acetic acid copolymer and its application to reagentless sensing of glucose. *Polymer*, 46, 8091-8097.

Kuwahara, T.; Oshima, K.; Shimomura, M.; Miyauchi, S. (2005) Glucose sensing with glucose oxidase immobilized covalently on the films of thiophene copolymers. *Synthetic Metals*, 152, pp. 29-32.

Lerch, K. (1988) Protein and active-site structure of tyrosinase. *Prog. Clin. Biol. Res.*, 256, pp. 85-98.

Li, G.; Mehl, G.; Hunnius, W.; Zhu, H.; Kautek, W.; Plieth, W.; Melsheimer, J.; Doblhofer, K. (2000) Reactive groups on polymer coated electrodes: 10. Electrogenerated conducting polyalkylthiophenes bearing activated ester groups. *Polymer*, 41, pp. 423-432.

Li, J. R.; Cai, M.; Chen, T. F. ; Jiang, L. (1989) Enzyme electrodes with conductive polymer membranes and Langmuir-Blodgett films. *Thin Solid Films*, 180, pp. 205-210.

Li, N.; Xue, M.-H.; Yao, H.; Zhu, J.-J. (2005), Reagentless biosensor for phenolic compounds based on tyrosinase entrapped within gelatine film. *Anal Bioanal. Chem.*, 383, pp. 1127–1132.

Liu, X. D.; Yu, W. T.; Wang, W.; Xiong, Y.; Ma X. J. and Yuan, Q. A. (2008) Polyelectrolyte microcapsules prepared by alginate and chitosan for biomedical application. *Prog. Chem.*, 20, pp. 126–139.

Lopez, M. S.-P.; Lopez-Cabarcos, E. and Lopez-Ruiz, B. (2006) Organic phase enzyme electrodes. *Biomol. Eng.*, 23(4), pp. 135–147.

Lourenco, J. M. C.; Ribeiro, P. A.; Botelho do Rego, A. M.; Raposo, M. (2007) Counterions in layer-by-layer films—Influence of the drying process. *J. Colloid Interface Sci.*, 313, pp. 26-33.

Macholan, L. (1990) Phenol-sensitive enzyme electrode with substrate cycling for quantification of certain inhibitory aromatic acids and thio compounds. *Collect. Czech. Chem. Commun.* 55, pp. 2152-2160.

Mai Anh, T.; Dzyadevych, S. V.; Soldatkin, A. P.; Duc Chien, N.; Jaffrezic-Renault, N. and Chovelon, J.-M. (2002) Development of tyrosinase biosensor based on pH-sensitive field-effect transistors for phenols determination in water solutions. *Talanta*, 56, pp. 627-634.

McQuade, D. T.; Pullen, A. E.; Swager, T. M. (2000) Conjugated Polymer-Based *Chemical Sensors. Chem. Rev.*, 100, 2537-2574.

Nakanishi, K.; Sakiyama, T. and Imamura, K. (2001) On the adsorption of proteins on solid surfaces, a common but very complicated phenomenon. *J. Biosci. Bioeng.*, 91, pp. 233-244.

Nakanishi, K.; Sakiyama, T.; Kumada, Y.; Imamura, K. and Imanaka, H. (2008) Recent Advances in Controlled Immobilization of Proteins onto the Surface of the Solid Substrate and Its Possible Application to Proteomics. *Cur. Proteom.*, 5, pp. 161-175.

Neto, J. M. G.; Leal Ferreira, G. F.; Ribeiro Santos, J.; da Cunha, H. N.; Dantas, I. F.; Bianchi, R. F. (2003) Complex conductance of carnauba wax/polyaniline composites. *Braz. J. Phys.*, 33, pp. 371-375.

Nicolini, C. *Molecular bioelectronics*, Worlds Scientific Publishing Co. Pte. Ltd.., Singapore, 1996.

Nicolini, C. (1997) Protein-monolayer engineering: principles and application to biocatalysis. *Trends Biotechnol.*, 15, pp. 395-401.

Ortega, F.; Dominguez, E.; Burestedt, E.; Emneus, J.; Gorton, L.; Marko-Varga, G. (1994) Phenol Oxidase Based Biosensors as Selective Detection Units in Column Liquid Chromatography for the Determination of Phenolic Compounds. *J. Chromatogr.*, 675, pp. 65-78.

Pałys, B.; Bokun, A.; Rogalski, J. (2007) Poly-o-fenylenediamine as redox mediator for laccase. *Electrochim. Acta*, 52, pp. 7075-7082.

Parellada J, Narváez A, Domínguez E, Katakis I. (1997) A new type of hydrophilic carbon paste electrodes for biosensor manufacturing: binder paste electrodes. *Biosens Bioelectron.* 12(4), pp. 267-75.

Pavinatto, F. J.; Caseli, L.; Pavinatto, A.; dos Santos, D. S.; Nobre, T. M.; Zaniquelli, M. E. D.; Silva, S. H.; Miranda, P. B.; de Oliveira, O. N. (2007) Probing Chitosan and Phospholipid Interactions Using Langmuir and Langmuir–Blodgett Films as Cell Membrane Models. *Langmuir*, 23, pp. 7666-7671.

Pechkova, E.; Nicolini, C. (2004) Protein nanocrystallography: a new approach to structural proteomics. *Trends Biotechnol.*, 22, 117-122.

Peng, H.; Soeller, C.; Vigar, N.; Kilmartin, P. A.; Cannell, M. B.; Bowmaker, G. A.; Cooney, R. P.; Travas-Sejdic, J. (2005) Label-free electrochemical DNA sensor based on functionalised conducting copolymer. *Biosens. Bioelectron.*, 20, 1821-1828.

Portnov, S.; Yashchenok, A.; Gubskii, A.; Gorin, D.; Neveshkin, A.; Klimov, B.; Nefedov, A.; Lomova, M. (2006) An automated setup for production of nanodimensional coatings by the polyelectrolyte self-assembly method. *Instruments and Experimental Techniques*, 49 (6), pp. 849-854.

Rahman, M.; Won, M. S.; Shim, Y. B. (2003) Characterization of an EDTA bonded conducting polymer modified electrode: its application for the simultaneous determination of heavy metal ions. *Anal. Chem.* 75, pp. 1123-1129.

Rahman, A.; Noh, H.-B.; Shim, Y.-B. (2008) Direct Electrochemistry of Laccase Immobilized on Au Nanoparticles Encapsulated-Dendrimer Bonded Conducting Polymer: Application for a Catechin Sensor. *Anal. Chem.* 80, pp. 8020–8027.

Ram, M. K.; Adami, M.; Paddeu, S.; Nicolini, C.; (2000) Nanoassembly of glucose oxidase on the in situ self-assembled electrochemical characterizations. *Nanotechnology*, 11, pp. 112-119.

Riccio, A.; Lanzi, M.; Antolini, C.; De Nitti, C.; Tavani, C.; Nicolini, C. (1996) Ordered Monolayer of Cytochrome c via Chemical Derivation of Its Outer Arginine. *Langmuir*, 12, pp. 1545-1549.

Riva, S. (2006) Laccases: blue enzymes for green chemistry. *Trends in Biotechnology*, 24, 219-226.

Rosatto, S. S.; Sotomayor, P. T.; Kubota, L. T.; Gushikem, Y. (2002) SiO_2/Nb_2O_5 sol–gel as a support for HRP immobilization in biosensor preparation for phenol detection. *Electrochim. Acta*, 47, pp. 4451-4458.

Ruzgas, T.; Csöregi, E.; Emnéus, J.; Gorton, L. and Marko-Varga, G. (1996) Peroxidase-modified electrodes: Fundamentals and application. *Analytica Chimica Acta*, 330, pp. 123-1138.

Scheller, W.; Schubert, F.; Fedrowitz, J. *Frontiers in Biosensorics I. Fundamental Aspects*, Birkhauser, Basel, 1997.

Serra, B.; Benito, B.; Agui, L.; Reviejo, A. J.; Pingarron, J. M. (2001) Graphite-Teflon-Peroxidase Composite Electrochemical Biosensors. A Tool for the Wide Detection of Phenolic Compounds. *Electroanal.*, 2001, *13*, 693-700

Shan, D.; Cosnier, S.; Mousty, C. (2003) Layered double hydroxides: an attractive material for electrochemical biosensor design. *Anal. Chem.*, 75, pp. 3872-3879.

Sharma, S. K.; Singhal, R.; Malhotra, B. D.; Sehgal, N.; Kumar, A. (2004) Langmuir–Blodgett film based biosensor for estimation of galactose in milk. *Electrochim. Acta*, 49, pp. 2479-2485.

Singhal, R.; Takashima, W.; Kaneto, K.; Samanta, S. B.; Annapoorni, S.; Malhotra, B. D. (2002) Langmuir-Blodgett film of poly-3-dodecylthiophene for application to glucose biosensor. *Sens. Actuat. B*, 86, pp. 42-48.

Singhal, R.; Chaubey, A.; Srikhirin, T.; Aphiwantrakul, S.; Pandey, S. S.; Malhotra, B. D. (2003) Immobilization of glucose oxidase onto Langmuir–Blodgett films of poly-3-hexylthiophene. *Curr. Appl. Phys.*, 3, pp. 75-279.

Siqueira Jr., J. R.; Caseli, L.; Crespilho, F. N.; Zucolotto, V.; Oliveira Jr., O. N. (2010) Immobilization of biomolecules on nanostructured film for biosensing. *Biosens. Bioelectron.*, 25, pp. 1254-1163.

Sivozhelezov, V.; Bruzzese, D.; Pastorino, L.; Pechkova, E.; Nicolini, C. (2009) Increase of catalytic activity of lipase towards olive oil by Langmuir film immobilization of lipase. *Enzyme Microb. Technol.*, 44, pp. 72-76.

Skompska, M.; Chmielewski, M. J.; Tarajko, A. (2007) Poly(1,8-diaminocarbazole) – A novel conducting polymer for sensor applications. *Electrochem. Commun.*, 9, pp. 540-544.

Sołoducho, J.; Cabaj, J. (2010) Biocatalysts Immobilized in Ultrathin Ordered Films. *Sensors*, *10*(11), pp. 10298-10313.

Timur, S.; Pazarlioglu, N.; Pilloton, R.; Telefoncu, A. (2004) Thick film sensors based on laccases from different sources immobilized in polyaniline matrix. *Sens. Actuat. B*, 97, pp. 132-136.

Vedrine, C.; Fabiano, S.; and Tran-Minh, C. (2003) Amperometric tyrosinase based biosensor using an electrogenerated polythiophene film as an entrapment support *Talanta*, 59(3), pp. 535–544.

Vianello, F.; Cambria, A.; Ragusa, S.; Cambria, M. T.; Zennaro, L.; Rigo, A. (2004) A high sensitivity amperometric biosensor using a monomolecular layer of laccase as biorecognition element. *Biosens. Bioelectron.*, 20, pp.315-321.

Vianello, F.; Ragusa, S.; Cambria, M.T.; Rogo, A. (2006) A high sensitivity amperometric biosensor using laccase as biorecognition element. *Biosens. Bioelectron.*, 21, pp. 2155–2160.

Yang, S.; Li, Y.; Jiang, X.; Chen, Z.; Lin, X. (2006) Horseradish peroxidase biosensor based on layerby-layer technique for the determination of phenolic compounds, *Sens. Actuat. B*, 114, pp. 774-780.

Yoshida, K.; Yoshimoto, M.; Sasaki, K.; Ohnishi, T.; Ushiki, T.; Hitomi, J.; Yamamoto, S. and Sigeno, M. (1998) Fabrication of a new substrate for atomic force microscopic observation of DNA molecules from an ultrasmooth sapphire plat. *Biophys. J.*, 74, pp. 1654-1657.

Yuan, C. and Johnston, L.J. (2000) Distribution of ganglioside GM1 in L-alpha-dipalmitoylphosphatidylcholine/cholesterol monolayers: a model for lipid rafts. *Biophys. J.*, 79, 2768–2781..

Zachariah, K. and Mottola, H.A. (1989) Continuous-flow determination of phenol with chemically immobilised polyphenol oxidase. *Anal. Lett.*, 22, pp. 1145-1158.

Zhu, H.; and Snyder, M. (2003) Protein chip technology. *Curr. Opin. Chem. Biol.*, 7, pp. 55-63

11

Application of Microwave Assisted Organic Synthesis to the Development of Near-IR Cyanine Dye Probes

Angela Winstead and Richard Williams
Morgan State University
USA

1. Introduction

Cyanine dyes are a class of fluorescent organic dyes that have been used extensively as the probe component of chemical and biological sensors. Generally, they are comprised of two nitrogen-containing heterocycles, one of which is positively charged. The heterocycles are linked by a conjugated polymethine chain with an odd number of carbons, common chain lengths are tri (n=1), penta (n=2) and heptamethine (n=3).

N	Absorbance (nm)	Emission (nm)
1	540-560	560-580
2	650-690	670-710
3	760-980	780-1000

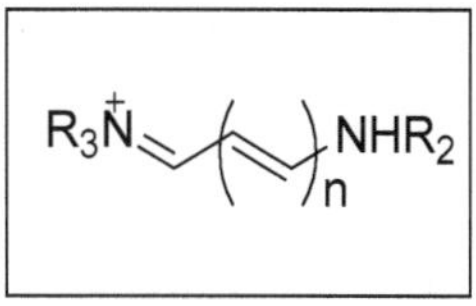

Fig. 1. Generic structure and properties of cyanine dyes

Pentamethine cyanine dyes (Cy5, n=2) are the universal probes for bio-analytical applications (Wehry 1976). They fluoresce at 670 nm and are compatible with laser and LED light sources that emit at 635 and 650 nm. A disadvantage of Cy5 probes is that most biomolecules also fluoresce in the same region, causing significant background interference. Heptamethine (Cy7, n=3) dyes, however, fluoresce in the near-IR (650-1000 nm) range with minimal background interference from biomolecules (Hammer, Owens et al. 2002). Although Cy5 technically falls in the near-IR (NIR) range, it fluoresces at relatively shorter wavelengths between 670 and 710 nm, a consequence of fewer methane carbons. In this region, some biological porphryrins, such as heme also exhibit emission spectra.

Microwave assisted organic synthesis (MAOS) has been used in the development of eco-friendly syntheses of biological and chemical sensor substrates. Often, sensor substrates designed to monitor and detect environmental pollutants are synthesized using reaction conditions and reagents that are not environmentally friendly and contribute to the problem they are designed to detect and or mitigate. Consequently, there is a need for a change in reaction conditions; increased atom efficiency, catalytic processes, and a decrease in solvent use in reactions, extractions, and purifications (Anastas 1998) are strategies used to improve reaction conditions. MAOS is widely accepted as a "green" technology; synthetic strategies enable solvent-free or minimal solvent use and increase atom efficiency in reactions (Hayes 2002). Further, because of the nature of MAOS, a variety of compounds can be synthesized that would not be feasible under conventional synthetic approaches (Lew, Krutzik et al. 2002).

The field of MAOS has changed dramatically from its first introduction in 1986 (Gedye, Smith et al. 1986; Giguere, Bray et al. 1986). In early work, microwave chemistry was often a last resource for an unsuccessful reaction or used to shorten reaction times from hours and days to minutes and seconds. The experiments were conducted in inexpensive domestic microwave ovens with time and power as the reaction conditions. Today, commercial microwave reactors control time, temperature, power, stirring and simultaneous cooling with real time monitoring of temperature, pressure, and power. Cutting edge instruments equipped with in situ analysis via raman spectroscopy (Leadbeater and Smith 2007), UV/VIS spectroscopy (Getvoldsen, Elander et al. 2002), and digital cameras and video that enable visual monitoring (Bowman, Leadbeater et al. 2008) have been developed capable of addressing these needs.

Our focus is the development of advanced microwave synthetic techniques that will aid in the synthesis of cyanine fluorophores and coupling agents needed in the development of biosensors with increased reaction efficiency, are environmentally benign, and cost effective.

1.1 MAOS of sensor probes

The rapid, cost-effective manner in which MAOS operates allows scientists access to a wide range and quantity of molecules that can be screened as potential biosensors. This section describes the application of MAOS to the synthesis of some fluorescent components in sensors. Perumal et al reported the microwave synthesis of an array of pyridinyl-1,2,4-triazine derivatives used as fluorescent sensors for ferric salts with reduced reaction times and comparable yields (66-85%) (Figure 2) (Perumal 2011). Bisaryl-3-pyrazine-1,2,4-triazine derivatives displayed good sensor property with Fe(III) ions in micro level concentrations.

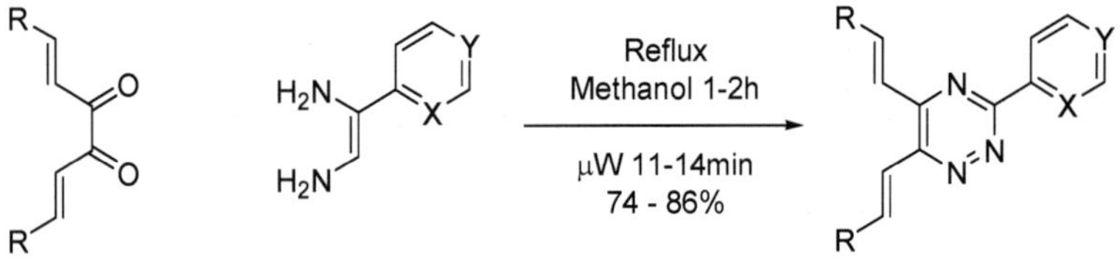

Fig. 2. Microwave synthesis of pyrazine derivatives

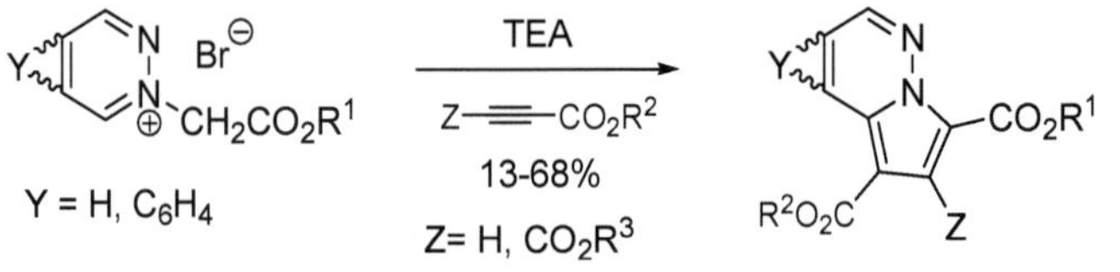

Fig. 3. Microwave synthesis of pyrrolodiazine derivatives

Pyrrolodiazine derivatives are highly fluorescent molecules that have potential applications in sensors and biosensors. The absorption and fluorescence of these N-heterocycles can be influenced by both solvent and substituent effects. MAOS has been used to rapidly and efficiently synthesize an array of pyrrolodiazine derivatives to study the relationship between structure and optical properties. The reactions were carried out using a monomode reactor varying temperature (by manipulating the power) and time with yields ranging from 13-68% (Figure 3).

Lamberto et al, studied the synthesis of symmetric and asymmetric viologen as well as a bis-viologen using microwave technology. Viologen derivatives are building blocks in biosensors for nitrite detection. A variety of solvents, reaction times, and molar ratios were manipulated. Seventeen viologen derivatives were successfully synthesized using microwave irradiation using a CEM LabMate microwave synthesizer (Figure 4).

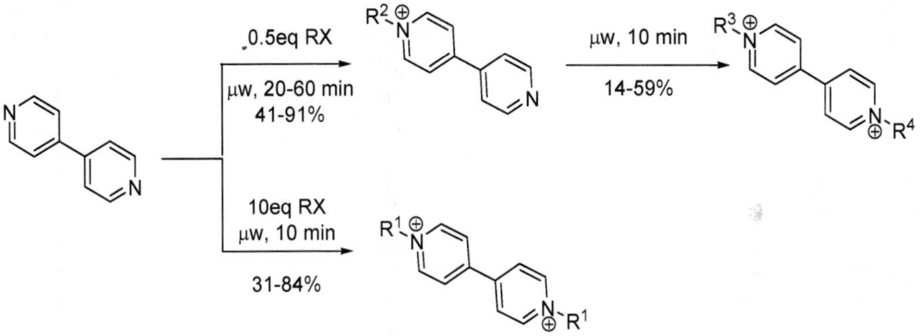

Fig. 4. Microwave synthesis of viologen derivatives

1.2 NIR cyanines

Applications that utilize NIR fluorescence technology are rapidly expanding. Heptamethine cyanines are an important class of NIR fluorescent molecules (Figure 5). Currently less than 1% of cyanine dye literature is concentrated on heptamethine derivatives. Consequently, progress in this area is limited by the lack of availability of suitable cyanine compounds that can be utilized as NIR labels and/or probes. More efficient ways of synthesizing these dyes are needed so that a variety of dyes with various spectral and chemical properties can be made available to thoroughly investigate their potential use as label and probes in analytical applications of NIR fluorescence.

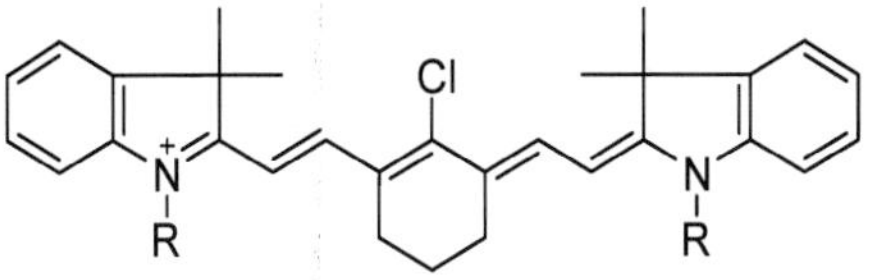

Fig. 5. Structure of NIR-1

Detection in the near-IR region demonstrates improved sensitivity over ultraviolet and visible (uv-vis) region detection because of the absence of background interference from competing analytes. This translates to a detection sensitivity based on the limitations of the detection instrument and not upon background interference. Based on this principle, shifting the fluorescence detection to longer wavelengths results in improved detection sensitivity. The increase in the polymethine chain increases the absorbance to the desired longer wavelengths and the inclusion of a six membered cyclic ring into the polymethine chain increases the photostability of the NIR Cy7 dye (Stoyanov). Near-IR fluorescent molecules also have other advantages including:

- higher molar absorptivities
- compatible with solid state excitation sources
- lower excitation energy requirements associated with the near-infrared region

1.2.1 NIR-cyanine synthesis

Cyanines can be described as either symmetrical (identical heterocyclic salts) or unsymmetrical (different heterocyclic salts). The general synthetic strategy used in the preparation of cyanine dyes is a multistep process (Figure 6). First, the nitrogen-containing heterocycle is quaternized with an alkyl halide to form a N-alkyl heterocyclic quaternary salt. The symmetrical dye is prepared through the addition of 2eq of the same heterocyclic salt to the an electrophile such as an imine, bisimine, or bisaldehyde and subjected to the reaction conditions.

Two published methods of synthesizing symmetrical cyanine dyes requires heating a mixture of a substituted quaternary salt (I) and either a bisimine or bisaldehyde to reflux in 1-butanol and benzene (Narayanan and Patonay 1995) or sodium acetate and ethanol (Jung and Kim 2006).

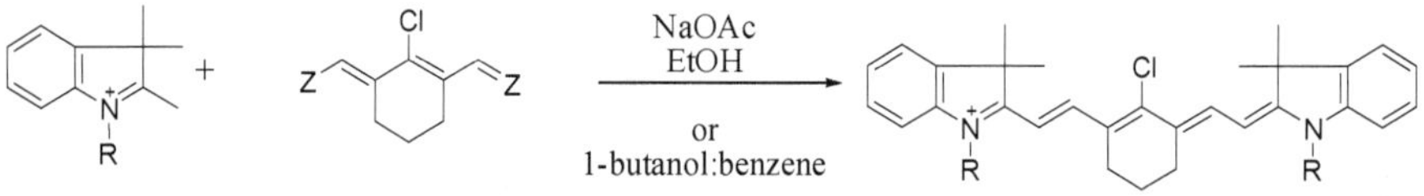

Fig. 6. Conventional synthesis of heptamethine cyanine dyes

The synthesis of unsymmetrical cyanines involves treatment of heterocyclic salt the desired electrophile to form the hemicyanine. The hemicyanine is reacted with a second heterocyclic quaternary salt to form an unsymmetrical cyanine dye. The major problem in the synthesis is the formation of the symmetric dye as a byproduct. A common solution is the isolation and rigorous purification of the hemicyanine, even though they are a challenge to isolate and purify (Mujumdar, Ernst et al. 1993; Mank, Van der Laan et al. 1995). This is a problem regardless of the methine chain length (Cy3, Cy5, Cy7), heterocyclic salt precursors, or water-solubility. The solid phase technique was applied to unsymmetrical cyanine synthesis by Mason et al (Mason, Hake et al. 2005) Solid phase synthesis of Cy3 and Cy5 unsymmetrical dyes is five and three steps respectively, in overall crude yields ranging from 3-60% with 50->95% purity (Figure 7). The synthesis requires extensive use of solvents as reaction media, and multiple purification techniques after each step with the amount of desired product only 2 – 11mg.

A second multi-step protocol that was developed by Caputo (Caputo 2002): 1) synthesized the acetylated hemicyanine with the less reactive heterocyclic salt 2) purified the hemicyanine with continuous extraction with ethyl acetate 3) reacted the pure hemicyanine with the more reactive heterocyclic salt. This strategy requires a large amount of solvent in the purification of the hemicyanine and the unsymmetrical dye is purified by column chromatography. The purity was not reported and the dye was not characterized or quantified prior to derivatization for several examples.

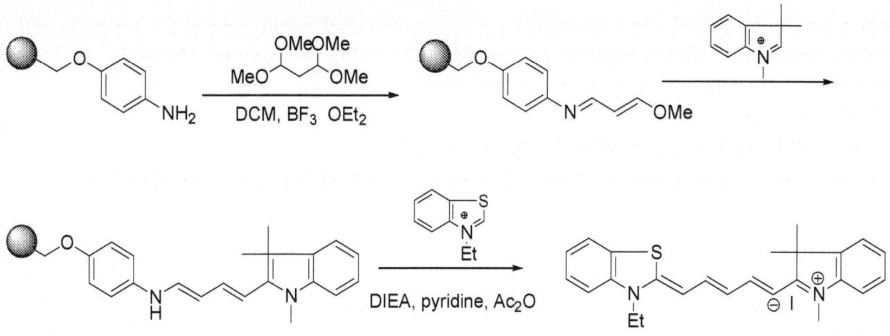

Fig. 7. Solid phase synthesis of cyanine dyes

2. Salt synthesis

N-Alkyl quaternary ammonium salts are used extensively as precursors of various cyanine dyes (Narayanan and Patonay 1995) and near-IR spiropyrans (Hirano, Osakada et al. 2002). The salts are synthesized by refluxing reagents with solvents such as chloroform, *o*-dichlorobenzene, acetonitrile and ethanol for 6 – 48h. One example requires refluxing in acetonitrile for 24 h, then treatment with diethyl ether followed by filtration. The combined filtrates were concentrated and refluxed for an additional 24 h, treated with diethyl ether and filtered (Pardal, Ramos et al. 2002). This process was repeated 1-3 times to achieve the published yields (25 – 78%). Another method heats the reagents at 80 °C for 21 h in an ampule tube sealed with a torch (Hirano, Osakada et al. 2002). Purification of the salts range from Soxhlet extraction with benzene for 24 h (Elizalde, Ledezma et al. 2005) to filtration with cold ether (Pardal, Ramos et al. 2002).

2.1 2,3,3-trimethylindolenine derivatives

The reaction of 2,3,3-trimethylindolenine with an array of alkyl halides with varied functionality were studied (Figure 8). The reactions were performed by charging each microwave reaction vial with of 2,3,3-trimethylindolenine and an alkyl halide. Our previously studied reaction of ethyl iodide with 2,3,3-trimethylindolenine served as the model system (Winstead 2008a). The microwave reaction conditions were determined using a single-mode microwave system. The temperature was monitored throughout each reaction. The optimized reaction condition was 130 °C, ramp time: 2:50 min, reaction time: 5:00 min giving a 95% yield.

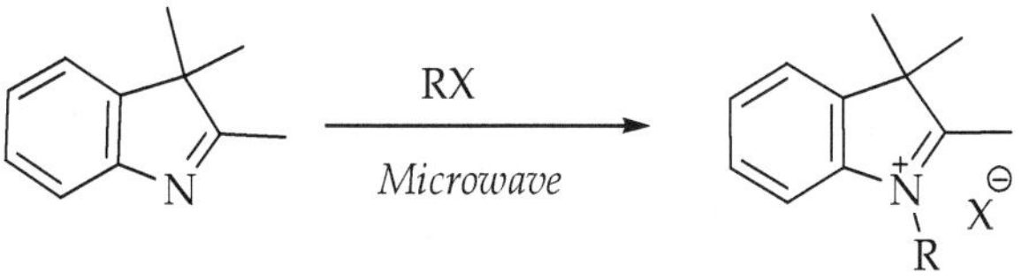

1 R = Me, **2** R = Et, **3** R = Pr, **4** R = -CH$_2$CH$_2$OH, **5** R = -(CH$_2$)$_5$CO$_2$H

Fig. 8. Synthesis of 2,3,3-trimethylindolenine quaternary salts

The scope of the reaction was examined with the coupling of 2,3,3-trimethylindolenine with iodomethane, iodopropane, bromoethanol and 6-bromohexanoic acid (Winstead 2008b). The hold time, ramp time, and temperature for each electrophile was studied. The optimized reaction conditions are presented in Table 1. In most cases, the yields were comparable or exceeded the published yields. Most significant is the substantially decreased reaction time and simplicity of the reaction procedure. The yields presented are the yields without resubjection of the filtrates.

Alkyl salts **1-3** were simply filtered and washed with cold ether. The products were pure by NMR analysis and no further purification was necessary. Salts **4** and **5** which all contain hydroxyl groups, did not crystallize right away. The reaction solution containing **4** was concentrated followed by the addition of hexanes. The solution was heated until crystals formed and then filtered. Similarly, salt **5** was recrystallized from acetone.

Salt	T (°C)	Time (min)	Yield (%)	Lit. Yield (%)	Lit Time (h)
1	130	5:00	95	59	48
2	110	2:30	93	75	21
3	110	7:00	83	44	24
4	110	7:00	73	69	24
5	110	7:00	59	67	12

Table 1. Reaction conditions for the synthesis of 2,3,3-trimethylindolenine salts

X ray chrystallography was taken of carboxylic acid derivative **5** (Figure 9) (Winstead 2010). The data revealed the hydrogen bonding of a water molecule to CO_2H moiety and the bromide ion is associated with water molecule.

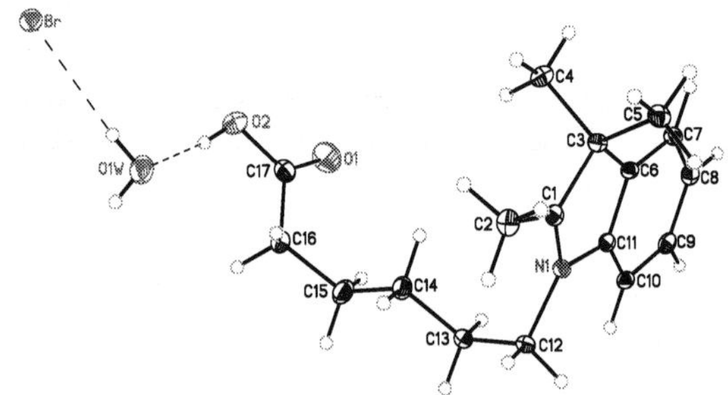

Fig. 9. X ray christallography

2.2 Benzothiazole derivatives

Benzothiazole derivatives were synthesized by reacting benzothiazole with various alkyl halides under microwave irradiation conditions to produce compounds 7-11 in good yields (Figure 10). Methyl indolenine salt 7 was synthesized in excellent yield without solvent. The reaction was complete in only 20 min (85%) compared to 7h and 60% yield under reflux conditions.

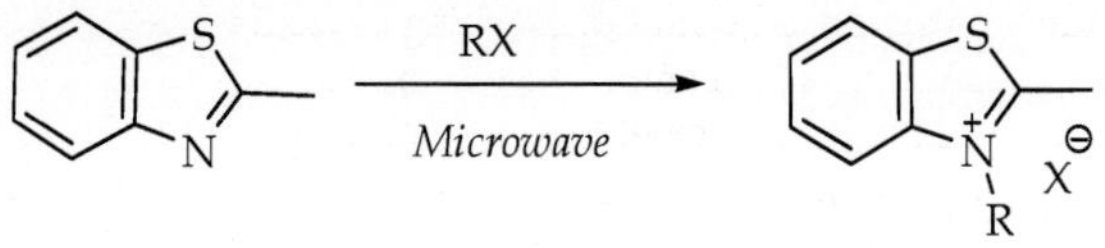

6 R = Me, **7** R = Et, **8** R = Pr, **9** R = -CH$_2$CH$_2$OH, **10** R = -(CH$_2$)$_5$CO$_2$H

Fig. 10. Synthesis of benzothiazole quaternary salts

The reaction ethyl indolenine **3** was synthesized in yields ranging from 4-80% (Table 2). The reaction conditions used for the methyl indolenine synthesis resulted in a 17% yield. The temperature was increased in 10 °C increments. The reaction was placed in a microwave vial and subjected to microwave irradiation at the set temperature (20 min). The vial was removed and checked for the presence of a solid. If there was small amount of solid, the reaction was placed back into the microwave oven for an addition 10 min. The yields reported are after the second reaction time and filtration. The reaction yield increased as the temperature increased to 170 °C. The reaction conditions for the synthesis of propyl indolenine **8** are the same as **7**.

T (°C)	Time (min)	Yield (%)
120	15	4
120	20	17
130	20 + 10	31
140	20 + 10	45
150	20 + 10	47
160	20 + 10	69
170	20	80

Table 2. Reaction conditions for ethyl indolenine **8**

2-ethanol indolenine **9** was synthesized in yield in 56% (Table 3). The temperatures studied were between 150 °C – 100 °C. 150 °C was selected as the starting point because it was in between the optimum reaction temperature for the methyl and the ethyl derivatives. The reaction time was 20 min. These initial conditions led to decomposition. A temperature reduction of 20 °C also provided decomposition (Entry 2), however a 30 °C decrease yielded starting material (Entry 3). When an increase in time produced decomposition (Entry 4), the temperature was reduced to 100 °C and held for 35 min. The reaction vial sat at room temperature for 2-4 h, after which time crystals formed and were filtered.

Entry	T (°C)	Time (min)	Yield (%)
1	150	20	Decomp
2	130	20	Decomp
3	120	20	SM
4	120	25	Decomp
5	100	35	56

Table 3. Reaction conditions for 2-ethanol indolenine **9**

It has been reported that the longer the alkyl chain length, the more difficult it is to perform the alkylation. Therefore, the higher 170 °C and 20 min was selected as the starting reaction conditions and yielded a low 30% yield (Table 4). The time was increased in 5 min increments to 35 min. Reaction times longer than 35 min lead to inconsistent results. The reaction vial sat at room temperature for 2-4 h, after which time crystals formed and were filtered.

T (°C)	Time (min)	Yield (%)
170	20	30
170	25	43
170	30	66
170	35	75

Table 4. Reaction conditions for 6-hexanoic acid indolenine **10**

The optimized reaction conditions for the synthesis of benzothiazole derivatives under microwave conditions and conventional heating methods in the literature are presented in Table 5. The reaction times are significantly shorter and the percent yields are higher in every example of the synthesis of these derivatives.

Salt	R	T (°C)	Time (min)	Yield (%)	Lit. Yield (%)	Lit Time (h)
6	Et	170	20	83	48	48
7	Me	120	20	85	60	7
8	Pr	170	20	65	5	7
9	-(CH$_2$)$_2$OH	100	35	58	N/A	6
10	-(CH$_2$)$_5$CO$_2$H	170	35	75	61	48

Table 5. Reaction conditions for benzothiazole quaternary salts

2.3 Other salt derivatives
2.3.1 Lepidine derviative

An interesting RNA antagonist precursor, 1-(6-ethoxy-6-oxohexyl)-4-methylquinolinium iodide quaternary ammonium salt has been synthesized. 1-(6-methoxy-6-oxohexyl)-4-methylquinolinium chloride quaternary salt has been synthesized in two steps in 37% overall yield (Figure 11) (Carreon, Stewart et al. 2007). The rapid solvent-free synthesis of the N-ethoxycarbonylhexyl quaternary salt from commercially available lepidine in one step in 56% yield with minimal purification is described.

Commercially available 6-bromohexanoic acid was converted to the iodide using Finkelstein reaction conditions. In these studies, the typically more reactive iodide alkyl halide yielded yields comparable to the commercially available bromide suggesting the extra step of halogen exhange is not necessary. The percent yields for the iodo and bromo salts were comparable (Table 6).

During the reaction time studies an increase in yield was observed when the reaction stopped midway, cooled to room temperature and resubjected to microwave irradiation for addional time. The difference was 10 points for both 9 and 7 min total reaction time (Table 7). This could be attributed to the additional time exposed to microwave irradiation due to a second 4 min ramp time during the resubjections.

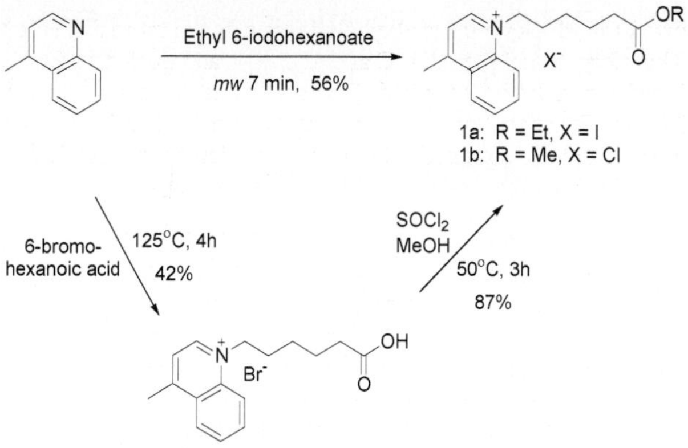

Fig. 11. Synthesis of lepidine quaternary salt

X	Temp	Time	Yield
Br	120	7	49
Br	120	7	54
I	120	7	52
I	120	7	56

Table 6. Reaction conditions for lepidine synthesis

Trial	Time	Yield
1	9	49
2	9	46
3	5 + 4	56
4	5 + 4	60
5	7	44
6	3 + 4	54

Table 7. Lepidine derivative time studies

2.3.2 Methylbenzothiazole acetonitrile

In some cases, cyanine dyes have a tendency to photobleach. The incorporation of α-cyano group has been shown to increase the photostability of the dye. Initial studies focus on the improvement of the synthesis of the α-cyano heterocyclic salt. Previous studies utilized conventional heating to synthesize α-cyano cyanine dyes.

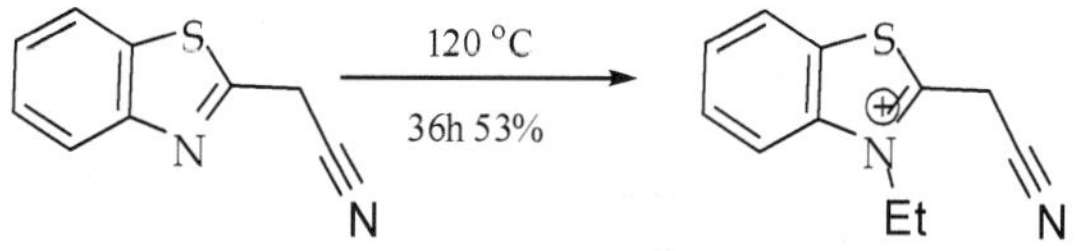

Fig. 12. Synthesis of methylbenzothiazole acetonitrile quaternary salt

Methybenzothiazole acetonitrile was treated with iodoethane; temperature, hold time, and mole ratio were manipulated. Based on NMR data, the reaction yielded a mixture of starting material and product. The reaction required heating at 120 °C for 36h to achieve the cyano salt in 53%. The sluggish nature of this reaction under conventional heating conditions was mimicked in our preliminary studies (Table 8). Higher temperatures lead to decomposition, while lower temperatures provide a mixture of starting material and product.

Entry	Temp (°C)	Time (min)	Outcome	Yield (%)
1	130 °C	40	Product	34%
2	130 °C	40	Product/ SM	36%
3	130 °C	50	Product/ SM	41%
5	140 °C	40	Product	21%
6	170 °C	10	decomposed	-
7	170 °C	20	decomposed	-

Table 8. Reaction conditions for methylbenzothiazole acetonitrile quaternary salts

3. Cyanine dye synthesis

3.1 Symmetrical cyanine dye synthesis

Herein, we report the microwave synthesis of five cyanine dyes derived from the aforementioned heterocyclic salts. The microwave reaction conditions were determined using the Biotage single-mode microwave system. The reaction of the N-((E)-(2-chloro-3-((E)-(phenylimino)methyl)cyclohex-2-enylidene)methyl)aniline (1eq) and 2,3,3-trimethyl-1-ethyl-3H-indolium iodide (2eq) served as the model system (Figure 13).

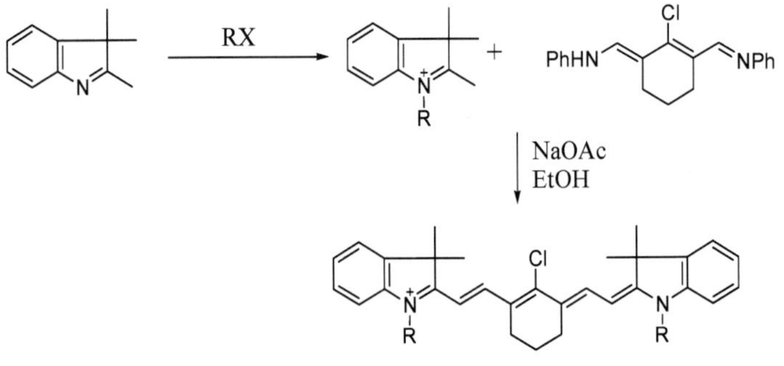

NIR-1: R = Me; **NIR-2:** R = Et; **NIR-3:** R = Pr; **NIR-4:** R = CH$_2$CH$_2$OH; **NIR-5:** R = (CH$_2$)$_5$CO$_2$H

Fig. 13. Synthesis of NIR dyes

The reaction was examined using a temperature range from 100 °C to 160 °C in 10 °C increments (Table 9). The vials were cooled to 0 °C, filtered and washed with diethyl ether to yield greenish-gold crystals in yields from 53 – 79%. The optimum temperature proved to be 120 °C (79%). The yield decreased by around 13% for each 10 °C increase from 140 °C to 160 °C. An increase in time from 20 min to 30 min did not lead to an improved percent yield. The purity of each sample produced from each trial was the same by ^{1}H and ^{13}C NMR.

Trial	Temp. (°C)	Time (min)	Yield (%)
1	100	20	61
2	110	20	63
3	120	20	79
4	130	20	70
5	140	20	72
6	150	20	61
7	160	20	53
8	120	30	77

Table 9. Temperature studies for NIR-2

The optimized reaction conditions for NIR-2 were examined using a variety of heterocyclic salts. The hold time, and temperature for each heterocyclic salt was studied. The optimized reaction conditions are presented in Table 2. NIR-5 provides two sites for protein labeling in biosensor applications. In most cases, the yield was comparable or exceeded the published yields (Table 10). Most significant is the substantially decreased reaction time and simplicity of reaction procedure. The yields presented are the yields without resubjection of the filtrates.

Dye	R	Yield (%)	Lit. Yield (%)
NIR-1	Me	72	66
NIR-2	Et	79	81
NIR-3	Pr	81	NR
NIR-4	$(CH_2)_2OH$	64	NR
NIR-5	$(CH_2)_5CO_2H$	83	NR

Table 10. Structure and yields for synthesis of symmetrical dyes

3.2 Unsymmetrical cyanine dye synthesis

Unsymmetrical cyanine dyes have been widely used in probes for bio-analytical applications. The growing range of applications has warranted the need for a convenient, reliable synthesis of these molecules. The synthesis of these compounds has been a challenge due to the production of the symmetrical dye alongside the desired unsymmetrical dye (Lin, Weissleder et al. 2002; Toutchkine, Nalbant et al. 2002; Kim, Kodagahally et al. 2005; Jiang, Dou et al. 2007). The purification of this mixture is nontrivial requiring multiple chromatographic steps that often lead to low yields. The reaction is a two step process that involves the synthesis of the hemicyanine intermediate followed by conversion to the unsymmetrical dye.

Our reaction protocol is: HET_1, bisimine, and NaOAc in ethanol are combined and irradiated for 15 min at 100 °C. Then HET_2 and NaOAc are added to the mixture and irradiated for 15 min at 100 °C. The reaction time is reduced from 2-4 hours at 120 °C to 15 min at 100 °C for initial step **A** (Figure 15). Isolation of the hemicyanine has been eliminated from the process, completely removing the opportunity for the hemicyanine to form symmetric dye during isolation. The absence of this also eliminates the use of large amounts of solvents during purification. The reduced reaction time substantially minimizes hemicyanine exposure to conditions that facilitate its reversibility. The same line of reasoning holds true for the second condensation reaction **C** between the hemicyanine and HET_2 to form the unsymmetrical dye. Although symmetric dye was not detected in the ^1H NMR, it cannot be stated that none was formed during the microwave process. The actual amount of symmetrical dye formed at each step must be quantified and will be the focus of future studies.

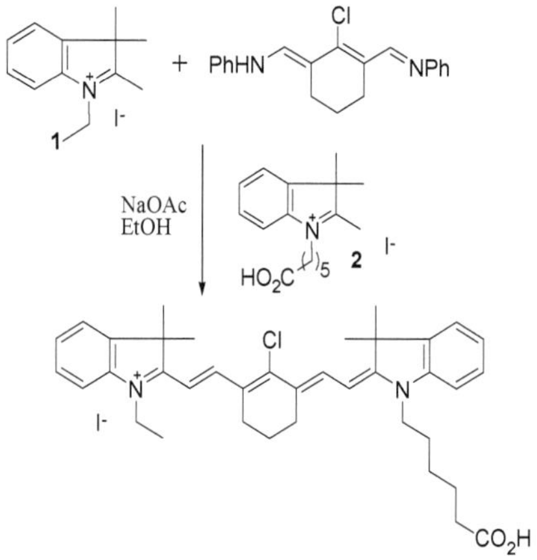

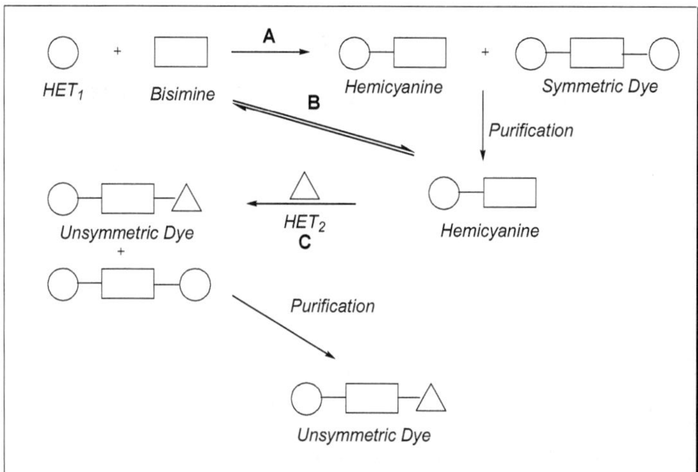

Fig. 14. Synthesis of unsymmetric dye

Fig. 15. Pathways of symmetrical dye formation

The treatment of ethyl salt **1** with bisimine and sodium acetate in ethanol for 10 min at 120 °C in a microwave oven affords the hemicyanine intermediate on literature precedence. The solution was allowed to cool to rt and salt **2** and 1 eq. of sodium acetate based was added to the vial and irradiated for an additional 10 min at 120 °C. The product was filtered and analyzed using ^{1}H NMR. These conditions were applied to the synthesis of three additional unsymmetrical cyanine dyes (Table 11).

Dye	R_1	R_2	Wavelength	Percent Yield
NIR-6	Et	Pr	780	58%
NIR-7	Et	Me	780	44%
NIR-8	$(CH_2)_2OH$	Et	782	75%
NIR-9	$(CH_2)_5CO_2H$	Me	780	69%

Table 11. Structure and yields for synthesis of unsymmetrical dyes

4. Properties

Fluorescence spectroscopy has become a key technique for the detection and elucidation of biological processes. Most fluorescence sensors for bio-analytical applications fluoresce in the visible region (400-650 nm) (Wehry 1976). A disadvantage of this technique is that most biomolecules also fluoresce in this region, causing significant background interference. Near infrared (NIR) dyes, however, fluoresce in the 650-1000 nm range with minimal background interference from biomolecules and high sensitivity (Hammer, Owens et al. 2002). Some of the advantages of NIR fluorescence based detection techniques include: increased detection sensitivity and selectivity due to the absence of background interference; increased photo stability and less photo bleaching effects because of lower excitation energies associated with the NIR region; good compatibility with cost effective light excitation sources and solid state detectors; and the easy adaptation to valuable visible fluorescence analytical techniques such as fluorescence energy transfer (FRET), two-photon excitation, and metal enhanced fluorescence (MEF).

In particular, cyanine dyes have widespread application as fluorescent probes. The spectral properties of fluorescent probes assist in the determination of how they are applied in a variety of analytical techniques. Properties such as wavelength absorbance and emission ranges, Stokes shifts, and spectral bandwidths are used to address the various requirements associated with different analytical techniques. One emerging area of interest for promising analysis applications that use cyanine dyes is metal enhanced fluorescence. The characteristic low quantum yields of these dyes suggest they have potential applications in metal enhanced fluorescence related techniques. Metal enhanced fluorescence is a form of fluorescence where plasmon waves generated by metals deposited on a glass substrate can increase the fluorescence of fluorophores within the influencing range of the wave. The phenomenon is demonstrated most effectively in fluorophores that have low quantum yields. Because of the energy associated with the longer wavelengths of the NIR, and the moderately low quantum yields associated with this class of compounds, the potential application for these compounds as MEF probes or labels are promising (Lakowicz, Parfenov et al. 2003; Lakowicz, Geddes et al. 2004). While these dyes are well-known, they are costly with limited commercial availability.

Figure 16 shows the absorbance spectrum for microwave synthesized NIR-2 along with its corresponding fluorescence spectrum. The fluorescence intensity was scaled to the absorbance scale for comparison purposes. The absorbance maximum was absorbed at 779 nm. A fluorescence maximum intensity was observed at 795 nm, demonstrating a Stokes shift of 16 nm. While the observed Stoke's shift is considered moderate, it is typical for the heptamethine cyanine class of NIR dyes.

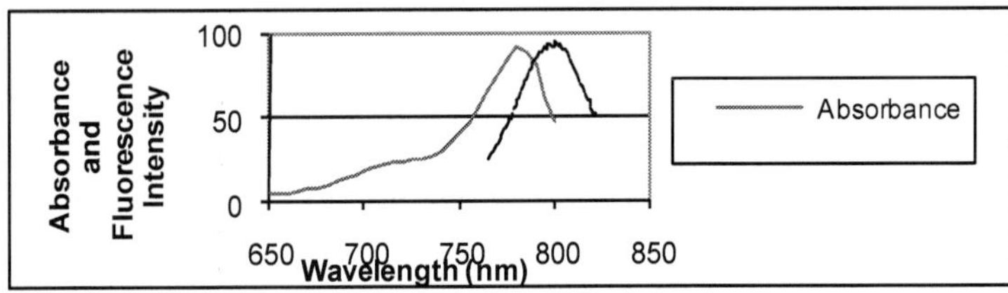

Fig. 16. A graph of the absorbance spectra along with the relative fluorescence intensity of the NIR dye, NIR-2 (in ethanol) demonstrates a Stokes shift of 16 nm

The spectral analysis of the five synthesized compounds demonstrates characteristics that are similar to what has been reported previously for this class of compounds (Table 12). They typically have large molar absorption extinction coefficients (molar absorptivities), moderate Stoke's shits, and small quantum yields. The overall detection sensitivities are a reflection of the moderate Stoke's shifts observed for the compounds as well as the limitation of the Hamatsu 928-PMT red sensitive photomultiplier tube used in the Cary Eclipse fluorescence spectrophotometer. This is expected as fluorescence detection moves to longer wavelengths because background interference from molecules with similar fluorescence properties is minimized. In the visible part of the spectrum, fluorescence detection sensitivity depends on both the responsivity of the detector and on background interference from molecules found in the detection medium with similar spectral properties. In the NIR spectral region, only the responsivity of the detector and the optical limitations of the instrument limit detection sensitivity. Because of the size of the Stoke's shift, detection sensitivity is influenced by scattered light effects that occur when light from the excitations source of the fluorescence spectrophotometer is scattered in the solvent medium and bleeds over into the instrument's detector. This reduces the signal-to-noise ratio (S/N) which limits the detection sensitivity. As the size of the Stoke's shift increases, the S/N increases and this improves the detection sensitivity. Increasing the size of the Stoke's shift is an important priority in future synthetic work as a way to directly improve detection sensitivity and create compounds that are effective labels and probes. Some promising work on increasing the Stoke's Shift of NIR heptamethine cyanine dyes has been reported (Peng, Song et al. 2005). Large Stoke's shifts greater than 100 nm can be obtained by substituting the chlorine with an alkyl amino group at the central position of the heptamethine cyanine dyes. This substitution process is very compatible with microwave assisted synthesis techniques and is expected to contribute greatly to future generations of NIR fluorescent probes. Increasing the Stoke's shifts of these dyes will translate into photo stable NIR fluorescent probes with significant increases in sensitivity. When used as the probe part of fluorescence based biosensors, novel NIR fluorescent dyes are expected to improve the sensitivity and dynamic range of current fluorescence biosensor analytical detection techniques.

Dye	Absorption λ_{abs} (nm)	Emission λ_{ems} (nm)	Stokes shift (nm)	Quantum yield	Detection sensitivity (M)	Molar absorptivity	E'_{ox} (V)
NIR-1	779	795	16	0.041	1E-08	2.8E+05	0.42
NIR-2	780	798	18	0.024	7.2E-09	1.6E+05	0.40
NIR-3	785	802	17	0.025	1.7E-08	1.2E+05	0.40
NIR-4	785	802	17	0.022	4E-09	4.3E+05	N/A
NIR-5	785	805	20	0.022	1.8E-08	2.4E+05	N/A

Table 12. Spectral characteristics

While the microwave assisted syntheses of unsymmetrical heptamethine cyanines remain a work in progress, preliminary spectral analyses of unsymmetrical compounds indicate no significant difference in spectral properties from their symmetrical counterparts (Table 13). While shifts were detected in absorption and emission wavelengths, the absorption wavelengths remained in the 780 nm range and the Stoke's shifts observed were considered insignificant. The quantum yields observed remained moderately low and very compatible for use in FRET and MEF applications. The slight increases observed for detection sensitivity were not significant, and the overall molar absorptivities remained in the range of symmetrical dyes with similar peripheral functional groups. The effect of substituting functional groups on the outer edges of the compounds in order to change their symmetry has no significant overall effect on the spectral properties of this class of dyes. This demonstrates the flexibility of these compounds in their ability to be used in fluorescence analytical applications. It is also a further indication of the potential of this class of compounds as probes for NIR fluorescence biosensor applications.

Dye	Absorption λ_{abs} (nm)	Emission λ_{ems} (nm)	Stokes shift (nm)	Quantum yield	Detection sensitivity (M)	Molar absorptivity
NIR-6	785	801	16	0.021	5.2E-09	1.9E+05
NIR-7	780	796	16	0.021	4E-09	3.1E+05
NIR-8	780	800	20	0.020	2.2E-09	2.1E+05
NIR-9	780	796	16	0.021	7.7E-09	2.3E+05

Table 13. Spectral characteristics of unsymmetrical dyes

In addition to Stoke's shift properties, the high molar absorptivity and molar extinction coefficients that characterize this class of dye are additional spectral properties that can be successfully utilized in the improvement of current fluorescence biosensor analytical techniques. The size of the molar extinction coefficients coupled with the relatively low quantum yields suggest several potentially useful application possibilities. A main concern for this class of NIR dyes are chemical photostability and photobleaching effects. Because the molar extinction coefficients are high compared to conventional visible fluorophores, larger amounts of energy can be pumped into these dyes and at longer wavelengths. When this effect is coupled with the presence of a stabilizing ring system in the center of the molecule, chemical photo stability is maximized. Typically, photostability is limited by the number of cycles that a single dye molecule can achieve before it decomposes. A cycle is defined as the process where a fluorophore is excited and then fluoresces as it returns to its ground state. More cycles suggest greater detection sensitivity because more photons are produced as photo stability increases. The combination of high molar absorptivity at longer wavelengths for the NIR cyanine class of fluorophores, coupled with the low excitation energy from NIR excitation light sources, insures maximized photo stability for dyes in this region.

Another useful application is related to the property of moderately low quantum yields that typically characterize cyanine dyes. Metal enhanced fluorescence is a form of fluorescence where plasmon waves generated by metals deposited on a glass substrate can increase the fluorescence of fluorophores within the influencing range of the wave. The phenomenon is dependent on the fraction of nonradiative decay being much greater than radiative decay. The lower the quantum yield, the greater the potential for the MEF effect to occur. Again, because of the energy associated with the longer wavelengths of the NIR, and the low quantum yields associated with this class of compounds, the potential application for these compounds as MEF probes or labels are promising.

5. Conclusions

MAOS has provided substantial decreased reaction times, simplicity of reaction procedure, and comparable or increased reaction yields for a variety of N-alkylheterocyclic salts and symmetrical and unsymmetrtical cyanines. The microwave synthesis is a straightforward synthesis that yields cyanine dyes that do not need rigorous purification via column chromatography. The percent yields are reflective of single runs without subjecting unreacted starting materials to the reaction conditions multiple times to achieve high percent yields.

6. Acknowledgments

This work was supported by NSF-HRD 0627276 and the Morgan State University's Department of Chemistry. The authors also acknowledge Department of Homeland Security, Office of University Programs, Science & Technology Directorate: DHS 2008-ST-062-000011 for support.

7. References

Anastas, P. T. W., T.C. (1998). Green Chemistry: Frontiers in Benign Chemical Synthesis and Processes. Oxford, UK, Oxford University Press.

Bowman, M. D., Leadbeater, N. E. and Barnard, T. M. (2008). "Watching microwave-promoted chemistry: reaction monitoring using a digital camera interfaced with a scientific microwave apparatus." Tetrahedron letters 49(1): 195-198.

Caputo, G., Della Ciana, L. (2002). Improved Process and Method for the Preparation of Asymmetric Monofunctionalised Indocyaine Labeling Reagents and Obtained Compounds. E. Patent.

Carreon, J. R., Stewart, K. M., Mahon Jr, K. P., Shin, S. and Kelley, S. O. (2007). "Cyanine dye conjugates as probes for live cell imaging." Bioorganic & medicinal chemistry letters 17(18): 5182-5185 %@ 0960-894X.

Elizalde, L. E., Ledezma, R. and Lopez, R. G. (2005). "Synthesis of Photochromic Monomers Derived from 1 -(2-Methacryloxyethyl)-3, 3-Dimethyl-2-[2H]-Spirobenzopyran Indoline." Synthetic Communications 35(4): 603-610 %@ 0039-7911.

Gedye, R., Smith, F., Westaway, K., Ali, H., Baldisera, L., Laberge, L. and Rousell, J. (1986). "The use of microwave ovens for rapid organic synthesis." Tetrahedron letters 27(3): 279-282.

Getvoldsen, G. S., Elander, N. and Stone-Elander, S. A. (2002). "UV Monitoring of Microwave-Heated Reactions-A Feasibility Study." CHEMISTRY-WEINHEIM-EUROPEAN JOURNAL- 8(10): 2255-2260.

Giguere, R. J., Bray, T. L., Duncan, S. M. and Majetich, G. (1986). "Application of commercial microwave ovens to organic synthesis." Tetrahedron letters 27(41): 4945-4948.

Hammer, R. P., Owens, C. V., Hwang, S. H., Sayes, C. M. and Soper, S. A. (2002). "asymmetrical, water-soluble Phthalocyanine dyes for covalent labeling of oligonucleotides." Bioconjugate Chem 13(6): 1244-1252.

Hayes, B. L. (2002). Microwave Synthesis Chemistry at the Speed of Light, CEM Publishing.

Hirano, M., Osakada, K., Nohira, H. and Miyashita, A. (2002). "Crystal and solution structures of photochromic spirobenzothiopyran. First full characterization of the meta-stable colored species." J. Org. Chem 67(2): 533-540.

Jiang, L. L., Dou, L. F. and Li, B. L. (2007). "An efficient approach to the synthesis of water-soluble cyanine dyes using poly (ethylene glycol) as a soluble support." Tetrahedron letters 48(33): 5825-5829 %@ 0040-4039.

Jung, M. E. and Kim, W. J. (2006). "Practical syntheses of dyes for difference gel electrophoresis." Bioorganic & medicinal chemistry 14(1): 92-97.

Kim, J. S., Kodagahally, R., Strekowski, L. and Patonay, G. (2005). "A study of intramolecular H-complexes of novel bis (heptamethine cyanine) dyes." Talanta 67(5): 947-954 %@ 0039-9140.

Lakowicz, J. R., Geddes, C. D., Gryczynski, I., Malicka, J., Gryczynski, Z., Aslan, K., Lukomska, J., Matveeva, E., Zhang, J. and Badugu, R. (2004). "Advances in surface-enhanced fluorescence." Journal of Fluorescence 14(4): 425-441 %@ 1053-0509.

Lakowicz, J. R., Parfenov, A., Gryczynski, I., Malicka, J. and Geddes, C. D. (2003). "Enhanced fluorescence from fluorophores on fractal silver surfaces." J. Phys. Chem. B 107(34): 8829-8833.

Leadbeater, N. E. and Smith, R. J. (2007). "In situ Raman spectroscopy as a probe for the effect of power on microwave-promoted Suzuki coupling reactions." Organic & Biomolecular Chemistry 5(17): 2770-2774.

Lew, A., Krutzik, P. O., Hart, M. E. and Chamberlin, A. R. (2002). "Increasing rates of reaction: microwave-assisted organic synthesis for combinatorial chemistry." J. Comb. Chem 4(2): 95-105.

Lin, Y., Weissleder, R. and Tung, C. H. (2002). "Novel near-infrared cyanine fluorochromes: synthesis, properties, and bioconjugation." Bioconjugate Chem 13(3): 605-610.

Mank, A. J. G., Van der Laan, H. T. C., Lingeman, H., Gooijer, C., Brinkman, U. A. T. and Velthorst, N. H. (1995). "Visible Diode Laser-Induced Fluorescence Detection in Liquid Chromatography after Precolumn Derivatization of Amines." Analytical Chemistry 67(10): 1742-1748.

Mason, S. J., Hake, J. L., Nairne, J., Cummins, W. J. and Balasubramanian, S. (2005). "Solid-phase methods for the synthesis of cyanine dyes." J. Org. Chem 70(8): 2939-2949.

Mujumdar, R. B., Ernst, L. A., Mujumdar, S. R., Lewis, C. J. and Waggoner, A. S. (1993). "Cyanine dye labeling reagents: sulfoindocyanine succinimidyl esters." Bioconjugate Chemistry 4(2): 105-111.

Narayanan, N. and Patonay, G. (1995). "A new method for the synthesis of heptamethine cyanine dyes: synthesis of new near-infrared fluorescent labels." The Journal of Organic Chemistry 60(8): 2391-2395 %@ 0022-3263.

Pardal, A. C., Ramos, S. S., Santos, P. F., Reis, L. V. and Almeida, P. (2002). "Synthesis and spectroscopic characterisation of n-Alkyl quaternary ammonium salts typical precursors of cyanines." Molecules 7: 320-330.

Peng, X., Song, F., Lu, E., Wang, Y., Zhou, W., Fan, J. and Gao, Y. (2005). "Heptamethine cyanine dyes with a large stokes shift and strong fluorescence: a paradigm for excited-state intramolecular charge transfer." J. Am. Chem. Soc 127(12): 4170-4171.

Perumal, P. T. a. P. T. (2011). "The synthesis and photophysical studies of pyridinyl-1,2,4-triazine derivatives and use as a fluorescent sensor for ferric salts." Dyes and Pigments 88: 403-412.

Stoyanov, S. "Probes: dyes fluorescing in the NIR region." Near-Infrared Applications in Biotechnology: 35–93.

Toutchkine, A., Nalbant, P. and Hahn, K. M. (2002). "Facile synthesis of thiol-reactive Cy3 and Cy5 derivatives with enhanced water solubility." Bioconjugate Chem 13(3): 387-391.

Wehry, E. (1976). Modern Fluorescence Spectroscopy 2, Plenum Press, New York.

Winstead, A., Hijji, Y., Hart, K., Jasinskib, J. and Butcher, R. (2010). "1-(5-carboxypentyl)-2,3,3-Trimethyl)-3H-Indolium Bromide." Acta Cryst E66: o171–o172.

Winstead, A.J., Hart, K., Fleming, N. Toney, D., *Microwave Synthesis of Quaternary Ammonium Salts*. Molecules, 2008. 13: p. 2107-2113.

Winstead, A.J., Williams, R., Hart, K., Fleming, N., Kennedy, A., *Microwave Synthesis of Near Infrared Heptamethine Cyanine Dye*. The J. of Microwave Power and Electromagnetic Energy, 2008. 42: p. 42-1-35 – 42-1-41.

Part 3

Detection and Monitoring

12

Biosensor for the Determination of Biochemical Oxygen Demand in Rivers

Gab-Joo Chee

Department of Biochemical Engineering, Dongyang Mirae University
South Korea

1. Introduction

Population growth and industrialization have caused serious environmental pollution. Environmental pollution has become global issues beyond a region or a country, and is the introduction of contaminants, which are synthetic chemicals, pesticides, and heavy metals etc, into environmental system. Biopersistent organic chemicals of them particularly cause instability, disorder, harm or discomfort to ecosystem. Biopersistent organic chemicals are present as pollutants in wastewater effluents from industrial manufacturers or normal households. Environmental problems with such organic pollutants are becoming progressively worse all over the world. They are increasingly found in groundwater wells, rivers, lakes, and seas. Our drinking water sources, in particular, have also become polluted. A rapid and online monitoring of organic pollutants in water systems is a process that is essential for not only health of human but also environmental protection and ecosystem. Environmental pollutants in water systems have been evaluated as indicators including biochemical oxygen demand (BOD), chemical oxygen demand (COD), and total organic carbon (TOC) etc. Unlike chemical analyses such as COD and TOC, BOD is a method evaluated by the microbial ecosystem by the American Public Health Association (APHA) (APHA, 1986) because the indicator is based on the metabolic activity of aerobic microorganisms. Therefore, BOD analysis directly represents some influences of organic pollutants on natural ecosystems. In 1884, the modern concept of biochemical oxidation aroses, when it was showed that the decrease in the dissolved oxygen (DO) content of incubated samples was caused by metabolic activity of the microorganisms present (Leblanc, 1974).

The 5-day BOD (BOD_5) test method, however, requires an incubation period of 5 days under specified standard conditions described by the American Public Health Association Standard Methods Committee. This method is not desirable not only for process control and environmental monitoring, but also for remediation it produces a quick feedback. Thus, a fast and simple estimation of BOD is required as an alternative method to circumvent the disadvantages of the conventional test (BOD_5). A rapid and reliable method for BOD estimation was first developed by Karube et al. (Karube et al., 1977b). Since then many kinds of microbial sensors instead of BOD_5 analysis have been reported (Kulys & Kadziauskiene, 1980; Lin et al., 2006; Riedel et al., 1990; Sakaguchi et al., 2003; Stand & Carlson, 1984; Tanaka et al., 1994; Yang et al., 1997; Yoshida et al., 2000). They generally consist of microorganisms immobilized on a porous membrane and an oxygen probe. In addition, bioelements have been used for over 17 species of microorganisms, and transducers have adopted ether amperometric or optical oxygen probes. These microbial sensors have been developed for BOD analysis in industrial effluents, which contain high concentrations of organic pollutants. Solutions containing glucose and glutamic acid (GGA), which are adopted to the BOD_5 test method by the APHA, have been used as standard for the calibration of BOD sensors.

On the other hand, rivers, which are drinking water sources, generally contain biopersistent organic compounds such as humic acid, lignin, tannic acid, and gum arabic. In river waters in Japan, the BOD values are generally <10 ppm, and especially drinking water sources < 3 ppm (River Bureau, 1992). The microbial biosensors described above are not suitable to evaluate low BOD in drinking water sources. Therefore, the biosensor for the evaluation of low BOD in rivers has been developed and applied.

2. Constituents of BOD sensor systems

BOD sensor systems consist of a bioelement and a transducer. Bioelements utilized in BOD sensors have shown high respiration rate to organic compounds in industrial effluents and wastewaters. Bioelements are as follows: *Arxula adeninivorans* LS3 (Chan et al., 1999; Lehmann et al., 1999), *Bacillus subtilis* (Riedel et al., 1988), *Bacillus subtilis* and *Bacillus licheniformis* (Tan et al., 1992), *Citrobacter* sp. and *Enterobacter* sp. (Galindo et al., 1992), *Enterobacter cloacae* (Villalobos et al., 2010), *Clostridium butyricum* (Karube et al., 1977a), *Hansenula anomala* (Kulys & Kadziauskiene, 1980), *Klebsiella oxytoca* (Ohki et al., 1994), *Lipomyces kononenkoae* (Reiss et al., 1993), *Photobacterium phosphoreum* (Hyun et al., 1993), *Pseudomonas putida* (Li & Chu, 1991), *Rhodococcus erythropolis* and *Issatchenkia orientalis* (Riedel, 1998), *Saccharomyces cerevisiae* (Seo et al., 2009), *Serratia marcescens* LSY4 (Kim & Kwon, 1999), *Torulopsis candida* (Sangeetha et al., 1996), *Trichosporon cutaneum* (Hikuma et al., 1979; Karube et al., 1977b; Yang et al., 1996), activated sludge (Sakai et al., 1995), mixture of microorganisms (Stand & Carlson, 1984), multi-species culture (BODSEED) (Tan & Wu, 1999), and thermophilic bacteria (Karube et al., 1989). Transducers are typically two oxygen probes, which are a Clark oxygen electrode described as an electrical current and an oxygen optrode displayed as fluorescence intensity. A Clark oxygen electrode (Karube et al., 1977b; Riedel et al., 1988; Seo et al., 2009; Tan & Wu, 1999; Yang ct al., 1996) has been widely adopted in BOD sensors, but recently an oxygen optrode (Chee et al., 2000; Jiang et al., 2006; Kwok et al., 2005; Lin et al., 2006) is also often applied. An oxygen electrode consists of a platinum cathode, a plumbum anode, a PTFE membrane and an electrolyte solution. While an oxygen optrode contains a fluorescer and an optical fiber. The presence of oxygen, acting as a quencher, reduces fluorescence intensity of the indicators.

2.1 Media and bioelement

The composition of the medium was selected according to the analysis of the secondary effluents from the municipal sewage treatment plants in Japan (Murakami et al., 1978; Tanaka et al., 1994). The limited medium involved humic acid, lignin, tannic acid, gum arabic, and surfactant as carbon sources. These organic compounds could not be easy biodegradable by microorganisms. For the isolation of microorganisms utilizing refractory organic compounds, the medium had the following composition per liter of distilled water:

nitrohumic acid, 2 g; gum arabic, 2 g; sodium ligninsulfonate (NaLS), 2 g; tannic acid, 2 g; linear alkylbenzene sulfonate (LAS), 2 g; $NaNO_3$, 2.2 g; NaH_2PO_4, 0.2 g; Na_2HPO_4, 0.3 g; $MgSO_4$, 0.2 g; KCl, 0.04 g; $CaCl_2$, 0.02 g; yeast extract, 0.02 g; $FeSO_4 \cdot 7H_2O$, 1 mg; MoO_3, 10 µg; $CuSO_4 \cdot 5H_2O$, 5 µg; H_3BO_3, 10 µg; $MnSO_4 \cdot 5H_2O$, 10 µg; $ZnSO_4 \cdot 7H_2O$, 7 µg; the final pH being 7.0. For the solid medium, 1.0% agarose was added. A serum bottle containing 20 mL of the medium was cultured at 30°C with shaking at 170 rpm. Subculturing was carried out every five days, each transfer involving the addition of 1 mL of broth to 20 mL fresh medium.

Many soils, muds and activated sludge samples were collected and screened as to their ability to biodegrade humic acid and lignin etc. *Pseudomonas putida* strain SG10 as an optimal bioelement was isolated under aerobic conditions in the limited medium (Chee et al., 1999b).

2.2 Measuring principles
2.2.1 The BOD₅ test method

The 5-day BOD (BOD_5) test method has been adopted in 1936 by the American Public Health Association Standard Methods Committee (APHA, 1986). The BOD_5 test method, however, requires not only many complicated procedures, including a 5-day incubation, but also experience and skill to get reproducible results. BOD_5 is defined as the biochemical oxygen demand of wastewaters, effluents, and polluted waters measured over 5 days at 20°C. Dissolved oxygen (DO) is measured initially and after incubation, and the BOD is computed from the difference bewteen initial and final DO. The parameter is widely used for the determination of biodegradable organic pollutants in water systems.

2.2.2 Biosensor method

BOD sensors consist of a biofilm and an oxygen electrode. A biofilm is a suitable microorganism immobilized on a porous cellulose membrane. Generally, the BOD levels by a biosensor are estimated with the steady-state method. A BOD sensor is immersed into a buffer solution saturated with an air, and in a few minutes the current output of an oxygen probe becomes a steady state, because the diffusion rate of oxygen into a biofilm from the solution in bulk reaches equilibrium with the consumption rate of oxygen by endogenous respiration of the immobilized microorganism. The current output values coincident with the DO in the solution in bulk are called an "initial or base current". When a BOD solution is injected into a biosensor system, biodegradable organics diffuse into a biofilm from the solution in bulk. Then, in several minutes the current output reaches another constant current value which is smaller than the initial current, because the diffusion rate of oxygen into a biofilm reaches equilibrium with the enhanced respiration rate of the biofilm by increasing organics. The current output values are called a "peak current". A difference between the initial current and the peak current is proportional to a concentration of immediately biodegradable organics in a sample. From this difference, unknown substrate concentrations are estimated. The determination time is normally 15–20 min followed by 15–60 min recovery time.

2.3 Standard solutions

Currently, the BOD_5 test method adopts glucose and glutamic acid (GGA) as a standard check solution. A standard solution is a mixture of 150 mg glucose/L and 150 mg glutamic acid/L, and is similar to that obtained with many municipal waste waters. For the 300 mg/L mixed solution, the BOD_5 would be 220 mg/L with standard deviation of 10 mg/L (JIS, 1993). GGA solution is used as a standard solution for calibration of most of previously reported BOD sensors. However, the components of river waters and secondary effluents differ greatly from GGA solution. River waters and effluents usually contain refractory organic compounds, such as humic acid, gum arabic, lignin, tannic acid, and surfactant. Especially, these organic compounds account for over 50% in compositions of secondary effluents according to the publication by Tanaka (Tanaka et al., 1994). Therefore, arbitrarily selected refractory organic compounds, as reported in previously papers (Murakami et al., 1978; Tanaka et al., 1994), were studied. The constituents of artificial wastewater (AWW) per liter of distilled water are as follows: nitrohumic acid, 4.246 mg; gum arabic, 4.696 mg; NaLS, 2.427 mg; tannic acid, 4.175 mg; LAS, 0.942 mg. In this study, The AWW is used as a standard solution for calibration of the *Pseudomonas putida* SG10 BOD sensor. The AWW solution is 3.7 mg/L of BOD_5, and 5.89 mg/L of COD_{MN}, respectively.

2.4 Biofilm

Cells in the stationary phase of growth were harvested by centrifugation at 6000 rpm for 10 min, washed twice with 50 mL of 10 mM phosphate buffer (pH 7.0), and were subsequently resuspended in the same buffer. The biofilm was prepared using an aspirator connected to a syringe filter holder (Advantec, Japan). Calculated amounts (wet cells 40 mg, OD_{660} = 1.7) of the pure culture broth were dropped on a porous cellulose nitrate membrane (20 mm diameter, 0.45 µm pore size, Advantec, Japan). Microorganisms were adsorbed on the membrane by suction, and then another similar membrane was placed on immobilized microorganism membrane and was re-adsorbed by using the equipment above, *i. e.* microorganisms were sandwiched between two porous membranes. The microorganism membrane was washed with 10 mM phosphate buffer. The biofilm was placed on an oxygen electrode, and fixed in place using 200 mesh nylon and an 'O'-ring.

3. Performance of the BOD sensors

3.1 Amperometic microbial BOD sensor

3.1.1 Chariterization and response

The oxygen electrode with the biofilm of *P. Putida* SG10 was inserted into the detection chamber containing 50 mL of 10 mM phosphate buffer saturated with air, while continuously stirring with a magnetic bar. The temperature of the detection chamber was maintained at 30°C using a constant temperature water bath. The current output of the oxygen electrode was measured using a digital multimeter (Model TR6840, TakedaRiken, Japan) and an electronic poly recorder (Model EPR-200A, TOA Electronics, Japan).

The values and linear correlation are shown in Fig. 1. Calibration was performed using the response data at the steady state. A linear relationship was observed ranging 0.5 to 10 mg/L BOD. The dectection limit was 0.5 mg/L BOD. The response time of the biosensor depended on the BOD level, taking between 2 and 15 min to reach the steady state. For BOD of 1 mg/L, reproducible responses could be obtained within ±10% relative error of the mean value; standard deviation was 0.078 mg/L (n=5). On the other hand, *Trichosporon cutaneum* typically employed in BOD sensors did not grow in AWW culture (data not shown).

The influences of pH and temperature on the response were investigated for BOD of 1 mg/L in 10 mM phosphate buffer. The optimum pH was investigated from pH 4.0 to 9.0. The response rapidly increased to give a maximum at pH 7.0, and then decreased (Fig. 2). It will be caused by the inactivation of *P. putida* at ether lower or higher pH values. The optimum temperature was evuluted in the range of 5 to 40°C, and found to be a maximum at 35°C (Fig. 3). Above 35°C, the response decreased slightly, which is probably also caused by the inactivation of the microorganism by heat. In order to prolong the lifetime of the bacteria, a temperature of 30°C was used in the biosensor.

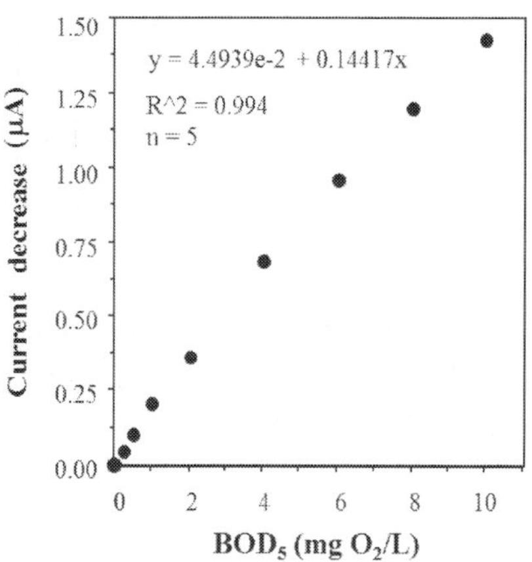

Fig. 1. Correlation between BOD concentration and current decrease using the AWW solution under the optimal conditions.

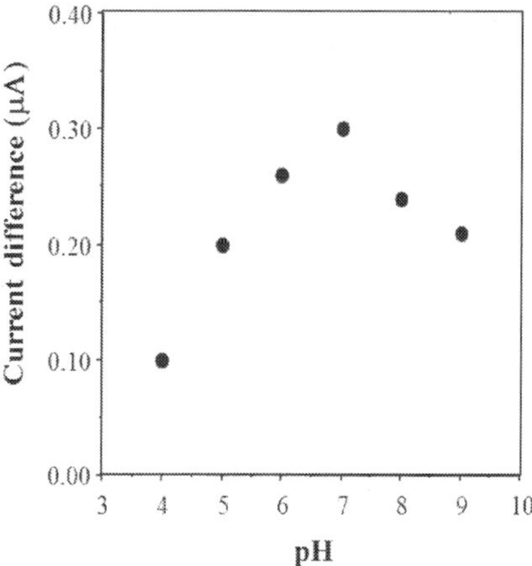

Fig. 2. Influence of pH on the sensor response at 30°C and AWW (1 mg/L BOD).

The stability of the biosensor was determined over 15 min after it had been immersed in the AWW solution (BOD of 1 mg/L), at pH 7.0 and 30°C. The biosensor response was found to be fairly constant over a period of 10 days, with about ±10% fluctuations. The storage stability of the biosensor was examined at 4°C. The biosensor could be stored in buffer solution for 5 months without significant deterioration, but the biosensor had to be pre-conditioned for 1-2 days in the AWW solution (BOD of 1 mg/L) before use, in order to achieve good sensitivity, stability, and reproducibility.

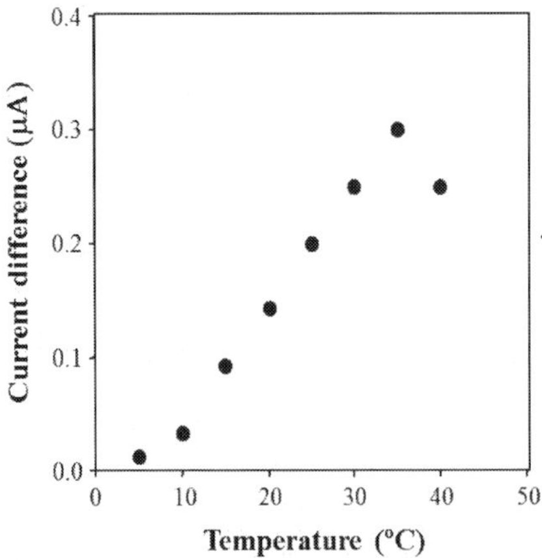

Fig. 3. Influence of temperature on the sensor response at pH 7.0 and AWW (1 mg/L BOD).

3.1.2 Interference on response

The influences of chloride ion and heavy metal ions on the biosensor response were examined in 10 mM phosphate buffer (pH 7.0), at BOD of 1 mg/L. Here, NaCl was used as the chloride ion source. The biosensor response was not dramatically affected by chloride ion concentration of up to 1000 mg/L. Therefore, the biosensor could be applied to analyze environmetal samples of high sodium chloride concentrations.

Most river waters contain various heavy metal ions like Fe^{3+}, Cu^{2+}, Mn^{2+}, Zn^{2+}, and Cr^{3+}, and the presence of these heavy metal ions in river waters may interfere with the activity of the microorganisms (Collins & Stotzky, 1989). The influence of heavy metal ions on the biosensor response was investigated at each 1 mg/L, because the highest concentration of these ions in polluted Japanese rivers should be below 1 mg/L (River Bureau, 1992). The results revealed that Fe^{3+}, Cu^{2+}, Mn^{2+}, Zn^{2+}, and Cr^{3+} have no effects on the response of the biosensor.

3.1.3 Application

The BOD sensor was used to determine the BOD of various river waters in Japan, and the values obtained were compared with the BOD_5 method. As shown in Figure 4, the results obtained by the biosensor are generally somewhat lower than the values obtained by the BOD_5 method. This behavior is attributable to the presence of compounds which are not easily assimilable to the biosensor in such a short time.

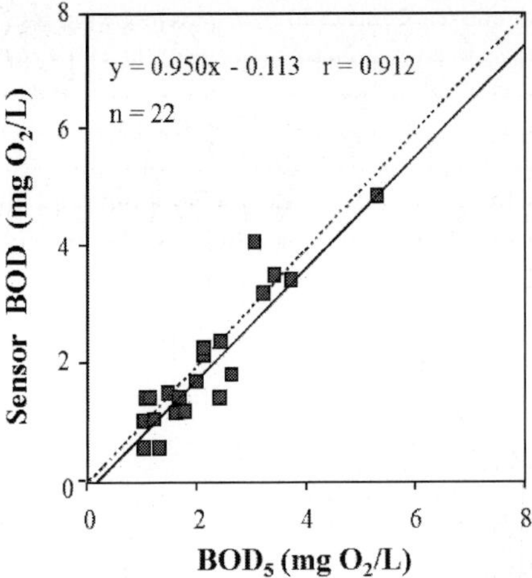

Fig. 4. Comparison of BOD values estimated by the sensor with those determined by the 5-day method for various river waters.

3.2 Highly sensitive BOD sensors

The BOD sensor showed the very good response in the range of AWW up to 10 mg/L. The biosensor with *P. putida* SG10 was used to evaluate BOD values of various enviromental samples. The biosensor could determine low BOD values, 1–10 mg/L in river waters. In analyses of river waters, however, this biosensor often dispalyed low values compared with the BOD_5 as well as other BOD sensors (Hikuma et al., 1979; Hyun et al., 1993; Ohki et al., 1994; Yang et al., 1996). The results would show that refractory organics in river waters are uneasily assimilable to the biofilm in such a short measuring time. To overcome this problem, pretreatment by photocatalytic oxidation or ozonation was introduced in the biosensor system (Chee et al., 1999a, 2001, 2005; Chee et al., 2007).

3.2.1 Photocatalytic BOD sensor

In 1972, Fujishima and Honda discorved the photocatalytic splitting of water which could be decomposed into hydrogen and oxygen over an illuminated titanium dioxide semiconductor electrode (Fujishima & Honda, 1972). With the help of this event, photoelectrochemistry has expanded into a formidable field encompassing solar energy conversion (Bard, 1982), photocatalysis (Fujihara et al., 1981; Hoffmann et al., 1995; Linsebigler et al., 1995), decomposition of agrochemical (Lu & Chen, 1997), air and water purification (Bolduc & Anderson, 1997; Rodriguez et al., 1996). In the photochemical oxidation method, short wavelength UV-C (< 280 nm) light is commonly employed (Prousek, 1996). On the other hand, photocatalytic experiments with titanium dioxide usually use long wavelength UV-A (> 315 nm) light (Bahnemann et al., 1991; Egerton & King, 1979; Hashimoto et al., 1984). Figure 5 shows a general schematic representation of the degradation of organic compound in aqueous solution by electron-hole (e⁻-h⁺) formation at the surface of an illuminated titanium dioxide particle (Rajeshwar, 1995). When titanium dioxide is illuminated with band gap energy of greater than 3.2 eV (380 nm), a photon excites an electron from the valence band (VB) to the conduction band (CB) and leaves an electronic vacancy commonly referred to as a hole in the VB. The electron in the CB can be transferred to adsorbed H^+, O_2 or the chlorinated pollutant initiating various reactions. The hole in the VB can react with surface-bound water, hydroxide groups, anions and organic substrate. Therefore organic compounds in aqueous solution are split to the formation of compounds of lower molecular weight by photocatalytic oxidation. The result will increase the sensitivity of the BOD sensor because organic compounds of lower molecular weight are more readily biodegradable through the biofilm.

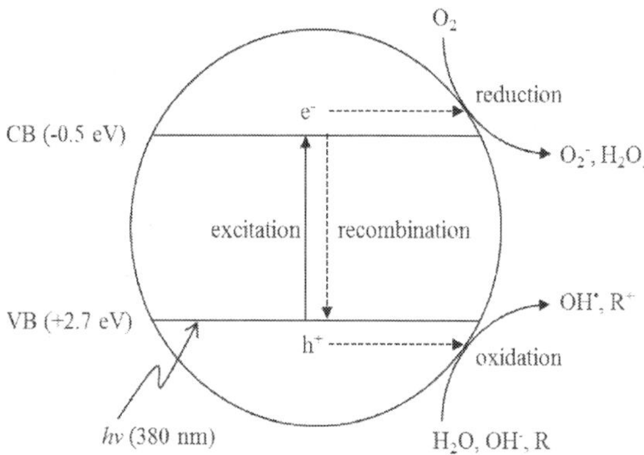

Fig. 5. Schematic diagram of simplified mechanism for the photoactivation of a titanium dioxide particle.

3.2.1.1 Flow system with semiconductor photocatalysis and response

The flow system was developed to monitor continuously in river waters, as schematically illustrated in Figure 6. The TiO_2-AWW solutions were illuminated from 0 to 5 min by black-light tube at room temperature. The slop of current was a maximum at irradiation time 4 min.

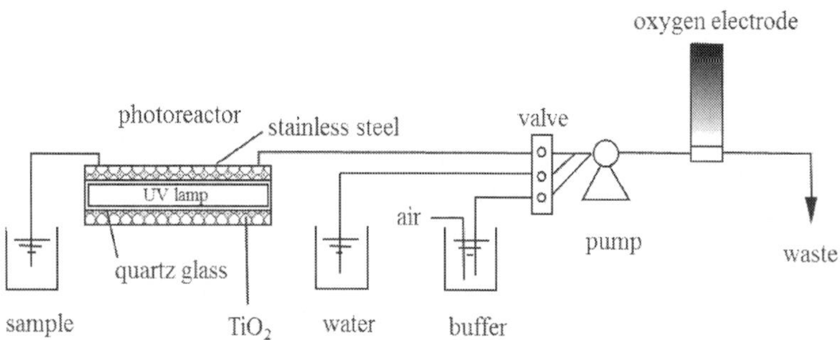

Fig. 6. The photocatalytic BOD sensor of the flow system using pretreatment by photocatalysis. A column size: ID 22 x OD 34 x L 205 mm, total volume: 63 mL, amount of TiO_2: 47 g, a UV lamp: a 6W black, flow rate: 3 mL/min.

The flow rate was 3 mL/min (Chee et al., 2005). Figure 7 showed the comparison of the biosensor responses with and without photocatalysis. At BOD of 1 mg/L, the biosensor responses obtained with and without photocatalysis were 0.20 and 0.15 µA, respectively. The slope with photocatalysis, up to 10 mg/L BOD, increased 1.39 fold that without photocatalysis. However, the biosensor response to BOD of 0.5 mg/L was hardly difference in bewteen with and without photocatalysis. The results would indicate that organic compounds in AWW solution would be adsorbed on surface of TiO_2 particle when the solution flowed into a photoreactor, and so some degraded organic compounds may stream to a cell on the biofilm. Consequently, the biosensor would give the low response, whereas, over 1 mg/L BOD, the solution from a photoreactor outlet after photocatalysis was fully to give high response to the biosensor. Relative standard deviations with and without photocatalysis at BOD of 2 mg/L were all both 12.0% (n=5).

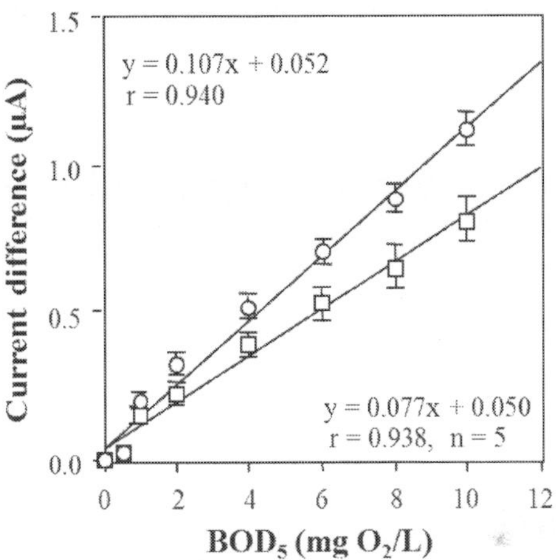

Fig. 7. Comparison of the sensor responses with and without photocatalytic oxidation using AWW in the flow system. □: without photocatalysis, ○: with photocatalysis.

The effect of the initial pH on the degradation of the photocatalytic AWW solution was examined at four different pH values between pH 5.0-8.0. The effectivity of the photocatalytic degradation is characterized by the current output. The current output is hardly changed to pH 6.0, but increased with rising pH over pH 6.0. It is well known that the surfaces of metal oxides in aqueous solution are covered with hydroxyl groups (Stumm & Morgan, 1981). Surface groups of a metal oxide are amphoteric and the surface acid-base equilibria are known as follows:

$$\equiv TiOH_2^+ \longleftrightarrow \equiv TiOH + H^+$$

$$\equiv TiOH \longleftrightarrow \equiv TiO^- + H^+$$

where $\equiv TiOH$ represents the "titanol" surface group. The neutral surface species, TiOH is predominant over a broad range of pH 3 to 10. At below the pH of zero point of charge, pH_{zpc}, the TiO_2 surface becomes a net positive charge because of the increasing fraction of total surface sites present as $\equiv TiOH_2^+$. On the other hand, at above pH_{zpc} the surface has a net negative charge because of a significant fraction of total surface sites present as $\equiv TiO^-$. The interaction with cationic electron donors and electron acceptors will be favored for heterogeneous photocatalytic activity at high pH ($> pH_{zpc}$), while anionic electron donors and acceptors will be favored at low pH ($< pH_{zpc}$). AWW solution is mixing compounds. The compounds may be significant photocatalytic activity at high pH. The organic compounds would be also decomposed by hydroxyl radicals which are very strong oxidant. The hydroxyl radicals mainly yield with hydroxide ions in the reaction of h^+ of the VB (Turchi & Ollis, 1990).

3.2.1.2 Lifetime of TiO$_2$

Lifetime of TiO$_2$ was investigated using AWW solution, BOD of 10 mg/L by TOC analyzer. As shown in Figure 8, TOC dramatically decreased until 40 times, and had 50% of the initiation at about 70 times. With 40 times or more, a decrease of TOC/TOC$_0$ with UV irradiation was observed. Because organic compounds streaming through a photoreactor may be quickly adsorbed by diffusing on the surface of TiO$_2$ particles (Bandala et al., 2002; Robert & Weber, 2000). The ratio of TOC/TOC$_0$ without UV irradiation also decreased 10-15%. Consequently, the degradation of organic compounds fell on the surface of TiO$_2$ with photocatalysis. Over 70 times, especially, the ratio of TOC/TOC$_0$ extremely decreased. It suggests that the surface of titanium dioxide particles might be saturate with organic compounds by repetition of the sample introduction. The detection time was 20 min, and sequent determination was carried out without washing. Determining real samples, half-lifetime of TiO$_2$ would be predicted longer because of a repetition of determining and washing.

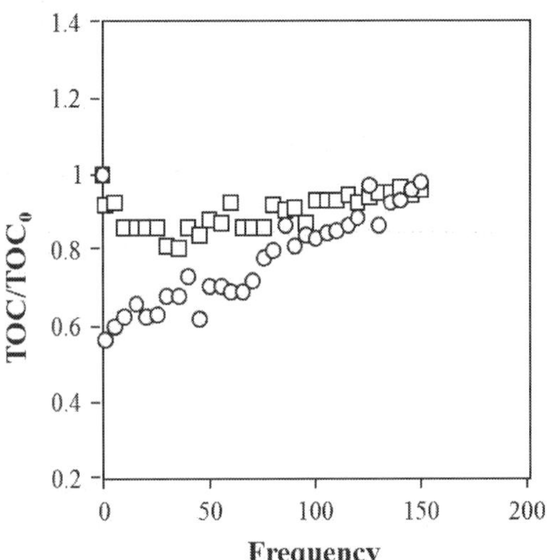

Fig. 8. Photocatalytic oxidation rates of TiO$_2$ using AWW in the flow system. TOC$_0$: AWW non-through a photoreactor, TOC: AWW through a photoreactor, □: without irradiation, ○: with irradiation (*i. e.* pretreatment by potocatalysis.).

3.2.1.3 Effects of free radicals and H_2O_2

The yield of hydrogen peroxide with photocatalysis was investigated in AWW solution (10 mg/L) that passed through a photoreactor. Organic compounds react on the surface of TiO_2 under UV irradiation, and many free radicals and hydrogen peroxide are produced. These were toxicity to microorganisms (Bilinski, 1991; Brandi et al., 1989a). Microorganisms exposed under hydrogen peroxide and/or oxy-radicals were killed or changed its morphology. Half-lifetime of hydroxyl radical, $HO^\bullet$ and superoxide, $O_2^{\bullet-}$ were 10^{-9} s at 1M and 2.5 s at 1 µM, respectively, while hydrogen peroxide was a relatively stable molecule in water (Pryor, 1986).

H_2O_2 yielded using AWW solution (10 mg/L) through a photoreactor under irradiating UV was quantitatively analyzed by spectrofluorometer (JASCO Co. Ltd, FP-770F, Japan). Excitation and emission of scopoletin were 366 and 460 nm, respectively. H_2O_2 of 3.56 µM was yielded in AWW solution that passed through a photoreactor under UV irradiation. H_2O_2 of 1.75 mM was approximately equitoxic in bacteria, as a previously described paper (Brandi et al., 1989b). Although bacteria are exposed to active oxygen or other radicals, they have evolved mechanisms of defense against oxidative stresses that can damage most cellular components, including proteins, lipids, and DNA. To overcome such oxidative stresses, bacteria have genes such as *soxR*, *soxS* for superoxide radical, and *oxyR*, *dps* for H_2O_2 and organic peroxides (Nair & Finkel, 2004). Dps protein, especially, can neutralize toxic peroxides through its ferroxidase activity. Bacteria also can repair oxidatively damaged DNA using genes such as *recA*, *polA*, and *xthA*. Accordingly, bacteria would be not affected in stress environmental of very low concentrations. Therefore, the concentration would not give the fluctuation in the sensor response. As shown in Figure 6, a photoreactor to the biofilm on the oxygen electrode takes 3 min when the flow rate was 3 mL/min. Other free radicals did not give the influences on the sensor response. The results showed that free radicals will not affect microorganisms in the biofilm on the oxygen electrode because their lifetime was extremely short.

3.2.2 Ozone catalytic BOD sensor

For the elimination of refractory organics from industrial wastewater and municipal effluents, ozone pretreatment has been studied by numerous investigators (Gulyas, 1997; Perkowski et al., 1996; Unkroth et al., 1997). Ozonolysis has been widely applied either to eliminate or to decompose refrectory organic compounds in industrial wastewater and municipal effluents (Gulyas et al., 1995; Widsten et al., 2004). During self-decomposition of ozone in aqueous solutions, free radicals yielded are used as powerful oxidants to cleave organic compounds in environmental samples (Staehelin & Holgné, 1982). The ozonation of organic compounds in aqueous solutions generates lower molecular weight than the parent compounds, and the decomposed organic compounds would assimilate faster into microbes immobilized on the membrane. The result will increase the sensitivity of the biosensor. Here, for higher sensitivity of the biosensor, ozone pretreatment was introduced in the biosensor (Chee et al., 1999a; Chee et al., 2007).

3.2.2.1 Stopped-flow system with ozonolysis and response

The stopped-flow BOD sensor consists of an ozonizer (ON-12, Nippon ozone, Japan), an oxygen probe (Able Co., Japan) with the biofilm, a digital meter (Model TR6840, TakedaRiken, Japan) and an electronic recorder (Model EPR-200A, TOA Electronics, Japan) (Fig. 9).

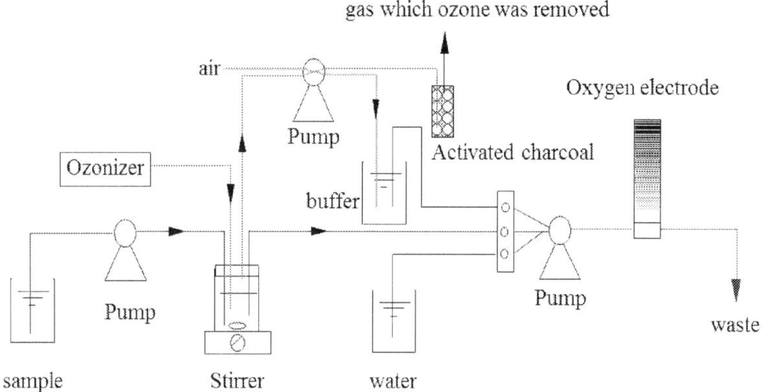

Fig. 9. Schematic diagram of the stopped-flow system of the BOD sensor using pretreatment with ozone.

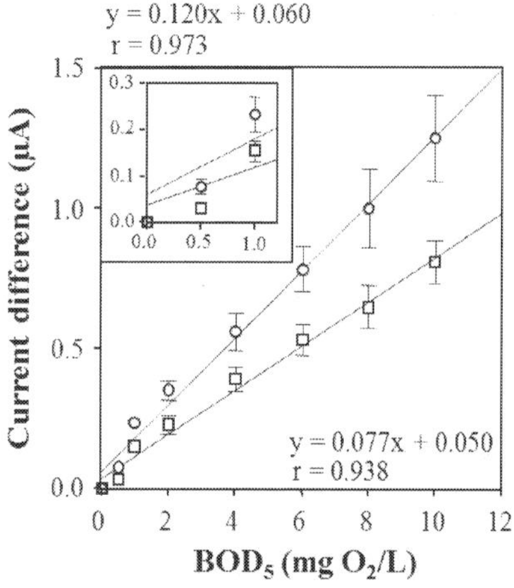

Fig. 10. The sensor responses before and after ozonation in the stopped-flow system under the optimal conditions. □: before ozonation, ○: after ozonation.

An ozonizer was installed to pretreat environmental samples. The biosensor responses before and after ozonation on AWW solutions were investigated in the range of 0 to 10 mg/L BOD (Fig. 10). At BOD of 1 mg/L, the biosensor responses obtained before and after ozonation were 0.15 and 0.24 µA, respectively. As shown in the window in Figure 10, the response before ozonation could be detected to 0.5 mg/L BOD, but the response had low reproducibility. On the other hand, the response after ozonation was 2.7-fold higher than that before; moreover, its reproducibility was increased. Up to 10 mg/L BOD, the slope of the biosensor responses after ozonation was 1.56-fold that before ozonation. Relative standard deviations with and without pretreatment at BOD of 2 mg/L were both 12.0%

(n = 5). The response time of the sensor was generally varied depending on the concentrations of BOD in the samples. In the range of the determined concentrations of BOD, however, the response time did not exceed 5 min. This indicates that the organic compounds in AWW solution degraded with ozone were easily assimilated by the microorganism on the biofilm. When the samples were treated with ozone in a pretreatment reactor, before pumping the samples into the flow cell with the biofilm, they were vigorously stirred using a magnetic bar over 20 min to completely remove the excess ozone.

3.2.2.2 Effect of ozonation time

Figure 11 showed the changes of TOC removal rates and pH with ozonation time in the range of 0 to 10 min in AWW solutions. The ozonation of AWW solutions was carried out by 25.9 g N^{-1} m^{-3} ozone at room temperature. TOC removal rates were defined as the ratio of TOC values before and after ozonation. TOC removal rates per time unit showed a maximum value at 3 min, and then slightly decreased. Results from Figure 11 indicate that there is an optimal ozonation time to degrade the maximum amount of AWW solution. Therefore, in subsequent experiments, ozonation times of the samples were adjusted to 3 min. During ozonation, a decrease in pH and a increase in variation followed by the steady state was observed, indicating the formation of ionic substances. Decolorization of the pale yellow AWW solution occurred after 3 minutes ozonation, and clear AWW solution was obtained in 10 minutes.

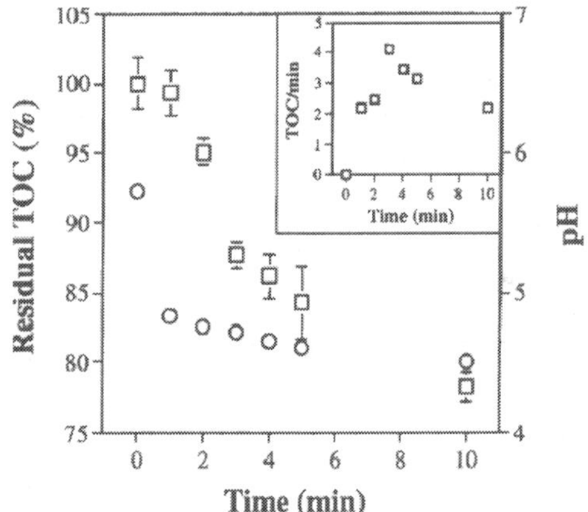

Fig. 11. Effect of ozonation time on pH and TOC removal rates in AWW solutions. AWW: 16.485 mg/L, concentration of ozone in the feed gas: 25.9 g N^{-1} m^{-3}, temperature: room temperature, □: TOC, ○: pH. Error bars represent the standard deviation of 6 expreriments.

3.2.2.3 Effect of pH and ozone concentration

The effects of initial pH on the ozonation rate of AWW solutions were investigated ranging 5.0 to 9.0. Owing to the strong effect of hydroxide ions on self-decomposition of ozone (Staehelin & Holgné, 1982), pH would be the most important parameter of ozonation process. TOC removal rates increased while rising initial pH. TOC removal rates at initial pH 7.0 were ca. 10 %, this value being twice that at pH 5.0. The ozonation rates are larger at higher pH values (Staehelin & Holgné, 1982), as predicted by the initial reaction of ozonation (Weiss, 1935):

$$O_3 + OH^- \quad \text{-------} \quad HO_2^{\bullet} + O_2^{\bullet -}$$

pH 7.0 was selected as working pH for further experiments. The reacting organic compounds can have different reactivities at different pH values because of their dissociation and electron activities. The reaction between ozone and organic compounds in aqueous solutions is known to be dependent on the pH value of the solution (Nadezhdin, 1988; Staehelin & Holgné, 1982). Ozone decomposed relatively well in the high pH region, where molecular ozone selectively reacts with unsaturated bonds in the molecule of organic compounds to produce ketones, carboxylic acid, alcohols, etc. Various free radicals yield from self-decomposition of ozone, like semiconductor photocatalysis, in aqueous solutions. It is well known that free radicals are toxic to bacteria since they can change the morphology of bacteria and damage most cellular components containing DNA, proteins, and lipids (Bilinski, 1991; Brandi et al., 1989a). In this work, however, bacteria on the biofilm will not be influenced by free radicals because only samples removed under controlled conditions of the ozone were pumped into the flow cell with the biofilm. Even if free radicals were residual in the samples pretreated with ozone, their functions immediately terminated due to their extremely short lifetimes, e.g., 10^{-9} sec for half-life of hydroxyl radical (Pryor, 1986), during the transfer of the samples from the pretreatment reactor to the flow cell with the biofilm.

Figure 12 showed the effect of ozone concentration on TOC removal rates. The feed ozone dose was investigated from 0 to 51.5 g N^{-1} m^{-3}, and the AWW solutions were adjusted to pH 7.0. As shown in Figure 12, TOC removal rates rapidly increased until reaching a plateau for 42.4 g N^{-1} m^{-3} ozone. TOC removal rates to 42.4 and 51.5 g N^{-1} m^{-3} ozone had values of about 17% and 18%, respectively.

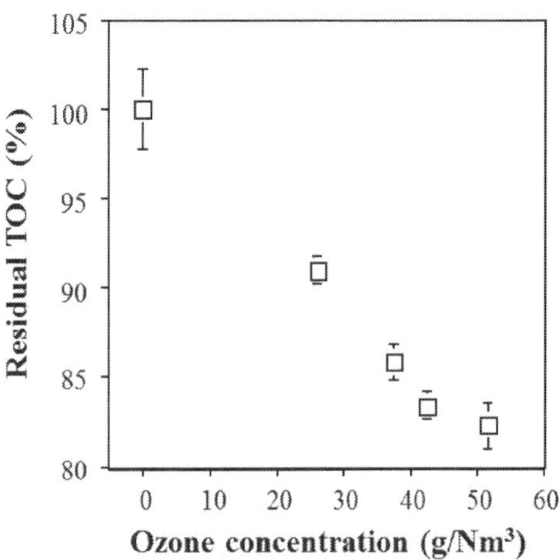

Fig. 12. Effect of ozone concentration on TOC removal in AWW solutions. Pretreatment time: 3 min, pH: 7.0, temperature: room temperature.

3.2.3 Application

The photocatalytic biosensor and stopped-flow system were used to evaluate BOD in environmental samples from various rivers before and after pretreatment. Environmental samples for the test were collected from 21 various rivers that were drinking water sources. The BOD levels obtained before and after pretreatment were compared with the conventional BOD_5 method. The BOD values evaluated without pretreatment showed lower than those obtained by the BOD_5, while the BOD values estimated with photocatalysis or ozonalysis were either a little low or the same to the BOD_5. The slope and correlation (r) between the photocatalytic biosensor and the conventional method were 0.908 and 0.983, respectively. The BOD levels obtained before and after ozonation were also compared with the BOD_5 method (Fig. 13). The slopes of the biosensor were 0.849 before ozonation and 0.933 after ozonation, respectively. The slope with pretreatment increased to approximately 17% in comparison with that without pretreatment. The correlation factor (r) between the stopped-flow system and the BOD_5 method was 0.989. The results indicate that the biosensors using pretreatments improved the estimation of BOD in environmental samples.

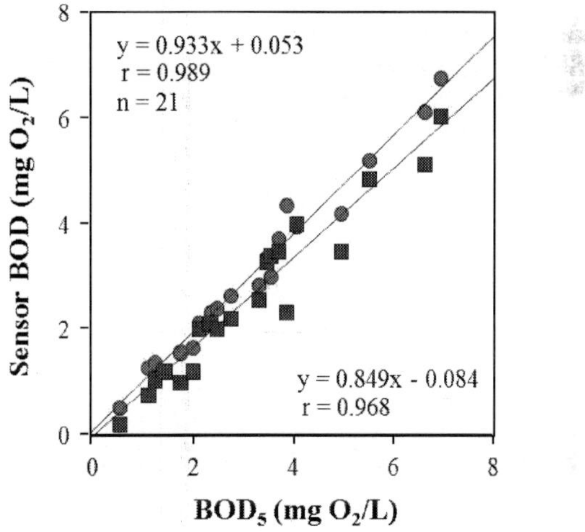

Fig. 13. Comparison of BOD values in river waters by with and without ozonation in the stopped-flow system. ■: without ozonation, ●: with ozonation.

4. Conclusions

The BOD sensor using *Trichosporon cutaneum* can be used in industrial wastewater, but not for river waters and secondary effluents due to no growth of *T. cutaneum* in AWW culture. The BOD sensor using *P. putida* SG10 was described to be suitable for determining low BOD values in river waters. BOD measurements could be determined at pH 7.0, and 30°C. The biosensor responses were observed a linear relationship in the range of 0.5 to 10 mg/L BOD. The detection limit and time were 0.5 mg/L BOD and less than 15 min, respectively, compared with the BOD_5 test method. Especially, the response time by ozonalysis was only 5 min. The biosensor system showed negligible response to chloride and heavy metal ions, and had good sensitivity, stability and reproducibility. Pretreatment by photocatalysis or ozonalysis was introduced to increase the sensitibity of the biosensor. When determining environmental samples by pretrements, the biosensor responses showed the increased levels, and the systems must be the advanced method for the evaluation of low BOD in rivers.

5. References

APHA. (1986). *Standard Methods for the Examination of Waters and Wastewater*, 16th ed. American Public Health Association, Washington, DC, pp. 525-531.

Bahnemann, D., Bockelmann, D. & Goslich, R. (1991). Mechanistic studies of water detoxification in illuminated TiO_2 suspensions, *Solar Energy Materials* Vol. 24: 564-583.

Bandala, E.R., Gelover, S., Leal, M.T., Arancibia-Bulnes, C., Jimenez, A. & Estrada, C.A. (2002). Solar photocatalytic degradation of aldrin, *Ctalysis Today* Vol. 76: 189-199.

Bard, A.J. (1982). Design of semiconductor photoelectrochemical system for solar energy conversion, *J. Phys. Chem.* Vol. 86: 172-177.

Bilinski, T. (1991). Oxygen toxicity and microbial evolution, *BioSystems* Vol. 24: 305-312.

Bolduc, L. & Anderson, W.A. (1997). Enhancement of the biodegradability of model wastewater containing recalcitrant or inhibitory chemical compounds by photocatalytic pre-oxidation, *Biodegradation* Vol. 8: 237-249.

Brandi, G., Cattabeni, F., Albano, A. & Cantoni, O. (1989a). Role of hydroxyl radicals in *Escherichica coli* killing induced by hydrogen peroxide, *Free Rad. Res. Comms.* Vol. 6: 47-55.

Brandi, G., Fiorani, M., Pierotti, C., Albano, A., Cattabeni, F. & Cantoni, O. (1989b). Morphological changes in *Escherichica coli* cells exposed to low or high concentrations of hydrogen peroxide, *Microbiol. Immunol.* Vol. 33: 991-1000.

Chan, C., Lehmann, M., Tag, K., Lung, M., Kunze, G., Riedel, K., Gruendig, B. & Renneberg, R. (1999). Measurement of biodegradable substances using the salt-tolerant yeast *Arxula adeninivorans* for a microbial sensor immobilized with poly (carbamoyl) sulfonate (PCS): Part I: construction and characterization of the microbial sensor, *Bioses. Bioelectron.* Vol. 14: 131-138.

Chee, G.-J., Nomura, Y., Ikebukuro, K. & Karube, I. (1999a). Development of highly sensitive BOD sensor and its evaluation using preozonation, *Anal. Chimi. Acta* Vol. 394: 65-71.

Chee, G.-J., Nomura, Y., Ikebukuro, K. & Karube, I. (2001). Biosensor for the evaluation of biochemical oxygen demand using photocatalytic pretreatment, *Sensors and Actuators B* Vol. 80: 15-20.

Chee, G.-J., Nomura, Y., Ikebukuro, K. & Karube, I. (2005). Development of photocatalytic biosensor for the evaluation of biochemical oxygen demand, *Biosens. Bioelectron.* Vol. 21: 67-73.

Chee, G.-J., Nomura, Y. & Karube, I. (1999b). Biosensor for the estimation of low biochemical oxygen demand, *Anal. Chim. Acta* Vol. 379: 185-191.

Chee, G.J., Nomura, Y., Ikebukuro, K. & Karube, I. (2000). Optical fiber biosensor for the determination of low biochemical oxygen demand, *Biosens. Bioelectron.* Vol. 15: 371-376.

Chee, G.J., Nomura, Y., Ikebukuro, K. & Karube, I. (2007). Stopped-flow system with ozonizer for the estimation of low biochemical oxygen demand in environmental samples, *Biosens. Bioelectron.* Vol. 22: 3092-3098.

Collins, Y.E. & Stotzky, G. (1989). Factors affecting the toxicity of heavy metals to microbes, In: *Metal ions and Bacteria*, Beveridge, T.J. & Doyle, R.J., (eds.), pp. 31-90, John Wiley & Sons, New York.

Egerton, T.A. & King, C.J. (1979). The influence of light intensity on photoactivity in TiO_2 pigmented systems, *J. Oil. Chem. Assoc.* Vol. 62: 386-391.

Fujihara, M., Satoh, Y. & Osa, T. (1981). Heterogeneous photocatalytic oxidation of aromatic compounds on TiO_2, *Nature* Vol. 293: 206-208.

Fujishima, A. & Honda, K. (1972). Electrochemical photolysis of water at a semiconductor electrode, *Nature* Vol. 238: 37-38.

Galindo, E., Garcia, J.L., Torres, L.G. & Quintero, R. (1992). Characterization of microbial membranes used for the estimation of biochemical oxygen demand with a biosensor, *Biotechnol. Tech.* Vol. 6: 399-404.

Gulyas, H. (1997). Processes for the Removal of Recalcitrant Organics from Industrial Wastewaters, *Wat. Sci. Tech.* Vol. 36: 9-16.

Gulyas, H., Bismarck, R.V. & Hemmerling, L. (1995). Treatment of Industrial Wastewater with Ozone/Hydrogen Peroxide, *Wat. Sci. Tech.* Vol. 32: 127-134.

Hashimoto, K., Kawai, T. & Sakata, T. (1984). Photocatalytic reactoins of hydrocarbons and fossil fuels with water: Hydrogen production and oxidation, *J. Phys. Chem.* Vol. 88: 4083-4088.

Hikuma, M., Suzuki, H., Yasuda, T., Karube, I. & Suzuki, S. (1979). Amperometric estimation of BOD by using living immobilized yeasts, *Eur. J. Appl. Microb. Biotechnol.* Vol. 8: 289-297.

Hoffmann, M.R., Martin, S.T., Choi, W. & Bahnemann, D.W. (1995). Environmental application of semiconductor photocatalysis, *Chem. Rev.* Vol. 95: 69-96.

Hyun, C.K., Tamiya, E., Takeuchi, T., Karube, I. & Inoue, N. (1993). A novel BOD sensor based on bacterial luminescence, *Biotechnol. Bioeng.* Vol. 41: 1107- 1111.

Jiang, Y., Xiao, L.L., Zhao, L., Chen, X., Wang, X. & Wong, K.Y. (2006). Optical biosensor for the determination of BOD in seawater, *Talanta* Vol. 70: 97-103.

JIS. (1993). *Japanese Industrial Standard Committee Testing Methods for Industrial Waste Water*, pp. 48-53 [in Japanese].

Karube, I., Matsunaga, T. & Suzuki, S. (1977a). A new microbial electrode for BOD estimation, *J. Solid-Phase Biochem.* Vol. 2: 97-104.

Karube, I., Mitsuda, S., Matsunaga, T. & Suzuki, S. (1977b). A Rapid Method for Estimation of BOD by using Immobilized Microbial Cells, *J. Ferment. Technol.* Vol. 55: 243-248.

Karube, I., Yokoyama, K., Sode, K. & Tamiya, E. (1989). Microbial BOD sensor utilizing thermophilic bacteria, *Anal. Lett.* Vol. 22: 791-801.

Kim, M.-N. & Kwon, H.-S. (1999). Biochemical oxygen demand sensor using *Serratia marcescens* LSY 4, *Biosens. Bioelectron.* Vol. 14: 1-7.

Kulys, J. & Kadziauskiene, K. (1980). Yeast BOD sensor, *Biotechnol. Bioeng.* Vol. 22: 221-226.

Kwok, N.-Y., Dong, S., Lo, W. & Wong, K.-Y. (2005). An optical biosensor for multi-sample determination of biochemical oxygen demand (BOD), *Sensors and Actuators B* Vol. 110 (2): 289-298.

Leblanc, P.J. (1974). Review of rapid BOD test methods, *J. WPCF* Vol. 46. 9: 2202-2208.

Lehmann, M., Chan, C., Lo, A., Lung, M., Tag, K., Kunze, G., Riedel, K., Gruendig, B. & Renneberg, R. (1999). Measurement of biodegradable substances using the salt-tolerant yeast *Arxula adeninivorans* for a microbial sensor immobilized with poly (carbamoyl) sulfonate (PCS): Part II: application of the novel biosensor to real samples from coastal and island regions, *Biosens. Bioelectron.* Vol. 14: 295-302.

Li, Y.-R. & Chu, J. (1991). Study of BOD microbial sensors for waste water treatment control, *Appl. Biochem. Biotechnol.* Vol. 28: 855-863.

Lin, L., Xiao, L.L., Huang, S., Zhao, L., Cui, J.S., Wang, X.H. & Chen, X. (2006). Novel BOD optical fiber biosensor based on co-immobilized microorganisms in ormosils matrix, *Biosens. Bioelectron.* Vol. 21: 1703-1709.

Linsebigler, A.L., Lu, G. & J. T. Yates, J. (1995). Photocatalysis on TiO_2 surfaces: principles, mechanism, and selected results, *Chem. Rev.* Vol. 95: 735-758.

Lu, M.-C. & Chen, J.-N. (1997). Pretreatment of pesticide wastewater by photocatalytic oxidation, *Wat. Sci. Tech.* Vol. 36: 117-122.

Murakami, K., Hasegawa, K., Watanabe, H. & Komori, K. (1978). Biodegradation of organic substances by biological treatment and in natural waters, *Public Works Research Institute Technical Memorandum* NO. 1421 [in Japanese with English abstract].

Nadezhdin, A.D. (1988). Mechanism of Ozone Decomposition in Water. The Role of Termination, *Ind. Eng. Chem. Res.* Vol. 27: 548-550.

Nair, S. & Finkel, S.E. (2004). Dps protects cells against multiple stresses during stationary phase, *J. Bacteriol.* Vol. 186: 4192-4198.

Ohki, A., Shinohara, K., Ito, O., Naka, K. & Maeda, S. (1994). A BOD sensor using *Klebsiella oxytoca* AS1, *J. Environ. Anal. Chem.* Vol. 56: 261-269.

Perkowski, J., Kos, L. & Ledakowicz, S. (1996). Application of Ozone in Textile Wastewater Treatment, *Ozone Sci. Eng.* Vol. 18: 73-85.

Prousek, J. (1996). Advanced oxidation processes for water treatment. Photochemical processes, *Chem. Listy* Vol. 90: 307-315.

Pryor, W.A. (1986). Oxy-radicals and related species: Their formation, lifetimes, and reactions, *Ann. Rev. Physiol.* Vol. 48: 657-667.

Rajeshwar, K. (1995). Photoelectrochemistry and the environment, *J. App. Electrochem.* Vol. 25: 1067-1082.

Reiss, M., Tari, A. & Hartmeier, W. (1993). BOD-biosensor basing on an amperometric oxygen electrode covered by *Lipomyces kononenkoae*, *Bioengineering* Vol. 9: 87.

Riedel, K. (1998). Application of biosensors to environmental samples, In: *Commercial biosensors: applications to clinical, bioprocess, and environmental samples*, Ramsay, G., (ed.), pp. 267-295, John Wiley & Sons Inc., New York.

Riedel, K., Lange, K.P., Stein, H.J., Kuhn, M., Ott, P. & Scheller. (1990). A microbial sensor for BOD, *Water Res.* Vol. 24: 883-887.

Riedel, K., Renneberg, R., Kuhn, M. & Scheller, F. (1988). A fast determination of BOD using microbial sensor, *Appl. Microb. Biotechnol.* Vol. 28: 316-318.

River Bureau. (1992). *Annual report on river water quality in Japan*. River Bureau, Japanese Ministry of Construction, Tokyo [in Japanese].

Robert, D. & Weber, J.V. (2000). Study of adsorption of dicarboxylic acids on titanium dioxide in aqueous solution, *Adsorption* Vol. 6: 175-178.

Rodriguez, S.M., Richter, C., Galvez, J.B. & Vincent, M. (1996). Photocatalytic degradation of industrial residual waters, *Solar Energy* Vol. 56: 401-410.

Sakaguchi, T., Kitagawa, K., Ando, T., Murakami, Y., Morita, Y., Yamamura, A., Yokoyama, K. & Tamiya, E. (2003). A rapid BOD sensing system using luminescent recombinants of *Escherichia coli*, *Biosens. Bioelectron.* Vol. 19: 115-121.

Sakai, Y., Abe, N., Takeuchi, S. & Takahashi, F. (1995). BOD sensor using magnetic activated sludge, *J. Fermen. Bioeng.* Vol. 80: 300-303.

Sangeetha, S., Sugandhi, G., Murugesan, M., Madhav, V.M., Berchmans, S., Rajasekar, R., Rajasekar, S., Jeyakumar, D. & Rao, G.P. (1996). *Torulopsis candida* based sensor for the estimation of biochemical oxygen demand and its evaluation, *Electroanalysis* Vol. 8: 698-701.

Seo, K.S., Choo, K.H., Chang, H.N. & Park, J.K. (2009). A flow injection analysis system with encapsulated high-density *Saccharomyces cerevisiae* cells for rapid determination of biochemical oxygen demand, *Appl. Microbiol. Biotechnol.* Vol. 83: 217-223.

Staehelin, J. & Holgné, J. (1982). Decomposition of Ozone in Water: Rate of Initiation By Hydroxide Ions and Hydrogen Peroxide, *Environ. Sci. Technol.* Vol. 16: 676-681.

Stand, S.E. & Carlson, D.A. (1984). Rapid BOD measurement for municipal wastewater samples using a biofilm electrode., *JWPCF* Vol. 56: 464-467.

Stumm, W. & Morgan, J.J. (1981). The surface chemistry of oxides, hydroxides, and oxide mineral, In: *Aquatic Chemistry*, 2nd ed. pp. 625-640, John Wiley & Sons, New York.

Tan, T.C., Neoh, K.G. & Li, F. (1992). Microbial membrane-modified dissolved oxygen probe for rapid biochemical oxygen demand measurement, *Sensors and Actuators B* Vol. 8: 167-172.

Tan, T.C. & Wu, C. (1999). BOD sensors using multi-species living or thermally killed cells of a BODSEED microbial culture, *Sensors and Actuators B* Vol. 54: 252-260.

Tanaka, H., Nakamura, E., Minamiyama, Y. & Toyoda, T. (1994). BOD biosensor for secondary effluent from wastewater treatment plants, *Wat. Sci. Tech.* Vol. 30: 215-227.

Turchi, C.S. & Ollis, D.F. (1990). Photocatalytic degradation of organic water contaminants: Mechanisms involving hydroxyl radical attack, *J. Catalysis* Vol. 122: 178-192.

Unkroth, A., Wagner, V. & Sauerbrey, R. (1997). Laser-Assisted Photochemical Wastewater Treatment, *Wat. Sci. Tech.* Vol. 35: 181-188.

Villalobos, P., Acevedo, C.A., Albornoz, F., Sanchez, E., Valdes, E., Galindo, R. & Young, M.E. (2010). A BOD monitoring disposable reactor with alginate-entrapped bacteria, *Bioprocess Biosyst. Eng.* Vol. 33: 961-970.

Weiss, J. (1935). Investigations on the Radical HO_2 in Solution, *Trans. Farad. Soc.* Vol. 31: 668-681.

Widsten, P., Hortling, B. & Poppius-Levlin, K. (2004). Ozonation of conventional kraft and SuperBatch residual lignins in methanol/water and water, *Holzforschung* Vol. 58: 363-368.

Yang, Z., Suzuki, H., Sasaki, S. & Karube, I. (1996). Disposable sensor for biochemical oxygen demand, *Appl. Microbiol. biotechnol.* Vol. 46: 10-14.

Yang, Z., Suzuki, H., Sasaki, S., McNiven, S. & Karube, I. (1997). Comparison of the dynamic transient- and steady-state measuring methods in a batch type BOD sensing system, *Sensor and Actuators B* Vol. 45: 217-222.

Yoshida, N., Yano, K., Morita, T., McNiven, S.J., Nakamura, H. & Karube, I. (2000). A mediator-type biosensor as a new approach to biochemical oxygen demand estimation, *Analyst* Vol. 125: 2280-2284.

13

Electrochemical Sensors for Pharmaceutical and Environmental Analysis

Robert Săndulescu, Cecilia Cristea, Veronica Hârceagă and Ede Bodoki
"Iuliu Hațieganu" University of Medicine and Pharmacy Cluj-Napoca,
Romania

1. Introduction

Improvement of life quality is one of the most important objectives of the global research efforts. Naturally, the quality of life is closely related to a better control of diseases, drug and food quality and safety, and last but not least, of the quality of our environment. In all these fields, a continuous, sensitive, fast and reliable monitoring is required to control key parameters (Castillo et al., 2004). In this context, the use of sensors and biosensors represent very promising tools. The interest in their application in the biomedical, pharmaceutical and environmental field increased lately as a result of their sensitivity, specificity and simple use, providing fast, cost-effective and repetitive measurements with miniaturized and portable devices.

2. How to construct and optimize an enzyme electrochemical biosensor?

A biosensor, according to the IUPAC definition, is a self-contained integrated device, which is capable of providing specific quantitative or semi-quantitative analytical information using a biological recognition element (biochemical receptor) which is retained in spatial contact with a physical transduction element (Thevenot et al., 2001). Actually they transform the reaction between the analyte and the biological component into a signal detected and monitored by the transducer that reflects the concentration of the analyte. Biosensors may be classified according to the bioactive component or the recognition event, the mode of signal transduction or according to the immobilization technique.

In the construction of the biosensors the biocomponent is selected depending on the application and the performance criteria requested. The most commonly applied bioactive components are the enzymes. They rapidly and cleanly form selective bonds with the substrates and convert them into products. One of the major issues when working with enzymes is their stability. Enzymes are very sensitive to their environment. Deactivation, inhibition or unfolding upon adsorption and chemical or thermal inactivation are common (Grieshaber et al. 2008). There are several advantages of the enzyme biosensors. These include their availability and the ability to modify the catalytic properties or substrate specificity by means of genetic engineering and catalytic amplification of the biosensor response by modulation of the enzyme activity with respect to the target analyte (Rogers, 2006). Among the enzymes commercially available, the oxidases: the glucose oxidase (GOD) and horseradish peroxidase (HRP) are the most often used. This type of enzyme offers the advantages of being stable and in some situations does not require coenzymes or cofactors. Other, less commonly used enzymes comprise beta-lactamase, urease, tyrosinase and acetylcholinesterase or choline oxidase.

Among different transducers employed in biosensors construction, the electrochemical transducers are the most frequently used. They occupy the first position as far as the disponibility on the market and they have already demonstrated their practical utility.

The most well known enzyme electrochemical biosensors reported in the literature are the biosensors with glucose oxidase for the detection of glucose (Ammam et al., 2010), the biosensors with HRP for the detection of compounds which can be peroxidate like clozapine, (Yu et al., 2006a), acetaminophen, (Sima et al., 2008) and of the thiolic products (gluthation, N-acetilcysteine) (Yu et al., 2006b) and the biosensors with acetylcholine and choline oxidase for detection of drugs used in Alzheimer treatment (Lenigk et al., 2000) and the organophosphorus pesticides (Andreescu & Marty, 2006).

The most important step in building a biosensor is the immobilization of the biomolecules. As a matter of fact the immobilization technique determines the biosensor performances. The selection of an appropriate immobilization method depends on the nature of the biological element, type of the transducer used, physicochemical properties of the analyte and operating conditions for the biosensor.

The purpose of any immobilization method is to get high stability and efficiency from the biosensor. The biocomponent should be stably immobilized in a physical environment close to the natural enzyme's environment in order to retain maximum functionality of its biological activity on the surface of the transducer. The immobilization might involve a number of disadvantages such as: worsening the performance of enzymatic system, lowering and even lost of activity, increase of Michaelis constant, modification of the optimal working temperature and pH of the enzyme in comparison with the free form. All these, however, may be reduced by applying an appropriate design.

The system should assure also good diffusion properties for substrates and should establish satisfactory electrical communication between the active component and the electrode surface (Xu et al., 2006) in order to give the electrochemical output signal required.

When the design of a biosensor is built, all the characteristics of an ideal analytical system: selectivity, limit of detection and limit of quantification, sensitivity, reliability, repetitively, robustness, throughput, reagentless, simplicity, costs, miniaturization and portability have to be taken into account and also the performances and the drawbacks of the systems already reported in the literature for the analyte of interest have to be deeply studied.

Enzyme biosensors can be categorized into those that monitor the catalytic transformation of a target analyte by a specific enzyme or the enzyme biological activity by measuring the producing or consumption of a given analyte and those that monitor the inhibition of a biochemical reaction due to the presence of target analyte. The inhibition-based biosensors can be designed in two different ways: the inhibitor interacts directly with the enzyme, blocking its activity, or the inhibitor interacts with a product of the enzymatic reaction. Only specific enzyme inhibitors should be regarded as "true inhibitors", which bind to the enzyme and inhibit its activity. However, the definition of "enzyme inhibitors" is sometimes misleading, for example in the case of tyrosinase based biosensors, the terminology being also used to refer to melanogenesis process inhibitors, whose action mainly reside in some interference in melanin formation, regardless of any direct inhibitor-enzyme interaction (Chang et al., 2009). "True tyrosinase inhibitors" can be considered kojic acid that links to the active site of tyrosinase like a copper chelator (Briganti et al., 1990), benzoic acid that functions as a monophenol analog, binding to one copper ion of the binuclear site via its carboxylate group (Menon et al., 1990) and azelaic acid that blocks the access of the substrate to the enzyme active center (Briganti et al.,1990) but not ascorbic acid that interrupts melanin synthesis process by reducing the enzymatically generated dopaquinone intermediate back to L-dopa.

When a biosensor applied to pharmaceutical or biomedical analysis is described the biological processes have to be studied in detail, taking in consideration their complexity, in order to clearly outline the field of application and the limitations of the device. For example the biosensor described by Sima et al. (2011) allowed the quantification and the characterization of inhibitors from the point of view of their interaction with tyrosinase and the enzymatic reaction in the presence of the natural substrate L-tyrosine. In the skin pigmentation process these compounds may interfere also in other biochemical steps having as final result a decrease in melanin formation. For example, azelaic acid inhibits melanin formation through other mechanism besides its interaction with the tyrosinase active site. It may also interfere with DNA synthesis and mitochondria activity in hyperactive and abnormal melanocytes (Briganti et al., 1990).

Once the biocomponent, the transducer and the immobilization technique are chosen the operation mode of the device should be established. For example in the article published by Sima et al., (2011), the obtained configuration was used to monitor the reduction current of the enzymatically generated oxidized species of L-tyrosine, i.e. the dopaquinone, in the presence of molecular oxygen (Fig. 1).

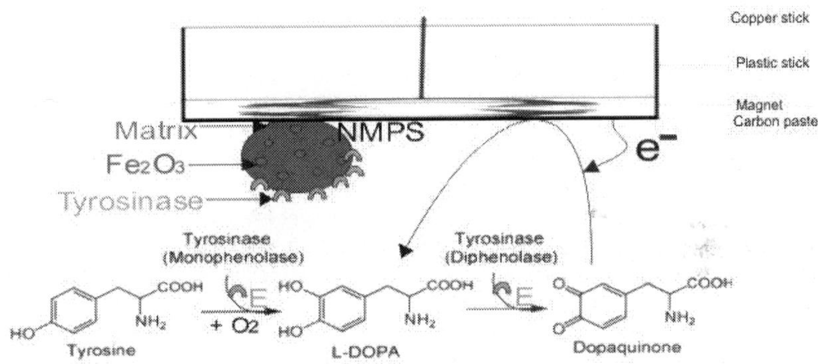

Fig. 1. Biosensor's mechanism of action (Sima et al., 2011)

The inhibition-based biosensors are preferred, if possible, in order to assure a good reproducibility of the results because in this case, when the analytes are evaluated, the biosensor inhibition signal is normalized to the initial signal of the substrate for each curve. Thus, the final results are not influenced by the variations of substrate's signal which can occur due to the loss of the enzyme activity or due to random errors that can appear during biosensor's preparation.

Also blank experiments have to be performed in order to check if "the biosensor" without enzyme or the inhibition based biosensor operating without addition of the enzymatic substrate does not give any response under the selected experimental conditions and in the presence of the studied compounds.

If the developed device is a novel analytical tool, in order to test its efficiency, it should be applied to the study of compounds that have already demonstrated their activity in clinical practice or by using other analytical methods, as it was the case in the article published by Sima et al. (2011).

When testing different compounds, because there are many factors which contribute to the final results (type of enzyme, quantity of immobilized enzyme, immobilization method, type of substrate, substrate concentration, time of contact between the enzyme, substrate and inhibitor, pH, temperature, applied potential, rate of solution stirring), in order to obtain a correct comparison of the studied compounds, the experiments have to be performed under the same experimental conditions (Sima et al., 2011).

First the working conditions for an optimal response to substrate of the enzyme based biosensor have to be tested. If the biosensor is designed for testing biologically active compounds it has to be taken into account that the device should operate as close as possible to the natural enzyme's environment.

The working applied potential, the type of enzyme substrate, the buffer pH, the quantity of immobilized enzyme, the substrate concentration in case of the inhibition studies, the presence of cofactors if necessary; all these have to be considered.

In order to select the applied potential for the amperometric testing, first of all, the biological process that produces electroactive compounds has to be demonstrated under the experimental conditions, even though it is well know in its natural environment. Normally cyclic voltammetric experiments are performed for detecting the electrochemical reduction or oxidation potential of the enzymatically generated electroactive species (Yu et al., 2006). It may be sometimes the case that if low reduction or oxidation currents are generated they might not be detected in cyclic voltammetric experiments, so amperometry is recommended, thanks to its low background currents (Shan et al., 2008). The applied potential has to be considered taking into account the magnitude of the electrochemical signals, the ratio between signal and background current, the steady-state of the response plateau, the possible interfering species at that applied potential and also, the working potential of other similar biosensors described in the literature.

When selecting the enzymatic substrate in case of inhibition based biosensors, this has to be the same as in the biological process, if possible, because it has been demonstrated that the potency of inhibitors varies considerably depending on the characteristics of the substrate (Shan et al., 2008). The sensitivity of the response has to be high enough and the steady-state of the response plateau has to be stable for assuring reliable results and the lag period of the enzymatic step sufficiently short for further readily implementation of the testing conditions.

It is well known that enzyme activity is highly pH dependent and that the optimum pH for an enzymatic assay must be determined empirically (Amine et al., 2006). The buffer pH considered optimal has to be chosen taking in consideration the magnitude of the amperometric signal, the steady-state of the response, the enzyme stability pH, the normal biological pH of the process and other similar biosensors reported in the literature.

The amount of enzyme immobilized at the electrode surface has to offer a good sensitivity and repeatability of the biosensor response taking in consideration also the costs of the assay, since some enzymes can be quite expensive.

The opinions about the substrate concentration to be considered in inhibition assays are quite contradictory. Kok et al. (2002), concluded, when measuring the inhibition potency of a competitive inhibitor with an acetylcholinesterase and a choline oxidase biosensor, that the inhibition percentage increased by raising the substrate concentration. Therefore they worked in enzyme saturated substrate conditions. Shan et al. (2008) demonstrated that for a tyrosinase biosensor for the determination of benzoic acid, the concentration of the catechol substrate did not affect the maximum inhibition percentage but it affected the sensitivity of the method. It must be noted, however, that the use of a high substrate concentration would not yield sensitive inhibition responses when the quantification of a competitive inhibitor is performed by simultaneous addition of the inhibitor and the substrate. Because in competitive inhibition the substrate competes with the inhibitor for the enzyme active site, and the inhibition, especially at low inhibitor concentrations, would likely not be detected (Kok et al., 2002). In the case of the biosensor made by Sima et al. (2011) as it was not designed for the quantification of inhibitors but for the screening and quantification of their inhibitory potency, it was considered that all the immobilized enzyme molecules must take part in the reaction and this could only be possible in substrate concentration corresponding to the saturation portion of the activity versus substrate curve. When the concentration of substrate is quite high, it has to be established if the concentration of other cofactors participating in the reaction is not a limiting factor and that the enzymatic process is saturated by the substrate.

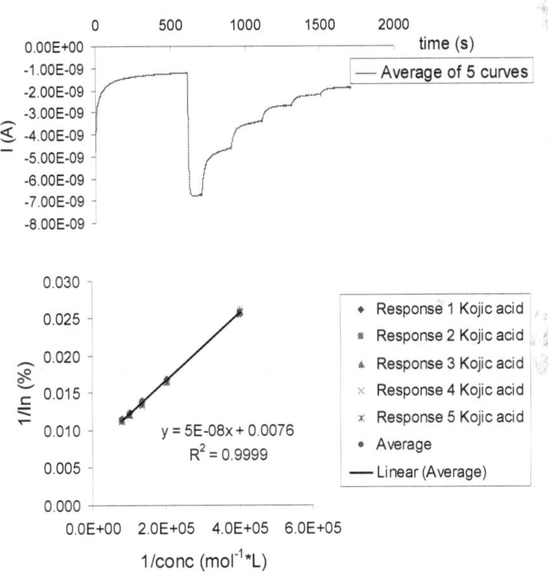

Fig. 2. Tyrosinase based biosensor amperometric response. L-tyrosine 3.33 ·10⁻⁴M, Kojic acid inhibitor between 2.50 ·10⁻⁶ and 1.24 ·10⁻⁵M, 0.1M phosphate buffer pH 6.5, Eapp -100mV, 10µL of a 1.25mg/mL tyrosinase-NMPS suspension spiked onto the mCPE. Insert: inhibition calibration curve (Sima et al., 2011)

In case of electrochemical biosensors parameters like apparent Michaelis-Menten constant (K_m^{app}), maximum current intensity, limit of detection and linear range for the analyte of interest have to be determined.

In case of electrochemical inhibition based biosensors parameters like IC_{50}, K_i^{app} (inhibition constant) and mechanism of inhibition of the studied compounds have to be determined.

In order to determine IC_{50}, the concentration of inhibitor which inhibits 50% of the substrate signal, the concentration of inhibitors can be correlated with the percentage of inhibition (In %) which can be calculated using the relationship: $In(\%)=[(I_0 - I_1)/ I_0]x100$; with I_0 and I_1 being the biosensor current signals before and after the addition of the inhibitor, respectively (Fig. 2).

The mechanism of inhibition can be established from the relationship between the biosensor response to substrate, K_m^{app} and I_{max}, in the absence and in the presence of different concentrations of inhibitor added in the initial testing solution (Fig. 3).

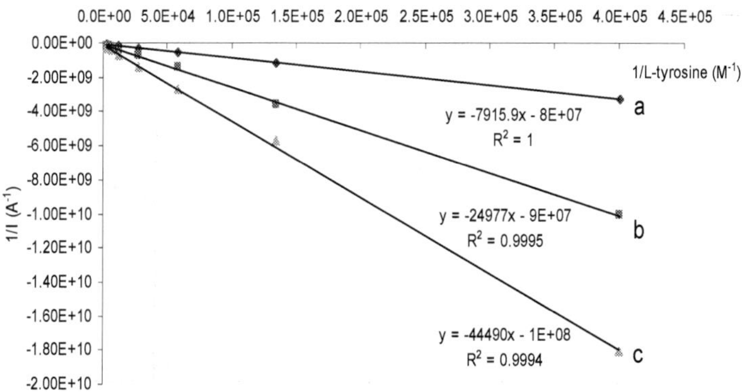

Fig. 3. Tyrosinase based biosensor amperometric data: 1/Signal vs 1/[L-tyrosine]. L-tyrosine between 2.48 ·10⁻⁶ - 4.39 ·10⁻⁴M a. without Kojic acid, b. in the presence of 2.09 ·10⁻⁶M Kojic acid, c. in the presence of 4.18 ·10⁻⁶M Kojic acid, 0.1M phosphate buffer pH 6.5, Eapp - 100mV, 10µL of a 1.25mg/mL tyrosinase-NMPS suspension deposited onto the mCPE (Sima et al., 2011)

From the primary plot of $1/I(A^{-1})$ versus 1/substrate concentration (M^{-1}) (Lineweaver-Burk), secondary plots can be generated, by plotting the slopes from the primary plots versus inhibitor concentration in order to determine the apparent inhibition constant, K_i^{app} (Fig. 4).

Finally the long-term stability of the immobilized enzyme has to be evaluated by measuring the biosensor response to substrate. Normally when not in use, the enzyme has to be maintained in hydrated conditions at around 4°C. In case of inhibition based biosensors IC_{50} of the tested inhibitors has to be determined at different time periods following enzyme immobilization to see if the obtained results are not affected by the loss of enzyme activity.

Biosensors have some major advantages over traditional analytical methods, which will certainly lead to their even more pronounced use in the pharmaceutical and biomedical field in the near future: they are prone to miniaturization which is of great importance because biological samples are available in small amounts and tissue damage must be minimized in case of in-vivo monitoring, the detection of the key substrate is very often made without prior separation, their sensitivity is usually in the order of ng/ml, they have a high selectivity, sometimes even specificity, they are characterized by short response times and quickness of data collection, they are easy to use (Castillo et al, 2004) and they have a high benefit/cost ratio.

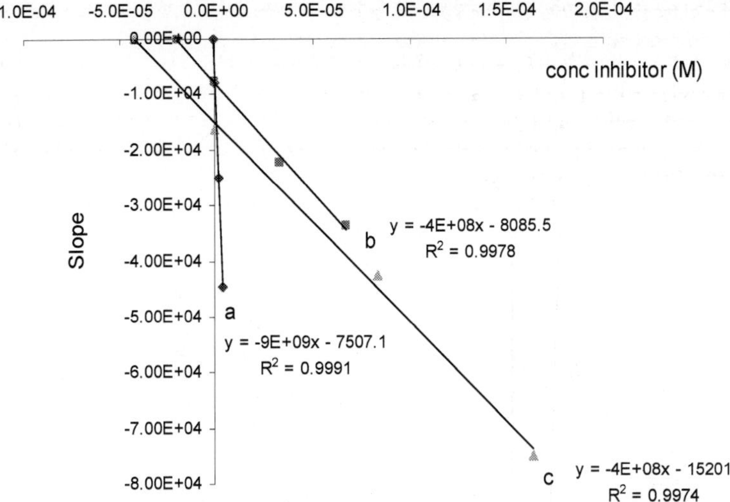

Fig. 4. Tyrosinase based biosensor data: slope of Fig. 3 data vs [In]. a. Kojic acid, b. Benzoic acid, c. Azelaic acid, 0.1M phosphate buffer pH 6.5, Eapp -100mV, 10µL of a 1.25mg/mL tyrosinase-NMPS suspension deposited onto the mCPE (Sima et al., 2011)

However, many enzyme-based amperometric biosensors described in the recent literature still display a few drawbacks when compared to other analytical methods. The most difficult problems to overcome for biosensors are: the low reproducibility of the response and the reduced stability due to the fact that the enzymes, removed from their natural environment, tend to rapidly loose their activity and thus limit the lifetime of the sensor, the electrochemical interferences in complex sample matrices and the biocompatibility and biofouling in case of in-vivo measurements (Castillo et al, 2004). Even if they do present some drawbacks these devices can be very efficacious used for adequate analysis. Moreover these disadvantages can be limited and the performances can be optimized by an adequate immobilization method of the biocomponent at the transducer surface.

While the concept of biosensors is simple, their commercialization is far from being simple. A lot of such devices are at scientific stage and only a few of the innovative ideas described in the scientific literature have reached the marketplace and they are only limited applied in clinical practice, as for example the well known glucose portable analyzer for home diagnostic (Wang, 1999); so still more efforts need to be undertaken.

3. Electrochemical sensors for health applications

3.1 Pharmaceutical analysis
The need for new sensitive devices allowing the simultaneous detection of several drugs showed a steadily increasing trend in the pharmaceutical research. Some examples of electroanalytical methods applied to ascorbic acid (Săndulescu et al, 2000; Marian et al., 2000), phenotiazines (Săndulescu et al, 2000; Blankert et al., 2005), vitamine B1 (Marian et al., 2001; Bonciocat et al., 2003), effervescent solid dosage forms (Săndulescu et al, 2000), benzodiazepines (Bănică et al., 2007), colchicine (Bodoki et al., 2007; Bodoki et al., 2007) and

paracetamol (Yu et al., 2006; Sima et al., 2008; Sima et al., 2010) were published in the last decade.

The electrochemical behaviour of some pharmaceuticals (caffeine, aminophylline, teophylline, Fig. 5) was investigated by cyclic voltammetry (CV), square-wave voltammetry (SWV) and differential pulse voltammetry (DPV), on different electrode materials: glassy carbon in native and electrochemically activated form, bare or 1,4-benzoquinone modified carbon paste. (Câmpean et al., 2011)

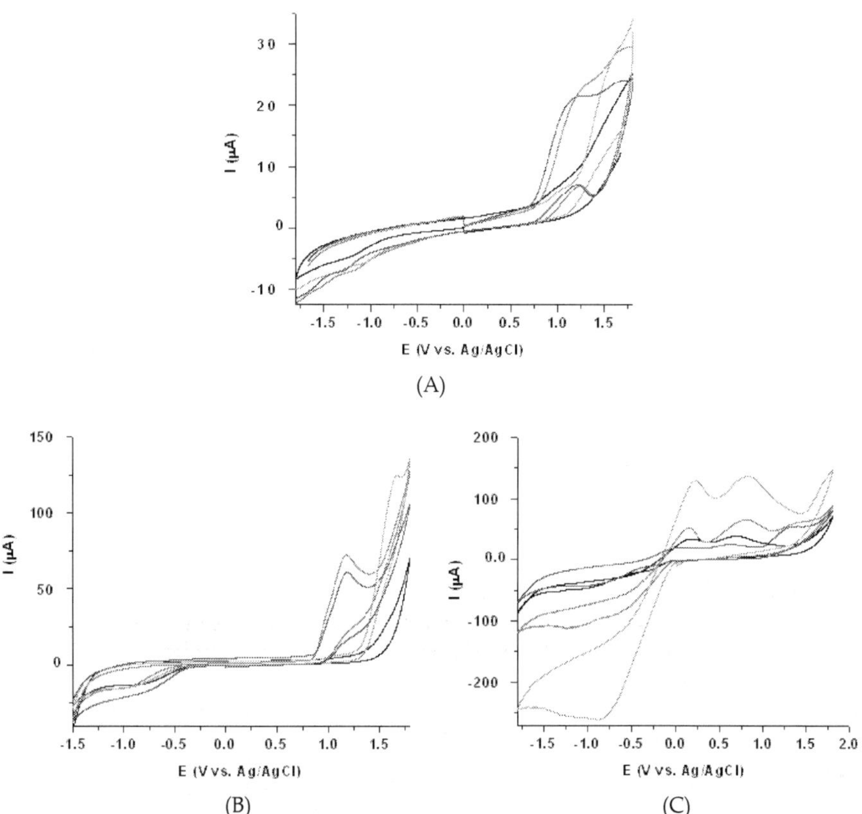

Fig. 5. Cyclic voltammograms of 10^{-3} M aminophylline (red line), caffeine (green line) and theophylline (blue line) solutions in Britton-Robinson buffer at pH 4 (black line), on:
(A) glassy carbon electrode; (B) carbon paste electrode and (C) 1,4-benzoquinine modified carbon paste electrode (Reference electrode Ag/ AgCl,(3 M KCl); auxilliary electrode Pt wire; scan rate 0.1 V s^{-1}; 20 °C)

The preliminary studies performed on traditional electrodes were extended on carbon based screen-printed electrodes (Fig. 6), simple and modified with multi-wall carbon nanotubes or cobalt-phtalocyanine. (Câmpean et al., 2011)

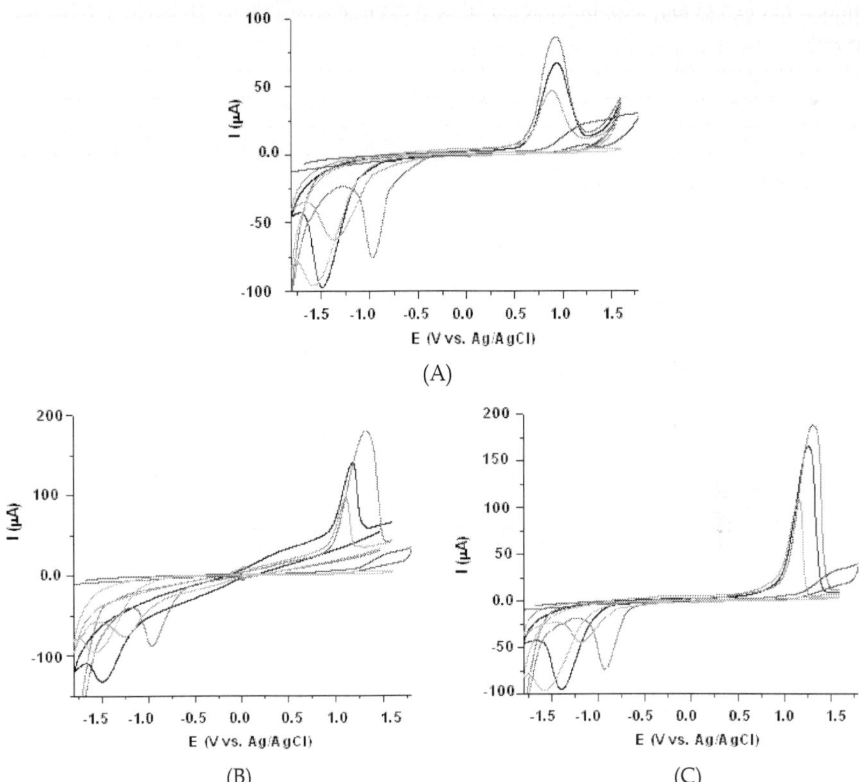

Fig. 6. Cyclic voltammograms obtained on screen-printed electrodes type: DS 110 with graphite (black line); DS 410 with graphite and cobalt phthalocyanine (red line); DS 110 with graphite modified with multiwall carbon nanotubes (green line); compared with the voltammograms obtained on glassy carbon electrode (blue line); recorded for 10^{-3} M: (A) aminophylline, (B) caffeine, (C) theophylline solutions in Britton-Robinson buffer at pH 4 (cyan line) (Reference electrode Ag; auxiliary electrode graphite; scan rate 0.1 V s^{-1}; 20 °C)

A horseradish peroxidase (HRP) - Zirconium alcoxide porous gel film biosensor for acetaminophen determination was obtained by enzyme immobilization on glassy carbon and carbon based screen-printed electrodes and has been reported by V. Sima et al. (2008).

The investigated zirconium alkoxide porous gel film represents an interesting way for biocomponent immobilization onto the glassy carbon electrode, conferring in the mean time a hydrated environment for the enzyme.

Two different zirconium alcoxide gels have been prepared, starting from different amounts of zirconium salt: 0.25 M and 0.4 M alcoholic solutions by refluxing for 2 hours at 90°C then allowed to cool at room temperature. 6.5 µl porous Zr alcoxide gel were added to 5 mg polyethyleneimine (PEI) in hydroalcoholic media. Equal amounts of the 0.03 and 0.06 mg/mL HRP enzyme solution and the above described alkoxide gel were mixed for 15 minutes and 20 µl of the resulting mixture was deposited on the glassy electrode surface and left over night at 4°C for drying. For another 24 hours it was left at 4°C in 5 ml phosphate buffer for hydratation.

The enzymatically generated reactive oxidized species of acetaminophen were electrochemically reduced and the amperometric signal was recorded. (Fig. 11)

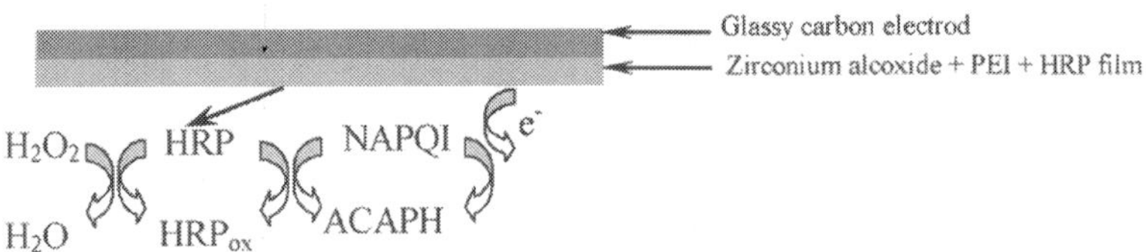

Fig. 11. The mechanism of biocatalytic peroxidation of acetaminophen (ACAPH) into N-acetylparabenzoquinonimine (NAPQI) by HRP immobilized at the surface of the electrode

The cyclic voltammetric assays showed that the zirconium alcoxide porous gel was a biocompatible material, capable to preserve the enzyme's bioactivity. The ZrO_2-PEI thin film exhibited good electrocatalytical and electroanalytical response towards acetaminophen and H_2O_2. The decrease of the oxidation current until about 50% between the electrodes modified with a film of PEI and a film containing Zr alcoxide shows the improved conductive properties of the alcoxide film.

Electrochemical differences were observed also between a hydrated and a dry film. To explain this behavior microscopic studies were conducted. Fig. 12 shows surface topography images of dry (a) and hydrated ZrO_2-PEI thin films (b), respectively. As shown in Fig. 12, particle size was in the ranges of micrometers. The surface topography changes considerably in the case of hydrated ZrO_2-PEI thin film (Fig. 12b). The dry film presented cracks and pores, whereas the hydrated film had a more homogenous structure.

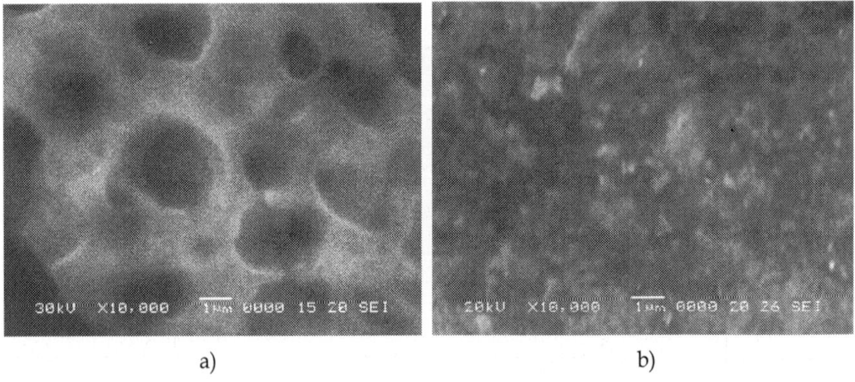

a)   b)

Fig. 12. Electron microscopy images of the dry (a) and hydrated film (b)

The biosensor was applied to the acetaminophen assay in drug formulation, Perdolan® and a generic solid dosage form made in our department. The standard addition method was used. A linear trend of current versus acetaminophen concentration was found for concentrations between $2.0 \cdot 10^{-5}$ M and $1.6 \cdot 10^{-4}$ M (R^2 = 0.9996, RSD of slope = 20%, n = 5) with a LOQ of $1.17 \cdot 10^{-7}$ M.

Another biosensor configuration based on the immobilization of horseradish peroxidase (HRP) within a zirconium alcoxide film on screen-printed electrodes (SPE) for the analysis of acetaminophen from Perdolan® tablets was also developed (Sima et al., 2010). The biosensor's operation mode was also based on monitoring the amperometric signal produced by the electrochemical reduction of the enzymatically generated electroactive oxidized species of acetaminophen in the presence of hydrogen peroxide.

In the case of SPEs modified with 10 µL film without HRP (5 mg PEI, 125 µL ethanol and 120 µL distilled water), in a control solution (phosphate buffer 0.1 M pH=7.4) at 0.57 V vs. Ag pseudoreference, an irreversible oxidation peak which becomes much smaller at the second cycle was observed. The compound formed during the first cycle was not further reduced, thus the oxidation current at the second cycle was much smaller. It is not clear which compound was oxidized, or whether this compound was part of the film, or an impurity in the film. The second cycle, which was stable for multiple scans, was taken to be the background level.

A higher background current in the control solution was observed in the case of film modified SPEs, probably due to specific interactions between the zirconium and the electroconductive graphite-based ink used to print the SPE working electrode.

After the deposition of the zirconium alcoxide-PEI film onto the glassy carbon working electrode, the electrochemical signal of paracetamol decreases, as a consequence of electroactive surface blockage of the of the working electrode by the PEI (Fig. 13).

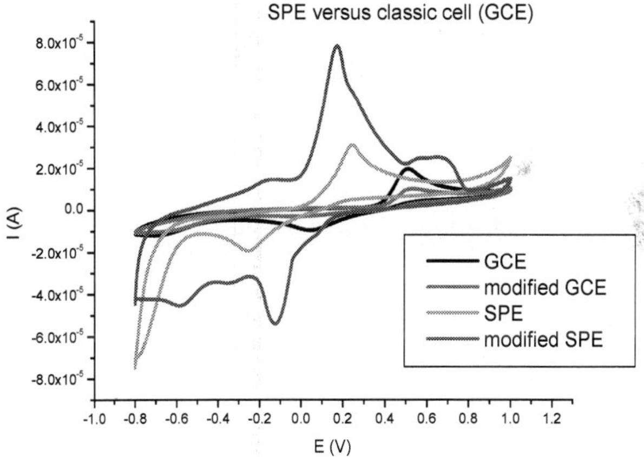

Fig. 13. Cyclic voltammograms for unmodified and 20 µL non enzymatic film (zirconium alcoxide-PEI gel) modified GC and SPE working electrodes in paracetamol $6.25 \cdot 10^{-4}$ M, 0.1 M phosphate buffer pH=7.4 – cyclic voltammetry, applied potential between -0.8 V and 1.0 V, scan rate 100 mV s^{-1}

After the deposition of the zirconium alcoxide-PEI film onto the surface of the SPE working electrode, the electrochemical signal of paracetamol increased, probably due to the catalytic activity of the zirconium alcoxide. Another difference between the unmodified and the film-modified SPE was the peak separation voltage (ΔEp): in the case of the unmodified SPE, ΔEp was 0.5 V, and in the case of the modified SPE, ΔEp was 0.29 V (Fig. 13). The smaller difference between the oxidation and reduction potential in the case of the film-modified SPE demonstrates that the zirconium alcoxide facilitated electron transfer between the analyte and the working electrode.

The sensitivity obtained with the SPE was significantly higher than the one obtained with glassy carbon electrode (GCE). The two experiments were carried out using the same fabrication technique for the entire biosensor (same composition of the enzymatic film) and using the same electrochemical parameters. Even if the electrochemical active surface area of the SPE was larger than that of the GCE, the difference of sensitivity between the two electrodes was much larger than what this difference can account for. This leaded us to believe that the improved sensitivity was due to factors other than the larger active area.

3.2 Biomedical analysis

The detection of a large spectrum of biological analytes has been performed during the last decades using inexpensive and sensitive electrochemical methods. The most commonly used electrodes were the carbon based electrodes because of their low cost, good electron transfer kinetics and biocompatibility. Recently, carbon nanotubes (CNTs) have also been incorporated into electrochemical sensors. The CNTs offer unique advantages including enhanced electronic properties, a large edge plane/basal plane ratio, and rapid electrode kinetics. Therefore, CNT-based sensors generally have higher sensitivities, lower limits of detection, and faster electron transfer kinetics than traditional carbon electrodes. To optimize a CNT sensor many variables need to be tested. Electrode performance depends on the synthesis method of the nanotube, CNT surface modification, the attachment method onto the electrode, and the addition of electron mediators. The physical and catalytic properties make CNTs ideal for use in sensors and biosensors. It is known that CNTs display high electrical conductivity, chemical stability, and mechanical strength. There are two main types of CNTs: single-walled CNTs (SWCNTs) and multi-walled carbon nanotubes (MWCNTs). SWCNTs are sp^2 hybridized carbon in a hexagonal honeycomb structure that is rolled into hollow tube morphology (Jacobs et al., 2010). MWCNTs are multiple concentric tubes encircling one another. A SWNT consists of a single graphite sheet rolled seamlessly, defining a cylinder of 1–2 nm diameter. Carbon nanotubes can behave as metals or semiconductors depending on the structure, mainly on the diameter and helicity.

Besides the incorporation of protein and antibodies which allowed their direct detection, the incorporation of enzymes onto the surface of the electrode is considered attractive for the direct detection of other non-electroactive biomolecules. The major properties of the enzymes are the specificity and high affinity for the target substrate. The CNTs were added to the enzyme biosensors for their electrocatalytic effect for the glucose or hydrogen peroxide.

In the development of the oxidase-based amperometric biosensors the glucose oxidase plays an important role. The glucose oxidase (GOx) is widely employed in most of the glucose biosensors due to its stability and high selectivity to glucose. The property of GOx with negative charge in neutral pH solutions also makes it feasible to immobilize GOx onto materials with positive charge by physical adsorption. It is formed by two flavine adenine dinucleotide (FAD) cofactors and catalyzes the oxidation of glucose according to the following reaction (Jacobs et al., 2010):

$$Glucose + O_2 \rightarrow Gluconolactone + H_2O_2$$

The amount of glucose is proportional to that of the produced H_2O_2, by consequence the glucose concentration can be directly determined through measuring the current derived from the electrochemical reaction of H_2O_2 (Kong et al., 1009).

The ability of carbon nanotubes (CNT) to promote the electron-transfer reaction of hydrogen peroxide suggests interesting applications of the CNTs for oxidase-based amperometric biosensors (Laschi et al. 2008).

Amperometric biosensors, which combine the bioselectivity of redox enzymes with the inherent sensitivity of amperometric transductions, have proven to be very useful for the detection of glucose.

The glucose biosensor is one of the most studied devices and is widely used as a clinical indicator of diabetes and in the food industry for quality control (Bakker, 2004; Prodromidis & Karayannis, 2002; Yang et al., 2004; Kong, et al., 2009; Liu et al., 2008). The glucose biosensors represent about 85% of the entire biosensor market. The considerable research and innovative detection strategies associated with glucose biosensors are justified by the interesting economic prospects. Amperometric glucose oxidase electrodes have played a leading role in the simple one-step blood sugar testing and continuous real time glucose monitoring. (Nenkova et al., 2010).

In the construction of an amperometric biosensor finding an appropriate matrix to immobilize the enzyme and to keep its activity for long-term application is another key issue. Many methods such as covalent binding (X.F. Yang et al., 2006) and cross-linking method have been used to immobilize the GOx onto different supporting materials. Silica gel (Jiang et al., 2006), polymers (Cosnier & Popescu, 1996), biomaterials (Wu et al., 2004) and their composites have been extensively explored to encapsulate enzymes and microorganism. Even that considerable progress has been made, some challenges still has to be overcome, such as leakage of enzyme from the matrix, long and complicated procedures for matrix preparation and short life time (Chang et al., 2010). Other examples of enzyme immobilization is represented by their inclusion in gels, cross-linked polymers, conductive salts or simply mixing them into carbon paste or carbon-organic polymer hosts. In general, the main requirement of support materials for enzyme immobilization is their adsorption ability and the preservation of the enzyme's catalytic properties during and after the immobilization process. Different polymeric materials have found wide application as enzyme immobilization matrices. In the literature are reported glucose biosensors developed by in situ electropolymerization of different monomers in the presence of glucose oxidase (Cosnier et al., 1999; Cosnier, 1999; Olea et al., 2008; Rubio Retama et al., 2004; Pan et al.; Ramanavicius et al., 2005).

The capture of CNTs in polymeric chains has been found useful for improving their solubility without impairing their physical properties. This strategy, based on wrapping water-soluble linear polymers around the tubes is robust and simple. Another advantage of this noncovalent attachment is that the structure of the nanotube is not altered, thus its mechanical properties remains unaltered. The advantages of dispersing multiwall carbon nanotubes in polyethylenimine for the development of electrochemical sensors for the detection of hydrogen peroxide, using screen-printed graphite-based electrodes as transducers and glucoseoxidase as bioreceptor has been reported (Rubianes & Rivas, 2007; Laschi et al., 2008).

An interesting approach was used for the elaboration of a new disposable electrochemical sensor for the detection of hydrogen peroxide, using screen-printed carbon-based electrodes (SPCEs) modified with multi-walled carbon nanotubes (MWCNs) dispersed in a polyethylenimine (PEI) mixture. The modified sensors showed an excellent electrocatalytic activity towards the analyte, with respect to the high overvoltage characterizing unmodified screen-printed sensors. The composition of the PEI/MWCNT dispersion was optimized in order to improve the sensitivity and reproducibility.

Different compositions of PEI/MWCNT suspension were tested. The sensitivity of the sensor (estimated on the basis of the slope of the calibration curve obtained for hydrogen peroxide) was used as a parameter to compare the performance of sensors obtained with different coatings. Each composition was analyzed in triplicate with the same sensor and the average sensitivity values lay in the range of 3.33 - 101.23 μAmM^{-1}. Not only the amounts of PEI and MWCNT were varied, but also the volume of the dispersion deposited onto the working electrode. The linear trend was always maintained, but an increase in sensitivity from 3.33 to 30.32 μAmM^{-1} was achieved by doubling the deposited volume from 5 to 10 μl. The optimized sensor showed good reproducibility (10% RSD calculated on three experiments repeated on the same electrode), whereas a reproducibility of 15% as RSD was calculated on electrodes from different preparations.

The amperometric response of a PEI/MWCNT/GOD modified screen-printed electrode was studied at +700mV in 0.05M phosphate buffer solution with pH 7.4. During successive additions of 0.5 mM glucose, a well-defined current response was observed. For each addition of glucose, a sharp rise in the current was observed, while no response was recorded in analogous measurements at the bare PEI/MWCNT electrode. The calibration plot was linear over a wide concentration range, 0.5–3.0 mM. The data interpolation in this concentration range had a slope of 22.18 μAmM^{-1}, with a correlation coefficient of 0.998.

Preliminary experiments carried out using glucose oxidase (GOD) as biorecognition element gave rise to promising results indicating that these new devices may represent interesting components for biosensor construction (Laschi et al., 2008).

3.3 Biorecognition and drug – receptor interaction studies

Molecular recognition or biorecognition is the essence of all biological interactions. These interactions generally involve noncovalent bonding (i.e. ionic, hydrogen bonding, hydrophobic interactions), where additionally shape complementarity plays an important role. During the last few years screen printed electrodes proved to be valuable tools in the study and development of new bioselectors. Their great versatility lies in the wide range of ways in which the electrodes may be modified, either by altering the composition of the printing inks or by depositing thin films of purposely designed supramolecular structures on the manufactured electrodes (Dominguez Renedo et al., 2007). Films of biomolecules have been used to tailor the properties of interfaces, providing solid surfaces to catalyze enzyme reactions, enabling fundamental biochemical and biophysical studies and nonetheless, serve in biosensors as biorecognition and biomimetic elements (Rusling, 2010).

The application field of such sensors is overwhelmingly broad, but the ones concerning clinical chemistry and analysis, drug and genetic testing stand out in particular. Electrochemical biosensors for the study of DNA interaction, hybridization and damage has become one of the most vibrant and dynamic areas of physical sciences and sensor engineering (Paleček et al., 1998; Teles & Fonseca, 2008). The molecule of DNA as carrier of genetic information is a major target for drug interaction. This phenomenon may interfere with transcription (gene expression and protein synthesis) and DNA replication, representing a fundamental issue in medical science. The evaluation of any interaction with DNA using biosensors helps to predict unwanted toxic side-effects and prevent DNA damage caused by therapeutic drugs.

Colchicine is a protoalkaloid, used as a specific antiinflammatory agent in acute attacks of gout by inhibiting the migration of leucocytes to inflammatory areas, thus interrupting the inflammatory response that sustains the acute attack. The drug also presents an antimitotic action, arresting dividing cells in metaphase by preventing normal function of mitotic spindle.

By studying colchicine's electrochemical signal obtained on graphite based screen printed electrode the influence of several bioselectors (calf thymus DNA, β-tubulin and bovine serum albumine) was investigated.

A disposable biosensor based on the immobilization of double – stranded calf thymus DNA (type XV, Sigma, Milan, Italy) was used to assess its possible interaction with colchicine.

The detailed protocol of production and use of the DNA biosensor was described earlier (Bagni et al., 2006), where the oxidation peak of guanine is used as the transduction signal to recognize DNA interacting agents. As the result of a specific or non-specific interaction (covalent binding, electrostatic interaction with the negatively charged nucleic acid sugar phosphate, hydrogen and/or van der Waals bonds) of the double stranded calf thymus DNA with a genotoxic agent, a decrease of the guanine peak (measured by square-wave voltammetry - SWV) is detected. Measurement of the target analyte with the DNA modified screen-printed electrode includes four main steps: (i) electrochemical conditioning of the electrode surface in order to oxidize the graphite impurities and to obtain a more hydrophilic surface to favor DNA immobilization, (ii) calf thymus ds-DNA immobilization, (iii) interaction with the colchicine solution and (iiii) electrode surface interrogation. In details, the experimental procedure was the following (Fig. 14) (Bodoki et al., 2009):

a. Electrode pre-treatment: applying potential of +1.6 V (vs. Ag-SPE pseudo-reference) for 120 s and +1.8 V (vs. Ag-SPE) for 60s; electrode in 5 mL of 0.25 M acetate buffer, containing 10 mM KCl (pH=4.75), under stirred conditions;

b. DNA-immobilization: electrode dipped in a solution of 50 ppm calf thymus ds-DNA in 0.25 M acetate buffer containing 10 mM KCl, applying a potential of +0.5 V (vs. Ag-SPE) for 5 minutes, under stirred conditions;

c. Blank or sample interaction: 10 uL of buffer or sample solution on the electrode's surface for 2 minutes;

d. Measurement: SWV scan in order to evaluate the oxidation of guanine residues on the electrode surface. The height of the guanine peak (at +0.95 V vs. Ag-SPE) was measured in 0.25 M acetate buffer, containing 10 mM KCl; SWV parameters: E_i +0.2V, E_f +1.45V, υ = 200 Hz, E_{step} 15 mV, $E_{amplitude}$ 40 mV.

Assessing the oxidation peak of guanine residues compared with the baseline, there is no current decrease, nor any potential shift, after the exposure to colchicine of the DNA biosensor, indicating an apparent lack of interaction.

However, at fairly higher concentrations of DNA (100 – 220 $\mu g \cdot L^{-1}$), based on the decrease of colchicine's peak current obtained on glassy carbon electrode by differential pulse voltammetry, the formation of DNA-colchicine complex was demonstrated with a combination ratio of 1:2 (Hui et al., 2011).

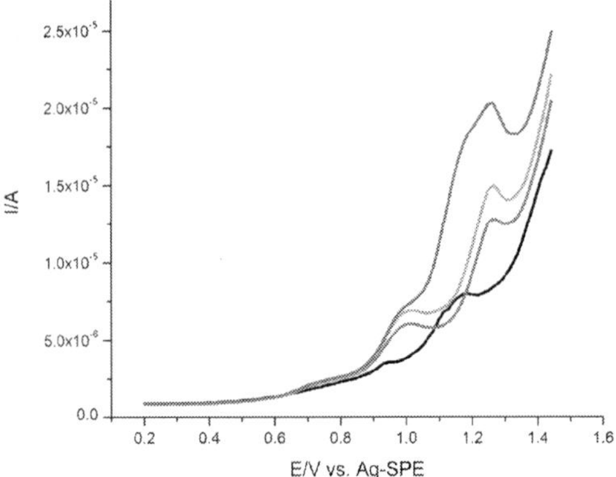

Fig. 14. Square wave voltammograms using screen-printed DNA biosensor. (black) without DNA immobilization, blank of colchicine ($2.6 \cdot 10^{-5}$ M); (red) with DNA immobilization, no colchicine; (green) with DNA immobilization, with colchicine ($2.6 \cdot 10^{-5}$ M); (magenta) with DNA immobilization, with higher concentration of colchicine ($2.6 \cdot 10^{-4}$ M).

Colchicine acts as an antimitotic agent by inhibiting microtubule polymerization by binding with high affinity to tubulin, a negatively charged protein (~120 kDa), formed by two globular polipeptidic subunits (α-tubulin and β-tubulin). Tubulin is one of the main constituents of the microtubules of the division spindle. Inhibitors of tubulin polymerization interacting at the colchicine binding site are potential anticancer agents, thus colchicine-tubulin interaction was very extensively studied in the last twenty years.

It has been demonstrated that the binding to the tubulin dimmer (α- and β-subunits) is slow and poorly reversible due to conformational changes in the protein's structure, whereas binding to β-tubulin monomers is reversible and 3-4 times faster (Banerjee et al., 1997). Therefore, the screen-printed electrodes were modified applying 5 μL of 25 ng mL^{-1} β-tubulin in 10 mM HEPES buffer on the electrode's surface, followed by drying at room temperature.

Due to the specific colchicine-β-tubulin interaction, an anodic shift of colchicine's oxidation peak can be observed (+1.06 V vs. Ag-SPE) accompanied by a decrease of the anodic current (Fig. 15). This decrease is not due to the electrode surface blockage by the deposited protein, since the same potential shift and current decrease is observed when β-tubulin is directly added in the voltammetric cell.

The literature describes also the formation of colchicine - bovine serum albumin complex, stabilized by van der Waals interactions and hydrogen bonds (Hu et al., 2005). Therefore, this interaction's influence on the electrochemical behavior of colchicine was also investigated using bovine serum albumin (BSA) surface modified SPE's.

A similar potential shift and anodic current decrease as in the case of β-tubulin was observed when the surface of screen-printed electrode was modified applying 7 μL of 0.025% m/v BSA followed by drying at room temperature or when the albumin was added to the bulk solution of the voltammetric cell. These data might be explained by the strong interaction of colchicine with the BSA determining a less favorable conformation of the colchicine molecule for the heterogeneous electron transfer.

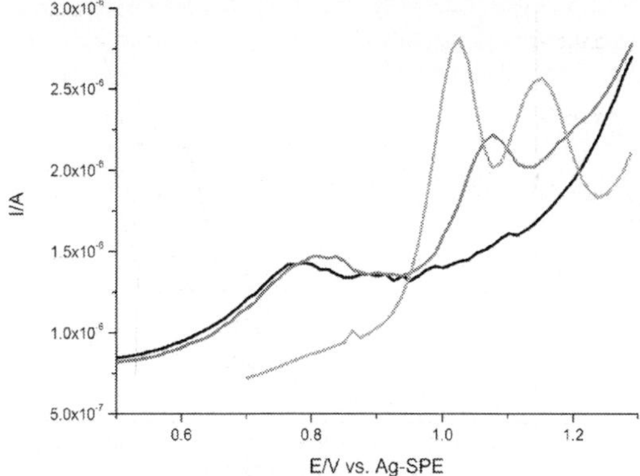

Fig. 15. Anodic differential pulse voltammograms using screen-printed electrodes modified with β-tubulin. (green) unmodified screen-printed electrode, 853.98 ng/mL colchicine in $H_3PO_4/HClO_4$ 0.01M; (black) β-tubulin modified screen-printed electrode, blank anodic DPV run in $H_3PO_4/HClO_4$ 0.01M; (red) β-tubulin modified screen-printed electrode, 853.98 ng/mL colchicine in $H_3PO_4/HClO_4$ 0.01M.

4. Electrochemical sensors for environmental applications

4.1 Heavy metal detection and removal

The importance of sensor development for the detection of pollutants from environmental samples lies in their large field of application. The detection and the removal of the heavy metals show a great interest because they are very persistent in the environment, generally without being decomposed by the bacteria, they are toxic even in traces and their accumulation in the living cells can cause severe pathologies.

A flow electroanalytical system for trace analysis, using modified graphite felt as working electrode was reported (Cristea et al., 2009). The graphite felt was chosen because of its numerous advantages like great specific surface, lack of toxicity, possibility to be used in flow systems, accessibility, low cost and ease of use (Cristea et al., 2005a; Cristea et al., 2005b). The immobilization of 1,8 diamino 3,6 dioxaoctane as receptor has been studied in order to increase the performances of the sensor in terms of selectivity (Geneste & Moinet, 2004; Geneste & Moinet, 2005; Nasraoui et al., 2009). The covalent attachment of the receptor at the surface of the porous electrode has been made by anodic oxidation in a cross-flow cell. The molecular structure of the receptor allows its attachment through either one or both amino groups leading to the formation of a cycle.

The modification of the graphite felt electrode was made by using a ligand previously tested in homogenous medium in reaction with some metallic ions (Loris et al., 1999). The literature reports the stability constants of the complexes obtained with several metallic ions (i.e. Pb(II), Cu(II), Ni(II), Cd(II), Zn(II), Hg(II)).

Graphite felt used as electrode in a flow cell was modified by anodic oxidation with 1,8-diamino-3,6-dioxaoctane in phosphate buffer at pH 7,2. Electrochemical and other experiments showed the double insertion of the compound by two amino groups giving a macrocycle at the surface of the graphite. In order to control the efficiency and the mode of grafting, part of the modified graphite was treated with a specific reagent for amino groups ($4\text{-}O_2NC_6H_4COCl$) bearing an electroactive nitro group that can be electrochemically detected after reaction. Grafting according reaction **4** was not detected; unfortunately, grafting according **5** was not detected too. In this last case, a steric effect can be invoked due to the surface of graphite fibber for explaining the non reactivity of the acyl chloride with the secondary amino groups bonded on the graphite:

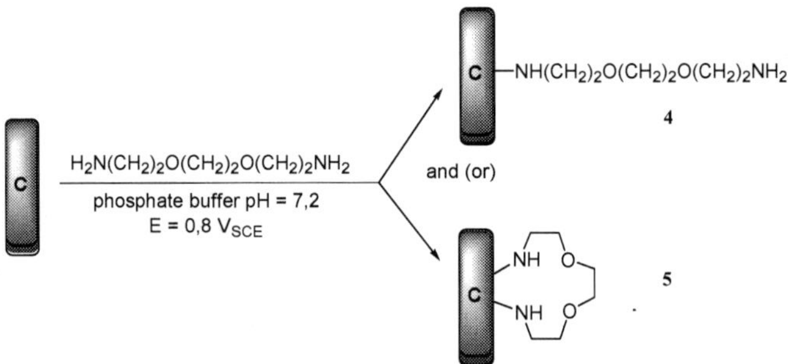

Different techniques were developed during the study, which allowed the grafting of the compound of interest in three distinct ways: predominantly as a cycle, predominantly as a one-end-attached molecule and as a mixture of the first two.

A formation mechanism of the macrocycle between the 1,8-diamino-3,6-dioxaoctane and the graphite felt was imagined. The proposed mechanism involves the two amino groups susceptible to oxidation at the same potential. It is probably a simultaneous radical attack of a double bond undergoing until the saturation of the surface sites. At this point, the oxidation driving force will disappear because of the lack of the radical coupling with the graphite and the oxidation of the two amino groups will become more difficult. It could be assumed that the macrocycle will be formed with two vicinal atoms of carbon:

The detection of the heavy metals (especially lead) has been performed in two steps: the preconcentration by complexation of the analyte by percolating the sample solution through the modified electrode and its analysis by linear sweep stripping voltammetry. By studying the influence of different experimental parameters (receptor's immobilization type, time of percolation, etc.) the best electrochemical signal for lead was obtained when the receptor was bonded to the surface of the electrode at only one end.

However, the electroanalytical verification of grafting according reaction **5** was obtained by Pb(II) trapping by ultrasonication, from a solution of $Pb(NO_3)_2$ ($5 \cdot 10^{-2}$ M), with prior rinsing with water:

The cyclic voltammogram using an electrode of 1 cm diameter confirms the presence of crypted Pb(II) which after the cathodic reduction shows a characteristic anodic peak denoting the dissolution of lead (Fig. 16).

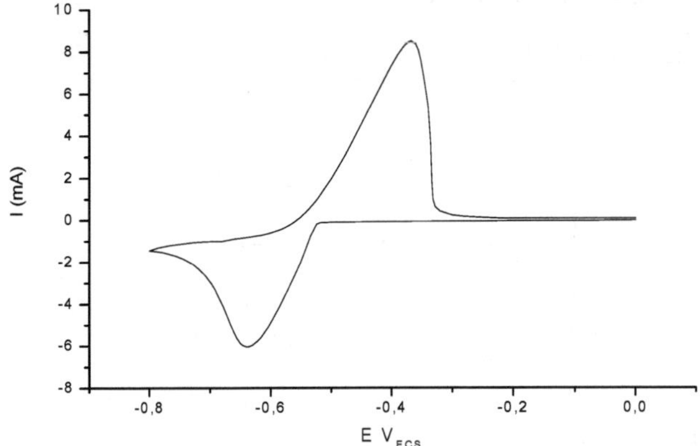

Fig. 16. Cyclic voltamogram on a modified graphite felt porous electrode immersed for 20 h $5 \cdot 10^{-2}$ M Pb(NO$_3$)$_2$; supporting electrolyte 0.1M LiClO$_4$, scan rate 20mVs^{-1}.

The reaction rate between the 1,8-diamino-3,6-dioxaoctane and the lead was also investigated, which turns out to be strongly time dependent. It can be concluded that the complexation reaction between the grafted ligand and the metal ion is relatively slow. The results show that that the complete trapping of the metallic ion is done after 3 hours at a flow rate of 0.6 mL/min of the metal ion solution. The signal decreases with the increase of the flow rate to 1.2, 2.4 and 4.8 mL/min for a percolation time of 30 minutes.

The trapping of other five heavy metals ions were investigated using the grafted porous graphite felt electrode. The obtained limits of detection expressed as concentration in mol/cm^3 of graphite felt are $1.6 \cdot 10^{-7}$ ($2.6 \cdot 10^{-10}$) for Pb^{2+}, $2 \cdot 10^{-8}$ ($3.3 \cdot 10^{-11}$) for Ni^{2+}, $1.2 \cdot 10^{-8}$ ($2 \cdot 10^{-11}$) for Cd^{2+}, $4.5 \cdot 10^{-9}$ ($7.4 \cdot 10^{-12}$) for Zn^{2+} and $3.9 \cdot 10^{-9}$ ($6.4 \cdot 10^{-12}$) for Cu^{2+}.

Other ligands were studied in order to obtain new electrode materials for the removal of the heavy metal ions from aqueous solutions. The covalent immobilization of cyclam on the graphite felt was carried out via a previously attached amino acid linker and the resulting sensor's performance for heavy metals detection was studied. The cyclam-modified electrode showed ability to detect Pb(II) ions without the interference of Cu(II) with a limit of detection down to $5 \cdot 10^{-8}$ mol L^{-1}. Its behavior exhibits several differences compared with the unmodified graphite felt: the influence of the flow rate on the complexation reaction and an improved selectivity for Pb(II) over Zn(II) (Nasraoui et al., 2010).

Another modified graphite felt electrode was obtained by using the carbamoyl-armed macrocycles which exhibit a good selectivity for Pb(II) ions. A flow-through electrochemical sensor was prepared by covalent attachment of 1,4,8-tri(carbamoylmethyl) hydroiodide (TETRAM) on a graphite felt electrode. This new sensor exhibits higher selectivity than the cyclam-modified electrode for Pb(II) ions, and does not show interference from other common metal ions. In the optimum analytical conditions a limit of detection of $2.5 \cdot 10^{-8}$ mol L^{-1} was assessed (Nasraoui et al., 2010).

4.2 Polyphenols detection

The phenolic compounds are widely used chemicals turning into toxic pollutants upon their release into the environment. Their environmental occurrence is a result of various production processes of man-made materials such as plastics, dyes, pesticides, paper and petrochemicals.

Phenols are easily absorbed through the skin and mucous membranes of animals and humans, representing a serious environmental issue.

The determination of phenolic compounds is of great importance due to their toxicity and persistency in the environment. Hence, the rapid determination of trace phenolic compounds is of great importance for evaluating the total toxicity of environmental or industrial samples.

One of the most effective and simplest ways to detect the phenolic compounds was by means of amperometric biosensors based on polyphenol oxidase (PPO) (Shan et al., 2007). Polyphenol oxydase (PPO), also called tyrosinase, is a metalloenzyme that contains a binuclear copper active site and catalyses, in the presence of dioxygen, the hydroxylation of monophenols to catechols (monooxygenase activity), which in turn are oxidized to *ortho*-quinone (catecholase activity) (Duckworth & Coleman, 1970). The phenol biosensor transduction is thus based on the amperometric detection of the enzymatically generated *o*-quinone.

The literature reports several configurations of amperometic biosensors for the detection of polyphenols in environmental and industrial samples. One of this was using for the first time a polycrystalline BiOx films for the entrapment of biomolecules. This procedure constituted an inexpensive, fast and easy method for the elaboration of enzyme electrodes, by the simple mixing of the colloidal suspension of bismuth oxide with proteins, followed by their deposition on the electrode surface (Shan et al., 2009).

Since PPO catalyses the formation of *ortho*-quinone and subsequent consumption of dioxygen, the majority of the PPO-based biosensors involves an amperometric transduction based on the quinone reduction. Actually the quinoid products inhibit PPO by undergoing an irreversible binding and passivate the electrode surface via the formation of insulating polymer films. Thus, the majority of the biosensor configurations were based on a mediated reduction of *orhto*-quinone. Different approaches involve conducting organic salt, metal microparticles and redox mediators immobilization together with the enzyme in a polymer matrix (Cosnier et al., 2001).

Recently, organic phase enzyme electrodes (OPEEs) have attracted considerable interest for their application in environmental and clinical monitoring (Cosnier & Innocent, 1993; Cosnier & Popescu, 1996; Mousty et al., 2001).

Electropolymerisation of functionalized polymers is an electrochemical method for the immobilization of biomolecules where the attachment of proteins is performed directly at the polymer–solution interface by chemical grafting or affinity. In comparison to the physical entrapment in polymer films, this approach preserves the catalytic activity and the molecular recognition properties of the immobilized biomolecules.

Several configurations of OPEE are reported in literature being recently reviewed (Lopez et al., 2006).

The use of organic solvents facilitates the detection of non-soluble or poorly water soluble compounds, prevents microbial contamination and may circumvent side reactions leading to enzyme deactivation or electrode fouling.

The key component of biosensors is the enzyme that is responsible for the specific and sensitive recognition of the analyte. The literature data show that the organic phase enzyme electrodes have been prepared with some proteins that after immobilization preserve functional activity. Some of such biosensors have been successfully applied for real samples working in organic media. Besides tyrosinase other enzymes were also tested in monoenzymatic or in bienzymatic systems. As working electrodes, graphite, glassy carbon, and oxygen gaseous diffusion electrodes were generally used. As for the solvents, acetonitrile, chloroform, dioxane and hexane have been tested (Reviejo et al., 1994). The activity of enzymes in organic media is strongly dependent on their hydration layer, which is essential for their conformational flexibility and for the sensor's stability.

A novel organic phase enzyme electrode (OPEE) via polyphenol oxidase (PPO) entrapment within a hydrophilic polypyrrole film electrogenerated from a new bispyrrolic derivative containing a long hydrophilic spacer was obtained. The pyrrole derivative used in the construction of the biosensor bears two pyrrole groups linked by water solubilizing ethoxy chain:

Its electrochemical behavior was investigated by cyclic voltammetry at glassy carbon electrode (GC) in aqueous solution containing 0.1M LiClO$_4$ as supporting electrolyte. During the oxidative potential scan up to 1.3V, the cyclic voltammogram of the monomer (5 mM) displays an irreversible anodic peak at 1.1V due to the oxidation of the pyrrole group into radical cation form. The repetitive potential scanning of the electrode between 0 and 0.8V induces the appearance and the growth of a quasi-reversible peak system, located around 0.4V. This evolution indicates the formation and growth of a polymer film on the glassy carbon surface. The polymerization can also be carried out by controlled potential electrolysis (E_{app} = 0.8V versus Ag/AgCl). It proceeds via a radical cation coupling, involving the release of two electrons and protons per pyrrole unit. Owing to the partly lipophilic character of the long alkyl chain linking the two polymerizable pyrrole groups, the monomer can be adsorbed onto the glassy carbon electrode prior its electropolymerization. This original method, developed in Serge Cosnier's laboratory (Cosnier et al. 2006) for amphiphilic polypyrrole films, allows the control of the monomer and enzyme amounts coated onto the electrode surface and, consequently, of the thickness of the polymer film. The immobilization of PPO was performed by electropolymerization of monomer /enzyme mixture deposited on the surface of glassy carbon electrode following the "adsorption step procedure". For this purpose, a controlled potential of 0.8 V versus Ag/AgCl was applied at glassy carbon electrodes, modified by PPO-monomer coatings, with different ratios of enzyme/monomer of PPO. Since the polymer films were formed from an adsorbed layer, the enzyme loss during the polymerization step can be determined by measuring the enzymatic activity of the surrounding electrolyte. Regardless of the initial amount of PPO in the coating, the quantity of actually immobilized PPO in the polymeric matrix has been estimated at ~99%. The PPO–polymer biosensors were calibrated with catechol, using the o-quinone reduction at an applied potential of −0.2 V versus Ag/AgCl.

The biosensor's sensitivity to catechol (incorporating 150 nmol monomer and 0.40 mg PPO), measured in aqueous buffer solution (pH 6.5), was found to be 116 mAM^{-1} cm^{-2}. In contact with anhydrous chloroform the biosensor still presents a rapid response (t_{90} ~30 s) to the changes in the catechol concentration. However, irrespective of the enzyme matrix composition, all sensitivities were significantly lower than that measured in the aqueous solution. It is worthy to mention that, the saturation of chloroform with water does not improve the biosensors sensitivity (Fig. 17).

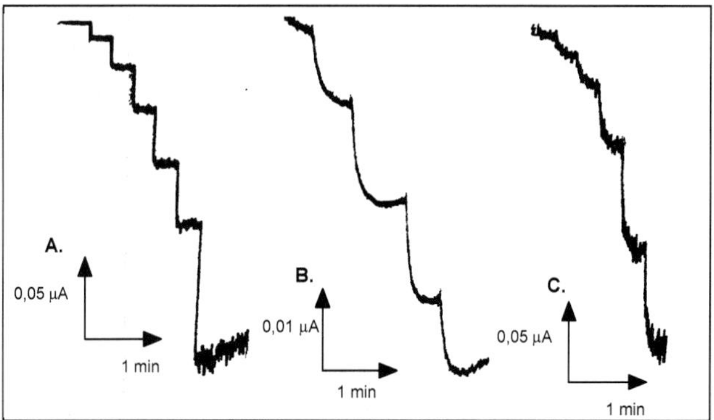

Fig. 17. Amperometric response of the modified electrode with GC-polymer-PPO (150 mmoles monomer and 0.40 mg PPO) in: A. 0.1 M phosphate buffer pH=6.5, successive addition of catechol ($10^{-6}<c<10^{-5}$ M), B. $CHCl_3$, 0.1 M TBAP and phosphate buffer pH=6.5 0.5% (v/v) to successive addition of cathecol ($2\cdot10^{-4}<c<8\cdot10^{-4}$ M), C. $CHCl_3$ and 0.1 M TBAP to successive additions of catehcol ($6\cdot10^{-6}<c<4\cdot10^{-5}$).

Consequently, the strong decrease of the biosensor sensitivity observed in chloroform may be due to the decrease of modified polymer matrix permeability but also to the modification of the enzyme microenvironment within the cross-linked polymer. In order to optimize the biosensor's construction, the effect of the composition of the PPO-monomer coatings on the biosensor's sensitivity was investigated. For the same enzyme loading (0.20 mg), an increase in the monomer's amount does not lead to an improvement of the biosensor sensitivity. As a matter of fact, an increasing polymer film thickness will increase the resistance to diffusion of the substrate from the polymer–solution interface to the enzyme located in proximity of the electrode surface. The optimum composition of the monomer–PPO coating was chosen as 150 nmol of monomer and 0.50 mg PPO. In order to examine the reproducibility of the biosensor's construction, four electrodes were prepared using the same mixture of enzyme and monomer. A relative standard deviation (RSD) of 7.6% was recorded for the biosensor response to catechol (5µM), this substrate concentration corresponding to the linear part of the calibration curve.

5. Conclusion

Electrochemical sensors are very attractive tools in pharmaceutical, biomedical and environmental analysis, combining all the advantages of the electrochemical methods, especially their high sensitivity allowing the detection of very low concentrations of analytes, with the extreme specificity of bioelements (i.e. enzymes, antibodies or DNA). Biosensors offer an extreme flexibility brought on not only by the wide range of biocomponents and composite electrode materials, but also by the inherent conformational and operational versatility that facilitates various configurations and geometrical shapes. Last but not least, their analytical value is further enhanced by the possibility of miniaturization (i.e. screen-printed electrodes and the so called „lab-on-a-chip" devices). Nevertheless there are still many drawbacks and problems more or less difficult to solve among which lack of reproducibility, very short service life (especially for enzyme-modified electrodes, either due to the immobilization or the loss of enzymatic activity), biocompatibility in the case of „in vivo" implanted sensors. However, all these issues represent an irresistible challenge for the scientists. It will be the task of this century to ensure the future development and discovery of new electrochemical sensors and their transfer from sophisticated laboratories to end-users from pharmaceutical industry, environmental agencies and hospitals.

6. References

Amine, A.; Mohammadi, H.; Bourais, I. & Palleschi, G. (2006). Enzyme inhibition-based biosensors for food safety and environmental monitoring. *Biosensors and Bioelectronics*, Vol. 21, No. 8, (February 2006), pp. 1405-1423

Ammam, M. & Fransaer, J. (2010). Glucose microbiosensor based on glucose oxidase immobilized by AC-EPD: Characteristics and performance in human serum and in blood of critically ill rabbits. *Sensors and Actuators B*, Vol. 145, No. 1, (March 2010), pp. 46-53

Andreescu, S. & Marty, J.L. (2006). Twenty years research in cholinesterase biosensors: From basic research to practical applications. *Biomolecular Engineering*, Vol. 23, No. 1, (March 2006), pp. 1-15

Bagni, G.; Osella, D.; Sturchio, E. & Mascini, M. (2006). Deoxyribonucleic acid (DNA) biosensors for environmental risk assessment and drug studies. *Analytica Chimica Acta*, Vol. 573-574, (2006), pp. 81-89

Bakker, E. (2004). Electrochemical sensors. *Analytical Chemistry*, Vol. 76, No. 12, (June, 2004), pp. 3285–3298

Banerjee, S.; Chakrabarti, G. & Bhattacharyya, B. (1997). Colchicine binding to tubulin monomers: A mechanistic study. *Biochemistry*, Vol. 36, No. 18, (1997), pp. 5600-5606

Bănică, F.; Bodoki, E.; Marian, E.; Vicaş, L. & Săndulescu, R. (2007). The construction and evaluation of an ion selective diazepam electrode for the potentiometric titration of diazepam. *Farmacia*, Vol. 55, No. 2, (March-April, 2007), pp. 207-216

Blankert, B.; Hayen, H.; van Leeuwen, S.M.; Karst, U.; Bodoki, E.; Lotrean, S.; Săndulescu, R.; Mora Diez, N.; Dominguez, O.; Arcos, J. & Kauffmann, J.-M. (2005). Electrochemical, chemical and enzymatic oxidations of phenothiazines. *Electroanalysis*, Vol. 17, No. 17, (September, 2005), pp. 1501-1510

Bodoki, E.; Laschi, S.; Palchetti, I.; Sandulescu, R. & Mascini, M. (2009). Electrochemical behavior of conchicine using graphite based screen-printed electrodes. *Talanta*, Vol. 76, No. 2, (2009), pp. 288-294

Bodoki, E.; Săndulescu, R. & Roman, L. (2007). Method validation in quantitative electrochemical analysis of colchicine using glassy carbon electrode. *Central European Journal of Chemistry*, Vol. 5, No. 3, (September, 2007), pp. 766-778

Bonciocat, N.; Marian, I.O.; Săndulescu, R.; Filip, C. & Lotrean, S. (2003). A new proposal for fast determination of vitamin B2 from aqueous pharmaceutical products. *Journal of*

Pharmaceutical and Biomedical Analysis, Vol. 32, No. 4-5, (August, 2003), pp. 1093-1098

Briganti, S.; Camera, E. & Picardo, M. (2003). Chemical and instrumental approaches to treat hyperpigmentation. *Pigment Cell Research*, Vol. 16, No. 2, (April 2003), pp. 101-110

Castillo, J.; Gaspar, S.; Leth, S.; Niculescu, M.; Mortari, A.; Bontidean, I. et al. (2004). Biosensors for life quality design, development and applications. *Sensors and Actuators B*, Vol. 102, No.2, (September 2004), pp. 179-194

Câmpean, A.; Tertiş, M. & Săndulescu, R. (2011) Electrochemical behavior of some purine derivatives on carbon based electrodes. *Central European Journal of Chemistry*, (2011), accepted for publication

Chang, G.; Tatsu, Y.; Goto, T.; Imaishi, H. & Morigaki, K. (2010). Glucose concentration determination based on silica sol–gel encapsulated glucose oxidase optical biosensor arrays. *Talanta*, Vol. 83, No. 1, (November, 2010), pp. 61-65

Chang, T.S. (2009). An Updated Review of Tyrosinase Inhibitors. *International Journal of Molecular Science*, Vol. 10, No. 6, (May 2009), pp. 2440-2475

Cosnier, S. & Innocent, C. (1993). A new strategy for the construction of a tyrosinase-based amperometric phenol and *o*-diphenol sensor *Bioelectrochemistty and Bioenergetics*, Vol. 31, No. 2, (July, 1993), pp. 147-160

Cosnier, S. & Popescu, I.C. (1996). Poly(amphiphilic pyrrole)-tyrosinase-peroxidase electrode for amplified flow injection-amperometric detection of phenol. *Analytica Chimica Acta*, Vol. 319, No. 1-2, (January, 1996), pp. 145-151

Cosnier, S. (1999). Biomolecule immobilization on electrode surfaces by entrapment or attachment to electrochemically polymerized films. A review. *Biosensors and Bioelectronics*, Vol. 14, No. 5, (May 1999), pp. 443-456

Cosnier, S.; Mousty, C.; Gondran, G. & Lepellec, A. (2006). Entrapment of enzyme within organic and inorganic materials for biosensor applications: Comparative study. *Materials Science and Engineering* C , Vol. 26, No 2-3, (March 2006), pp. 442–447

Cosnier, S.; Senillou, A.; Grätzel, M.; Comte, P.; Vlachopoulos, N.; Jaffrezic-Renault, N. & Martelet, C. (1999). A glucose biosensor based on enzyme entrapment within polypyrrole films electrodeposited on mesoporous titanium dioxide. *Journal of Electroanalytical Chemistry*, Vol. 469, No. 2, (July 1999), pp. 176-181

Cosnier, S.; Szunerits, S.; Marks, R.S.; Lellouche, J.P. & Perie, K. (2001). Mediated electrochemical detection of catechol by tyrosinase-based poly(dicarbazole) electrodes. *Journal of Biochemical and Biophysical Methods*, Vol. 50, No. 1, (December, 2001), pp. 65–77

Cristea, C.; Feier, B.; Geneste, F.; Săndulescu, R. & Moinet, C. (2009). New modified porous electrodes for the removal of heavy metal ions from aqueos solutions. *Journal of Environmental Protection and Ecology*, No. 3, (2009), pp. 633-640

Cristea, C.; Moinet, C.; Jitaru, M. & Dărăbanţu, M. (2005). Electroreduction of (1S,2S)-2-amino-1-(4-nitrophenyl)-propane-1,3-diol derivatives. Behaviour of electrogenerated species and applications to organic synthesis. *Journal of Applied Electrochemistry*, Vol. 35, No. 9, (2005), pp. 845-849

Cristea, C.; Moinet, C; Jitaru, M. & Popescu, I.C. (2005). Electrosynthesis of nitroso compounds from (1S, 2S)-2-amino-1-(4-nitrophenyl)-propane-1,3-diol derivatives. *Journal of Applied Electrochemistry*, Vol. 35, No.9, (2005), pp. 851-855

Dominguez Renedo, O.; Alonso-Lomillo, M.A.; & Arcos Martinez, M.J. (2007). Recent developments in the field of screen-printed electrodes and their related applications. Vol. 73, No. 2, (2007), pp.202-219

Duckworth, H. & Coleman, J.E. (1970). Determination of metals in 100 μl sample volumes by atomic absorption analysis using the "spike-height" method of absorbance measurement. *Analytical Biochemistry*, Vol. 34, No. 2, (April 1970), pp. 382-386

Geneste, F. & Moinet, C. (2004). Electrocatalytic activity of a polypyridyl ruthenium-oxo complex covalently attached to a graphite felt electrode. *New Journal of Chemistry*, Vol. 28, No. 6, (June, 2004), pp. 722-726

Geneste, F. & Moinet, C. (2005). Electrochemically linking TEMPO to carbon via amine bridges. *New Journal of Chemistry*, Vol. 29, No. 2, (February, 2005), pp. 269-271

Grieshaber, D.; MacKenzie, R.; Voros, J. & Reimhult, E. (2008). Electrochemical biosensors - sensor principles and architectures. *Sensors*, Vol. 8, No. 3, (April 2008), pp. 1400-1458

Hu, Y.J.; Liu, Y.; Zhang, L.X.; Zhao, R.M. & Qu, S.Sh. (2005). Studies of interaction between colchicine and bovine serum albumin by fluorescence quenching method. *Journal of Molecular Structure*, Vol. 750, No. 1-3, (2005), pp. 174-178

Jacobs, C.B.; Peairs, M.J. & Venton, B.J. (2010). Carbon nanotube based electrochemical sensors for biomolecules. *Analytica Chimica Acta*, Vol. 662, No. 2, (March, 2010), pp. 105-127

Jiang, Y.; Xiao, L.; Zhao, L.; Chen, X.; Wang, X. & Wong, K. (2006). Optical biosensor for the determination of BOD in seawater, *Talanta* Vol. 70, No. 1, (August, 2006), pp. 97-103

Kok, F.N.; Bozoglu, F. & Hasirci, V. (2002). Construction of an acetylcholinesterase–choline oxidase biosensor for aldicarb determination. *Biosensors and Bioelectronics*, Vol. 17, No. 6-7, (June 2002), pp. 531-539

Kong, T.; Chen, Y.; Ye, Y.; Zhang, K.; Wang, Z. & Wang, X. (2009). An amperometric glucose biosensor based on the immobilization of glucose oxidase on the ZnO nanotubes. *Sensors and Actuators B*, Vol. 138, No. 1, (April, 2009), pp. 344–350

Laschi, S.; Bulukin, E.; Palchetti, I,; Cristea, C. & Mascini, M. (2008). Disposable electrodes modified with multi-wall carbon nanotubes for biosensor applications. *IRBM* , Vol. 29, No. 2-3, (April-May 2008), pp. 202-207

Lenigk, R.; Lam, E.; Lai, A.; Wang, H.; Han, Y.; Carlier, P. & Renneberg, R. (2000). Enzyme biosensor for studying therapeutics of Alzheimer's disease. *Biosensors and Bioelectronics*, Vol. 15, No. 9-10, (November 2000), pp. 541-547

Liu, X.; Shi, L.; Niu, W.; Li, H. & Xu, G. (2008). Amperometric glucose biosensor based on single-walled carbon nanohorns. *Biosensors and Bioelectronics*, Vol. 23, No. 12, (July, 2008), pp. 1887–1890

Loris, J.M.; Martinez-Manez, R.; Padilla-Tosta, M.E.; Pardo, T.; Soto, J.; Beer, P. D.; Cadman, J. & Smith, D. K. (1999). Cyclic and open-chain aza–oxa ferrocene-functionalized derivatives as receptors for the selective electrochemical sensing of toxic heavy metal ions in aqueous environments, *Journal of the Chemical Society, Dalton Transactions*, No. 14, (1999), pp. 2359-2370

Marian, I.O.; Bonciocat, N.; Săndulescu, R. & Filip, C. (2001). Direct voltammetry for vitamin B_2 determination in aqueous solutions by using a glassy carbon electrode. *Journal of Pharmaceutical and Biomedical Analysis*, Vol. 24, No. 5-6, (March, 2001), pp. 1175-1179

Marian, I.O.; Săndulescu, R. & Bonciocat, N. (2000). Diffusion coefficient and concentration determination of ascorbic acid in the Fredholm alternative using a carbon paste, *Journal of Pharmaceutical and Biomedical Analysis*, Vol. 23, No. 1, (August, 2000), pp. 227-230

Menon, S.; Fleck, R.W.; Yong, G. & Strothkamp, K.G. (1990). Benzoic acid inhibition of the alpha, beta and gama isozymes of Agaricus bisporus tyrosinase. *Archives of Biochemistry and Biophysics*, Vol. 280, No. 1, (July 1990), pp. 27-32

Mousty, C.; Lepellec, A.; Cosnier, S.; Novoa, A. & Marks, R. (2991). Fabrication of organic phase biosensors based on multilayered polyphenol oxidase protected by an alginate coating. *Electrochemistry Communication*, Vol. 3, No.12, (December 2001), pp. 727-732

Nasraoui, R.; Floner, D. & Geneste, F. (2009). Analytical performances of a flow electrochemical sensor for preconcentration and stripping voltammetry of metal ions. *Journal of Electroanalytical Chemistry*, Vol. 629, No. 1-2, (April, 2009), pp. 30–34

Nasraoui, R.; Floner, D. & Geneste, F. (2010). Improvement in performance of a flow electrochemical sensor by using carbamoyl-arms polyazamacrocycle for the preconcentration of lead ions onto the electrode. *Electrochemistry Communications*, Vol. 12, No. 1, (January, 2010), pp. 98–100

Nasraoui, R.; Floner, D.; Paul-Roth, C. & Geneste, F. (2010). Flow electroanalytical system based on cyclam-modified graphite felt electrodes for lead detection. *Journal of Electroanalytical Chemistry*, Vol. 638, No. 1, (January, 2010), pp. 9–14

Nenkova, R.; Ivanova, D.; Vladimirova, J. & Godjevargova, T. (2010). New amperometric glucose biosensor based on cross-linking of glucose oxidase on silica gel/multiwalled carbon nanotubes/polyacrylonitrile nanocomposite film. *Sensors and Actuators B*, Vol. 148, No. 1, (June, 2010), pp. 59–65

Olea, D.; Viratelle, O. & Faure, C. (2008). Polypyrrole-glucose oxidase biosensor: Effect of enzyme encapsulation in multilamellar vesicles on analytical properties. *Biosensors and Bioelectronics*, Vol. 23, No. 6, (January, 2008), pp. 788–794

Paleček, E.; Fojta, M.; Tomschik, M. & Wang, J. (1998). Electrochemical biosensors for DNA hybridization and DNA damage. *Biosensors and Bioelectronics*, Vol. 13, No. 6, (1998), pp. 621-628

Pan, D.; Chen, J.; Yao, Sh.; Nie, L.; Xia, J. & Tao, W. (2005). Amperometric glucose biosensor based on immobilization of glucose oxidase in electropolymerized o-aminophenol film at copper-modified gold electrode. *Sensors and Actuators B*, Vol. 104, No. 1, (January, 2005), pp. 68–74

Prodromidis, M.I. & Karayannis, M.I. (2002). Enzyme based amperometric biosensors for food analysis, *Electroanalysis* Vol. 14, No. 2 (2002) pp. 241–261

Ramanavicius, A.; Kausaite, A. & Ramanaviciene, A. (2005). Polypyrrole-coated glucose oxidase nanoparticles for biosensor design. *Sensors and Actuators* B, Vol. 111–112, (November, 2005), pp. 532–539

Reviejo, A.J.; Liu, F. & Pingarron, J.M. (1994). Amperometric biosensors in reversed micelles. *Journal of Electroanalytical Chemistry*, Vol. 374, No. 1-2, (August 1994), pp.133–139

Rogers, K.R. (2006). Recent advances in biosensor techniques for environmental monitoring. *Analytica Chimica Acta*, Vol. 568, No. 1-2, (May 2006), pp. 222-231

Rubianes, M.D. & Rivas, G.A. (2007). Dispersion of multi-wall carbon nanotubes in polyethylenimine: A new alternative for preparing electrochemical sensors. *Electrochemistry Communications*, Vol. 9, No. 3, (March, 2007), pp. 480–484

Rubio Retama, J.; Lopez Cabarcos, E.; Mecerreyesc, D. & Lopez-Ruiz, B. (2004). Design of an amperometric biosensor using polypyrrole-microgel composites containing glucose oxidase. *Biosensors and Bioelectronics*, Vol. 20, No. 6, (December, 2004), pp. 1111–1117

Rusling, J.F. (2010). Biomolecular Films. Design, Function, and Applications, Surfactant Science Series, Vol.111, Marcel Decker, Inc., 2003, New York; Pier Andrea Serra, Biosensors, InTech, 2010

Sanchez-Paniagua Lopez, M.; Lopez-Cabarcos, E. & Lopez-Ruiz, B. (2006). Organic phase enzyme electrodes. *Biomolecular Engineering*, Vol. 23, No. 4, (September 2006), pp. 135–147

Săndulescu, R.; Mirel, S. & Oprean, R. (2000). The development of spectrophotometric and electrochemical methods for ascorbic acid and acetaminophen determination and their applications in the analysis of effervescent dosage forms. *Journal of Pharmaceutical and Biomedical Analysis*, Vol. 23, No. 1, (August, 2000), pp. 77-87

Săndulescu, R.; Mirel, S.; Oprean, R. & Lotrean, S. (2000). Comparative electrochemical study of some phenothiazines with carbon paste, solid carbon paste and glassy carbon electrodes. *Collection of Czechoslovak Chemical Communications*, Vol. 65, No. 6, (June, 2000), pp. 1014-1028

Shan, D.; Li, Q.; Xue, H. & Cosnier, S. (2008). A highly reversible and sensitive tyrosinase inhibition-based amperometric biosensor for benzoic acid monitoring. *Sensors and Actuators B*, Vol. 134, No. 2, (September 2008), pp. 1016-1021

Shan, D.; Zhang, J.; Xue, H.; Zhang, Y.; Cosnier, S. & Ding, Sh. (2009). Polycrystalline bismuth oxide films for development of amperometric biosensor for phenolic compounds. *Biosensors and Bioelectronics*, Vol. 24, No. 12, (August, 2009), pp. 3671–3676

Shan, D.; Zhu, M.; Han, E., Xue, H. & Cosnier. (2007). Calcium carbonate nanoparticles: A host matrix for the construction of highly sensitive amperometric phenol biosensor. *Biosensors and Bioelectronics*, Vol. 23, No. 5, (December, 2007), pp. 648–654

Sima, V.; Cristea, C.; Bodoki, E.; Duţu, G. & Săndulescu, R. (2010). Screen-printed electrodes modified with HRP– zirconium alcoxide film for the development of a biosensor for acetaminophen detection. *Central European Journal of Chemistry*, Vol. 8, No. 5, (October 2010), pp. 1034-1040

Sima, V.; Cristea, C.; Lapadus, F.; Marian, I.O.; Marian, A. & Sandulescu, R. (2010). Electroanalytical properties of a novel biosensor modified with zirconium alcoxide porous gels for the detection of acetaminophen. *Journal of Pharmaceutical and Biomedical Analysis*, Vol. 48, No. 4, (December 2008), pp. 1195-1200

Sima, V.H.; Patris, S.; Aydogmus, Z.; Sarakbi, A.; Sandulescu, R. & Kauffmann, J.M. (2011). Tyrosinase immobilized magnetic nanobeads for the amperometric assay of enzyme inhibitors: Application to the skin whitening agents. *Talanta*, Vol. 83, No. 3, (January 2011), pp. 980-987

Teles, F.R.R. & Fonseca, L.P. Trends in DNA biosensors. *Talanta*, Vol.77, No. 2, (2008), pp. 606-623

Thevenot, D.R.; Toth, K.; Durst, R.A. & Wilson, G.S. (2001). Electrochemical biosensors: recommended definitions and classification. *Biosensors and Bioelectronics*, Vol. 16, No. 1-2, (January 2001), pp. 121-131

Wang, J. (1999). Amperometric biosensors for clinical and therapeutic drug monitoring: a review. *Journal of Pharmaceutical and Biomedical Analysis*, Vol. 19, No. 1-2, (February 1999), pp. 47-53

Wu, B.; Zhang, G.; Shuang, Sh & Choi, M.M.F. (2004). Biosensors for determination of glucose with glucose oxidase immobilized on an eggshell membrane. *Talanta*, Vol. 64, No. 2, (Octomber, 2004), pp. 546-553

Xu, H.; Hou F. & Zhang, X. (2011). Electrochemical investigation on the interaction between colchicine and DNA. *Current Physical Chemistry*, Vol.1, No. 1, (2011), pp.11-15

Xu, Z.; Chen, X. & Dong, S. (2006). Electrochemical biosensors based on advanced bioimmobilization matrices. *Trends in Analytical Chemistry*, Vol. 25, No. 9, (October 2006), pp. 899-908

Yang, X.; Zhou, Z.; Xiao, D. & Choi, M.M.F. (2006). A fluorescent glucose biosensor based on immobilized glucose oxidase on bamboo inner shell membrane. *Biosenosors and Bioelectronics*, Vol. 21, No. 8, (February, 2006), pp. 1613-1620

Yang, Y.; Yang, H.; Yang, M; Liu, Y.; Shen, G. & Yu. R. (2004). Amperometric glucose biosensor based on a surface treated nanoporous ZrO_2/Chitosan composite film as immobilization matrix, *Analitica Chimica Acta*, Vol. 525, No. 2, (November, 2004), pp. 213–220

Yu, D.; Blankert, B.; Bodoki, E.; Bollo, S.; Vire, J.C.; Sandulescu, R.; Nomura, A. & Kauffmann, J.M. (2006). Amperometric biosensor based on horseradish peroxidase-immobilised magnetic microparticles. *Sensors and Actuators B*, Vol. 113, No. 2, (February 2006), pp. 749-754

Yu, D.; Dominguez, R.O.; Blankert, B.; Sima, V.; Sandulescu, R.; Arcos, J. & Kauffmman, J.M. (2006). A peroxidase-based biosensor supported by nanoporous magnetic silica microparticles for acetaminophen biotransformation and inhibition studies. *Electroanalysis*, Vol. 18, No. 17, (September 2006), pp. 1637-1642

14

Ultra-sensitive Detection Using Integrated Waveguide Technologies

Shuhong Li[1], Lynda Kiefer[2], Peixuan Zhu[1],
Daniel Adams[1], Steingrimur Stefansson[3],
Daniel R. Shelton[2], Platte Amstutz[1] and Cha-Mei Tang[1]
[1]Creatv MicroTech, Inc. 11609 Lake Potomac Drive, Potomac, MD 20854,
[2]Environmental Microbial and Food Safety Laboratory, USDA, Beltsville, MD 20705,
[3]Fuzbien Research Institute, 9700 Great Seneca Hwy. Rockville MD 20850,
USA

1. Introduction

There is a pressing need to detect analytes at very low concentrations, such as food- and water-borne pathogens (e.g. *E. coli* O157:H7) and biothreat agents (e.g., anthrax, toxins). Common fluorescence detection methods, such as 96 well plate readers, are not sufficiently sensitive for low concentrations. We describe here a novel detection principle---integrating waveguide technology (IWT)---that allows for greater sensitivity, and report on the sensitivity of a new instrument (Signalyte™-II) based on IWT, and relevant assays.

When fluorescent labels emit light, the emission is typically in all directions such that only a small fraction of the light is collected by the detector as signal. This is the case with 96 well plate readers. Simultaneously, light from the excitation source, auto fluorescence from the sample, Raman emission from water, and electronic noise of the detector contribute to background noise. Consequently, a detection technology that captures the majority of emitted light while eliminating background is inherently more sensitive.

2. Integrating Waveguide Technology (IWT)

Creatv MicroTech, Inc. has developed assays based on IWT which achieves high sensitivity by maximizing the signal while minimizing background noise. IWT assays can be conducted either in a solid-phase or liquid-phase format. In the solid-phase IWT format, analytes are first captured on the inner surface of the capillary tube or cuvette; while in the liquid-phase IWT format, the analyte is in solution inside of the cuvette.

The basic principle of liquid-phase IWT detection is shown in Figure 1. In this configuration, the sample containing fluorescent dye is placed in a glass capillary cuvette. The closed end of the capillary cuvette forms a half-ball lens that focuses the emitted light. Detection and quantitation are achieved by illuminating the cuvette at a 90-degree angle relative to the length of the cuvette. The glass walls of the tube together with the sample act as a waveguide, efficiently gathering and propagating the fluorescent signal from the entire sample to the end of the cuvette and exits through the half-ball lens. The half-ball lens focuses the signal and together with additional optics, allows efficient application of bandpass or longpass filters before sending the signal to the detection.

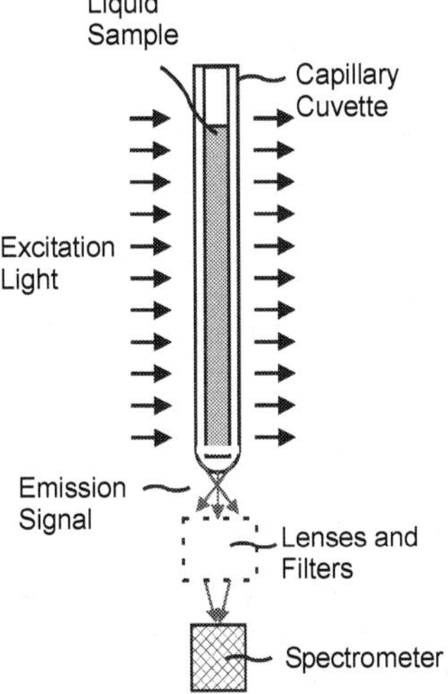

Fig. 1. Integrating Waveguide Technology detection concept.

We have previously reported applications using solid-phase IWT, where analytes are captured on the inner surface of an open ended capillary tube and analyzed using an instrument called Signalyte™ (Li, 2005a). This format has been used to detect and quantify *Escherichia coli* O157 (Zhu et al., 2005) and Bacillus anthracis spores (Hang et al., 2008), where dection limits were 10 cells and 1000 spores, respectively.

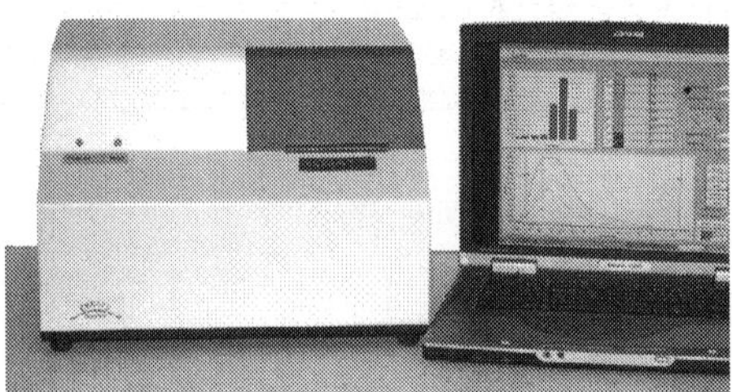

Fig. 2. Signalyte™-II instrument.

Based on these previous studies, a new generation intrument was developed to optimize assays based on liquid phase IWT, called Signalyte™-II; the instrument is shown in Figure 2. Signalyte™-II allows for multiplex assays by providing up to four fluorescence excitation wavelengths from 365 nm – 635 nm. This is achieved by a combination of the excitation source and band pass filters. To allow for maximum flexibility of high power excitation source, LEDs are used. High power LEDs are typically available at peak wavelengths of 365 nm, 470 nm, 530 nm, 590 nm and 635 nm. Because their bandwidths overlaps, these LEDs are applicable for most commercially available fluorescent dyes. The excitation wavelength can be altered, depending on the application.

Bandpass filters for the exciation source and long pass filters for the emission signal are selected to match the absorption and emission spectra for each specific dye. These filters are easily installed or changed. A band pass filter is needed to select the excitation LED light in the wavelength region of the emission signal of interest. This filtered excitation light impinges on the sample. Under ideal situations, all the excitation light leaves. Because of scattering of the excitation light by the sample and curvette, a long pass filter is needed to eliminate or minimize the signal caused by the excitation source.

The detection system for Signalyte™-II is a spectrometer that is sensitive from 350 nm – 800 nm. This spectrum range is applicable for most common fluorescence applications. A spectrometer, rather than photomultiplier tubes (PMT), is used because the spectrum provides information about the background noise, thereby allowing it to be eliminated from data analysis.

To achieve maximum sensitivity, the signal to noise ratio has to be as high as possible. The signal is increased in two ways. First, the cuvette is efficient in gathering and tramsmitting the emitted light to the detector. Numerical simulation shows that about 13 percent of the emission is collected by the cuvette. This percentage is much larger than other standard collection methods such as the use of a lens or optical fiber. Secondly, the duration of time the spectrometer can collect the light can be adjusted over four orders of magnitude with the maximum of 65 seconds. This integration of the signal is important for low fluorescence.

The Signalyte™-II has been designed to minimize noise from the following sources:

- **Excitation source.** Background noise from the excitation light is minimized, because illumination is perpendicular to the waveguide and the majority of the excitation passes through the curvette. Only small amount of scattered light is trapped by the cuvette.

- **Electronic noise.** The spectrumeter is cooled providing low electronic noise. Thus allowing integration of the signal in time without raising the noise and long signal collection time is good for low concentrations.

- **Other instrument related noises.** Other noises in the instrument are primarily associated with optics. Appropriate optics and optical filters reduces this noise.

- **Non-specific binding of fluorescent dyes in sample.** Non-specific binding comes from fluorescent dyes that are not captured on purpose. This can be reduced by optimizing the assay.

Sample size is up to 35 microliters. The instrument simultaneously tests up to eight samples that are loaded using a standard multi-channel pipetter, plus an additional control to measure background Raman and Raleigh scattering. The sample holder including a reference is shown in Figure 3.

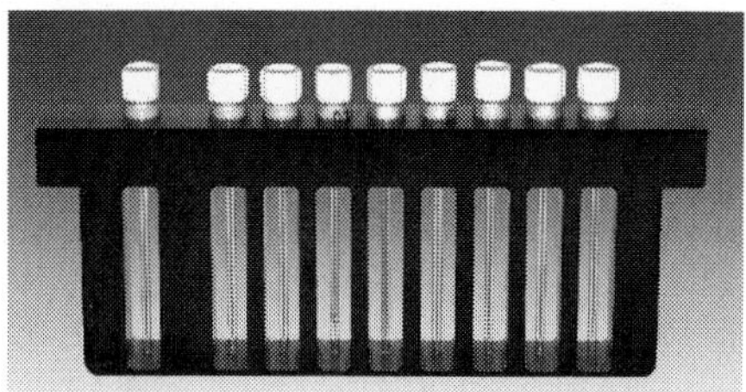

Fig. 3. Cuvette holder allows testing of 8 samples and a reference.

"Signalyte-II Control" is the software that operates Signalyte™-II. The interface of the software is shown in Figure 4.

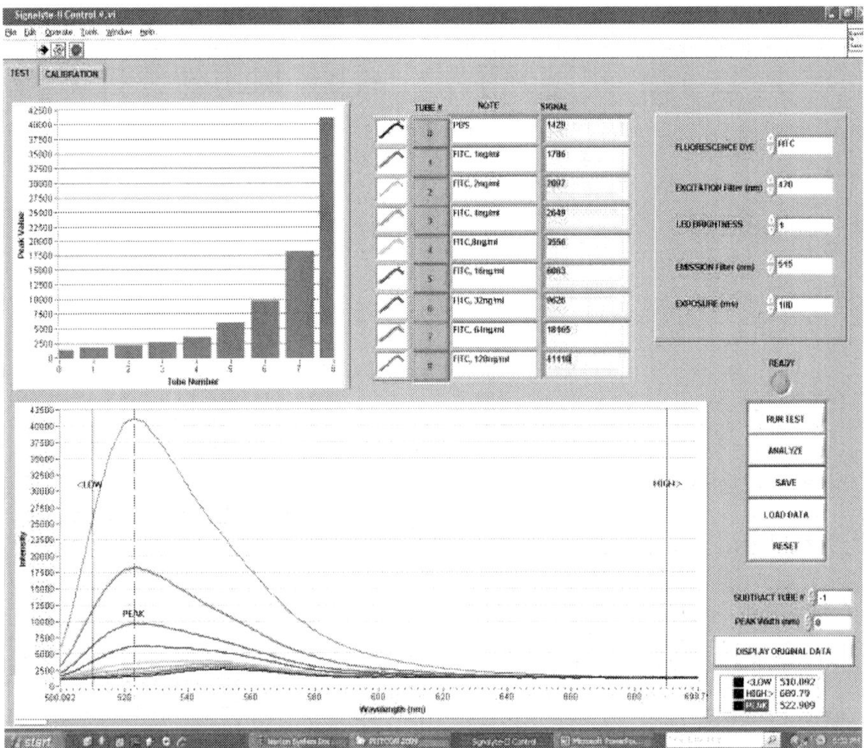

Fig. 4. Signalyte™-II user software interface

Signalyte™-II can test 8 samples plus a reference (tube #0). When testing the samples, the exposure time can be set from 5 milliseconds to 65 seconds. With long exposure time, weak signal can be integrated over a longer period of time, but background noise does not increase as much as the signal does because of the integrating waveguide technology and the fact that the detector is a cooled CCD. A typital exposure time for weak signal is 1 to 3 seconds.

High-power light-emitting diode (LED) is used as illumination light source in Signaly™-II. Comparing to broad spectrum lamps, such as mercury or xenon arc lamps, which are widely used in fluorescence detection instruments, LEDs possess all of the desirable features they lack. Although arc lamps can generate a broad spectrum, only a small percentage of the projected light is utilized for fluorescence detection. Another problem for arc lampd is the excessive heat they generate. In contrast, LEDs are cooler, smaller, and provide a far more convenient mechanism to cycle the source on and off.

Compared to laser light, the wider bandwidth featured by LEDs is more useful for exciting a variety of fluorescent probes. The diverse spectral output afforded by LEDs can supply the optimum excitation wavelength band for fluorophores spanning the ultraviolet, visible, and near-infrared regions. Furthermore, high-power LEDs generate sufficient intensity to provide a useful illumination source for a wide spectrum for fluorescence detection.

Up to four LEDs can be installed in Signalyte™-II. In the following example, there are four LEDs, including red, amber, green and blue LEDs. The fluorescence dyes for the corresponding LEDs are listed accordingly as Cy5, ROX, Cy3, and FITC. When choosing one of these dyes, the excitation filter and emission filter are set at predetermined wavelengths that optimal for the chosen dye. excitation filter shows the center wavelength of the excitation light, and emission filter shows the cut-on wavelength of the long pass emission filter. Table 1 shows the instrument settings for different dyes.

Fluorescence Name	Excitation Wavelength	Long Pass Filter
Cy5	635 nm	LP665
ROX	590 nm	LP630
Cy3	530 nm	LP570
FITC	470 nm	LP515

Table 1. An example of Signalyte™-II setting for fluorescence dyes testing

After the samples are tested, the software automatically measures peak intensity of the signal spectrum, integrates signal strength into a bar chart display, and displays the relevant spectrum.

The user software interface shows three graphs of the testing result. The one on the lower left corner of the user software interface, Figure 4, is SPECTRUM. The one on the upper left corner is BAR CHART. On the upper right side above the SPECTRUM is SIGNAL. After RUN TEST, the software automatically finds the peak intensity of the signal spectrum. The full spectrum (350-800 nm) of each tested sample is displayed in SPECTRUM DISPLAY with colors assigned by SPECTRUM COLOR. The integrated signal strength is displayed in BAR CHART, and is displayed as integer values in BAR CHART.

The original data can be analyzed by dragging the three cursors, including LOW, HIGH, and PEAK, on the SPECTRUM graph. For example, you can expand and investigate a sub spectrum between wavelength range of LOW to HIGH by dragging the two cursors. You can also choose your own PEAK wavelength by changing the PEAK cursor location.

The background noise can be illeminated from the emission spectrum by subtracting tube # 0 (reference) to provide a net fluorescence signal. This feature allows the Signalyte™-II instrument to provide both highly sensitive and precise quantitative data.

Data can be saved as an Excel spreadsheet on a computer connected by USB cable. You can also upload previously saved data to review it.

The limits of detection for three fluorescent dyes and three fluorescent particles using Signalyte™-II are shown in Table 2.

The Signalyte™-II enables numerous assay applications, such as multiplex immunoassays, FRET- based assays, chemiluminescence detection, end point PCR, and DNA and RNA detections without using thermal cycling amplification, multiplex quantum dots assays, polarization assays, etc.

Number of Detected Dye Molecules	Number of Detected Particles
Cy™5: 2.5 pM (5×10^7 dyes)	Purple: 10 nanoparticles
FITC: 25 pM (5×10^8 dyes)	Sky Blue:12 nanoparticles
ROX: 25 pM (5×10^8 dyes)	FluoSphere: 1.3×10^4 particles

Table 2. Signalyte™-II is typically a factor of 100 to 1000 times more sensitive than plate readers.

3. Ultra-sensitive immunoassays using Signalyte™-II

Escherichia coli **O157:H7 Immunoassay.** *E. coli* O157:H7, the most common serotype of enterohemorrhagic *E. coli* (EHEC), is responsible for numerous food-borne and water-borne infections worldwide. Consequently, rapid and sensitive assays for detection of *E. coli* O157 are highly desirable. We have investigated two approaches for the detection of *E. coli* O157 based immunomagnetic separation and on filtration.

Immunomagnetic separation (IMS) techniques are routinely used for isolation, and detection, of EHEC O157:H7 from enriched food and water samples. The assay protocol is as follows. Antibody-coated magnetic beads were first added to the test samples to specifically capture *E. coli* O157:H7 bacteria. Other bacteria and contaminants in the water samples were removed by subsequent washes. The bead-bound *E. coli* O157:H7 were recognized by a Cy5-conjugated anti-O157:H7 polyclonal antibody to form an immunosandwich complex. After washing to remove the excess antibodies, the antibody/cell complex was dissociated from the beads into the liquid phase. The Cy5 fluorescence intensity of the supernatant, which was related to the target concentration, was measured using the Signalyte™-II.

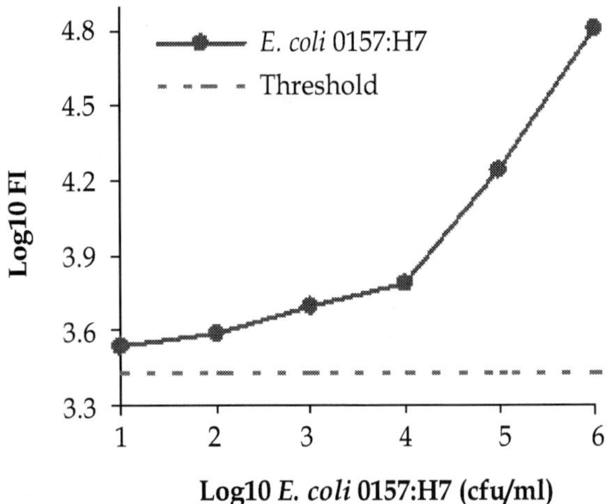

Fig. 5. Results from Creatv's immunoassay kit for detection of 10-fold serial dilutions of *E. coli* O157:H7 read with the Signalyte™-II. For IMS detection format, the limit of detection is 10 cfu/ml.

Our data demonstrate that it is possible to detect as few as 10 cfu/ml of *E. coli* in a 1 ml sample using an IMS format (Figure 5), thus allowing for rapid detection of low concentrations of bacterial cells. The same assay read on a 96-well fluorescent plate reader BMG FLUOstar Omega (BMG Labtech) (Figure 6) had limit of detection of 10^5 cfu/ml.

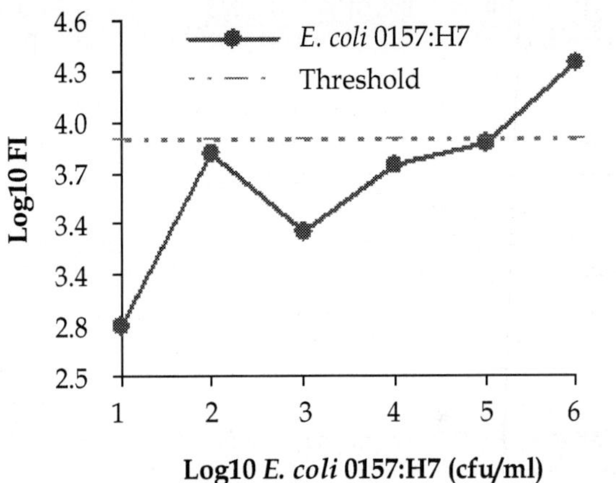

Fig. 6. Results from Creatv's fluorescence immunoassay kit for detection of 10-fold serial dilutions of *E. coli* O157:H7 read with the BMG FLUOstar Omega. The limit of detection is 100,000 cfu/ml.

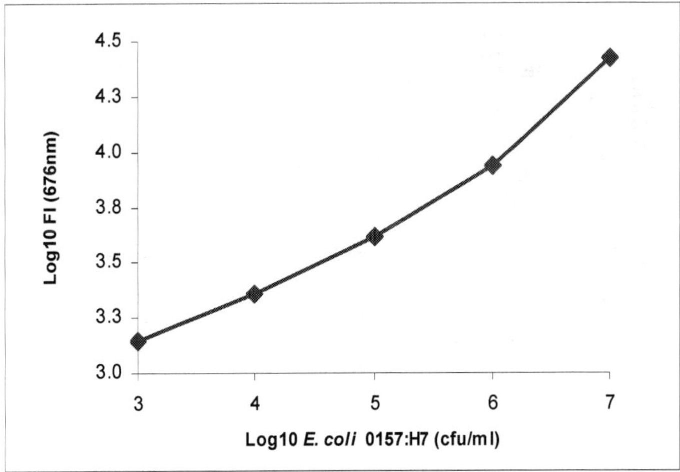

Fig. 7. Results from polyclonal antibody fluorescence immunoassay for detection of 10-fold serial dilutions of *E. coli* 0157:H7 demonstrating 1000 cfu/ml sensitivity. For filtration-based format, the limit of detection was 1000 cfu/ml.

Alternatively, *Escherichia coli* O157:H7 may be detected using a filtration-based format. The assay protocol is as follows. *E. coli* 0157:H7 cells were grown to an approximate cell density of 10^9 cfu/ml. Cells were serially diluted by 10-fold increments in phosphate buffered saline and contacted for 1 hour by tumbling at 37° C with Dylight 649 conjugated *E. coli* 0157:H7 polyclonal antibody. Cell suspensions were filtered through Durapore PDVF (0.1 um pore size) spin filters and the filters exhaustively washed to remove excess antibody. The antibody/cell complex trapped on spin filters was then dissociated with a low pH buffer and the fluorescent antibodies quantified with the Signalyte™-II.

As shown in Figure 7, there was an approximate linear relationship between signal versus log cell concentration; the limit of detection was 1000 cfu *E. coli* 0157:H7 in a 1 ml suspension with polyclonal antibody.

The sensitivity using this format is lower due to non-specific binding of antibodies to the filter polymer, which was eluted with the low pH buffer. We are currently evaluating alternative filter polymers to eliminate this non-specific binding, which will result in lower detection limits. Note that since larger volumes of sample can be incubated with antibody and filtered, the ultimate detection limit is dependent on sample size.

Further, this technology is applicable to any pathogen for which suitable antibodies are available.

4. Integrating waveguide technology in flow through format

IWT is also applicable to the analysis of multiple samples in a flow through format that would be suitable for continuous monitoring. A schematic of the concept is shown in Figure 8, where the sample enters from one end of the capillary and exits from the other end. Again, the capillary along with the sample enclosed by the capillary together acts as a waveguide. The excitation light is incident perpendicular to the capillary, while the fluorescent emission is gathered by a lens, passing through filters and lenses to a detector. The advantage of a flow-through format is the ability to couple high analyte sensitivity with automated detection. A prototype system has been fabricated and is currently in development.

4.1 Flow through format demonstration by using Cy5 fluorescence solution

Proof-of-concept has been demonstrated by passing through a serial dilution of Cy5 fluorescence dye ranging from 1 picomolar to 100 picomolar using a flow through format. Different concentrations of Cy5 were placed in vials respectively. The sample was pumped into the top of the capillary by a parastatic pump at a flow rate of 0.5 ml/min. The parastatic pump had two tubings, input and output. The input tubing was immersed into a vial with Cy5 while the output tubing was connected to the top of the capillary so that the sample cirulated through the capillary. The sample and the capillary formed a waveguide which was illuminated by a 635 nm laser, and signal was detected by a PMT.

The tubing was initially immersed into PBS buffer, followed sequentially by Cy5 solutions of 1 picomolar, 10 picomolar, and 100 picomolar. Finally, the input tubing was left exposed to the air. A software was used to monitor the signal on a real time basis. Figure 9 shows the PMT signal changing over time, which correlated with the Cy5 sample concentration. The signal dropped to background level when there was no liquid (air only) in the flow cell.

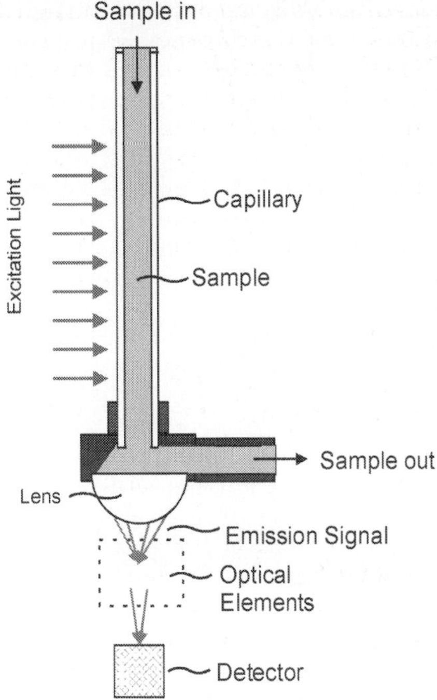

Fig. 8. Integrating Waveguide Technology in a flow through format.

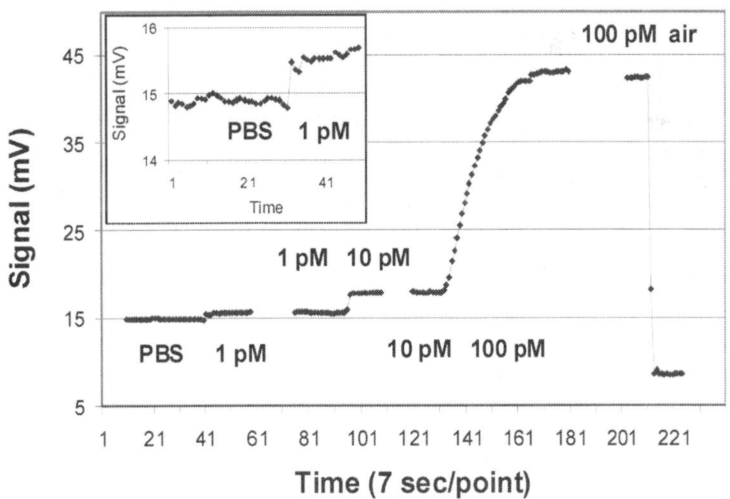

Fig. 9. Cy5 flow through data with a flow rate of 0.5 ml/min.

4.2 In-line detection of treated waste water using ATP assay

As fresh water supplies dwindle, it has become apparent that all water sources must be available for supply to both irrigation and consumption uses. Treating and supplying municipal water is well established throughout the world. However, the reclaiming of water from human and animal waste, most notably for reuse in crop irrigation, is coming to a front as water supplies become scarce.

A number of companies are studying the possibility of treating animal waste to remove chemical and bacterial contaminants, while retaining the water to use on crops. This water quality should conform with applicable national and international regulations and at a minimum follow the World Health Organization's guideline of 1000 coliforms/100 mL for irrigation water. This presents a problem for most water specialists, in that coliform determination is usually done off site and by specialized technicians using testing that requires several days for verification. Here is the point at which Creatv's Integrating Waveguide Technology flow-through biosensor will simplify and streamline the process. Combined with Creatv's buoyant silica microbubbles which are capable of capturing and concentrating various coliforms with high binding efficiency, this technologies can provide the means to concentrate bacteria in real time while providing an automated detection platform to determine coliform contamination. Creatv is collaborating with a developer of a waste treatment instrument, Spiralcat of Maryland, which has the need to test water reclaimed from animal waste using an in-line system.

Creatv has developed poly-L-Lysine coated buoyant silica microbubbles which can bind and retain gram negative and gram positive bacteria cells. The binding efficiencies of Poly-L-Lysine coated microbubbles are 90% for *E. coli* O111, 60% for *E. coli* O26, 58% for *Enterobacter* and 48% for *Salmonella*. To evaluate the use of microbubbles in a flow-though in-line testing environment, they used a cartridge system, as shown in Figure 10. The sample flows in from the top and leaves from the bottom. After the coliform are captured on the surface of microbubbles, an adenosine 5′-triphosphate (ATP) assay using Promega Bactiter-Glo™ reagents has been adapted to gernerate luminescence signal. There are two steps in the process, reaction and detection. After all input sample was flowed through the system, the cartridge was washed by PBS. The "capping" frit was then removed and an ATP Luminescence solution was added to the cartridge, the solution was incubated for 5 minutes at room temperature. After incubation, the samples are transferred to the in-line biosensor, where luminescence signal was tested and coliform concentration was determined.

The in-line biosensor shown schematically in Figure 11 is a portable instrument that can be used for onsite testing. After the ATP incubation was finished, the sample was injected into the testing cartridge, in this prototype a syringe was used to inject the sample. The glass tube and the sample in the testing cartridge act as a waveguide for gathering and guiding the luminescence signal to the PMT detector. Computer software controls the PMT, reads out the signal, and displays the result. When testing is done, sample is removed from the outlet by a syringe. The syringes are used to demonstrate the feasability of the in-line testing concept, and are to be replaced by an automated pumping system, in future designs.

Creatv made several different testing cartridges, and tested them with ATP assays. The optimized cartridge used in the final design was built with a flat glass window at the exit. The testing cartridge consists of a sample chamber, a glass tube that holds the sample, an inlet, an outlet, and a flat glass exit window, as shown in Figure 12. The sample is injected into the chamber from the inlet located at the bottom of the testing cartridge. Sample and the glass tube act as a waveguide for gathering and guiding luminescence to the exit window. The exit window is mounted on top of the input window of the PMT. Computer software controls the PMT, reads out the signal, and displays the result. When testing is done, sample is sucked up from the outlet through a capillary tube. The testing cartridge is reusable by washing the chamber several times with buffer. Buffer can be injected from the inlet and removed from the outlet.

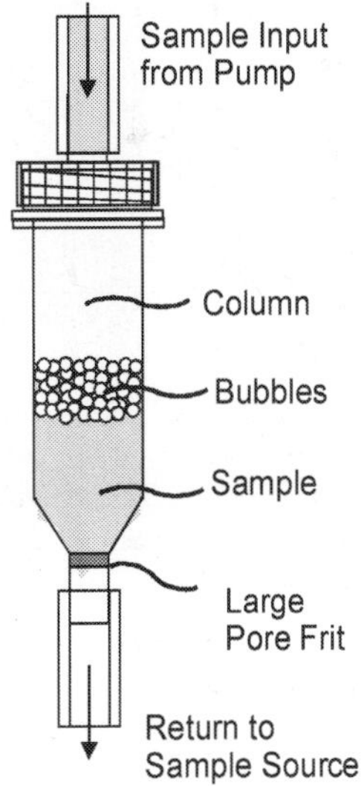

Fig. 10. Flow through cartridge with bubbles.

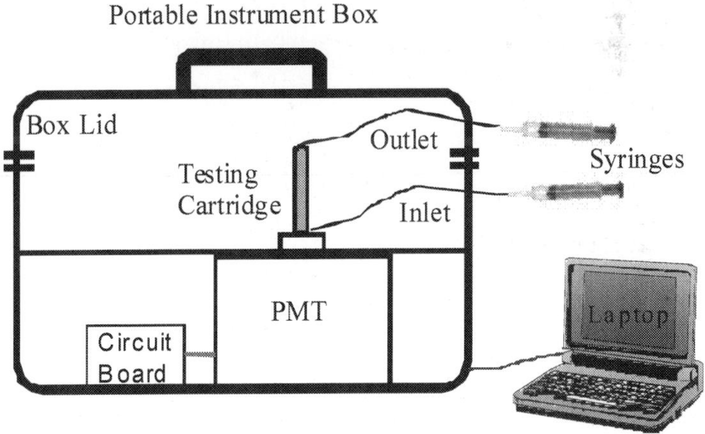

Fig. 11. Schematic of the detection instrument

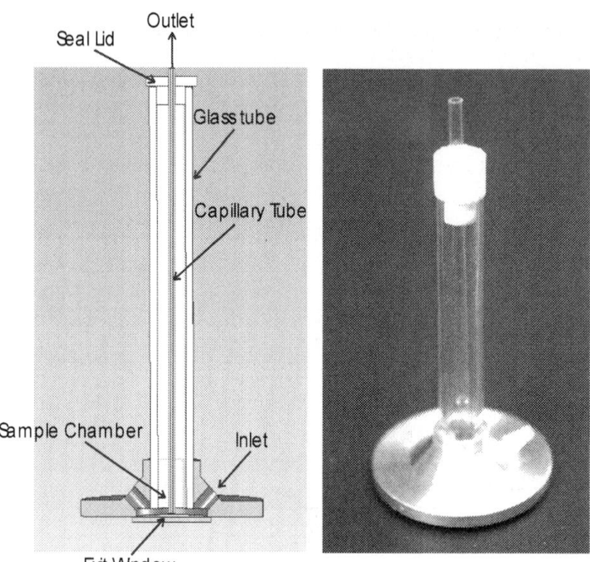

Fig. 12. Illustration of disposable cartridge construction and the photograph of the cartridge.

The full bubble and luminescence assay was run using sterile water to obtain a positive control. A summary of the developed protocol is:

1. Add 250 uL microbubbles to a column with a large pore frit and a capping frit (Cartridge picture) and attach to a peristaltic pump.
2. Flow 100 mL sample at 2mL/min through the column.
3. Wash the column with 5 mL PBS.
4. Add 100 uL PBS and 100 uL Luminescent Reagent to column, Incubate 5 minutes.
5. Remove 200 uL of solution to a 96 well plate for detection.

The positive controls were run in 100 mL sample inputs, with cultured *E. coli* O157 serial diluted from 1:10 concentrations of 1 cfu/mL to 10^6 cfu/mL.

Using the samples from the various batch runs from Spiralcat, this experiment was repeated by spiking *E. coli* into the Spiralcat reclaimed water sample. In Figure 12, the above protocol was used and every concentration was done in triplicate. The "zero" concentration was run 26 times in order to determine limit of detection (LOD). As seen in Figure 13, R values of the pure culture using the bioluminescent reagent was +99%, the R value for *E. coli* spiked in water was 98%, and the R value of the full assay using a "real" output sample from Spiralcat was 91%. Importantly, the capture rate of *E. coli* did seem to be lower in the Spiralcat samples, but additionally the negative control of the Spiralcat sample had a 40% lower background signal with a lower deviation between runs. These results show us that the Luminescent Bubbles assay is highly reproducible over time using "real" samples sampled on different days. Further, the assay is in a format which is adaptable into the prototype instrument we designed with no technician input needed to run a liquid sample.

Test results in Figure 13 show that the in-line luminescent detection system can detect 100 cells of *E. coli* in a 100 mL sample of Spiralcat reclaimed water.

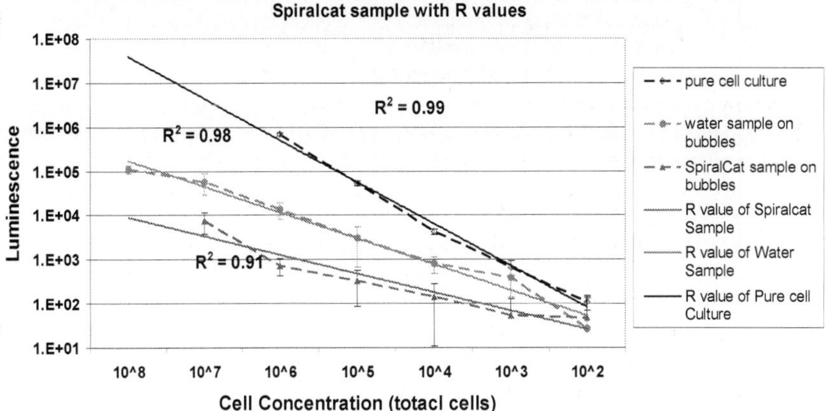

Fig. 13. Comparison of experiment results from the pure cell culture using the bioluminescent reagent, *E. coli* spiked in water using Creatv's microbubble assay, and the same assay using a reclaimed water sample from Spiralcat.

5. Conclusion

The ultra-sensitive fluorescence detection instrument, Signalyte™-II, is based on Integrating Waveguide Technology. The sensitivity is achieved by maximizing the signal while minimizing background noise. A very sensitive *E. coli* O157:H7 detection assay based on IMS techniques was developed and fluorescent signal was tested on Signalyte™-II. For IMS detection format, testing data demonstrate that as few as 10 cfu/ml of *E. coli* in a 1 ml sample is detectable. Another application of Integrating Waveguide Technology is in flow-through format. We demonstrate a real time fluorescence detection system using Cy5 dye. The flow through format can also be adapted to luminescence detection. We developed an in-line detection system of treated waste water using ATP assay. Testing result shows that 100 cells of *E. coli* in a 100 mL sample of reclaimed waste water is detectable.

6. Acknowledgement

The authors would like to thank our collaborators from US Department of Agriculture for their team work and using of their facilities. This work was supported by a grant from the National Institutes of Health Phase II SBIR Grant No. R44 CA094430, and a grant from National Science Foundation Phase I SBIR Grant No. IIP-0945747.

7. Reference

Li, S.; Amstutz, III, P.; Tang, C.-M.; Hang, J.; Zhu, P.; Zhang, Y.; Shelton, D. R., & Karns, J. D. (2009a). Integrating Waveguide Biosensor, In: *Biosensors and Biodetection Methods and Protocols, Vol. 1: Optical-Based Detectors,* A. Rasooly & K. E. Herold, (Ed.), pp, 289-401, Humana Press, ISBN 978-1-60327-566-8, New York, NY

Li, S.; Zhang, Y.; Amstutz, III, P. & Tang, C.-M. (2009b). Multiplex Integrating Waveguide Sensor – SignalyteTM-II, In: *Biosensors and Biodetection Methods and Protocols, Vol. 1: Optical-Based Detectors,* Avraham Rasooly and Keith. E. Herold (Ed.) pp. 423-434, Humana Press, ISBN 978-1-60327-566-8, New York, NY

Hang, J.; Sundaram, A. K.; Zhu, P.; Shelton, D. R.; Karns, J. S.; Martin, P. A. W.; Li, S.; Amstutz, P., & Tang, C.-M. (2008). Development of a Rapid and Sensitive Immunoassay for Detection and Subsequent Recovery of *Bacillus anthracis* Spores in Environmental Samples. *Journal of Microbiological Methods*, Vol.73, No.3, (June 2008), pp. 242-246, ISSN 0167-7012.

Zhu, P.; Shelton, D. R.; Karns, J. S.; Sundaram, A.; Li, S.; Amstutz, P., & Tang, C.-M., (2005). Detection of Water-Borne *E. coli* O157 Using the Integrated Waveguide Biosensor. *Biosensor and Bioelectronics*, Vol.21, No. 4, (October 2005), pp. 678-683, ISSN 0956-5663.

15

Enzyme Based Phenol Biosensors

Seyda Korkut Ozoner[1], Elif Erhan[1] and Faruk Yilmaz[2]
[1]Department of Environmental Engineering,
[2]Department of Chemistry,
Gebze Institute of Technology, Gebze, Kocaeli
Turkey

1. Introduction

Phenol and its derivatives is one of the most important parameters which should be monitored in environmental engineering. They are present in many wastewater streams of the oil, paint, paper, polymer and pharmaceutical industries. Phenolic compounds reach into the food chain by wastewaters then lead to dangerous and toxic effect on aquatic organisms. Principal standard methods for quantitative phenol measurement are high performance liquid chromatography (HPLC), electrochemical capillary electrophoresis (CE), gas chromatography (GC) and colorimetric spectrophotometry. Although, these methods are analytically capable, generally they require pretreatment processes such as extraction, cleaning, dilution of the samples as well as additional chemicals. Owing to those disadvantages, researchers have focused on enzyme based amperometric biosensors for measuring phenolic compounds due to their advantages such as good selectivity, working possibility in aqueous medium, fast responding, relatively low cost of realization and storage and the potential for miniaturization and automation. Amperometric biosensors, have been developing for phenol and its derivatives, are usually prepared with working electrodes which include polyphenol oxidases (PPO) (tyrosinase and laccase) and enzyme horseradish peroxidase (HRP). HRP reaction with phenols is faster than PPO enzyme reactions, and HRP-based working electrodes show higher sensitivity in comparison to PPO-based electrodes. Thus, the usage of HRP on working electrodes can be advised for fast and effective phenol measurements.

The design of a support matrix that binds the enzyme and bare electrode can be target specific providing efficient electron transport via added functional groups or nanoparticles into the composite structure of the electrode. Conducting polymers as supporting matrix are usually used as copolymers or composite films in biosensor systems since mechanical and processing properties of their homopolymers are weak (Tsai & Chui, 2007; Heras et al., 2005; Carvalho et al., 2007; Serra et al., 2001; Mailley et al., 2003). Copolymerization does not require rigorous experimental conditions, and can be employed for the polymerization of a large variety of monomers leading to the formation of new advantageous materials (Böyükbayram et al., 2006; Kuwahara et al., 2005; Yilmaz et al., 2004; Yilmaz et al., 2005). Nanomaterials have also been used to improve the operational characteristics of biosensors (Yang et al., 2006; Zhou et al., 2007; Rajesh et al, 2005; Shan et al., 2007). This improvement results from both increased surface area and increased catalytic activity. Carbon nanotubes (CNTs) have emerged as a new class nanomaterials that are receiving considerable interest owing to their ability to promote electron transportation (Zhao et al., 2006; Chen et al, 2007; Zeng et al., 2007; Liu et al., 2006; Vega et al., 2007; Santos et al., 2007). The high conductivity of this carbon material leading to a level of 10^2 $\Omega^{-1}cm^{-1}$ improves electrochemical signal transduction, while its nano-architecture imposes the electron contact between redox centers, deeply inlaid in enzyme structure, and the smooth surface of the electrode.

In this chapter, we reported HRP-based amperometric phenol biosensors, which were comprised of working electrodes prepared in various designs, developed in order to get reliable, selective, sensitive and fast detection of phenol and its derivatives. Various compositions of polymeric/composite films were synthesized onto the surface of the electrodes. Various supporting matrix, designed target specific, were used for the fabrication of some of these polymeric/composite films. We are planning to cover a detailed investigation and discussion of the enzyme based working electrodes with regard to the response dependences as well as their amperometric characteristics including sensitivity, linear range, detection limit, relative standard deviation and reproducibility of the composite film electrodes.

2. General principle of enzyme-based amperometric biosensors

Amperometric biosensors are analytical devices in which a biological material is used as a biological catalyst in combination with an electrical transducer. A biosensor responds to an analyte in a sample and interprets its concentration as an electrical signal via a biological recognition system and the electrochemical transducer. Amperometric biosensors possess linear concentration dependence, compared to a logarithmic relationship in potentiometric systems and measure change in the current on the working electrode due to the direct oxidation of the products of a biochemical reaction. Electrochemical biosensors have been under development for 40 years, and over this time a wide variety of sensors has been developed. The overriding theme of biosensors is the ability to perform selective biological recognition of the target analyte in a complex sample matrix and couple this to sensitivity of electrochemical detection. The magnitude of the response of amperometric biosensors depends on a number of factors, including the kinetics of the enzymatic reaction, the construction, and the operation mode of the enzyme electrode. The response from the electrode can either be diffusionaly or kinetically controlled. With kinetically controlled enzyme electrodes, the enzyme loading is sufficiently low that the response depends on the enzyme concentration and the kinetics of the enzymatic reaction. Such behaviour has limited analytical utility as response saturation occurs at low substrate concentration. The diffusionaly controlled electrode possesses very high enzyme loadings such that the current is independent of small changes in enzyme concentration; consequently, the current response is a function of analyte concentration and diffusion. Enzyme electrodes can be operated in several measurement modes: dynamic steady-state, potential step, and flow-injection mode. The steady-state mode allows reaction equilibrium to be reached before the analytical signal is obtained, whereas in dynamic measurement the signal is obtained quickly as a predetermined timepoint after introduction of the sample. Both potential step and flow-injection measurements are transient responses due to the transient nature of the techniques (Diamond, 1998).

2.1 HRP-based amperometric phenol biosensors

Amperometric biosensors for the detection of phenolic compounds have been introduced as a mono-enzyme system using tyrosinase, laccase or HRP. Tyrosinase biosensors are restricted to the monitoring of phenolic compounds having at least one *ortho*-position free. On the other hand, laccase biosensors give response to phenolic compounds with free *para*- and *meta*-position with a complicated catalytic cycle. HRP having less selectivity to phenolics is capable of giving response to a large number of phenol derivatives, and shows a high stability and efficiency for different biosensor designs. HRP was oxidized by hydrogen peroxide and re-reduced by phenols. Phenoxy radicals, formed during the enzymatic oxidation of phenolic compounds in the presence of hydrogen peroxide, were reduced electrochemically on the electrode surface; the reduction current is proportional to concentration of phenolic compound (Korkut el al., 2008; Korkut Ozoner et al., 2011).

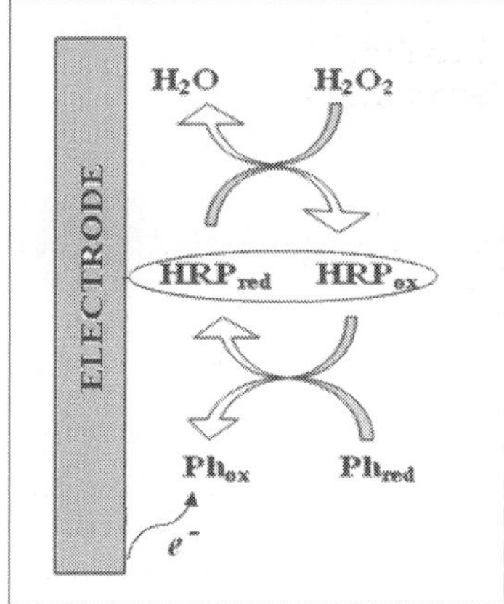

Fig. 1. The electrochemical reaction between HRP and phenol on electrode surface.

2.2 Polymers for working electrodes

To achieve high biosensor performance, it is very necessary to fabricate excellent electrode support materials for both effective immobilization of enzyme and fast electron transport between enzyme and metallic electrode. Electropolymerizable conducting polymers are generally used as supporting matrix for working electrodes. Among the conducting polymers, polythiophene has a special place due to their electrical properties, rich synthetic flexibility, and environmental stability in doped and undoped states, non-linear optical properties, and highly reversible redox switching. Synthesis of a thiophene-functionalized methacrylate monomer [3-methylthienylmethacrylate (MTM)] via the esterification of 3-thiophene methanol with methacryloyl chloride can be prepared. Thus, the MTM monomer obtained has two polymerizable groups: the vinyl group is useful for radical polymerization while the thiophene ring, with substitution at the 3-position, can be employed in both oxidative polymerization and electropolymerization. It is also possible to prepare block and random copolymers of MTM with other acrylic or vinyl monomers at different compositions. Subsequently, constant-potential electrolyses can be employed for the synthesis of the graft copolymers of the side chain thiophene (Depoli et al., 1985).

Copolymerization is the most effective and successful way among the existing polymerization techniques for incorporation of systematic changes in polymer properties. It does not require rigorous experimental conditions, and can be employed for the polymerization of a large variety of monomers leading to the formation of new materials. Reactive functional polymers can be prepared by incorporation of acrylates and methacrylates monomers containing side chain reactive functional groups into polymers. Various architectures of epoxy group possessing polymers have been developed in the literature. Copolymers of glycidyl methacrylate (GMA), an epoxy group containing methacrylate monomer, have received great interest. Epoxide is a three-membered cyclic ether and very reactive due to the large strain energy (about 25 kcal mol^{-1}) associated with the three-membered ring. Therefore, it can be employed into a large number of chemical reactions by ring opening. Various applications of chemically modified pendant copolymers, such as immobilization of enzymes, DNA, catalysts, and biomolecules, were reported (Hradil & Svec, 1985; Lukas & Kalal, 1978).

2.3 Working electrode fabrications

Polymeric coatings can be applied to a wide range of electrode support materials. Electrodes covered with polymeric coatings have thicker layers which, besides increasing the flexibility for choosing the coating material, permit to obtain a higher surface coverage and therefore to increase the amount of electroactive material attached to the surface. Polymeric coatings can be formed by electropolymerization of monomers or by solution casting of preformed polymers. Electropolymerization, a recent focus among immobilization strategies, is an electrochemical route to form polymeric coatings by entrapment of biomolecules and involves the application of an appropriate potential to a working electrode immersed in an aqueous solution containing the electropolymerizable monomer and enzyme, which is homogeneously incorporated in the growing polymer. The enzyme/polymer interaction is of paramount importance to improve the fundamental knowledge about the biological interface of the biosensor. So far, difficulties to understand the exact mechanism of entrapment and the dynamic effects on biosensors partially result from a scarcity of reports comparing different polymer matrix for immobilization of the same enzyme. In the entrapment technique which is easy and rapid one-step procedure, enzyme does not link onto the polymeric structure, and can act as it is in its free form in the pores during/after electropolimerization process. However, biological activity of the entrapped enzyme decreases probably due to the hydrophobic character of polymers and the steric hindrances caused by the surrounding polymer, which drastically reduces the accessibility to the immobilized biomolecules. Enzymes can also be chemically immobilized to a polymer matrix basically in a two-step process: a polymer film containing functional groups for enzyme immobilization is formed on electrode surface, then the electrode is dipped in enzyme solution or the enzyme solution is dropped onto the surface of the electrode. Covalent bindings are stronger and therefore less prone to biomolecule detachment, thus increasing the stability of the linkage. As an alternative to electropolymerization, polymer coatings can also be formed by casting of films from solution using preformed polymers. In this way, the amount of material on the electrode surface can be controlled by the concentration and the amount of polymer solution applied. The actual layer thickness is, however, less well defined than in electropolymerization, an important issue as often the analytical signal will depend on the thickness of the modifying layer.

3. Experimental procedures

3.1 Chemicals

Horseradish peroxidase (E.C.1.11.1.7) with an activity of 10 000U vial[-1] (according to pyrogallol method performed by the supplier), aqueous solution of hydrogen peroxide (30%), glutaraldehyde (25%), lithium chloride, dichloromethane (DCM), N,N-dimethyl formamide (DMF), α,α'-Azobisisobutyronitrile (AIBN), di-potassium hydrogen phosphate, citric acid, tri-sodium citrate, acetic acid (96%), sodium acetate tri-hydrate and potassium di-hydrogen phosphate were purchased from Merck. Tetrahydrofuran (THF) was obtained from Riedel. Phenol, p-benzoquinone, hydroquinone, 2,6-dimethoxyphenol, 2-chlorophenol, 3-chlorophenol, 4-chlorophenol, 2-aminophenol, 4-methoxyphenol, pyrocatechol, guaiacol, m-cresol, o-cresol, p-cresol, catechol, 4-acetamidophenol, pyrogallol, 2,4-dimethylphenol, pyrrole monomer (99%), sodium dodecyl sulfate (SDS) and 1-cyclohexyl-3(2-morpholinoethyl)carbodiimide metho-p-tolueno-sulfonate were obtained from Sigma. Stock solutions of various phenols were daily prepared in 0.1 M, pH 7 phosphate buffer solution. Multiwalled carbon nanotubes (MWCNTs) were obtained from Nanocs. Inc., Newyork, USA.

3.2 Synthesis of Poly(glycidyl methacrylate-*co*-3-thienylmethyl methacrylate) {Poly(GMA-*co*-MTM)}

Side chain thiophene containing monomer, 3-thienylmethylmethacrylate (MTM) was synthesized according to the previously reported papers (Yilmaz et al., 2004; 2005).

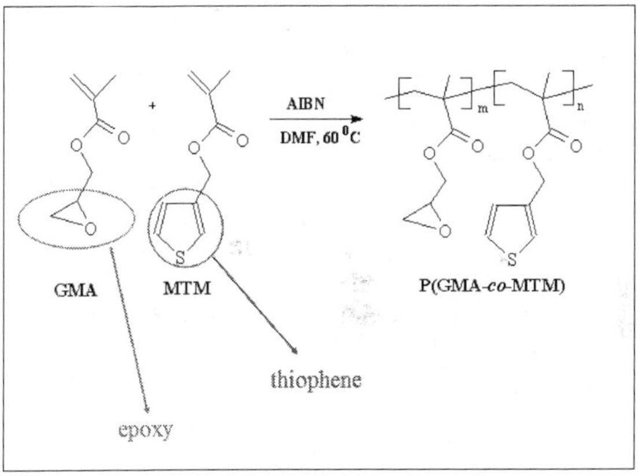

Fig. 2. Synthesis of Poly(GMA-*co*-MTM).

We have previously reported the copolymers of MTM with glycidyl methacrylate (GMA) and monomer reactivity ratios were determined for low conversion using Fineman Ross (FR) (r_{GMA}= 0:9795; r_{MTM}= 0:5641) and Kelen Tüdös (KT) (r_{GMA} = 0:9796; r_{MTM} =0:5771) graphical methods (Gunaydın & Yilmaz, 2007). Poly(GMA-*co*-MTM) was synthesized via radical polymerization of appropriate GMA/MTM feed mixture in the presence of AIBN as an initiator. Predetermined quantities of MTM, GMA and AIBN (1% of total weight of monomers) in DMF with a volume of 1.5 mL were placed in a Pyrex tube. The mixture was deoxygenated by flushing with oxygen-free argon for at least 15 min. The tube was tightly sealed and immersed in a thermostated oil bath at 60 ± 1°C. The conversion was determined by gravimetric measurements. After the reaction, copolymer was precipitated in methanol, filtered off, and purified by reprecipitation from DCM solution into methanol and finally dried in vacuo for 24 h. The solution of Poly(GMA-*co*-MTM) was prepared in THF solvent.

3.3 Amperometric measurements

Amperometric measurements were performed by using a CHI Model 840B electrochemical analyzer. A gold working electrode (2 mm diameter), a glassy carbon working electrode with a diameter of 3 mm for batch measurements, 2 mm for flow injection analyses (FIA), a Platinum wire counter electrode, a Ag/AgCl (3M NaCl) reference electrode, and a conventional three-electrode electrochemical cell were used in the experiments. Measurements of phenolic compounds were carried out in 0.1 M, pH 7 phosphate buffer in the presence of 0.7 mg mL⁻¹ lithium chloride with an applied working potential of -50 mV.

3.4 Experimental setup
3.4.1 FIA system

FIA system was set up with an HPLC pump (GBC LC1120), an injection valve (Shimadzu) and a flow cell including three-electrode system. The HPLC pump was adjusted to deliver a carrier solution at a constant flow rate. Potassium phosphate buffer was used as carrier solution. Samples were injected into the carrier solution, passing from the flow cell, by an injection valve with a volume of 1 mL. A sharp current peak was formed for each phenolic injection at a working potential of -50 mV.

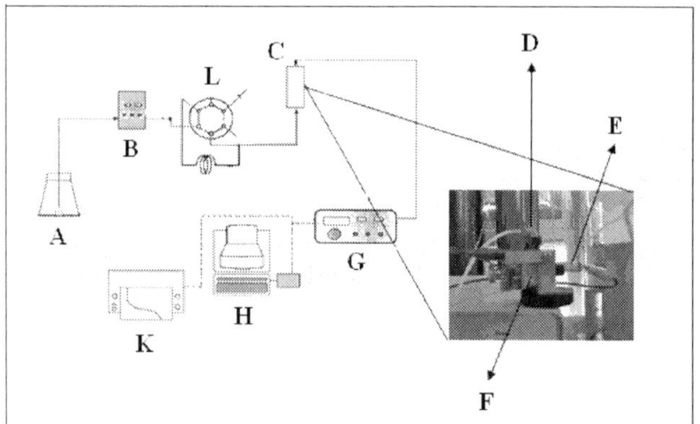

Fig. 3. Schematic diagram of the FIA system: carrier solution (A), HPLC pump (B), injection valve (L), flow cell (C), Pt counter electrode (D), Ag/AgCl reference electrode (E), glassy carbon working electrode (F), potentiostat (G), computer (H) and data recorder (K).

3.4.2 Batch system

Electrochemical batch measurements were carried out in 10 mL of potassium phosphate buffer with a continuous stirring at 600 rpm in three-electrode cell. Three-electrode system was immersed into the electrochemical cell, a working potential of -50 mV was applied and current was allowed to reach a steady-state value then, various concentrations of phenolic compounds were added into the cell to produce i-t curves of amperometric measurements.

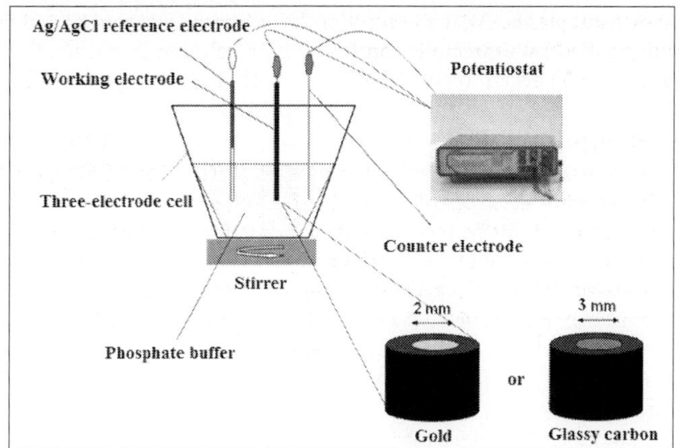

Fig. 4. Schematic diagram of the batch system.

3.5 Fabrication of the working electrodes

3.5.1 Poly(glutaraldehyde-*co*-pyrrole)/HRP {Poly(GA-*co*-Py)/HRP} composite film electrode

The composite film electrode was used in FIA system. Poly(glutaraldehyde) solution was prepared by adding 2 mL of 0.1 M NaOH and 2 mL of 25% glutaraldehyde to 10 mL of distilled water. The solution was stirred at 600 rpm for 30 minutes in order to polymerize glutaraldehyde. The final pH must be 9-10. The copolymerization medium was comprised of 0.01 M pyrrole and 0.6 mg mL^{-1} SDS in 10 mL of prepared PGA solution. This medium was circulated through the flow cell using the HPLC pump under a potential scan between 0 and +1.2 V with the scan rate of 100 mV s^{-1}. Then the copolymerized film coated electrode was immersed into 25% glutaraldehyde solution to increase the number of aldehyde groups in the composite film of Poly(glutaraldehyde-*co*-pyrrole) {Poly(GA-*co*-Py)}, and stored at +4°C overnight. The electrode was washed potassium phosphate buffer, and immersed in 0.3 mg mL^{-1} HRP solution for 20 hours. Finally, the electrode was washed again with buffer to remove excess HRP.

3.5.2 Carbon nanotube/Polypyrrole/HRP (CNT/PPy/HRP) nanocomposite film electrode

A modified acid oxidative method was used for preparation of water-soluble CNTs (Zhao et al., 2002). 14 mg of MWCNTs were added into 5 mL of a 9:1 concentrated H_2SO_4/H_2O_2 (30%) solution and stirred for 30 min for CNTs oxidation. After the reaction, 15 mL of the 9:1 concentrated H_2SO_4/H_2O_2 solution was added into the mixture. The mixture was placed in an ultrasonic bath and sonicated for 5 min. Resulting CNTs dispersion was diluted using 1 L of distilled water and filtered through a 0.45 µM cellulose membrane. The filtrate was washed with 0.01 M NaOH solution and distilled water till the pH level reaching to 7 and dispersed in distilled water (0.03 mg L^{-1}). The resulting CNTs solution was sonicated for 2 min to obtain a homogeneous CNTs solution. Nanocomposite film was formed onto the surface of the gold electrode by immersing the electrode to an electropolymerization medium contained 5 mL of oxidized CNTs solution, 5 mL of 0.05 M pH 6.5 citrate buffer,

0.01 M pyrrole, 0.6 mg L⁻¹ SDS and 0.3 mg L⁻¹ HRP under a potential scan between 0 and +1.2 V for 4 minutes at a scan rate of 100 mVs⁻¹.

3.5.3 Poly(glycidyl methacrylate-*co*-3-thienylmethyl methacrylate)-Polypyrrole-Carbon nanotube-HRP {Poly(GMA-*co*-MTM)/PPy/CNT/HRP} composite film electrode

6 mg of Poly(GMA-*co*-MTM) was dissolved in 10 mL THF. The polymer solution with a volume of 20 µL was directly spread onto the surface of a gold electrode. The electrode was then allowed to dry for solvent evaporation at room temperature. Poly(GMA-*co*-MTM) coated electrode was dipped into the electropolymerization medium contained 5 mL of oxidized CNTs solution, 5 mL of 0.05 M pH 6.5 citrate buffer, 0.01 M pyrrole, and 0.6 mg L⁻¹ SDS under a potential scan between (-1.2) – (+1.2) V for 4 minutes with a scan rate of 100 mVs⁻¹. Poly(GMA-*co*-MTM)/PPy/CNT electrode was pre-treated at a potential of +2 V vs Ag/AgCl for 5 minutes in 0.1 M, pH 7 phosphate buffer. The electrode was then allowed to react for 3.5 hours within the solution of 5 mg L⁻¹ 1-cyclohexyl-3(2-morpholinoethyl) carbodiimide metho-*p*-tolueno-sulfonate with a continuous stirring at 200 rpm. The electrode was dipped into a solution of 0.3 mg L⁻¹ HRP, dissolved in 0.1 M, pH 7 phosphate buffer, and stored at +4°C overnight.

3.5.4 Poly(glycidyl methacrylate-*co*-3-thienylmethyl methacrylate)-Polypyrrole-HRP {Poly(GMA-*co*-MTM)/PPy/HRP} composite film electrode

20 µL of 0.6 mg L⁻¹ of Poly(GMA-*co*-MTM) was directly spread onto the surface of the polished glassy carbon electrode. After the solvent evaporation polymer coated electrode electropolymerized with polypyrrole in a polymerization medium contained 10 mL of 50 mM pH 6.5 citrate buffer including 0.01 M pyrrole, 0.6 mg mL⁻¹ SDS and 0.6 mg mL⁻¹ of HRP at a potential scan between (-1.2) – (+1.2) V for 4 minutes at a scan rate of 100 mVs⁻¹.

4. Results and discussions

4.1 FIA of phenols by using Poly(GA-*co*-Py)/HRP composite film electrode

Various concentrations of *p*-benzoquinone, catechol and phenol ranging between 75 µM and 750 µM with 1.5 mM hydrogen peroxide were injected to the carrier solution at a flow rate of 1 mL min⁻¹ (Fig. 5). No reproducible response was obtained for phenol from Poly(GA-*co*-Py)/HRP composite film electrode, and the best response was observed for catechol as model phenolic. The difference in response among the phenolic compounds depends on the different affinity of HRP towards its substrates, and the formation of *o*-quinones during the enzymatic reaction for each phenolic (Tsai & Cheng-Chui, 2007).

Effect of flow rate was investigated on the FIA system response to a series of *p*-benzoquinone injections at different flow rates ranged between 0.25-6 mL min⁻¹ (Fig. 6). It was observed that the obtained peak currents decreased as a consequence of the short retention time of the substrate with the enzyme depending on the increase of the flow rate. In addition, unstable signals were shaped with the increasing flow rate due to the unsteady flow conditions through the composite film. The flow rate in the FIA system affected the sample throughput, detection limit and accuracy. The disadvantages of lower flow rates were low sample throughput and an increase in dispersion. By using a higher flow rate, a greater number of samples could be analyzed, and the peaks became narrower, but the detection limit increased and unstable responses were observed.

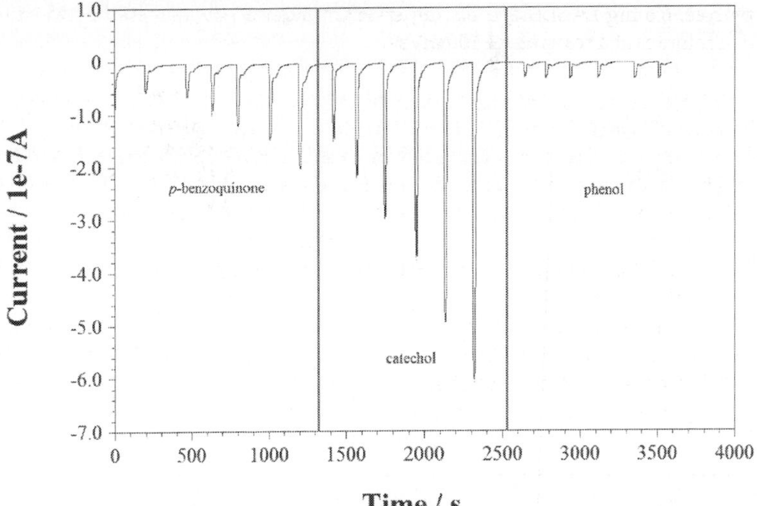

Fig. 5. Poly(GA-*co*-Py)/HRP composite film electrode response to 75-125-200-300-500-750 μM *p*-benzoquinone, catechol and phenol injections. Applied potential was -50 mV (vs. Ag/AgCl, 3 M NaCl).

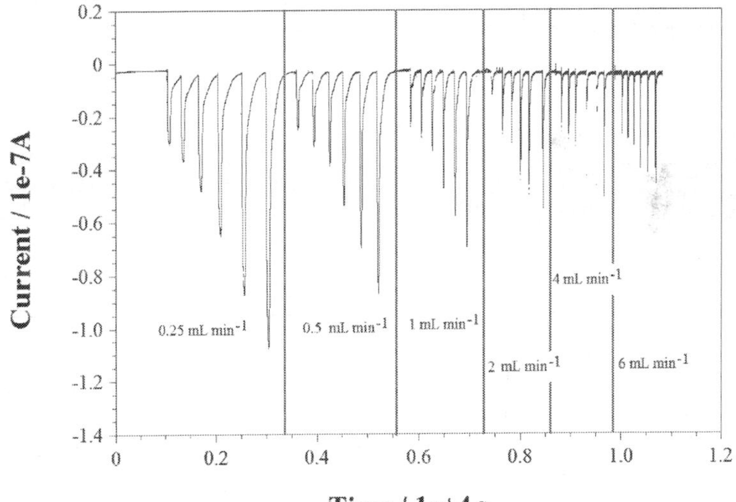

Fig. 6. Effect of 0.25-0.5-1-2-4-6 mL min⁻¹ flow rate on Poly(GA-*co*-Py)/HRP composite film electrode response to 75-125-200-300-500-750 μM *p*-benzoquinone injections. Applied potential was -50 mV (vs. Ag/AgCl, 3 M NaCl).

Stability of Poly(glutaraldehyde-*co*-pyrrole)/HRP composite film electrode was evaluated by the 20 repetitive analyses of *p*-benzoquinone at a concentration of 2.5 µM recorded at 1 min intervals over a prolonged period. Well-defined reduction responses were obtained with a standard deviation of ±0.23 nA. Poly(GA-*co*-Py)/HRP composite film electrode could be used for one month without loosing its initial response. The high operational and storage stability of the electrode can be a result of removing of the enzymatic phenolic products by continuous flow. Owing to the copolymerization of pyrrole with glutaraldehyde, a sufficient electron transfer was provided between enzyme and the electrode since PGA contains conjugated electroactive aldehyde groups. These active aldehyde groups incorporated to the conductive polymeric backbone by the copolymerization with pyrrole. Therefore, strong chemical bonds were formed between HRP and the copolymeric film via the aldehyde groups of the copolymer. There have been a few reports of FIA phenol biosensor in literature. Poly(GA-*co*-Py)/HRP composite film electrode showed lower detection limit and wider linear range in comparison to those reports (Table 1). This can be attributed to the electrocopolymerization of glutaraldehyde with pyrrole monomer since the aldehyde groups of PGA both electroactive and capable to bind the enzyme chemically.

Analyte	Biosensor	Detection limit (µM)	Linear range (µM)	Reference
p-Cresol	Laccase/Graphite	39	10-1000	(Wilkolaza et al., 2005)
4-Chlorophenol	Laccase/Graphite	346	1000-10000	(Wilkolaza et al., 2005)
Hydroquinone	Laccase/Graphite	0.58	1-10	(Wilkolaza et al., 2005)
Hydroquinone	Laccase/ECH Sepharose	-	0-500	(Vianello et al., 2006)
4-Aminophenol	Laccase/Graphite	0.61	1-10	(Wilkolazka et al., 2005)
4-Methoxyphenol	Laccase/Graphite	7.9	1-100	(Wilkolazka et al., 2005)
p-Benzoquinone *(at a flow rate of 1mL min⁻¹)*	Poly(GA-*co*-Py)/HRP	2	2.5-750	This study

Table 1. Analytical parameters of some FIA biosensors for phenolic compounds.

4.2 Amperometric detection of phenolic compounds in batch operation
4.2.1 CNT/PPy/HRP nanocomposite film electrode

Electropolymerization CVs of CNT/PPy/HRP nanocomposite film and PPy/HRP (without CNT) electrode were shown in Fig. 7. Typical polypyrrole voltammograms were obtained for both electrodes. Oxidation current of pyrrole was much higher for CNT/PPy/HRP nanocomposite film electrode than the other electrode fabricated without CNT. This can be attributed to the enhanced pyrrole oxidation process since the electron transfer mechanism was facilitated by the incorporation of CNT into the PPy film structure.

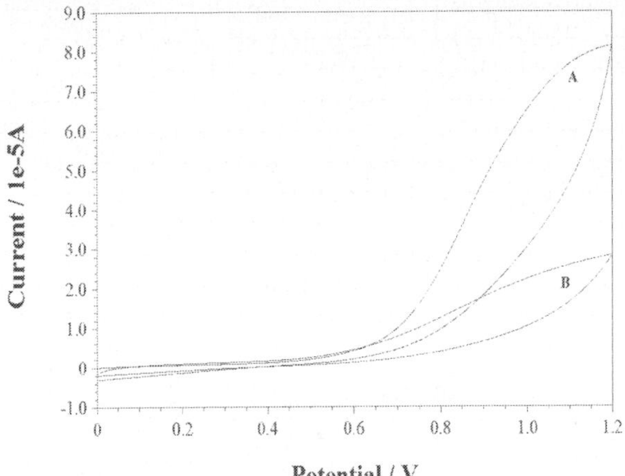

Fig. 7. Cyclic voltammogram of CNT/PPy/HRP (A), and PPy/HRP (B) electrode at a scan rate of 100 mV s^{-1}.

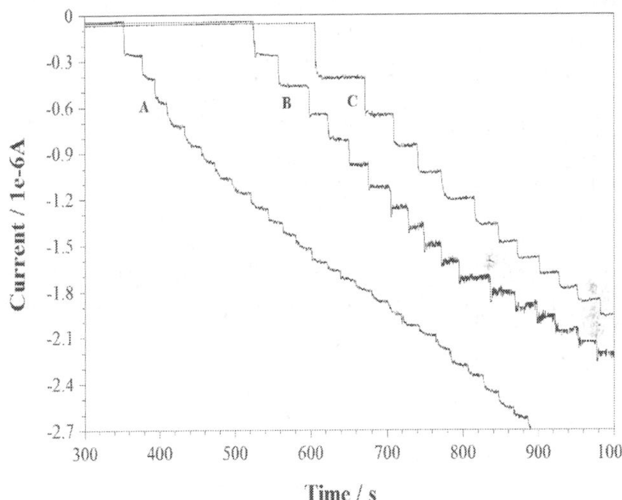

Fig. 8. Amperometric response of CNT/PPy/HRP electrode to 4-methoxyphenol (A), hydroquinone (B) and 2-aminophenol (C) additions.

Eighteen phenolics were tested for CNT/PPy/HRP nanocomposite film electrode with an applied potential of -50 mV. Fig. 8 illustrates typical amperometric responses of the electrode after the addition of successive aliquots of some of the phenolic compounds under a constant stirring at batch operation. Table 2 summarizes the characteristics of the calibration plots obtained for the phenol derivatives, as well as the corresponding limits of detection calculated according to the $3s_b/m$ criteria where m is the slope of the linear range of the respective calibration plot, and s_b is estimated as the standard deviation of the signals from different solutions of the phenolics at the concentration level corresponding to the lowest concentration of the calibration plot. The lowest detection limit was found to be 0.027 μM (S/N=3) for *p*-benzoquinone and the highest detection limit was found to be 27.9 μM (S/N=3) for 2,4-dimethylphenol among the tested phenolics. It was previously reported that the phenolic compounds with electron-donor substituents in an *ortho*-position gave no response (Kane & Iwuoha, 1998). CNT/PPy/HRP nanobiocomposite film electrode did not give any response to *o*-cresol. The highest sensitivity was obtained for 4-methoxyphenol since presence of –OCH₃ group of 4-methoxyphenol allows HRP to oxidize more efficiently. A lower sensitivity was observed for 2,4-dimethylphenol, as expected, for the one having the *ortho*-position occupied by a methyl group. The sensitivity ranges between 1-50 nA μM⁻¹ for the phenolics tested.

Analyte	r	Sensitivity (nA μM⁻¹)	Linear range (μM)	LOD (μM)	%RSD
Phenol	0.99	1	16-144	3.52	2.89
p-Benzoquinone	0.99	3	0.02-0.16	0.027	4.43
Hydroquinone	0.99	8	16-240	6.42	6.5
2,6-Dimethoxyphenol	0.99	7	1.6-19.2	0.29	1.8
2-Chlorophenol	0.99	8	1.6-8	0.26	1.7
3-Chlorophenol	0.99	6	1.6-12.8	0.2	1.1
4-Chlorophenol	0.99	8	1.6-14.4	0.3	1.87
2-Aminophenol	0.99	40	8-60.8	1.53	5.4
4-Methoxyphenol	0.99	50	1.6-81.6	1.06	2.8
Pyrocatechol	0.99	8	1.6-446.4	6.27	6.7
Guaiacol	0.98	9	1.6-9.6	0.3	1.92
m-Cresol	0.99	9	8-20.8	1.5	2.84
o-Cresol			no response		
p-Cresol	0.98	5	128-832	24	2.5
Catechol	0.98	2	1.6-8	0.93	3.8
4-Acetamidophenol	0.99	3	1.6-16	1.11	2.57
Pyrogallol	0.98	1	1.6-22.4	1.24	1.2
2,4-Dimethylphenol	0.98	1	64-240	27.9	2.2

Table 2. Analytical characteristics of CNT/PPy/HRP nanobiocomposite film electrode for various phenolic compounds. Applied potential; -50 mV, 0.1 M phosphate buffer (pH 7) containing 16 μM hydrogen peroxide.

Fig. 9 illustrates the amperometric responses of CNT/PPy/HRP and PPy/HRP working electrodes to increasing concentrations of hydroquinone additions into the 0.1 M, pH 7 phosphate buffer at an applied potential of -50 mV (vs. Ag/AgCl). No reproducible signals were observed for PPy/HRP electrodes. CNT/PPy/HRP nanobiocomposite film electrode achieved to produce measurable responses by the regular growth of reduction currents. CNTs were thought to impose the electron transfer of the mediated reaction. It was previously reported that peroxidases were able to do direct electron transfer between enzyme molecules and electrode thus they did not need electron mediators for electron transfer (Gorton et al., 1992). However, in this study, the available responses could only be obtained by CNTs-based electrode due to its ability to promote electron transfer reaction with HRP. Furthermore, the amount of active immobilized enzyme in CNT/PPy/HRP nanobiocomposite film and PPy/HRP biocomposite film was found to be 6.1 and 2.7 μg, respectively. The immobilized enzyme quantity was measured by using the enzyme activity assay according to the previously reported procedure (Vojinovic et al., 2004). Nanobiocomposite film, involving CNTs, attached higher amount of enzyme than the composite film without CNTs due to their unique structure having activated large surface area. The nanostructure of the biocomposite could intensify the surface for higher biocatalytic activity. HRP was mainly entrapped into the polymeric film structure during the pyrrole electropolymerization process, and chemically linked via the carboxylated groups of CNTs.

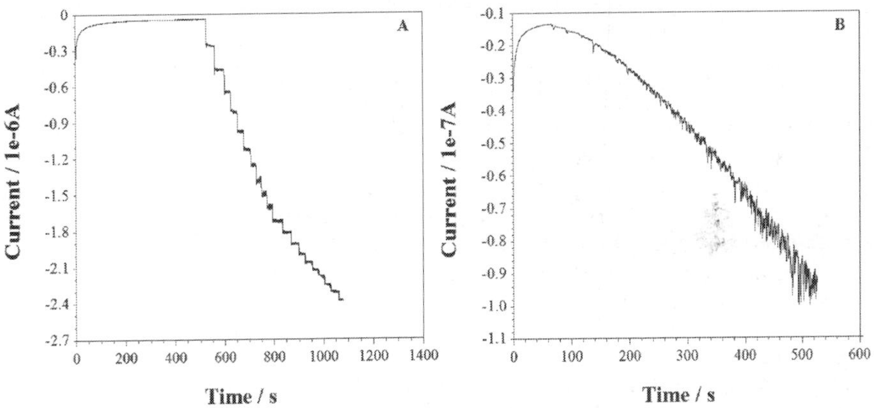

Fig. 9. Amperometric responses of CNT/PPy/HRP (A) and PPy/HRP working electrode (B) to the successive additions of hydroquinone.

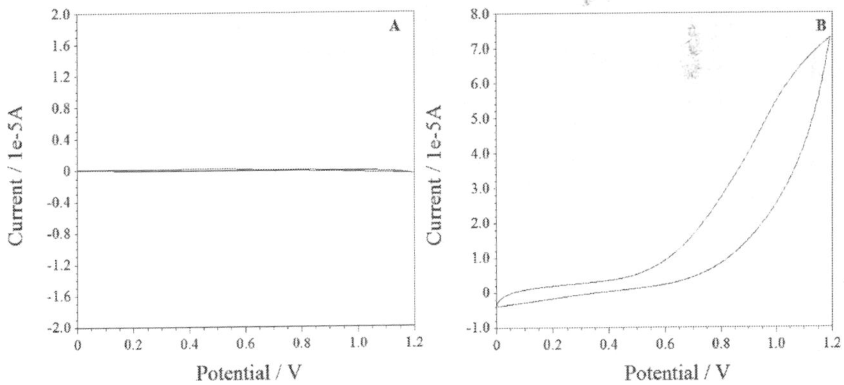

Fig. 10. Cyclic voltammogram of Poly(GMA-*co*-MTM) film electrode (A) and Poly(GMA-*co*-MTM)/PPy/CNT/HRP composite film electrode (B) in 0.1 M, pH 7 phosphate buffer at a scan rate of 100 mVs⁻¹.

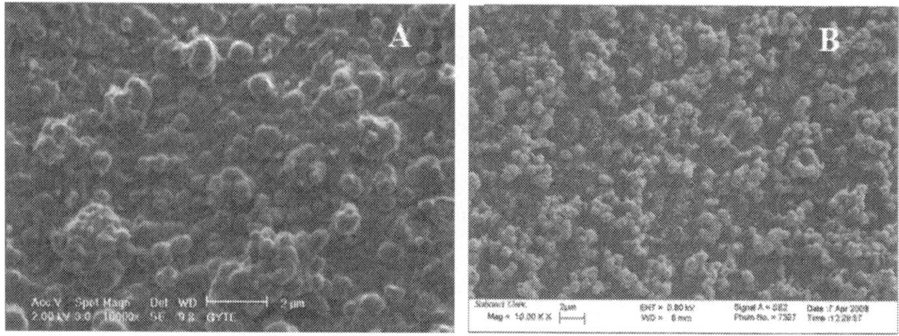

Fig. 11. SEM images of Poly(GMA-*co*-MTM) film (A) and Poly(GMA-*co*-MTM)/PPy/CNT composite film (B).

4.2.2 Poly(GMA-*co*-MTM)/PPy/CNT/HRP composite film electrode

The electrochemical properties of the composite electrodes of Poly(GMA-*co*-MTM) (Fig. 10A) and Poly(GMA-*co*-MTM)/PPy/CNT (Fig. 10B) were evaluated through cyclic voltammetry in 10 mL of 0.1 M phosphate buffer solution (pH 7) contained 0.7 mg mL^{-1} of lithium chloride. CV obtained with Poly(GMA-*co*-MTM) film electrode revealed, in both scan directions, that no voltametric peak in the scanning potential range (0 to 1.2 V vs. Ag/AgCl) is obtained. It means that the thiophene groups on the copolymer did not show any electroactivity. In the case of pyrrole present in the system, the usual pyrrole polymerization peaks were drastically shifted (Fig. 10B). It is an indication for the electropolymerization reaction between pyrrole and the thiophene moiety of the copolymer.

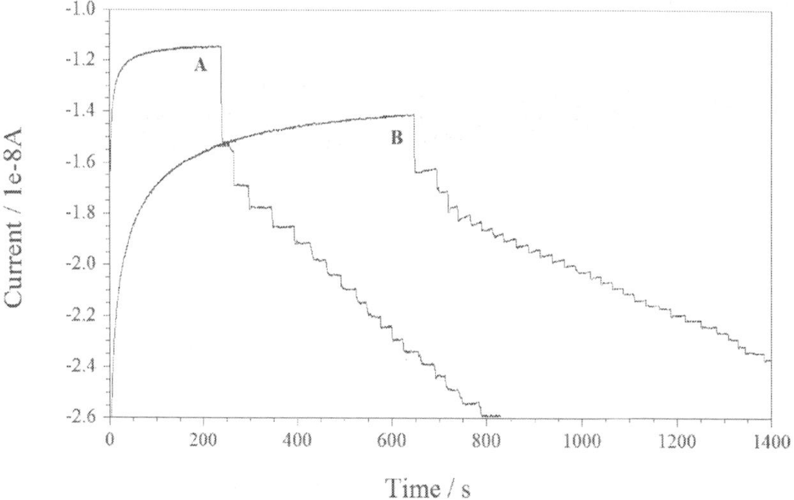

Fig. 12. Amperometric response of Poly(GMA-*co*-MTM)/PPy/CNT/HRP composite film electrode to the successive additions of guaiacol (A) and pyrogallol (B).

Fig. 12 shows amperometric response of Poly(GMA-*co*-MTM)/PPy/CNT composite film electrode to some of the phenolics. The biosensor responded rapidly to the concentration increments of the various phenolics. Rapid response indicates a fast electron exchange between HRP and its substrates, indicating that the catalytic properties of the enzyme were not hindered by Poly(GMA-*co*-MTM)/PPy/CNT/HRP composite film. Analytical parameters were presented in Table 3. Wide linear range was observed for 2-chlorophenol (1.6-68.8 µM), 3-chlorophenol (1.6-81.6 µM) and 4-chlorophenol (1.6-86.4 µM) with the correlation coefficient of 0.999. The biosensor was also tested by the phenolics recovery experiments, which showed satisfactory results, with recoveries from 95% to 107% for the all tested phenolics. The available responses could only be obtained by the copolymeric film of Poly(GMA-*co*-MTM)/PPy. No reproducible response was observed by the electrode only coated with Poly(GMA-*co*-MTM) since it was not an electroactive polymer.

Analyte	r	Sensitivity (nA µM⁻¹)	Linear range (µM)	LOD (µM)	%RSD
Phenol	0.99	0.7	1.6-72	0.732	7.5
p-Benzoquinone	0.99	5	1.6-25.6	0.409	13
Hydroquinone	0.99	9	1.6-25.6	0.336	8.8
2,6-Dimethoxyphenol	0.99	0.8	1.6-36.8	0.382	8.38
2-Chlorophenol	0.99	1	1.6-68.8	0.249	4.7
3-Chlorophenol	0.99	1	1.6-81.6	0.441	9.9
4-Chlorophenol	0.99	1	1.6-86.4	0.336	6.9
2-Aminophenol	0.99	2	1.6-44.8	0.247	6.6
4-Methoxyphenol	0.99	2	1.6-35.2	0.312	6.35
Pyrocatechol	0.99	1	1.6-49.6	0.516	11
Guaiacol	0.99	0.3	3.2-52.8	0.490	10
m-Cresol			no response		
o-Cresol			no response		
p-Cresol			no response		
Catechol	0.99	2	1.6-44.8	0.304	7.5
4-Acetamidophenol	0.99	3	1.6-22.4	0.624	12
Pyrogallol	0.99	0.1	4.8-48	0.660	11
2,4-Dimethylphenol	0.99	0.4	1.6-40	0.382	7.8

Table 3. Analytical characteristics of Poly(GMA-*co*-MTM)/PPy/CNT/HRP composite film electrode for various phenolic compounds. Applied potential; -50 mV, 0.1 M phosphate buffer (pH 7) containing 16 µM hydrogen peroxide.

The most possible linkages between HRP and the functional groups of the composite film were C-N bonds. The enzyme HRP was chemically immobilized via the epoxy groups of the Poly(GMA-*co*-MTM) and the carboxyl groups of the CNTs. The bonding mechanisms are illustrated in Fig. 13. Theoretically, it is possible for an enzyme molecule to bind to the composite film through the two different mechanisms simultaneously. Such multiple linkages might be resulted an increased steric hindrance on the enzyme molecule (Korkut Ozoner et al., 2010; Bayramoğlu & Yakup Arıca, 2008). Moreover, Kobayashi et al. 2005 reported that their results suggested the magnitude of the effect of steric hindrance depended on the disubstitution of phenol derivatives (Kobayashi et al., 2005). It is the fact that the enzyme molecules directly bond onto the CNTs, acting as an electron transferring bridge or a wire, may stabilize the microenvironmental conditions for the desired electrochemical reaction. Hence, the enzyme was also immobilized chemically to the composite film of Poly(GMA-*co*-MTM)/PPy/CNT supported by a conductive copolymer.

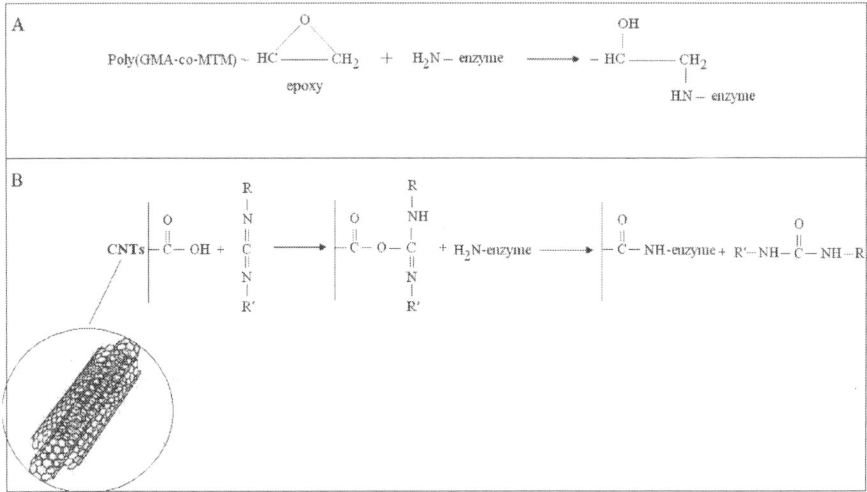

Fig. 13. The proposed immobilization mechanism of HRP on Poly(GMA-*co*-MTM)/PPy/CNT composite film via the epoxy groups of Poly(GMA-*co*-MTM) (A) and via the carboxyl groups of CNTs (B).

4.2.3 Poly(GMA-*co*-MTM)/PPy/HRP composite film electrode

The typical amperometric responses and the calibration curves of the electrode F are illustrated in Fig. 14 and Fig. 15, respectively after the addition of successive aliquots of phenolic compounds at an applied potential of -50 mV under continuous stirring at 600 rpm. Poly(GMA-*co*-MTM)/PPy/HRP composite film electrode reached to the steady-state current of 95% in less than 3 s.

Table 4 summarizes the characteristics of the calibration plots obtained from the current-time recordings of phenol derivatives. The detection limit ranged between 0.13 and 1.87 µM for the tested phenol derivatives. The different sensitivities varied between 3-200 nA µM⁻¹ for the tested phenolics can be related to the formation of *o*-quinones during the enzymatic reaction. The maximum sensitivity was found to be 200 nA µM⁻¹ for hydroquinone. In addition to this, 4-methoxyphenol and 4-acetamidophenol showed higher sensitivity than the other phenolics. This can be dialed with the presence of $-OCH_3$ group of 4-

methoxyphenol which enhances oxidation of the phenolic by HRP. Due to the strong ability of electron-donor conjugation of hydroquinone and 4-acetamidophenol, the corresponding conjugation structure could be easily formed. The higher sensitivity can be attributed to the favorable microenvironment of the immobilization matrix and enzyme immobilization procedure, which was performed by both chemical bonding via the epoxy groups and entrapment during the electropolymerization step. However, the type of the electrode material played an important role on the value of the sensitivity. Glassy carbon electrodes (GCEs) have been widely used compared with metal electrodes due to its biocompatibility with tissue, having low residual current over a wide potential range and minimal propensity to show a deteriorated response as a result of electrode fouling (Jin et al., 2008). Recently reported papers have stated that HRP is more compatible with carbon electrode materials (Santos et al., 2007; Carvalho et al., 2007; Huang et al., 2008). Rabinovich and Lev have claimed that the response of a phenol biosensor is usually limited by the electrochemical back reduction of the quinone leading to the diphenolic compound. Carbon electrode material affects significantly the sensitivity of the biosensor, because the limiting electrochemical back reduction of the enzymatic products takes place on the grain of the carbon materials (Rabinovich & Lev, 2001).

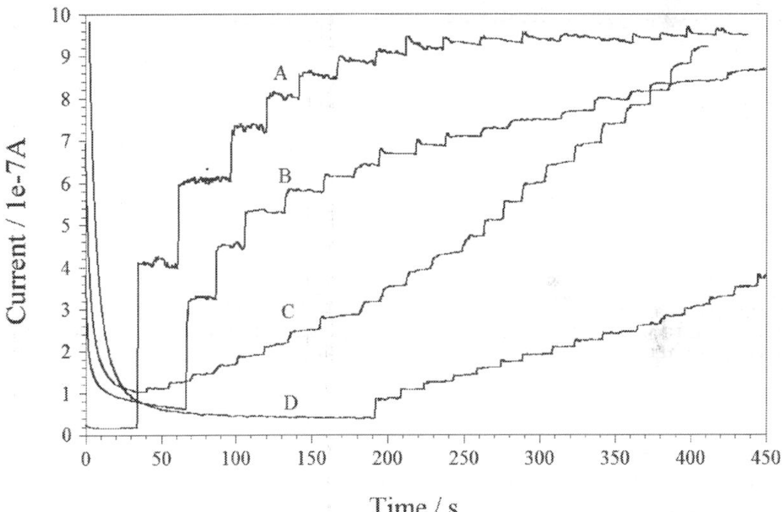

Fig. 14. Amperometric response of Poly(GMA-*co*-MTM)/PPy/HRP composite film electrode to the successive additions of catechol (A), *p*-benzoquinone (B), *p*-cresol (C) and *m*-cresol (D).

No response was obtained for 2,4-dimethylphenol, as expected, for the one having the *ortho*-position occupied by a methyl group. Not only o-cresol and 2,4-dimethylphenol but also 2-aminophenol, pyrogallol and 2,6-dimethoxyphenol gave no response. The operational stability of the electrode was monitored for a series of 20 succesive additions of 2 μM phenolic compounds. High operational stability was observed with relative standard deviations (RSD) ranging between 2% and 5.1% as seen in Table 4.

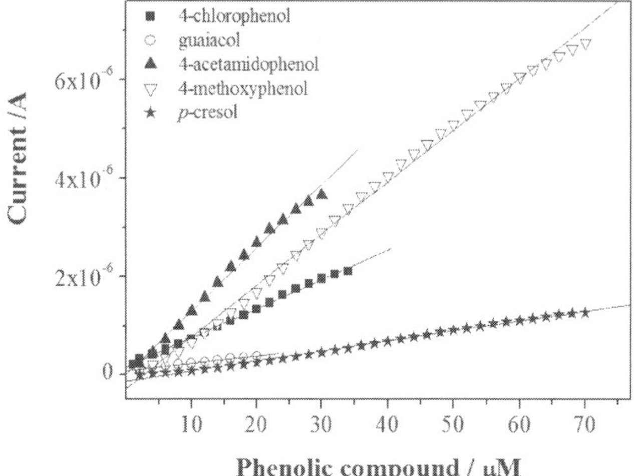

Fig. 15. Calibration curves of Poly(GMA-*co*-MTM)/PPy/HRP composite film electrode to increasing phenolic concentrations (initial phenolic concentration is 2 μM).

Analyte	r	Sensitivity (nA μM⁻¹)	Linear range (μM)	LOD (μM)	%RSD
Hydroquinone	0.99	200	2-34	0.13	2.3
Catechol	0.92	30	2-12	0.87	4.5
p-Benzoquinone	0.92	30	2-10	0.85	5
2-Chlorophenol	0.97	10	4-10	1.62	4.1
3-Chlorophenol	0.98	20	2-12	1.31	5
4-Chlorophenol	0.99	60	1-34	0.55	2
2-Aminophenol			no response		
Phenol	0.98	90	2-12	0.3	2.1
Guaiacol	0.99	10	2-20	1.2	3.8
2,6-Dimethoxyphenol			no response		
4-Acetamidophenol	0.99	100	2-30	0.21	2.3
4-Methoxyphenol	0.99	100	2-70	0.25	3.2
2,4-Dimethylphenol			no response		
Pyrogallol			no response		
Pyrocatechol	0.98	3	2-22	1.87	2.8
m-Cresol	0.99	10	2-88	1.43	3.8
o-Cresol			no response		
p-Cresol	0.99	20	2-70	1.28	5.1

Table 4. Analytical characteristics of Poly(GMA-*co*-MTM)/PPy/HRP composite film electrode for various phenolic compounds. Applied potential; -50 mV, 0.1 M phosphate buffer (pH 7) containing 20 μM hydrogen peroxide.

5. Conclusion

In this study a series of working electrode was fabricated for the amperometric detection of different phenolic compounds. For the fabrication of the working electrodes Poly(GMA-*co*-MTM) was synthesized as target spesific regarding to its chemically enzyme immobilization capacity and electropolymerizable thiophene groups with a conductive polymer such as polypyrrole. Different electrode designs conducted by using the same polymers of Poly(GMA-*co*-MTM) and polypyrrole showed different measurement results for the tested phenolics due to the differences of enzyme immobilization techniques, film electroactivity and variety of composite/copolymeric film structures of the fabricated electrodes. Electron transfer promoting effect of CNTs was distinctly observed for some of the fabricated electrodes.

6. Acknowledgement

The authors are grateful to "The Scientific & Technological Research Council of Turkey" (TUBITAK) for financial support of this project under grant Number 105Y130.

7. References

Bayramoğlu, G. & Yakup Arıca, G. (2008). Enzymatic removal of phenol and *p*-chlorophenol in enzyme reactor: Horseradish peroxidase immobilized on magnetic beads. *Journal of Hazardous Materials*, Vol.156, pp. 148-155.

Böyükbayram, A. E., Kıralp, S., Toppare, L. & Yagci, Y. (2006). Preparation of biosensors by immobilization of polyphenol oxidase in conducting copolymers and their use in determination of phenolic compounds in red wine. *Bioelectrochem*, Vol.69, pp. 164-171.

Carvalho, R. H., Lemos, F., Lemos, M. A. N. D. A., Cabral, J. M. S. & Ramoa Ribeiro, F. (2007). Electro-oxidation of phenol on a new type of zeolite/graphite biocomposite electrode with horseradish peroxidase. *J Mol. Cat. A: Chemical*, Vol.278, pp. 47-52.

Chen, S., Yuan, R., Chai, Y., Zhang, L., Wang, N. & Li, X. (2007). Amperometric third-generation hydrogen peroxide biosensor based on the immobilization of hemoglobin on multiwall carbon nanotubes and gold colloidal nanoparticles. *Biosens. Bioelectron*, Vol.22, pp. 1268-1274.

Depoli, M. A., Waltman, R. J., Diaz, A. F. & Bargon, J. (1985). An electrically conductive plastic composite derived from polypyrrole and polyvinyl-chloride. *J.Polym Sci Polym Chem Ed*, Vol. 23, pp. 1687-1985.

Diamond D., (Ed(s).)). (1998). *Principles of Chemical and Biological Sensors* (first edition), John Wiley & Sons Ltd., ISBN 0-471-54619-4, England.

Gorton, L., Jönsson-Petersson, G., Csöregi, E., Johansson, K., Dominguez, E. & Marko-Varga, G. (1992). Amperometric biosensors based on an apparent direct electron transfer between electrodes and immobilized peroxidases. *Analyst*, Vol.117, pp. 1235-1241.

Gunaydin, O. & Yilmaz, F. (2007). Copolymers of glycidyl methacrylate with 3-methylthienyl methacrylate: synthesis, characterization and reactivity ratios. *Polymer Journal*, Vol.39, pp. 579-588.

Heras, M. A., Lupu, S., Pigani, L., Pirvu, C., Seeber, R., Terzi, F. & Zanardi, C. (2005). A poly(3,4-ethylenedioxythiophene)-poly(styrene sulphonate) composite electrode coating in the electrooxidation of phenol. *Electrochim. Acta,* Vol.50, pp. 1685-1691.

Hradil, J. & Svec, F. (1985). Changes in the porous structure of macroporous copolymers due to successive effects of solvents and temperature. *Makromol. Chem,* Vol.130, pp. 81-90.

Huang, S., Qu, Y., Li, R., Shen, J. & Zhu, L. (2008). Biosensor based on horseradish peroxidase modified carbon nanotubes for determination of 2,4-dichlorophenol. *Microchim. Acta,* Vol.162, pp. 261-268.

Jin, G., Huang, F., Li, W., Yu, S., Zhang, S. & Kong, J. (2008). Sensitive detection of trifluoperazine using a poly-ABSA/SWNTs film-modified glassy carbon electrode. *Talanta,* Vol.74, pp. 815-820.

Kane, S. A., Iwuoha, E. I. & Smyth, M. R. (1998). Development of a sol-gel based amperometric biosensor for the determination of phenolics. *Analyst,* Vol.123, pp. 2001-2006.

Kobayashi, T., Taguchi, H., Shigematsu, M. & Tanahashi, M. (2005). Substituent effects of 3,5-disubstituted p-coumaryl alcohols on their oxidation using horseradish peroxidase-H_2O_2 as the oxidant. *Journal of Wood Science,* Vol.51, pp. 607-614.

Korkut Ozoner, S., Erhan, E., Yilmaz, F., Celik, A. & Keskinler, B. (2010). Newly synthesized poly(glycidylmethacrylate-co-3-thienylmethylmethacrylate)-based electrode designs for phenol biosensors. *Talanta,* Vol.81, pp. 82-87.

Korkut Ozoner, S., Yilmaz, F., Celik, A., Keskinler, B. & Erhan E. (2011). A novel poly(glycidly methacrylate-co-3-thienylmethyl methacrylate)-polypyrrole-carbon nanotube-horseradish peroxidase composite film electrode for the detection of phenolic compounds. *Curr. Appl. Phy,-* article in press.

Korkut, S., Keskinler, B. & Erhan E. (2008). An amperometric biosensor based on multiwalled carbon nanotube-poly(pyrrole)-horseradish peroxidase nanobiocomposite film for determination of phenol derivatives. *Talanta,* Vol.76, pp. 1147-1152.

Kuwahara, T., Oshima, K., Shimomura, M. & Miyauchi, S. (2005). Glucose sensing with glucose oxidase immobilized covalently on the films of thiophene copolymers. *Synthetic Metals,* Vol.152, pp. 29-32.

Liu, Y., Qu, X., Guo, H., Chen, H., Liu, B. & Dong, S. (2006). Facile preparation of amperometric laccase biosensor with multifunction based on the matrix of carbon nanotubes–chitosan composite. *Biosens. Bioelectron,* Vol.21, pp. 2195-2201.

Lukas, J. & Kalal, T. (1978). Reactive polymers: XV. polar polymeric sorbents based on glydicyl methoacrylate copolymers. *J. Chromatogr. A,* Vol.153, pp. 15-22.

Mailley, P., Cummings, E. A., Mailley, S. C., Eggins, B. R., McAdams, E. & Cosnier, S. (2003). Composite carbon paste biosensor for phenolic derivatives based on in situ electrogenerated polypyrrole binder. *Anal. Chem,* Vol.75, pp. 5422-5428.

Rabinovich, L. & Lev, O. (2001). Sol-gel derived composite ceramic carbon electrodes. *Electroanalysis,* Vol.13, pp. 265-275.

Rajesh, R., Pandey, S. S., Takashima, W. & Kaneto, K. (2005). Simultaneous co-immobilization of enzyme and a redox mediator in polypyrrole film for the fabrication of an amperometric phenol biosensor. *Curr. Appl. Phy,* Vol.5, pp. 184-188.

Santos, A. S., Pereira, A. C., Sotomayor,M. D. P. T., Tarley, C. R. T., Duran, N. & Kubota, L. T. (2007). Determination of phenolic compounds based on co-immobilization of methylene blue and HRP on multi-wall carbon nanotubes. *Electroanalysis,* Vol.19, pp. 549-554.

Serra, B., Benito, B., Agüi, L., Reviejo, A. J. & Pingarron, J. M. (2001). Graphite-teflon-peroxidase composite electrochemical biosensors. A tool for the wide detection of phenolic compounds. *Electroanalysis,* Vol.13, pp. 693-700.

Shan, D., Zhu, M., Han, E., Xue, H. & Cosnier, S. (2007). Calcium carbonate nanoparticles: a host matrix for the construction of highly sensitive amperometric phenol biosensor. *Biosens. Bioelectron,* Vol.23, pp. 648-654.

Tsai, Y. C. & Cheng-Chiu, C. (2007). Amperometric biosensors based on multiwalled carbon nanotube-Nafion-tyrosinase nanobiocomposites for the determination of phenolic compounds. *Sens. Actuators B: Chem,* Vol.125, pp. 10-16.

Vega, D., Agüi, L., Gonzalez-Cortes, A., Yanez-Sedeno, P. & Pingarron, J. M. (2007). Electrochemical detection of phenolic estrogenic compounds at carbon nanotube-modified electrodes. *Talanta,* Vol.71, pp. 1031-1038.

Vianello, F., Ragusa, S., Cambria, M. T. & Rigo, A. (2006). A high sensitivity amperometric biosensor using laccase as biorecognition element. *Biosens. Bioelectron,* Vol.21, pp. 2155-2160.

Vojinovic, V., Azevedo, A. M., Martins, V. C. B., Cabral, J. M. S., Gibson,T. D. & Fonseca, L. P. (2004). Assay of H2O2 by HRP catalysed co-oxidation of phenol-4-sulphonic acid and 4-aminoantipyrine: characterisation and optimisation. *J. Mol. Catal. B: Enzym,* Vol.28, pp. 129-135.

Wilkolazka, A. J., Ruzgas, T. & Gorton, L. (2005). Amperometric detection of mono- and diphenols at Cerrena unicolor laccase-modified graphite electrode: correlation between sensitivity and substrate structure. *Talanta,* Vol.66, pp. 1219-1224.

Yang, M., Jiang, J., Yang, Y., Chen, X., Shen, G. & Yu, R. (2006). Carbon nanotube cobalt hexacyanoferrate nanoparticle-biopolymer system for the fabrication of biosensors. *Biosens. Bioelectron,* Vol.21, pp. 1791-1797.

Yilmaz, F., Kasapoglu, F., Hepuzer, Y., Yagci, Y., Toppare, L., Fernandes, E. G. & Galli, G. (2005). Synthesis and mesophase properties of block and random copolymers of electroactive and liquid crystalline monomers. *Designed Monomers and Polymers,* Vol.8, pp. 223-236.

Yilmaz, F., Sel, O., Guner, Y., Toppare, L., Hepuzer, Y. & Yagci, Y. (2004). Controlled synthesis of block copolymers containing side chain thiophene units and their use in electrocopolymerization with thiophene and pyrrole. *Journal of Macromolecular Science Part A,* Vol.41, pp. 401-418.

Zeng, Y. L., Huang, Y. F., Jiang, J. H., Zhang, X. B., Tang, C. R., Shen, G. L. & Yu, R. Q. (2007). Functionalization of multi-walled carbon nanotubes with poly(amidoamine) dendrimer for mediator-free glucose biosensor. *Electrochem. Commun,* Vol.9, pp. 185-190.

Zhao, L., Liu, H. & Hu, N. (2006). Electroactive films of heme protein-coated multiwalled carbon nanotubes. *Journal of Colloid and Interface Science,* Vol.296, pp. 204-211.

Zhao, W., Song, C. & Pehrsson, P. E. (2002). Water-soluble and optically pH sensitive single-walled carbon nanotubes from surface modification. *J. Am. Chem. Soc,* Vol.124, pp. 12418-12419.

Zhou, Y. L., Tian, R. H. & Zhi, J. F. (2007). Amperometric biosensor based on tyrosinase immobilized on a boron-doped diamond electrode. *Biosens. Bioelectron,* Vol.22, pp. 822-828.

16

Flow Injection Biosensor System for 2,4-Dichlorophenoxyacetate Based on a Microbial Reactor and Tyrosinase Modified Electrode

Mitsuhiro Shimojo, Kei Amada, Hidekazu Koya and Mitsuyasu Kawakami

Fukuoka Institute of Technology

Japan

1. Introduction

Synthetic chlorinated organic compounds have been used extensively as herbicides and pesticides and the contamination of ecosystems with these compounds has stimulated great interest in investigation of these frequently toxic or bioaccumulatable compounds. 2,4-Dichlorophenoxyacetate (2,4-D), one of these compounds, is a synthetic phytohormone that has been widely used as a herbicide for controlling broadleaf weeds, and huge amount of 2,4-D has been released into the environment. 2,4-D is known, on the other hand, to be susceptible to rapid biological degradation in natural environments, and has been used as a model for genetic and biochemical study on the chloroaromatic degradation. A number of its degrading bacteria have been isolated worldwide from a variety of environments and extensively examined on a molecular basis (Amy et al., 1985; Sinton et al., 1986; Perkins et al., 1990; Fulthorpe et al., 1992, 1995; Ka et al., 1994; Tonso et al., 1995; Top et al., 1995; Maltseva et al., 1996; Suwa et al., 1996; Vallaeys et al., 1996; Kamagata et al., 1997; Cavalca et al., 1999; Laemmli et al., 2000; Itoh et al., 2002). One of the most extensively studied strain is *Ralstonia eutrophus* JMP134, which carries a 2,4-D-degrading gene cluster on the transmissible plasmid pJP4 (Don & Pemberton, 1981; Neilson et al., 1992; Fukumori & Hausinger, 1993a; Laemmli et al., 2000).

On the other hand, reliable determination of 2,4-D is indispensable to investigate its biological degradation. A biosensor is a device utilizing a biological sensing element, and a variety of biosensors for 2,4-D detection have been developed (Table 1). Most of sensors reported so far, however, can be classified into either immunoassay or immunoenzymatic assay. They are based on the specific interaction between 2,4-D (antigen) and its antibody. These biosensors have been demonstrated to show very high sensitivities (*e.g.* the lower detection limit of less than 1 µg/L), while these assays are generally said to be somewhat cumbersome to perform and require considerably expensive reagents. A biosensor based on the inhibition of catalytic activity of enzyme alkaline phosphatase in the presence of 2,4-D has been also proposed and a detection limit of 0.5-6 µg/L has been obtained. The enzyme inhibition effect has been also observed for another pesticide, indicating the analyte selectivity of the sensor to be not expected. On the other hand, very few attempts have been made on the biosensor employing microorganisms as the sensing component.

Bioelement	Transducer	Principle/Comment	Reference
antibody	SPR	immunoassay/ resonance angle	Starodub et al., 2005
	electrode	affinity sensor /conductance	Hianik et al., 1998,1999
	luminometer	immunoenzymatic assay/ chemiluminescence	Rubtsova et al., 1997
	QCM	immunoassay/ piezoelectricity	Horacek et al., 1997, 2000
	TIRF[a] fibre	immunoassay/optical	Mosiello et al., 1997
	electrode	immunoassay/ bi-enzyme amplifier, amperometry	Bier et al., 1997
	electrode	immunoassay/ enzymatic inhibition, amperometry	Medyantseva et al., 1997
	electrode	immunoenzymatic assay/ amperometry	Trau et al., 1997
	electrode	immunoenzymatic assay/ potentiometry	Piras et al., 1996
	electrode	immunoenzymatic assay/ amperometry	Skladal et al., 1995
	electrode	immunoenzymatic assay/ amperometry	Kalab et al., 1995
enzyme	electrode	enzymatic inhibition/ amperometry, voltammometry	Mazzei et al., 2004
micro-organism	O_2 electrode	respiration activity/ *A.eutrophus*	Beyersdorf-Radeck et al., 1993

([a]: total internal reflection fluorescence)

Table 1. Biosensors for 2,4-D analysis.

2. Principle of the present sensor system

2.1 2,4-D degrading microorganisms

As described above, a number of 2,4-D degrading bacteria have been isolated worldwide from varying environments. Bacteria capable of degrading 2,4-D are of general importance for the preservation of the environment, and microbial degradation with such bacteria has been extensively studied as a model for decomposition of hazardous chloroaromatic compounds (Häggblom,1990, 1992; van der Meer et al., 1992).

A microbial 2,4-D biosensor has been developed by employing a layer of *Alcaligenes eutrophus* (*Ralstonia eutrophus*) JMP134 (Table 1). Concerning to this sensor, the concentration of 2,4-D has been determined by monitoring the oxygen consumption with an oxygen electrode since the bacterium requires oxygen for the degradation of the xenobiotic. However, this approach can be disturbed by the presence of other oxygenases in a microbial cell.

Recent physiological and evolutional studies have demonstrated that these 2,4-D-degrading bacteria could be categorized into three groups (Ka et al., 1994; Kamagata et al., 1997; Kitagawa et al., 2002; Itoh et al., 2004). Most of the 2,4-D-degrading bacteria isolated from human-disturbed sites include copiotrophic and fast-growing genera in the β and γ subdivisions of *Proteobacteria*, which has been classified as class I 2,4-D degraders. The degradation pathway of 2,4-D has been extensively characterized with *Ralstonia eutrophus* JMP134, one of the class I degraders (Don et al., 1985; Streber et al., 1987; Perkins et al., 1990). Degradation of 2,4-D in this class of degraders is considered to be initiated by cleavage of the ether linkage to yield 2,4-dichlorophenol (2,4-DCP), which is then hydroxylated to 3,5-dichlorocatechol (3,5-DCC), followed by ring cleavage (Fig. 1).

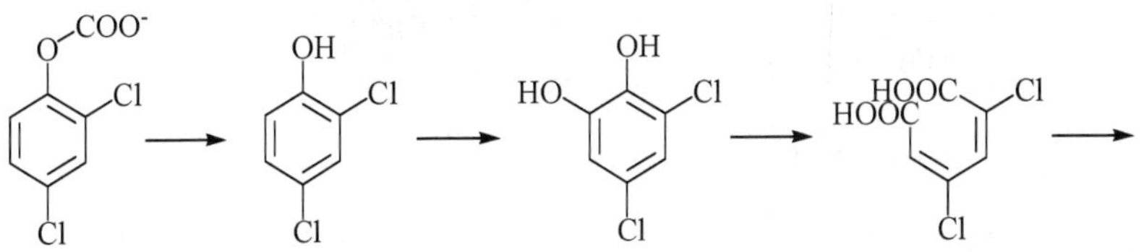

2,4-D   **2,4-DCP**   **3,5-DCC**   **2,4-Dichloromuconate**

Fig. 1. Aerobic degradation pathway of 2,4-D for class I degraders.

2.2 Biosensors detecting metabolic intermediates

It is noteworthy that the phenolic and the catecholic compounds are produced as the first and the second degradation products, respectively, in the pathway described above, while these compounds are metabolized finally to carbon dioxide. This implies that the concentration of 2,4-D might be estimated by determining these intermediates with a sensing device. Microbial biosensors based on the determination of the metabolic intermediate have not been reported for 2,4-D, while such a type of whole cell amperometric biosensors have been reported for *p*-nitrophenol (Lei et al., 2004; Mulchandani et al., 2005). A tyrosinase-modified electrode is one of effective transducers for phenols or catechols detection, for which active studies have been made on the development of highly efficient tyrosinase electrodes (for example, Notsu & Tatsuma, 2004; Mailley et al., 2004; Stanca & Popescu, 2004; Gutés et al., 2005; Carralero Sanz et al., 2005; Liu et al., 2005). Developments of flow injection analysis with tyrosinase electrode have been also attempted (Dall' Orto et al., 1999; Li & Tan, 2000; Notsu et al., 2002; Serra et al., 2003).

Then, in the present work a novel biosensor system was constructed with a fixed-bed reactor packed with immobilized microbes and a tyrosinase-modified graphite electrode. This sensor system monitors phenolic and catecholic compounds produced in the first two steps by the microbial degradation of 2,4-D.

3. Construction of sensor system

3.1 Isolation of 2,4-D-degrading bacteria and confirmation of degradation

In order to isolate 2,4-D-degrading bacteria, soil samples were collected from various environments such as mountains, farmlands, riversides, and industrial areas in Northern Kyushu. They were mixed with a mineral salt (MS) medium (0.5 g K_2HPO_4, 0.5 g KH_2PO_4,

0.25 g NH_4NO_3, 0.002 g $NaMoO_4 \cdot 2H_2O$, 0.001 g $FeSO_4 \cdot 7H_2O$, and 0.001 g $MnSO_4 \cdot 7H_2O$ per liter) containing 2,4-D (0.005 g) as the sole carbon source, and enrichment culture was performed at 30 °C for 7 d. The culture broth was centrifuged, and the concentration of 2,4-D in the supernatant was measured using a spectrophotometer at 284 nm and a high-performance liquid chromatography (HPLC) (LaChrom series, Hitachi, Tokyo, Japan). The culture broths with decreased 2,4-D concentration were then transferred to basal agar plates containing 2,4-D. Finally, we could obtain 11 isolates. 2,4-D degradation ability of the strains was confirmed by the decrease in the amount of remaining substrate, which was measured with HPLC. Identification of the strains was performed by 16S rDNA sequence analysis. Among them the bacterium identified as *Ralstonia* sp., which showed relatively high degradation capability and rapid growth in a medium containing 2,4-D as a sole carbon source, was employed for the sensor system. The strain was found to have 97% sequence similarity to *Ralstonia* sp. JMP134.

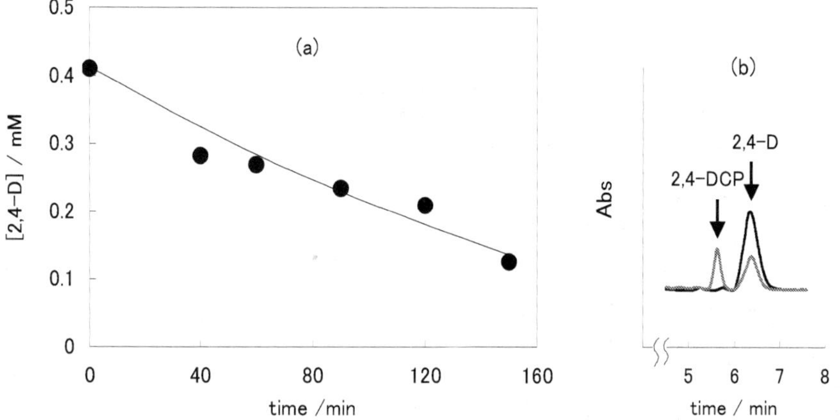

Fig. 2. Time course of 2,4-D concentration (a) and an example of chromatogram (b) obtained for degradation experiment with cell suspension; reaction time —: 0 min, ┉┉: 150 min.

The 2,4-D degradation activity of the strain was examined by monitoring the reduction of the amount of remaining substrate in a culture. The microorganism was grown aerobically in Luria-Bertani (LB) medium (NaCl, 10; Tryptone, 10; Yeast Extract, 5 g/L) containing 0.10 g/L (0.4 mM) 2,4-D at 30°C for 20 h to a late-exponential phase. The cells were harvested by centrifugation and washed thoroughly with sterile saline, which was then resuspended in the same solution. Degradation experiment initiated by transferring a fixed amount of the cell suspension into MS medium supplemented with 2,4-D (0.10 g/L), was performed at 30°C on a reciprocating shaker. The 2,4-D content was measured periodically with HPLC.

The time course of decrease in 2,4-D concentration and an example of HPLC chromatogram are shown in Fig. 2(a) and (b), respectively. It is seen that the 2,4-D content decreases smoothly with increasing the incubation time. Further, a decrease in peak-height for 2,4-D and an appearance of 2,4-DCP peak were confirmed by comparing the chromatograms obtained at the incubation time of 0 and 150 min, while variation of the amount of 2,4-DCP could not be followed accurately since most of their peaks were too small.

3.2 immobilization of microbes

Silica gel particles (Wako Gel type G, 300-600 µm, Wako Pure Chemical Industries Ltd., Tokyo, Japan) were used as the carrier for immobilization of bacteria. LB medium containing 0.05 g/L 2,4-D was used for cultivation and immobilization. A continuous flow reactor was constructed with a glass column (13 mmϕ × 80 mm) in which silica gel particles were packed. A sterile medium (100 mL) prepared in a reservoir (250 mL- glass bottle) was circulated with a peristaltic pump at a flow rate of 1 mL/min, followed by inoculation with the pre-incubated culture (1 mL). After the incubation at 30°C for 72 h the cells were harvested and washed thoroughly with 0.1 M sodium phosphate buffer (PBS; pH 7.0). Then, the immobilized microbes particles were suspended in sterile 0.9 % NaCl solution and transferred into a centrifuge tube. The immobilized microbes particles thus obtained were stored in a refrigerator for at least 24 h before use.

3.3 Instrumentation

A schematic diagram of the reactor type biosensor system is illustrated in Fig. 3. The system consists of a peristaltic pump (model 7553-80, Masterflex, Cole-Parmer Instrument Co., Vernon Hills, IL, USA), a sample injector with a 50 µL loop, a 6-way valve, a microbial reactor, and a flow cell. Two PTFE rotary valves (Rheodyne, Rhonert Park, CA, USA) were used as the injector and the valve. The 6-way valve was switched to the reactor port when the response to 2,4-D was measured. On the other hand, the valve was switched to the flow cell port when the sensitivity of enzyme electrode was checked. The microbial reactor, flow cell, and carrier reservoir were placed in an incubator maintained at 30°C.

Enlarged schematics of the microbial reactor and the flow cell are illustrated in Figs. 4(a) and 4(b), respectively. The parts of the reactor body were made of a PTFE rod (60 mmϕ). Silica gel particles with immobilized microbes were packed in the cell holder (20 mmϕ) and supported with two pieces of SUS mesh screen. In the present work the cell holders of 3 and 12 mm in height (H) were used. The flow cell was made of a PTFE sheet and a 3 mm- thick silicone gasket. The cell volume was estimated to be about 0.7 mL. A platinum plate (20×10×0.5 mm) and an Ag/AgCl electrode (1 mmϕ) were used as the auxiliary and the reference electrode, respectively. The microbial reactor and the flow cell were kept at 30°C in an incubator.

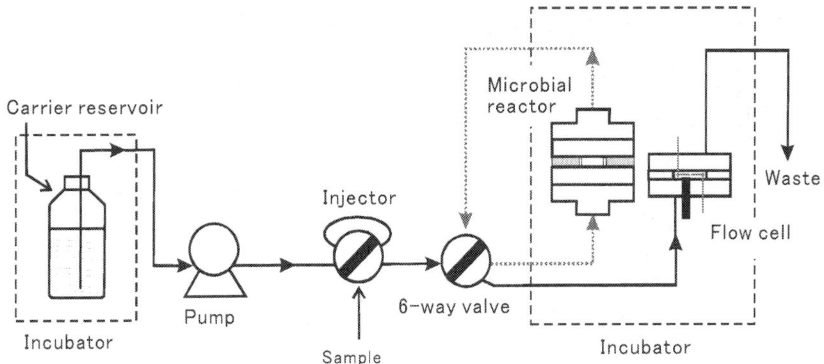

Fig. 3. Schematic diagram of the 2,4-D sensor system.

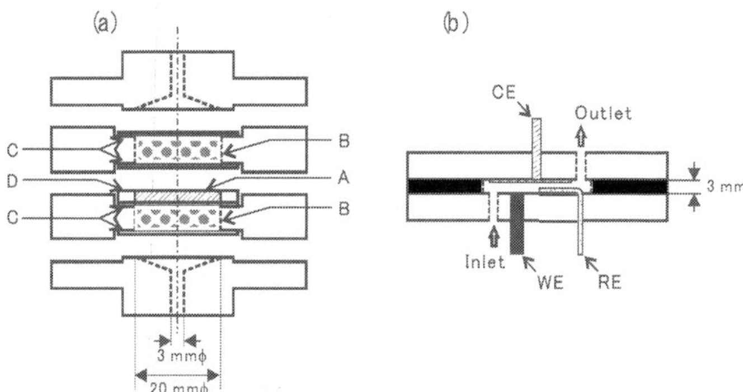

Fig. 4. Schematics of the microbial reactor (a) and the flow cell (b). A: immobilized microbe, B: glass beads, C: SUS mesh screen, D: immobilized microbes holder.

Graphite electrodes were polished with 0.1 μm alumina powder, and rinsed thoroughly with deionized water. Then, these electrodes were sonicated in acetone and deionized water successively, and allowed to dry at room temperature. Tyrosinase (from mushroom, E.C. 1.14.18.1) was purchased from Sigma (St. Louis, MO, USA) and used without further purification. The enzyme activity was determined by using tyrosine as a substrate. An enzyme solution was prepared by dissolving 3.0 mg tyrosinase in 1.0 mL PBS (0.1 M; pH 7.0). The 20 μL of the tyrosinase solution (about 36 units) were deposited with a microsyringe on a graphite electrode surface, and allowed to dry at 4°C. Finally, the 20 μL of Nafion solution (Sigma-Aldrich, 0.5%) were deposited on the tyrosinase-coated electrode surface, and allowed to dry at 4°C for an overnight. The enzyme electrodes were stored in PBS (0.1 M, pH 7.0) at 4°C when not in use.

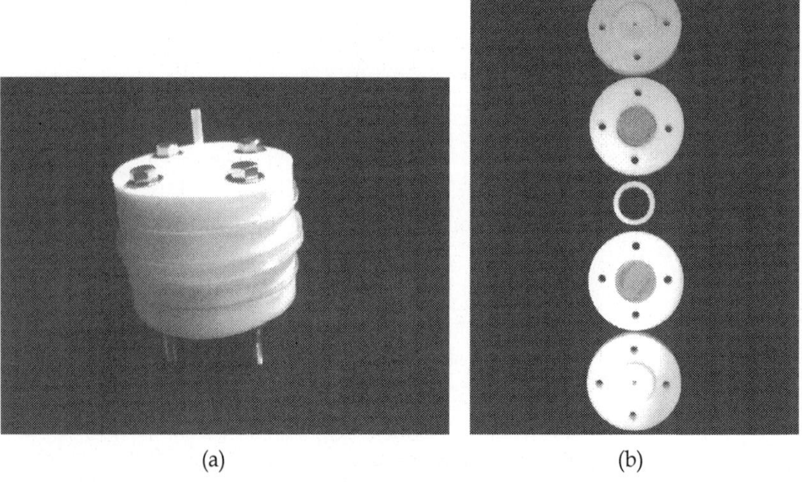

Fig. 5. Photographs of the reactor (a) and its parts (b).

3.4 Response measurement

0.1 M PBS (pH 6.0, 6.5, and 7.0) containing 0.1 M NaCl was used as the mobile phase. The carrier reservoir was hold in an incubator kept at 30°C, and the carrier solution was made to flow through the system at a constant flow rate. Amperometric measurements for the sensor system were made by applying a given potential on the working electrode (enzyme electrode) with a potentiostat (model HECS 318C, Huso Electro Chemical Systems, Kawasaki, Japan) connected to a personal computer. 2,4-D, 2,4-DCP and 3,5-DCC standard solutions for flow injection measurements were prepared with the PBS used as the mobile phase. The microbial reactor was filled with the mobile phase solution employed and left at 30 °C when not used for the measurements.

4. Performance and characteristics of sensor system

4.1 Sensor response of the tyrosinase-modified electrode

The enzyme tyrosinase catalyses two reactions; the hydroxylation of phenols to give catechols and the oxidation of catechols to o-quinones in the presence of oxygen. The determination of phenolic compounds with the tyrosinase modified electrodes can be based either on the electrochemical oxidation of catechols or the electrochemical reduction of quinones.

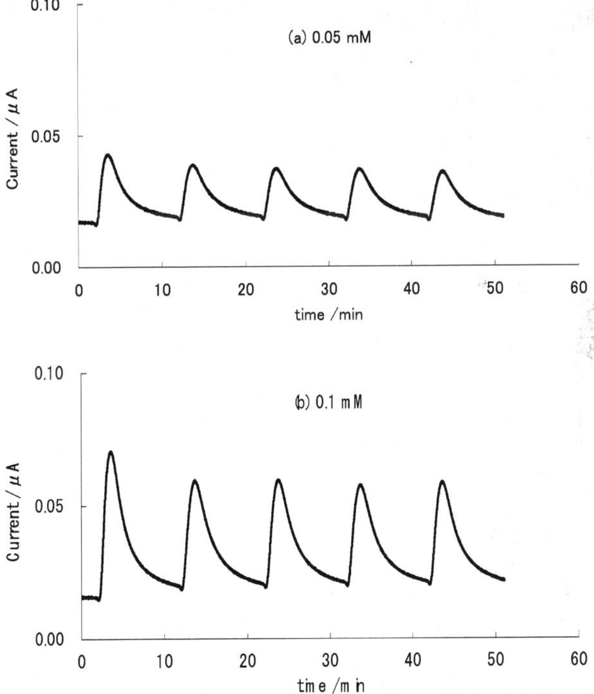

Fig. 6. Diagrams of amperometric responses obtained with the flow cell for successive injections of (a) 0.05 mM and (b) 0.1 mM 2,4-DCP (pH 6.5; H = 3 mm; flow rate 1.0 mL/min).

The oxidation on the electrode occurs at a relatively high potential, while the reduction at a low potential. Then, effects of the applied potential on the amperometric response for the tyrosinase electrode were first examined by applying a potential of ±0, +0.2, and +0.5 V (*vs.* Ag/AgCl). In this case the microbial reactor was bypassed and the substrate was sent directly to the flow cell. Consequently, the magnitudes of both the response and the residual currents were found to be affected seriously by the electrode potential, and the best sensitivity and baseline stability were obtained at +0.5 V as compared with those at ±0 and +0.2 V. Then the potential was fixed at +0.5 V and anodic output currents were monitored throughout the measurements.

Examples of amperometric responses obtained with the tyrosinase-modified electrode for successive injections of 2,4-DCP standard solutions with PBS (pH 7.0) containing 0.1 M NaCl as the mobile phase at the applied potential of 0.5 V and the flow rate of 1.0 mL/min are shown in Fig. 6(a) and 6(b). With an injection of the substrate, as it can be seen, an anodic peak was obtained as the electrochemical response. Both the peak current (height) and the peak area were found to increase with the concentration of 2,4-DCP injected. The average peak area determined for quintuplicate output signals was used to depict the calibration curve (Fig. 7). It can be seen that there exists not appreciable difference between the sensitivity for 2,4-DCP and that for 3,5-DCC. The lower detection limit for both compounds was 0.01 mM.

On the other hand, appreciable response current could not be observed when 1.0 mM 2,4-D standard solutions injected did not passthrough the microbial reactor and was sent directly to the tyrosinase electrode. This implies that 2,4-D is not detected electrochemically with the enzyme-modified electrode and either 2,4-DCP or 3,5-DCP is not significantly contained in the 2,4-D standard solution as an impurity.

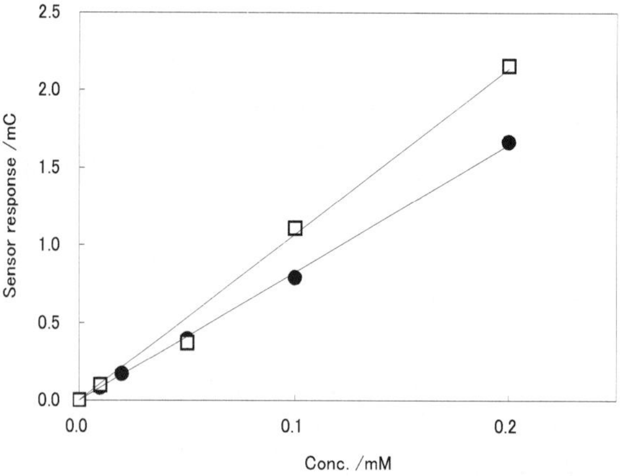

Fig. 7. Calibration plots for 2,4-DCP (•) and 3,5-DCC (□) determined with the flow cell.

4.2 Performance of 2,4-D biosensor system

The 2,4-D degradation activity of the strain was examined, as described before, using suspension of *Ralstonia* sp. in MS medium containing 2,4-D as a sole carbon source. In the case of immobilized cells, however, it is necessary to take into account the possibility of adsorption of the substrate by carrier particles employed for immobilization. Then, a fixed-bed reactor was constructed by replacing the glass column (8 mmϕ × 50 mm) of the system utilized for cell immobilization. After the immobilized cells or bare silica gel particles were packed into the column, a sterile MS medium (100 mL) containing 2,4-D prepared in a reservoir was circulated at a flow rate of 1.5 mL/min at 30°C for 12 h. The 2,4-D content was analyzed periodically with HPLC.

The time course of decrease in 2,4-D concentration observed for the immobilized microbial reactor is shown in Fig. 8. It is seen that 2,4-D is appreciably adsorbed by silica gel particles and its concentration is reduced to almost one-half its initial value for 12 h. The amount of adsorbed 2,4-D would decrease when microorganisms are immobilized onto the carrier particles. In the assay of the culture, on the other hand, the signal of 2,4-DCP could not be observed in the chromatogram. This may be due to the possibilities of adsorption of 2,4-DCP by the solid carrier and/or limitation of the detector of HPLC. Moreover, the initial rates of the decrease in 2,4-D content for both cases could be considered to be almost the same, suggesting that the degradation and the adsorption of 2,4-D might occur competitively when the substrate and the immobilized microbe particles come into contact in the reactor. Then, in order to eliminate the effect of adsorption on the silica gel surface, successive injections of the substrate was continued until stable response was obtained in the amperometric measurements.

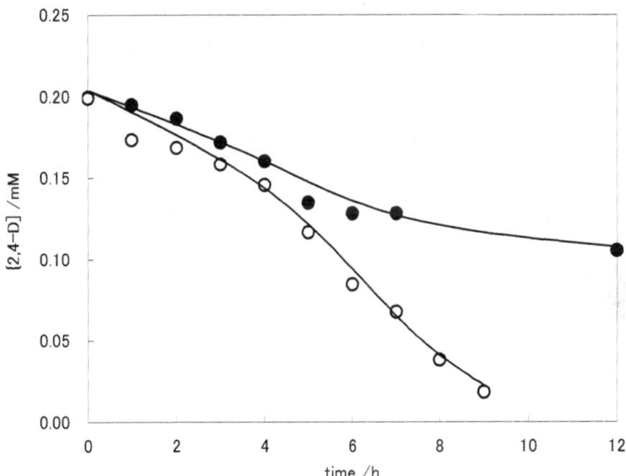

Fig. 8. Time course of 2,4-D concentration obtained for the fixed bed reactor packed with immobilized cells (○) and carrier particles alone (•).

Fig. 9 shows an example of amperometric response obtained with the sensor system employing *Ralstonia* sp. for which the immobilized cell holder of 3 mm in height was used. Using 0.1 M PBS (pH 6.5) containing 0.1 M NaCl as the mobile phase, 1.0 mM 2,4-D was injected successively at the flow rate of 1.0 mL/min. An anodic peak was observed each time with an injection of the substrate. It can be also seen not to take more than 10 min for each measurement. Response current could not be observed, as described before, when the substrate did not pass through the microbial reactor. This indicates the response of the sensor system to be induced by the electrode reaction of electrochemically active substances such as phenolic and/or catecholic compounds, which could be produced by the microbial degradation of 2,4-D in the reactor. Then, the average peak area determined for triplicate output signals was used as the response of the sensor system.

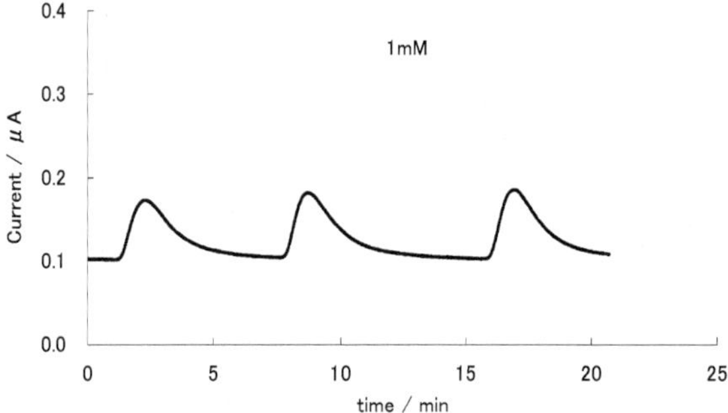

Fig. 9. Diagram of amperometric response obtained with the sensor system for successive injections of 2,4-D standard solution (pH 6.5; H = 3mm; flow rate 1.0 mL/min).

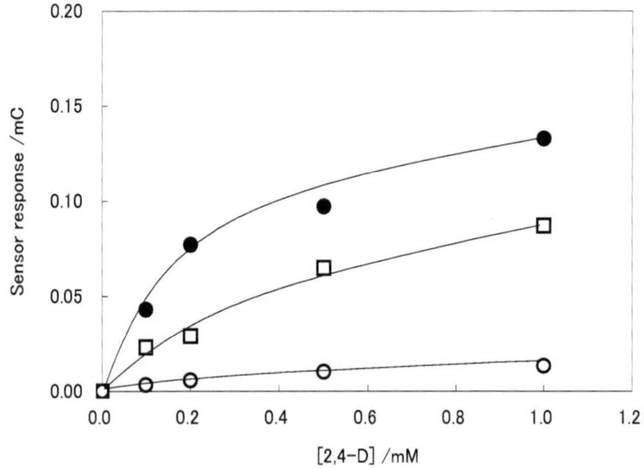

Fig. 10. Effect of the pH of carrier solution on the response of sensor system for 2,4-D; (●) pH 7.0, (□) pH 6.5, (○) pH 6.0 (H = 3 mm; flow rate 1.0 mL/min).

The effect of the pH of mobile phase solution on the response of sensor system was investigated using carrier solutions of pH 6.0, 6.5 and 7.0. The results are depicted in Fig. 10. The sensor response can be seen to increase with increasing pH of carrier solution in the range applied here. It has been reported that 2,4-dichlorophenoxyacetate/α-ketoglutarate dioxygenase, the enzyme responsible for the first step of degradation pathway for 2,4-D exhibits maximum activity at pH 6.5-7 (Fukumori & Hausinger, 1993b). On the other hand, the sensitivity of the tyrosinase-modified electrode for 2,4-DCP has been also observed to exibit almost the same pH dependence (the data are not shown). The pH dependence of the responsibility of present sensor system would result from the synergetic effect of the pH dependencies of enzyme activities for 2,4-D degradation and product detection.

Effect of the height of immobilized microbes bed on the responsibility of sensor system is shown in Fig. 11. It is seen that the response obtained by using the bed of 12 mm in height is considerably enhanced when compared with that observed with the bed of 3 mm in height. As the immobilized microbes bed becomes higher both the residence time and the amount of biocatalyst in the reactor increase. It is reasonable that the amount of degradation product increase with increasing the height of biocatalyst bed. Thus, employment of higher bed in the reactor is considerably effective to enhance the sensitivity of the sensor system, while it has been found to require longer time for a measurement (about 20 min at flow rate of 1.0 mL/min).

The flow rate of carrier solution has been also found to considerably affect the sensor response. Calibration plots obtained by the sensor system for 2,4-D with different flow rates are shown in Fig. 12. As can be seen from the figure, the sensitivity of the sensor system was appreciably enhanced by lowering the flow rate. The residence time in the reactor increases with decreasing the flow rate, which exerts almost the same effect as that of the height of biocatalyst bed. It is also confirmed that lowering the flow rate enhances the sensitivity of tyrosinase-modified electrode in the flow cell. The result obtained here is considered to be induced by combined effects of the flow rate on the reactor and the flow cell. In any event, it is evident that the sensitivity of sensor system can be improved by making the carrier solution to flow at a reduced rate, while it takes longer time for a measurement.

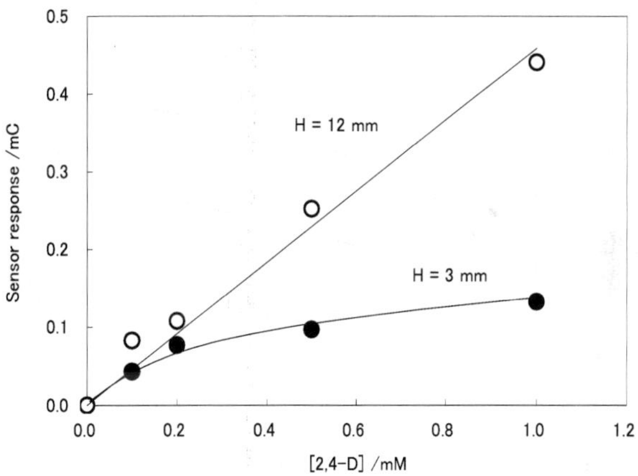

Fig. 11. Effect of the height of immobilized microorganisms bed on the response of sensor system for 2,4-D (pH 7.0; flow rate 1.0 mL/min).

The sensitivity of the sensor system was found to be almost unchanged during the measurements for one week when the reactor and the tyrosinase electrode were stored as described before. The lower detection limit and the detection range at the present state were 0.1 mM and 0.1 – 1 mM, respectively for 2,4-D. There is a possibility, however, that the sensitivity of the present sensor system can be enhanced by improving the performance of biocatalyst and/or by applying more effective detection method.

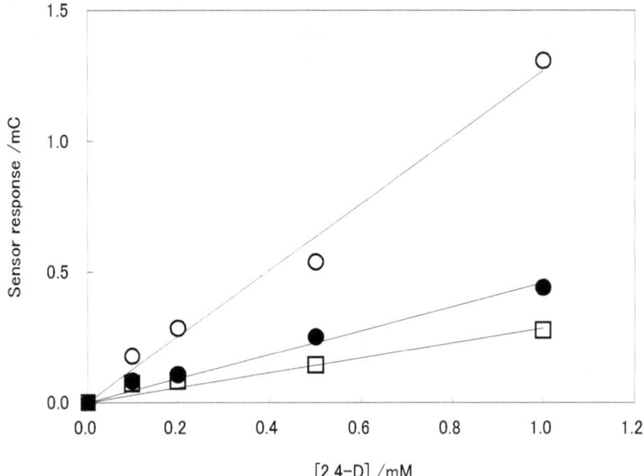

Fig. 12. Effect of the flow rate of carrier solution on sensitivities of the sensor system for 2,4-D; (○) 0.5, (●) 1.0, (□) 2.0 mL/min. (H = 12 mm, pH 7.0).

5. Conclusions

A novel flow-injection biosensor system for 2,4-D detection consisting of the microbial reactor for substrate degradation and the enzyme electrode for product detection was demonstrated. The most remarkable feature of the sensor system is to utilize a 2,4-D-degrading microorganism such as *Ralstonia* sp. which produces phenolic compounds at initial stages in the degradation pathway. The 2,4-D degrader was immobilized on silica gel particles and a fixed-bed bioreactor was constructed. The resulting 2,4-DCP and/or 3,5-DCC were detected amperometricaly with the tyrosinase-modified electrode. The sensitivity of the system was found to be considerably affected by pH and the flow rate of carrier solution. The height of immobilized microorganism bed also exerted reasonable effect on the responsibility. Although the sensitivity is not high enough for microanalysis at the present stage, further improvement of the responsibility is possible. The strategy proposed here can be easily extended to other biosensor development.

6. References

Amy, P.S.; Schulke, J.W.; Frazier, L.M.; Seidler, R.J. (1985). Characterization of aquatic bacteria and cloning of genes specifying partial degradation of 2,4-dichlorophenoxyacetic acid. *Appl. Environ. Microbiol.* 49,1237–1245.

Beyersdorf-Radeck, B.; Schmid, R. D. & Malik, K. A. (1993). Long-term storage of bacterial inoculum for direct use in a microbial biosensor. *J. Microbiol. Methods*, 18, 33-39.

Bier, F.F.; Ehrentreich-Förster, E.; Dölling, R.; Eremenko, A.V.; Scheller, F.W. (1997). A redox-label immunosensor on basis of a bi-enzyme electrode. *Anal. Chim. Acta*, 344, 119-124.

Carralero Sanz, V.; Luz Mena, M.; González-Cortés, A.; Yáñez-Sedeño, P. & Pingarrón, J. M. (2005). Development of a tyrosinase biosensor based on gold nanoparticles-modified glassy carbon electrodes. Application to the measurement of a bioelectrochemical polyphenols index in wines. *Anal. Chim. Acta*, 528, 1-8.

Cavalca, L.; Hartmann, A.; Rouard, N.; Soulas, G. (1999). Diversity of tfdC genes: distribution and polymorphism among 2,4-dichlorophenoxyacetic acid degrading soil bacteria. *FEMS Microb. Ecol.* 29,45–58.

Dall' Orto, V.C.; Danilowicz, C.; Rezzano, I.; Del Carlo, M.; Mascini, M. (1999). Comparison between three amperometric sensors for phenol determination in olive oil samples. *Anal. Lett.*, 32, 1981-1990.

Don, R.H. & Pemberton, J.M. (1981). Properties of six pesticide degradation plasmids isolated from *Alcaligenes paradoxus* and *Alcaligenes eutrophus*. *J. Bacteriol.* 145, 681–686.

Don, R.H.; Weightman, A.J.; Knackmuss, H.J.; Timmis, K.N. (1985). Transposon mutagenesis and cloning analysis of the pathways for degradation of 2,4-dichlorophenoxyacetic acid and 3-chlorobenzoate in *Alcaligenes eutrophus* JMP134(pJP4). *J. Bacteriol.* 161, 85–90.

Fukumori, F. & Hausinger, R.P. (1993a). *Alcaligenes eutrophus* JMP134 "2,4-dichlorophenoxyacetate monooxygenase" is an a-ketoglutarate-dependent dioxygenase. *J. Bacteriol.* 175, 2083–2086.

Fukumori, F. & Hausinger, R.P. (1993b). Purification and characterization of 2,4-dichlorophenoxyacetate/ a-ketoglutarate dioxygenase. *J. Biol. Chem.*, 268, 24311-24317.

Fulthorpe, R.R., & Wyndham, R.C. (1992). Involvement of a chlorobenzoate catabolic transposon, Tn5271, in community adaptation to chlorobiphenyl, chloroaniline, and 2,4-dichlorophenoxyacetic acid in a freshwater ecosystem. *Appl. Environ. Microbiol.* 58, 314–325.

Fulthorpe, R.R.; McGowan, C.; Maltseva, O.V.; Holben, W.H.; Tiedje, J.M. (1995). 2,4-Dichlorophenoxyacetic acid-degrading bacteria contain mosaics of catabolic genes. *Appl. Environ. Microbiol.* 61, 3274–3281.

Gutés, A.; Céspedes, F.; Alegret, S. & del Valle, M. (2005). Determination of phenolic compounds by a polyphenol oxidase amperometric biosensor and artificial neural network analysis. *Biosens. Bioelectron.*, 20, 1668-1673.

Häggblom, M. M. (1990). Mechanism of bacterial degradation and transformation of chlorinated monoaromatic compounds. *J. Basic. Microbiol.*, 30, 115-141.

Häggblom, M. M. (1992). Microbial breakdown of halogenated aromatic pesticides and related compounds. *FEMS Microbiol. Rev.*, 103, 29-72.

Hianik, T.; Passechnik, V. I. ; Sokolíková, L.; Šnejdárková, M.; Sivák, B.; Fajkus, M.; A.Ivanov, S. & Fránek, M. (1998). Affinity biosensors based on solid supported lipid membranes. Their structure, physical properties and dynamics. *Bioelectrochem. Bioenerg.*, 47, 47-55.

Hianik, T.; Šnejdárková, M.; Sokolíková, L.; Meszár, E.; Krivánek, R.; Tvarožek, V.; Novotný, I. & Wang, J. (1999). Immunosensors based on supported lipid membranes, protein films and liposomes modified by antibodies. *Sensors and Actuators B*, 57, 201-212.

Horáček, J. & Skládal, P. (1997). Improved direct piezoelectric biosensors operating in liquid solution for the competitive label-free immunoassay of 2,4-dichlorophenoxyacetic acid. *Anal. Chim. Acta*, 347, 43-50.

Horáček, J. & Skládal, P. (2000). Effect of organic solvents on immunoassays of environmental pollutants studied using a piezoelectric biosensor. *Anal. Chim. Acta*, 412, 37-45.

Itoh, K.; Kanda, R.; Sumita, Y.; Kim, H.; Kamagata, Y.; Suyama, K.; Yamamoto, H.; Hausinger, R.P.; Tiedje, J.M. (2002). tfdA-like genes in 2,4-dichlorophenoxyacetic acid-degrading bacteria belonging to the Bradyrhizobium- Agromonas-Nitrobacter-Afipia cluster in α-Proteobacteria. *Appl. Environ. Microbiol.* 68, 3449-3454.

Itoh, K.; Tashiro, Y.; Uobe, K.; Kamagata, Y.; Suyama, K. & Yamamoto, H. (2004). Root nodule *Bradyrhizobium* spp. harbor *tfdAa* and *cadA*, homologous with genes encoding 2,4-dichlorophenoxyacetic acid – degrading proteins. *Appl. Environ. Microbiol.*, 70, 2110-2118.

Ka, J.O.; Holben, W. E. & Tiedje, J. M. (1994). Genetic and phenotypic diversity of 2,4-dichlorophenoxyacetic acd (2,4-D)- degrading bacteria isolated from 2,4-D- treated field soils. *Appl. Environ. Microbiol.*, 60, 1106-1115.

Kaláb, T. & Skládal, P. (1995). A disposable amperometric immunosensor for 2,4-dichlorophenoxyacetic acid. *Anal. Chim. Acta*, 304, 361-368.

Kamagata, Y.; Fulthorpe, R. R.; Tamura, K.; Takami, H.; Forney, L. J. & Tiedje, J. M. (1997). Pristine environments harbor a new group of oligotrophic 2,4-dichlorophenoxyacetic acid – degrading bacteria. *Appl. Environ. Microbiol.*, 63, 2266-2272.

Kitagawa, W.; Takami, S.; Miyauchi, K.; Masai, E.; Kamagata, Y.; Tiedje, J. M. & Fukuda, M. (2002). Novel 2,4-dichlorophenoxyacetic acid degradation genes from oligotrophic *Bradyrhizobium* sp. strain HW13 isolated from a pristine environment. *J. Bacteriol.*, 184, 509-518.

Laemmli, C.M.; Leveau, J.H.J.; Zehnder, A.J.B.; van der Meer, J.R. (2000). Characterization of a second tfd gene cluster for chlorophenol and chlorocatechol metabolism on plasmid pJP4 in *Ralstonia eutropha* JMP134(pJP4). *J. Bacteriol.* 182, 4165-4172.

Lei, Y.; Mulchandani, P.; Chen, W.; Wang, J. & Mulchandani, A. (2004). *Arthrobacter* sp. JS443 – based whole cell amperometric biosensor for p-nitrophenol, *Electroanalysis*, 16, 2030-2034.

Li, J. & Tan, S.N. (2000). Applications of amperometric silica sol-gel modified enzyme biosensors. *Anal. Lett.*, 33, 1467-1477.

Liu, Z.; Liu, Y.; Yang, H.; Yang, Y.; Shen, G. & Yu, R. (2005). A phenol biosensor based on immobilizing tyrosinase to modified core-shell magnetic nanoparticles supported at a carbon paste electrode, *Anal. Chim. Acta*, 533, 3-9.

Mailley, P.; Cummings, E.A.; Mailley, S.; Cosnier, S.; Eggins, B.R. & McAdams, E. (2004). Amperometric detection of phenolic compounds by polypyrrole-based composite carbon paste electrodes, *Bioelectrochemistry*, 63, 291-296.

Maltseva, O.; McGowan, C.; Fulthorpe, R.; Oriel, P. (1996). Degradation of 2,4-dichlorophenoxyacetic acid by haloalkaliphilic bacteria. *Microbiology*, 142, 1115-1122.

Mazzei, F.; Botrè, F.; Montilla, S.; Pilloton, R.; Podestà, E. & Botrè, C. (2004). Alkaline phosphatase inhibition based electrochemical sensors for the detection of pesticides. *J. Electroanal. Chem.*, 574, 95-100.

Medyantseva, E. P.; Vertlib, M. G.; Kutyreva, M. P.; Khaldeeva, E. I.; Budnikov, G. K. & Eremin, S. A. (1997). The specific immunochemical detection of 2,4-dichlorophenoxyacetic acid and 2,4,5-trichlorophenoxyacetic acid pesticides by amperometric cholinesterase biosensors. *Anal. Chim. Acta*, 347, 71-78.

Mosiello, L.; Nencini, L.; Segre, L. & Spanò, M. (1997). A fibre-optic immunosensor for 2,4-dichlorophenoxyacetic acid detection. *Sensors and Actuators B*, 39, 353-359.

Mulchandani, P.; Hangarter, C. M.; Lei, Y.; Chen, W. & Mulchandani, A. (2005). Amperometric microbial biosensor for *p*-nitrophenol using *Moraxella* sp.-modified carbon paste electrode, *Biosens. Bioelectron.*, 21, 523-527.

Notsu, H., Tatsuma, T., Fujishima, A. (2002). Tyrosinase-modified boron–doped diamond electrodes for the determination of phenol derivatives. *J. Electroanal. Chem.*, 523, 86-92.

Neilson, J.W.; Josephson, K.L.; Pepper, I.L.; Arnold, R.B.; DiGiovanni, G.D.; Sinclair, N.A. (1994). Frequency of horizontal gene transfer of a large catabolic plasmid (pJP4) in soil. *Appl. Environ. Microbiol.* 60, 4053–4058.

Notsu, H. & Tatsuma, T. (2004). Simultaneous determination of phenolic compounds by using a dual enzyme electrodes system, *J. Electroanal. Chem.*, 566, 379-384.

Perkins, E.J.; Gordon, M.P.; Caceres, O.; Lurquin, P.F. (1990). Organization and sequence analysis of the 2,4-dichlorophenol hydroxylase and dichlorocatechol oxidative operons of plasmid pJP4. *J. Bacteriol.* 172, 2351–2359.

Piras, L.; Adami, M.; Fenu, S.; Dovis, M. & Nicolini, C. (1996). Immunoenzymatic application of a redox potential biosensor. *Anal. Chim. Acta*, 335, 127-135.

Rubtsova, M.Y.; Kovba, G. V. & Egorov, A. M. (1998). Chemiluminescent biosensors based on porous supports with immobilized peroxidase. *Biosens. Bioelectron.*, 13, 75-85.

Serra, B., Reviejo, A.J., Pingarron, J.M. (2003). Flow injection amperometric detection of phenolic compounds at enzyme composite biosensors, Applicatio to their monitoring during industrial waste waters purification processes. *Anal. Lett.*, 36, 1965-1986.

Skládal, P. & Kaláb, T. (1995). A multichannel immunochemical sensor for determination of 2,4-dichlorophenoxyacetic acid. *Anal. Chim. Acta*, 316, 73-78.

Suwa, Y.; Wright, A.D.; Fukumori, F.; Nummy, K.A.; Hausinger, R.P.; Holben, W.E.; Forney, L.J. (1996). Characterization of a chromosomally encoded 2,4-dichlorophenoxyacetic acid/ -ketoglutarate dioxygenase from *Burkholderia* sp. strain RASC. *Appl. Environ. Microbiol.* 62, 2464–2469.

Stanca, S. E. & Popescu, I. C. (2004). Phenols monitoring and Hill coefficient evaluation using tyrosinase-based amperometric biosensors, *Bioelectrochemistry*, 64, 47-52.

Starodub, N. F.; Pirogova, L.V.; Demchenko, A. & Nabok, A. V. (2005). Antibody immobilisation on the metal and silicon surfaces. The use of self-assembled layers and specific receptors. *Bioelectrochemistry*, 66, 111-115.

Tonso, N.L.; Matheson, V.G.; Holben, W.E. (1995). Polyphasic characterization of a suite of bacterial isolates capable of degrading 2,4-D. *Microb. Ecol.* 30, 3–24.

Top, E.M.; Holben, W.E.; Forney, L.J. (1995). Characterization of diverse 2,4-dichlorophenoxyacetic acid-degradative plasmids isolated from soil by complementation. *Appl. Environ. Microbiol.* 61, 1691–1698.

Trau, D.; Theuerl, T.; Wilmer, M.; Meusel, M. & Spener, F. (1997). Development of an amperometric flow injection immunoanalysis system for the determination of the herbicide 2,4-dichlorophenoxyacetic acid in water. *Biosens. Bioelectron.*, 12, 499-510.

Vallaeys, T.; Fulthorpe, R.R.; Wright, A.M.; Soulas, G. (1996). The metabolic pathway of 2,4-dichlorophenoxyacetic acid degradation involves different families of tfdA and tfdB genes according to PCR-RFLP analysis. *FEMS Microb. Ecol.* 20, 163–172.

van der Meer, J. R.; de Vos, W. M.; Harayama, S. & Zehnder, A. J. B. (1992). Molecular Mechanism of genetic adaptation to xenobiotic compounds. *Microbiol. Rev.*, 56, 677-694.